Animal Cell Biotechnology
Methods and Protocols
SECOND EDITION

METHODS IN BIOTECHNOLOGY™

John M. Walker, *Series Editor*

24. **Animal Cell Biotechology: *Methods and Protocols, Second Edition,*** edited by *Ralf Pörtner, 2007.*
23. **Phytoremediation: *Methods and Reviews,*** edited by *Neil Willey, 2007.*
22. **Immobilization of Enzymes and Cells**, ***Second Edition***, edited by *Jose M. Guisan, 2006*
21. **Food-Borne Pathogens**: ***Methods and Protocols,*** edited by *Catherine Adley, 2006*
20. **Natural Products Isolation**, ***Second Edition***, edited by *Satyajit D. Sarker, Zahid Latif, and Alexander I. Gray, 2006*
19. **Pesticide Protocols**, edited by *José L. Martínez Vidal and Antonia Garrido Frenich, 2006*
18. **Microbial Processes and Products**, edited by *Jose Luis Barredo, 2005*
17. **Microbial Enzymes and Biotransformations**, edited by *Jose Luis Barredo, 2005*
16. **Environmental Microbiology: *Methods and Protocols***, edited by *John F. T. Spencer and Alicia L. Ragout de Spencer, 2004*
15. **Enzymes in Nonaqueous Solvents: *Methods and Protocols***, edited by *Evgeny N. Vulfson, Peter J. Halling, and Herbert L. Holland, 2001*
14. **Food Microbiology Protocols**, edited by *J. F. T. Spencer and Alicia Leonor Ragout de Spencer, 2000*
13. **Supercritical Fluid Methods and Protocols**, edited by *John R. Williams and Anthony A. Clifford, 2000*
12. **Environmental Monitoring of Bacteria**, edited by *Clive Edwards, 1999*
11. **Aqueous Two-Phase Systems**, edited by *Rajni Hatti-Kaul, 2000*
10. **Carbohydrate Biotechnology Protocols**, edited by *Christopher Bucke, 1999*
9. **Downstream Processing Methods**, edited by *Mohamed A. Desai, 2000*
8. **Animal Cell Biotechnology**, edited by *Nigel Jenkins, 1999*
7. **Affinity Biosensors: *Techniques and Protocols***, edited by *Kim R. Rogers and Ashok Mulchandani, 1998*
6. **Enzyme and Microbial Biosensors: *Techniques and Protocols***, edited by *Ashok Mulchandani and Kim R. Rogers, 1998*
5. **Biopesticides: *Use and Delivery***, edited by *Franklin R. Hall and Julius J. Menn, 1999*
4. **Natural Products Isolation**, edited by *Richard J. P. Cannell, 1998*
3. **Recombinant Proteins from Plants: *Production and Isolation of Clinically Useful Compounds***, edited by *Charles Cunningham and Andrew J. R. Porter, 1998*

METHODS IN BIOTECHNOLOGY™

Animal Cell Biotechnology

Methods and Protocols

SECOND EDITION

Edited by

Ralf Pörtner

Institute for Bioprocess and Biochemical Engineering
Hamburg University of Technology
Hamburg, Germany

HUMANA PRESS ✷ TOTOWA, NEW JERSEY

999 Riverview Drive, Suite 208
Totowa, New Jersey 07512

www.humanapress.com

This publication is printed on acid-free paper. ∞
ANSI Z39.48-1984 (American Standards Institute) Permanence of Paper for Printed Library Materials.

Cover design by Nancy K. Fallatt

Cover illustration: *(Background)* AM12-cells grown on Fibra-Cel® macroporous carrier. *(Foreground)* CHO-cells grown on macroporous Sponceram® disc.

For additional copies, pricing for bulk purchases, and/or information about other Humana titles, contact Humana at the above address or at any of the following numbers: Tel.: 973-256-1699; Fax: 973-256-8341; E-mail: orders@humanapr.com; or visit our Website: www.humanapress.com

Printed in the United States of America. 10 9 8 7 6 5 4 3 2 1

eISBN-10: 1-59745-399-4

eISBN-13: 978-1-59745-399-8

Library of Congress Cataloging-in-Publication Data

Animal cell biotechnology : methods and protocols / edited by Ralf Pörtner.--2nd ed.
p. cm.-- (Methods in biotechnology)
Includes bibliographical references and index.
ISBN 1-58829-660-1 (alk. paper)
1. Animal cell biotechnology. 2. Cell culture. I. Pörtner, Ralf.
TP248.27.A53A542 2007
660.6--dc22

2006049541

Preface

Mammalian cells are used in industry as well as in research for a variety of applications. Examples are the production of monoclonal antibodies with hybridoma cells or proteins for diagnostic or therapeutic use with recombinant cells, production of viral vaccines, as well as cultivation of tissue cells for artificial organs or for gene therapy. Beside the techniques required for establishing specific cell lines are techniques required for optimized cultivation in small and large scale cell characterization and analysis and purification of biopharmaceuticals and vaccines.

Animal Cell Biotechnology: Methods and Protocols is divided into parts that reflect the processes required for different stages of production. In Part I basic techniques for establishment of production cell lines are addressed. Safe and reliable transfer of genetic information at high frequencies into desired target cells remains one of the major challenges in therapeutic life sciences. Multiply attenuated viral vector systems, which are able to transfer heterologous DNA or RNA into almost any cell phenotype as well as in vivo, emerged as the most efficient transgene delivery systems in modern biology.

The treatment of genetic diseases and of several acquired diseases can only reasonably be performed by using gene or cell therapy approaches. Chapter 1 provides a comprehensive technical overview of the most commonly used viral transduction systems and compares their specific characteristics in order for researchers to make the best choice in a particular scientific setting. In Chapter 2, in addition to protocols for the establishment of producer cells and the production of viral vectors, some basic titration methods are presented.

Monoclonal antibodies (mAbs) are unique and versatile molecules that have been applied in research, as well as to in vitro and in vivo diagnostics and treatment of several diseases. Chapter 3 is an overview of the state of the art of antibody engineering and production methods and describes standard protocols for producing a hybridoma cell line the building up of transfectomas. A suitable bacterial expression system for the production of antibodies and antibody fragments is also given.

The history of the culture of animal cell lines is littered with published and much unpublished experience with cell lines that have become switched, mislabelled, or cross-contaminated during laboratory handling. To deliver valid and good quality research and to avoid expenditure of time and resources on such rogue lines, it is vital to perform some kind of qualification for the provenance of cell lines used in research and particularly in the development of biomedical products. DNA fingerprinting provides a valuable tool to compare different sources of the same cells and, where original material or tissue is available, to confirm the correct identity of a cell line. Chapter 4 provides a review of some of the most useful techniques to test the identity of cells in the cell culture laboratory and presents methods used in the authentication of cell lines.

In Part II basic cultivation techniques are described. Microcarrier culture introduces new possibilities and, for the first time, allows practical high-yield culture of anchorage-dependent cells. In microcarrier culture, cells grow as monolayers on the surface of small spheres or as multilayers in the pores of macroporous structures that are usually suspended in culture medium by gentle stirring. The microcarrier examples described in Chapter 5 summarize the procedures for the major types existing. Encapsulation of cell cultures is a universally applicable tool in cell culture technology. Basically it can be used to protect cells against hazardous environmental conditions as well as the environment against disadvantageous effects triggered by the immobilized cultures. The three basic encapsulation systems existing and described in Chapter 6 are the bead, the coated bead, and the membrane-coated hollow sphere. In addition, two other universally applicable methods to quantify the number of living cells immobilized in those encapsulation systems could be used.

There is an increasing need for high-throughput scaled-down models that are representative of commercial bioprocesses, particularly in the field of animal cell biotechnology. Chapter 6 describes two protocols for small-scale disposable plastic vessels: 125-mL Erlenmeyer shake flasks, and 50-mL centrifuge tubes. We also describe two common applications for these scale-down platforms: (1) satellite cultures derived from conventional bioreactor runs for optimizing the feed strategy for fed-batch cultures; and (2) testing multifactorial experimental designs for designing the components of a base medium.

Part III covers cell characterization and analysis. The accurate determination of cell growth is pivotal to monitoring a bioprocess. Direct methods are described in Chapter 8 to determine the cells in a bioprocess, including microscopic counting, electronic particle counting, biomass monitoring, and image analysis. These methods work most simply when a fixed volume sample can be taken from a suspension culture. Indirect methods of cell determination involve thc chemical analysis of a culture component or a measure of metabolic activity. These methods are most useful when it is difficult to obtain intact cell samples. However, the relationship between these parameters and the cell number may not be linear through the phases of a cell culture.

The development of flow cytometric techniques has greatly advanced our ability to identify and characterize the morphological and biochemical heterogeneity of cell populations, permitting a rapid and sensitive cell analysis. In Chapter 9, various flow cytometric applications in cell culture are described for cell lines grown on a laboratory scale, but the principles can also be employed for the monitoring of cell lines up to the large industrial scale.

Magnetic resonance imaging (MRI) and spectroscopy (MRS) are powerful noninvasive techniques that can be used to monitor the behavior of cells in intensive bioreactor systems. In Chapter 10 a number of NMR-based techniques that have been used successfully to investigate cell growth and distribution, cellular energetics, and the porosity to medium flow and linear flow velocity profiles around and across cell layers in perfusion bioreactors are discussed. Deermining these parameters are important when designing bioreactor systems and operation protocols that optimize cell productivity.

For an established culture it is of great importance to know the demands of the cells as exactly as possible. Apart from global parameters like temperature, pH, oxygen tension, and osmolarity the availability of substrates and the concentration of metabolic waste products play a major role in the performance of the cell culture. In order to monitor the concentrations of these substances and to evaluate the state of the culture, some different methods have to be applied. These are described in Chapter 11.

Stoichiometric and kinetic analyses are critical tools for developing efficient processes for the production of animal cell culture products such as therapeutic proteins or vaccines. Chapter 12 describes the common analytical methods for quantifying animal cell growth and metabolism. It also presents sample calculations to illustrate the utility and limitations of various analytical methods. The application of calculated stoichiometric and kinetic parameters in mathematical models is also discussed. Hence, this chapter provides the basic tools for obtaining a quantitative description of current cell culture state as well as applying the information to optimize the process.

Apoptosis is a genetically regulated process by which cells can be eliminated in vivo in response to a wide range of physiological and toxicological signals. Cells in vitro may spontaneously or be induced to die by apoptosis, e.g., depletion of nutrients or survival factors from the culture media. In Chapter 13 several biochemical and cytometric techniques of varying complexity that have been developed to detect cellular and subcellular changes occurring during apoptosis are described. For all procedures, cells can be obtained from either monolayer or suspension culture.

Metabolic flux analysis with its ability to quantify cellular metabolism is an attractive tool for accelerating cell line selection, medium optimization, and other bioprocess development activities. In the stoichiometric flux estimation approach, unknown fluxes are determined using intracellular metabolite mass balance expressions and measured extra-cellular rates. The simplicity of the stoichiometric approach extends its application to most cell culture systems and the steps involved in metabolic flux estimation by the stoichiometric method are presented in detail in Chapter 14.

In Part IV special cultivation techniques are addressed. Although glass-stirred or stainless-steel-stirred reactors dominate in the area of animal cell cultivation, users are increasingly trying to integrate disposable bioreactors wherever possible. Today applications range from inoculum to glycoprotein production processes, which are most commonly performed in various membrane bioreactors and the Wave in small- and middle-scale.

For this reason Chapter 15 presents protocols instructing for handling the CeLLine, the miniPerm, and the Wave. In our experience, the methods described for a CHO suspension cell line can also be successfully applied to other animal suspension cells, such as insect cells (Sf-9) and human embryogenic kidney cells (HEK-293 EBNA cells).

The use of hollow fiber culture is a convenient method for making moderately large quantities of high-molecular-weight products secreted by human and animal cells, at high concentrations and with a higher ratio of product to medium-derived impurities than is generally achieved using homogeneous (e.g., stirred-tank) culture. Methods and options for operating and optimizing such systems are described in Chapter 16, with special reference to the Biovest AcuSyst systems.

Fixed-bed reactors for cultivation of mammalian cells can be used for a variety of applications, such as the production of mAb with hybridoma cells or proteins for diagnostic or therapeutic use with recombinant, mostly anchorage-dependent animal cells, production of retroviral vectors, as well as for cultivation of tissue cells for artificial organs. Chapter 17 describes the basic characteristics of fixed-bed reactor systems and macroporous carriers, such as selection of suitable carriers, procedures for cultivation in lab-scale fixed-bed. and additional requirements for larger reactor volumes.

For basic research on animal cells and in process development and troubleshooting, lab-scale bioreactors (0.2–20 L) are used extensively. In Chapter 18 setting up the bioreactor system for different configurations is described and two special set-ups (acceleration stat and reactors in series) are treated. In addition, on-line measurement and control of bioreactor parameters is described, with special attention to controller settings (PID) and on-line measurement of oxygen consumption and carbon dioxide production. Finally, methods for determining the oxygen transfer coefficient are described.

Part V covers downstream techniques. In Chapter 19 membrane filtration techniques are addressed. Membrane filtration is frequently used in animal cell culture for bioreactor harvesting, protein concentration, buffer exchange, virus filtration, and sterile filtration. A variety of membrane materials and pore sizes ranging from loose microfiltration membranes, to tight ultrafiltration membranes which reject small proteins, are frequently found in a purification train. While all of these operations make use of the same-size-based separation principle, the actual methods of operation vary significantly.

The purification of a (recombinant) protein produced by animal cell culture, the so-called downstream process (DSP), tends to be one of the most costly aspects of bioproduction. Chromatography is still the major tool on all levels of the DSP from the first capture to the final polishing step. In Chapter 20 first the commonly used methods and their set-up, in particular ion exchange chromatography (IEX), hydrophobic interaction chromatography (HIC), affinity chromatography (AC), and gel filtration (GPC, SEC), but also some lesser known alternatives, are outlined. Then the rational design of a downstream process, which usually comprises three orthogonal chromatographic steps, is discussed. Finally process variants deviating from the usual batch column/gradient elution approach is presented, including expanded bed, displacement, and continuous annular chromatography, but also affinity precipitation.

In Part VI special applications are described. The production of viral vaccines in animal cell culture can be accomplished with primary, diploid, or continuous (transformed) cell lines. Each cell line, each virus type, and each vaccine definition requires a specific production and purification process. Media have to be selected as well as the production vessel, production conditions, and the type of process. In Chapter 21, emphasis is put on different issues that have to be considered during virus production processes by discussing the influenza virus production in a microcarrier system in detail as an example. The use of serum-containing as well as serum-free media, but also the use of stirred tank bioreactors or wave bioreactors is considered.

Retroviral vectors are involved in more than 25% of the gene therapy trials, being the pioneering vector in this field (Chapter 22). The production of retroviral vectors still poses some challenges owing mainly to the relatively low cell-specific productivity and the low vector stability. Having clinical applications in mind, it is clear that robust production and purification processes are necessary, along with a proper vector characterization. The determination of vector titer in terms of infectious particles and transduction efficiency, combined with product quality analysis in terms of total particles and proteins present in the final preparation, are essential for a complete assessment and definition of the final product specifications.

Viral transduction of eukaryotic cell lines is one possibility for efficient generation of recombinant proteins, and among all viral-based expression systems the Baculovirus/insect cell expression system (BEVS) is certainly the best known and applied. Chapter 22 delineates the individual process steps of the system: maintenance of insect cell cultures, transfection of recombinant DNA into different insect cell lines, amplification of virus stocks, virus titer determination and accessory experiments required to give rise to optimal production conditions.

In summary, this volume constitutes a comprehensive manual of state-of-the-art and new techniques for setting up mammalian cell lines for production of biopharmaceuticals, and for optimizing critical parameters for cell culture throughout the whole cascade from lab to final production. Inevitably, some omissions will occur in the test, but the authors have sought to avoid duplications by extensive cross-referencing to chapters in other volumes of this series and elsewhere. We hope *Animal Cell Biotechnology: Methods and Protocols, Second Edition* provides a useful compendium of techniques for scientists in industrial and research laboratories that use mammalian cells for biotechnology purposes. The editor is grateful for the support of all the contributors, the series editor Prof. John Walker, Hertfordsire, UK, and the publisher, who have made this volume possible.

Ralf Pörtner

Contents

Contributors

DEREK ADAMS • *Alexion Pharmaceuticals, Inc., Cheshire, United Kingdom*
MOHAMED AL-RUBEAI • *Department of Chemical and Bioprocess Engineering and Centre for Synthesis and Chemical Biology, University College Dublin, Belfield, Dublin, Ireland*
PAULA ALVES • *Instituto Di Biologia Experimental E Tecnológica (IBET), Oeiras, Portugal*
RICHARD BIENER • *Department of Natural Sciences, University of Applied Sciences Esslingen, Esslingen, Germany*
GERALD BLÜML • *Department of Biotechnology, GE Healthcare, Vienna, Austria*
KEVIN M. BRINDLE • *Biochemistry Department, University of Cambridge, Cambridge, United Kingdom*
HEINO BÜNTEMEYER • *Institute of Cell Culture Technology, University of Bielefeld, Bielefeld, Germany*
MICHAEL BUTLER • *Department of Microbiology, University of Manitoba, Winnipeg, Manitoba, Canada*
ED BYRNE • *National Institute for Biological Standards and Control, South Mimms, United Kingdom*
MARLENE CARMO • *Instituto Di Biologia Experimental E Tecnológica (IBET), Oeiras, Portugal*
SILVIA CARROLL • *Department of Chemical and Bioprocess Engineering and Centre for Synthesis and Chemical Biology, University College Dublin, Belfield, Dublin, Ireland*
MIGUEL CHILLON • *Universitat Autònoma de Barcelona, Unitat Vectors Virals, Bellaterra, Spain*
PEDRO E. CRUZ • *Instituto Di Biologia Experimental E Tecnológica (IBET), Oeiras, Portugal*
PETER CZERMAK • *Institute of Biopharmaceutical Technology, University of Applied Sciences Giessen-Friedberg, Giessen, Germany; Department of Chemical Engineering, Kansas State University, Manhattan,* KS
JOHN M. DAVIS • *Research and Development Department, Bio-Products Laboratory, Elstree, Hertfordshire, United Kingdom*
CHRISTOPHE DELENDA • *Généthon, Evry, France*
ALEXANDRA DORN-BEINEKE • *Institute for Clinical Chemistry, Faculty for Clinical Medicine Mannheim, University of Heidelberg, Heidelberg, Germany*
ANNE-MARIE DOUAR • *Généthon, Evry, France*
DIETER EIBL • *Department of Biotechnology, University of Applied Sciences Wädenswil, Wädenswil, Switzerland*
REGINE EIBL • *Department of Biotechnology, University of Applied Sciences Wädenswil, Wädenswil, Switzerland*

DAVID A. FLURI • *Institute for Chemical and Bio-Engineering (ICB), Swiss Federal Institute of Technology, ETH Hoenggerberg, Zurich, Switzerland*
RUTH FREITAG • *Process Biotechnology, University of Bayreuth, Bayreuth, Germany*
MARTIN FUSSENEGGER • *Institute for Chemical and Bio-Engineering (ICB), Swiss Federal Institute of Technology, ETH Hoenggerberg, Zurich, Switzerland*
SABINE GEISSE • *Novartis Institutes for BioMedical Research, Basel, Switzerland*
YVONNE GENZEL • *Max Planck Institute for Dynamics of Complex Technical Systems, Magdeburg, Germany*
CHETAN T. GOUDAR • *Research and Development, Process Sciences, Bayer HealthCare, Biological Products Division, Berkeley*
J. ROSS HAWKINS • *National Institute for Biological Standards and Control, South Mimms, United Kingdom*
WEI-SHOU HU • *Department of Chemical Engineering and Materials Science, University of Minnesota, Minneapolis,* MN
HOLGER HÜBNER • *Institute of Bioprocess Engineering, Friedrich-Alexander-University of Erlangen-Nuremberg, Erlangen, Germany*
ADIBA ISHAQUE • *Process Sciences, Bayer Healthcare, Berkeley, CA*
NIGEL JENKINS • *Bioprocess Development Department, Laboratoires Serono S.A., Fenil-sur-Corsier, Switzerland*
MARTIN JORDAN • *Bioprocess Development Department, Laboratoires Serono S.A., Fenil-sur-Corsier, Switzerland*
KONSTANTIN B. KONSTANTINOV • *Research and Development, Process Sciences, Bayer HealthCare, Biological Products Division, Berkeley, CA*
RASHMI KORKE • *Biogen Idec, Cambridge,* MA
DIRK E. MARTENS • *Food and Bioprocess Engineering Group, Wageningen University, Wageningen, The Netherlands*
OTTO-WILHELM MERTEN • *Généthon, Evry, France*
MARIAM NAEIRI • *Department of Chemical and Bioprocess Engineering and Centre for Synthesis and Chemical Biology, University College Dublin, Belfield, Dublin, Ireland*
DIRK NEHRING • *Institute of Biopharmaceutical Technology, University of Applied Sciences Giessen-Friedberg, Giessen, Germany*
MICHAEL NEUMAIER • *Institute for Clinical Chemistry, Faculty for Clinical Medicine Mannheim, University of Heidelberg, Heidelberg, Germany*
ANDRÉ A. NEVES • *Biochemistry Department, University of Cambridge, Cambridge, United Kingdom*
STEFANIE NITTKA • *Institute for Clinical Chemistry, Faculty for Clinical Medicine Mannheim, University of Heidelberg, Heidelberg, Germany*
JAMES M. PIRET • *Michael Smith Laboratories and Department of Chemical and Biological Engineering, University of British Columbia, Vancouver, British Columbia, Canada*
OSCAR B. J. PLATAS BARRADAS • *Institute for Bioprocess and Biochemical Engineering, Hamburg University of Technology (TUHH), Hamburg, Germany*
RALF PÖRTNER • *Institute for Bioprocess and Biochemical Engineering, Hamburg University of Technology (TUHH), Hamburg, Germany*

Udo Reichl • *Max Planck Institute for Dynamics of Complex Technical Systems; Otto-von-Guericke-Universität Magdeburg; Institute for Bioprocess Technology, Magdeburg, Germany*

Teresa Rodrigues • *Instituto Di Biologia Experimental E Tecnológica (IBET), Oeiras, Portugal*

Maureen Spearman • *Department of Microbiology, University of Manitoba, Winnipeg, Manitoba, Canada*

Glyn N. Stacey • *National Institute for Biological Standards and Control, South Mimms, United Kingdom*

Evert Jan van den End • *Food and Bioprocess Engineering Group, Wageningen University, Wageningen, The Netherlands*

Ranil Wickramasinghe • *Department of Chemical and Biological Engineering, Colorado State University, Fort Collins, CO*

I

Basic Techniques for Establishing Production Cell Lines

1

Transduction Technologies

David A. Fluri and Martin Fussenegger

Summary

Safe and reliable transfer of genetic information at high frequencies into desired target cells remains one of the major challenges in therapeutic life sciences. Multiply attenuated viral vector systems, which are able to transfer heterologous DNA or RNA into almost any cell phenotype as well as in vivo, emerged as the most efficient transgene delivery systems in modern biology. We provide a comprehensive technical overview on the most commonly used viral transduction systems and compare their specific characteristics in order for researchers to make the best choice in a particular scientific setting.

Key Words: Adenovirus; adeno-associated virus; E.REX system; gene therapy; lentivirus; oncoretrovirus.

1. Introduction

The successful transfer of genetic material to target cells still represents a major experimental challenge in all aspects of molecular life sciences, including biotechnology and therapeutic initiatives. A promising delivery method involves the use of engineered animal and human viruses. Such vectors are designed to transfer a desired transgene by exploiting the efficient delivery machinery of the wild-type virus while at the same time minimizing the risk of pathogenicity.

Different systems, based on animal as well as human viruses, have recently been developed and, in some cases, advanced to clinical application. Worldwide more than 400 clinical trials using viral vectors have been, or are currently being, conducted *(1)*. The majority of these trials have been performed using retroviral vectors, followed by adenoviral and other viral vector systems, among them adeno-associated virus (AAV) vectors. Because of the high transfer efficiency and reasonably straightforward production protocols, viral vectors have also become increasingly popular for routine applications in various fields of life sciences.

The majority of viral vector systems have been designed to retain, and even extend, the viruses' cell tropism while ensuring high biosafety characteristics.

From: *Methods in Biotechnology, Vol. 2: Animal Cell Biotechnology: Methods and Protocols, 2nd Ed.* Edited by: R. Pörtner © Humana Press Inc., Totowa, NJ

For some viruses the ability to transduce many cell types is intrinsically present (e.g., adenovirus or AAV). Other viruses have a far narrower host range (e.g., lentiviruses, which are mainly restricted to CD4-expressing cells). Such viruses must be altered through pseudotyping with envelope or capsid proteins from other compatible virus species, thereby enabling the recombinant vector to transduce a broader variety of cell types.

To ensure biosafety, most vectors are rendered replication incompetent through selective deletion of wild-type viral genetic information, thereby preventing productive viral life cycles within the target cell *(2)*. In addition to the removal of nonessential sequences from the wild-type genome, a fundamental principle underlying the production of recombinant viral particles is the redistribution of indispensable elements to either helper plasmids or producer cell lines. The resulting free space flanked by the viral inverted/long terminal repeats (ITRs/LTRs) is then employed to accommodate the heterologous DNA to be delivered. For a number of vector types only one viral open reading frame (ORF) is replaced and delivered *in trans* (e.g., first-generation adenovirus) *(3)*, whereas for other types the entire viral coding sequence is replaced by foreign DNA with only minimal viral elements retained (e.g., adeno-associated virus, "gutless" adenovirus) *(4,5)*.

The architecture of a viral vector has significant impact on the safety of the transduction system (*see* **Fig. 1**). Typically, the more the viral genome is segregated and lacks overlapping sequences, the smaller is the probability for recombination-based generation of replication-competent revertants during the production process. Because wild-type virus is on the whole highly optimized and adapted to its environment, altering its organization and expression patterns, as is done for the construction of viral vectors, is likely to decrease production efficiency. This often manifests in lower titers, poor encapsidation of viral genomes, or changes in transduction efficiency of recombinant viral particles. It is therefore crucial to find an optimal balance between high-safety profiles while retaining high efficiency and quality of viral particle production.

Every viral transduction system has its unique characteristics that render it more suitable for one or another application. Crucial parameters influencing selection of appropriate transduction systems are outlined in **Table 1**. For every application one should carefully consider the advantages and drawbacks of each system and decide according to the experimental requirements.

This chapter provides detailed protocols and guidelines for the production and purification of the four most common viral transduction systems; namely, oncoretroviral, lentiviral, adenoviral, and AAV vectors. At the end of the chapter we provide an example in which AAV vectors have been engineered to deliver adjustable reporter genes in vitro to exemplify the convenience of viral delivery systems for various applications.

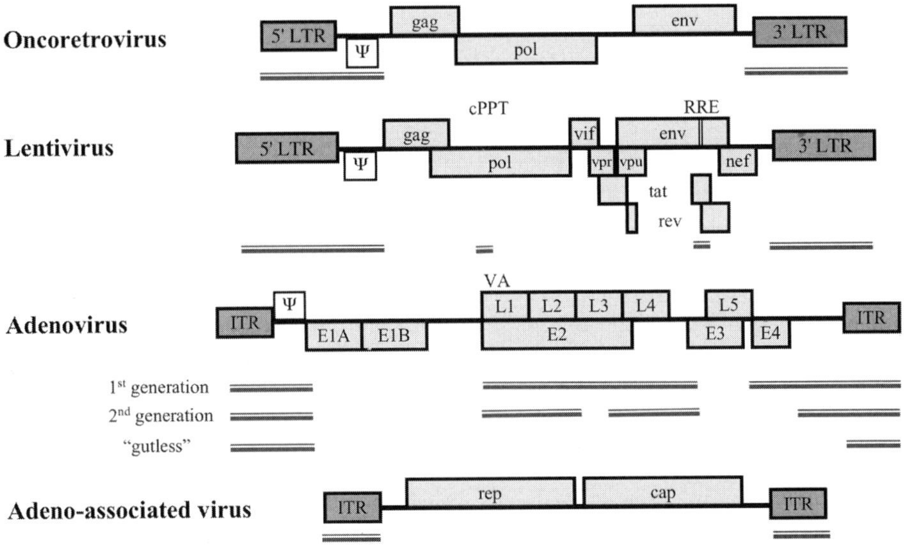

Fig. 1. Schematic representation of the most popular viral transduction systems. Elements originating from the wild-type virus, which are retained in transfer vectors, are underlined. Only essential viral elements are shown and not drawn to scale. Abbreviations: cap, gene coding for structural adeno-associated virus (AAV) proteins; cPPT, central polypurine tract; E1-E4, adenovirus early transcription units 1–4; env, gene encoding envelope protein; gag, gene encoding core proteins (capsid, matrix, nucleocapsid); ITR, inverted terminal repeat; pol, gene encoding viral reverse transcriptase and integrase proteins; L1–L5, adenovirus late transcription units 1–5; LTR, long terminal repeat; rep, gene coding for replication and multiple other AAV functions; RRE, rev responsive element; nef, rev, tat, vif, vpr, vpu, genes encoding for accessory lenitviral proteins; VA, gene encoding viral associated RNA; ψ, packaging signal.

2. Materials

2.1. General

1. HEK293-T cells ***(6)***.
2. Dulbecco's modified Eagle's medium (DMEM) complete medium: DMEM medium (DMEM, Invitrogen, Carlsbad, CA) containing 10% heat-inactivated fetal calf serum (FCS; PAN Biotech GmbH, Aidenbach, Germany, cat. no. 3302-P231902) and 1% penicillin–streptomycin solution (Sigma, St. Louis, MO, cat. no. P4458-100).
3. Trypsin solution: trypsin–ethylenediaminetetraacetic acid (EDTA) solution in phosphate-buffered saline (PBS; PAN Biotech GmbH, Aidenbach, Germany, cat. no. P10-023100).
4. Sterile $CaCl_2$ (1 *M*) solution.
5. Sterile 2X HeBS solution (HeBS: 0.28 *M* NaCl, 0.05 *M* hydroxyethyl piperazine ethane sulfonate [HEPES], 1.5 m*M* Na_2HPO_4, pH 7.08).

Table 1
Selection Criteria for Viral Transduction Systems

	Oncoretroviral vectors	Lentiviral vectors	Adenoviral vectors	Adeno-associated viral vectors
Transduction of nondividing cells	–	+	+	+
Stability of viral particle	+/–	+/–	+/–	+
Episomal maintenance	–	–	+	+/–
Integration into the chromosome	+	+	–	+/–
Extensive packaging capacity	–	–	+	–
Fast production protocols	+	+	–	+
High safety profile	+/–	+/–	+/–	+
Low immunogenicity	+	+	+/–	+/–
Clinical data available	+	–	+	+

6. PBS solution without magnesium and calcium, pH 7.2 (PBS, Sigma, cat. no. D-5652).
7. Cell culture-certified disposable plastic ware: Petri dishes, 6-well plates, 24-well plates (TPP, Trasadingen, Switzerland), pipets, Falcon tubes 15 mL and Eppendorf tubes 1.5 mL (Greiner bio-one, Kremsmuenster, Austria).
8. Ultracentrifuge tubes (e.g., Beckman Quick-Seal tubes [Beckman cat. no. 342 413] or Sorvall centrifuge tubes [Sorvall, cat. no. 03141]).
9. 0.45-µm filters (Schleicher & Schuell GmbH, Dassel, Germany).
10. Plasmid DNA purification kits (Jet-Star 2.0 Midiprep, Genomed, AG, Bad Oyenhausen, Germany; Wizard mini-prep kit, Promega, Madison, WI).
11. Cell counting device (Casy1® counter, Schaerfe System, Reutlingen, Germany); alternatively, a hemacytometer can be used.
12. Incubator (e.g., Hereaus, HERAcell, Langenselbold, Germany) for cultivation of mammalian cells at 37°C in a humidified atmosphere containing 5% CO_2.
13. Ultracentrifuge equipped with swing-out rotor (e.g., Sorvall, Hanau, Germany).

2.2. Oncoretroviral/Lentiviral Vectors

1. GP-293 cells (Clontech, Palo Alto, CA, cat. no. K1063-1).
2. pVSV-G (Clontech).
3. pLEGFP-N1 (Clontech).
4. pMF365 ***(6)***.
5. pCD/NL-BH* ***(7)***.
6. pLTR-G ***(8)***.
7. Advanced DMEM (Gibco, cat. no. 12491-015).
8. Chemically defined lipid concentrate (Gibco, cat. no. 11905-031).

9. Egg lecithin (Serva Electrophoresis GmbH, cat. no. 27608).
10. Cholesterol (Sigma, cat. no. C-4951, diluted in PBS).

2.3. Adenoviral Vectors

1. Buffer A (10 m*M* Tris-HCl, 1 m*M* $MgCl_2$, 135 m*M* NaCl, pH7.5).
2. Buffer B (buffer A + 10% glycerol).
3. Cesium chloride (Gibco, cat. no. 15507-023).
4. Agarose (Brunschwig, Basel, Switzerland, cat. no. 8008).
5. Phenol/chloroform/isoamylalcohol 25:24:1 saturated with 10 m*M* Tris/1 m*M* EDTA (Sigma, cat. no. P-3803).
6. Ethanol 96%.
7. TE buffer (10 m*M* Tris, 1 m*M* EDTA).
8. Glycerol (Fluka, cat. no. 49767).
9. 16G needles (Becton Dickinson, Franklin Lakes, NJ).
10. Dialysis tubing (Sigma, cat. no. D9777-100FT).
11. Dialysis closures (Sigma, cat. no. Z370959-10EA).
12. HEK293 (Microbix, Toronto, Canada).
13. Adenoviral type 5 genomic vector pJM17 (Microbix).
14. Adenovector encoding desired transgene (e.g., pVN31 *[9]*).
15. Proteinase K (Qiagen GmbH, Hilden, Germany, cat. no. 19131).

2.4. Adeno-Associated Viral Vectors

1. pDG *(10)*.
2. Spin columns MWCO 30,000 (Vivaspin, Vivascience, Germany).
3. Heparin affinity columns (HiTrap heparin HP, Amersham Biosciences, Sweden).
4. Pump, allowing flow rates of 5 mL/min.
5. Liquid nitrogen or dry ice/ethanol bath.
6. Binding buffer (10 m*M* Na_2PO_4, pH 7).
7. Elution buffer (10 m*M* Na_2PO_4, 1 *M* NaCl, pH 7).
8. Storage buffer (binding buffer containing 20% ethanol).
9. PBS-MK buffer (PBS 1x containing 1 m*M* $MgCl_2$ and 2.5 m*M* KCl).
10. Benzonase (Sigma, cat. no. E1014).
11. Optiprep™ iodixanol solution (Axis-Shield, Rodelokka, Norway).
12. Phenol red solution 0.5% (Sigma, cat. no. P0290).

2.5. Trigger Inducible Transgene Expression

1. HT-1080 cells (ATCC, CCL-121).
2. AAV vector encoding desired transgene (e.g., pDF141; **Fig. 2**).
3. Erythromycin (Fluka, Buchs, Switzerland, cat. no. 4573) stock solution of 1 mg/mL in 96% ethanol.
4. 2x Secreted alkaline phosphatase (SEAP) buffer: (20 m*M* homoarginine, 1 m*M* $MgCl_2$, and 21% (v/v) diethanolamine/HCl, pH 9.8.).
5. pNPP solution (120 m*M* *para*-nitrophenylphosphate [pNPP hexahydrate, Chemie Brunschwig, Basel, Switzerland, cat. no. 12886-0100]) in 2x SEAP buffer.

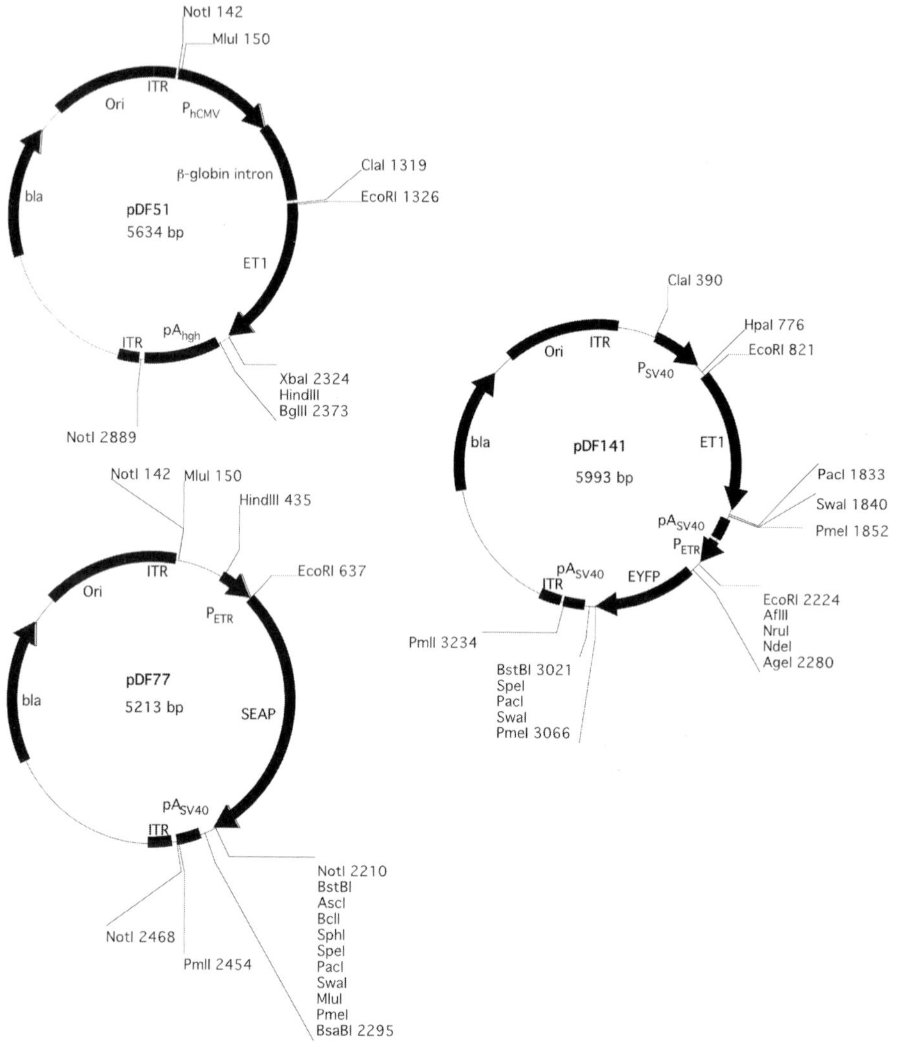

Fig. 2. Plasmid maps of adeno-associated viral vectors engineered for macrolide-responsive expression of transduced transgenes. Vectors encoding (1) the macrolide-dependent transactivator ET1 driven by a constitutive promoter P_{hCMV} (pDF51), (2) SEAP driven by the erythromycin-responsive promoter P_{ETR} (pDF77), and (3) a macrolide-responsive self-regulated one-vector configuration combining P_{SV40}- driven ET1 and P_{ETR}-driven EYFP expression (pDF141). Abbreviations: bla, β-lactamase; ET1, erythromycin transactivator; EYFP, enhanced yellow fluorescent protein; ITR, inverted terminal repeat; ORI, origin of replication; pA_{hgh}, human growth hormone polyadenylation signal; pA_{SV40}, simian virus 40 polyadenylation signal; P_{hCMV}, human immediate early cytomegalovirus promoter; P_{ETR}, erythromycin-responsive promoter; P_{SV40}, simian virus 40 promoter; SEAP, secreted alkaline phosphatase.

6. Microplate reader (GeniusPro, Tecan Group Ltd., Maennedorf, Switzerland).
7. Fluorescence microscope (DMI6000B, Leica Inc., Heerbrugg, Switzerland) equipped with CFP- and YFP-specific filters.

3. Methods

All of the viral vectors introduced in this manual are classified as biosafety level 2 according to the National Institutes of Health (NIH). Working with these systems therefore requires appropriate safety equipment and governmental regulations may apply. Additional precautions must be followed if working with vectors hosting toxic or oncogenic inserts.

The production protocols for all of the following viral vectors are based on comparable techniques. Nevertheless, the protocols for each viral vector system have been separately provided (apart from general procedures which are discussed in the first chapter) to ensure simplicity for the user and to provide comprehensive instructions for each virus type.

3.1. General Procedures

All plasmids were propagated in *Escherichia coli* DH5α. Standard cloning techniques were performed to engineer plasmids ***(11)***. The DNA used for virus production was purified using a silica-based anion-exchange DNA purification kit (Genomed) or with "Wizard" mini-prep kits (Promega) (*see* **Note 1**), according to the manufacturers' protocols. DNA concentrations were measured with an ultraviolet (UV) spectrometer (BioPhotometer, Eppendorf, Hamburg, Germany), and only DNA with a 260/280 coefficient of greater than 1.75 was used for transfections. Unless otherwise specified, HEK293 cells were cultivated in DMEM supplemented with 10% FCS and 1% penicillin/streptomycin solution in a humidified atmosphere at 37°C containing 5% CO_2 (*see* **Note 2**).

3.2. Oncoretroviral/Lentiviral Vectors

In this chapter oncoretroviral and lentiviral vectors are introduced together because the two systems are closely related and share numerous features. However, separate production protocols are provided for each vector system to simplify their utilization.

Both oncoretroviral and lentiviral vectors are enveloped RNA viruses packaging two copies of the viral RNA genome. Oncoretroviral vectors have been derived from different animal viruses such as murine leukemia virus (MLV), Rous sarcoma virus (RSV), or avian leukosis virus (ALV) and were among the first viral transduction systems used in gene therapy. The most commonly used lentiviral vectors are derived from human immunodeficiency virus type 1 (HIV-1) ***(6,7)***. However, members of the lentivirus family, such as simian immunodeficiency viruses (SIVs), feline immunodeficiency virus (FIV), or equine infectious anemia virus (EIAV), also serve as parental viruses for the generation of vectors for gene therapy purposes ***(12)***. Oncoretroviruses and lentiviruses randomly integrate as proviruses into target chromosomes. In contrast to oncoretroviruses, lentiviruses encode accessory genes, which facilitates transduction of mitotically inert target cells.

Oncoretroviral vector design is based upon the replacement of the viral gag, pol, and env ORFs by the desired transgene under the control of either a heterologous promoter or the viral long terminal repeat (LTR). The only elements required *in cis* are the extended packaging signal (Ψ^+) and the viral LTRs together with adjacent regions essential for reverse transcription and integration ***(13,14)***.

The design of lentivectors is very similar to oncoretroviral vectors but requires additional viral elements provided *in trans*, such as the Rev and Tat proteins. Besides the LTRs and the packaging signal, lentiviral vectors also contain the Rev-responsive element (RRE) and the central polypurine tract (cPPT) to enable efficient nuclear transport and second DNA strand priming, respectively.

Both oncoretroviral and lentiviral vectors can be efficiently pseudotyped with envelope proteins from other virus species, enabling the relatively narrow host-range of the wild-type virus to be broadened to a wider variety of cell types. The envelope protein most frequently used for pseudotyping is derived from the vesicular stomatitis virus glycoprotein (VSV-G), which shows fusogenic activity and uses ubiquitous cell-surface receptors ***(8,15)***.

Retroviral vectors have become very attractive for gene therapy initiatives because of their stable integration of genetic information combined with low immunogenicity, improved safety profiles through use of self-inactivating third-generation vectors ***(16)***, and simple production protocols. Nonetheless, safety concerns remain because of random transgene integration resulting in the risk of proto-oncogene activation or insertional mutagenesis.

In this chapter we cover the production of small-scale crude oncoretroviral and lentiviral particle preparations by transient transfection methods and provide simple purification and concentration protocols using ultracentrifugation.

3.2.1. Retroviral Vector Production

Day 1: Seeding of cells for transfection:

1. Seed GP-293 cells at a concentration of 300,000 cells per well of a 6-well plate and cultivate at 37°C in DMEM supplemented with 10% FCS, 1% penicillin/streptomycin solution.

Day 2: Calcium phosphate transfection of GP-293:

1. Prepare Eppendorf tube containing 80 µL of HeBS solution.
2. Prepare Eppendorf tube containing 2 µg pVSV-G and 2 µg of plasmid encoding the retroviral expression construct (e.g., pLEGFP-N1) in 60 µL ddH_2O. Add 20 µL of 1M $CaCl_2$ solution.

3. Add DNA mix dropwise to HeBS solution while bubbling with a pipet, vortex 5 s, and let sit for 2 min. Meanwhile, add chloroquine at a final concentration of 25 μM to the cells. Add precipitates to the cells and let sit for 6 h.
4. Replace medium with fresh complete DMEM and incubate for another 48 h.

Day 4: Virus harvest:

1. Collect virus-containing supernatant from GP-293 culture and filter through a 0.45-μm filter (Schleicher & Schuell).
2. Aliquot the viral particle preparation and freeze at –80°C.
3. Concentrate viral stocks by ultracentrifugation for 3.5 h at 4°C and 70,000g in an ultracentrifuge equipped with a swing-out rotor and optiseal tubes (Beckman Instruments Inc., Fullerton, CA; cat. no. 361625).
4. Discard supernatant and resuspend the pellet in an appropriate volume of PBS and store aliquots at –80°C or in liquid nitrogen (*see* **Note 5**).

3.2.2. Lentiviral Vector Production

Day 1: Seeding of cells for transfection:

1. Seed 500,000 HEK293-T cells in one well of a 6-well plate and cultivate at 37°C in advanced DMEM supplemented with 2% FCS, 0.01 m*M* cholesterol, 0.01 m*M* egg lecithin, and 1x chemically defined lipid concentrate (*see* **Note 3**).

Day 2: Calcium phosphate transfection of HEK293-T:

1. Prepare Eppendorf tube containing 80 μL of HeBS solution.
2. Prepare Eppendorf tube containing 1 μg of pCD-NL (helper construct), 1 μg of pLTR-G (construct encoding the VSV-G gene), and 1 μg of the vector encoding the lentiviral expression construct (e.g., pMF365) in 60 μL ddH_2O. Add 20 μL of 1 *M* $CaCl_2$ solution.
3. Add DNA mix dropwise to HeBS solution while bubbling with a pipet, vortex 5 s, and let sit for 2 min. Meanwhile, add chloroquine at a final concentration of 25 μ*M* to the cells. Add precipitates and let sit for 5 h.
4. Replace medium with fresh advanced DMEM containing 2% FCS, cholesterol, egg lecithin, and chemically defined lipid concentrate. Let sit for a further 48 h.

Day 4: Virus harvest:

1. Collect virus-containing medium from HEK293-T cells and filter through a 0.45-μm filter (Schleicher & Schuell) (*see* **Note 4**).
2. Aliquot the viral particle preparation and freeze at –80°C or in liquid nitrogen (*see* **Note 5**).
3. Concentration of viral stocks: lentiviral particles can be purified using the protocol for oncoretroviral vector ultracentrifugation.

3.3. Adenovirus

Adenoviral vectors are among the oldest and most widely used vectors for gene therapy applications. The advantages of this vector type are the extensive packaging capacity, the ability to transduce a variety of different cell types (including nondividing cells), and its potential for high-titer preparations. Adenovirus is a nonenveloped icosahedral structure and for its genome contains a double-stranded DNA molecule of 30–40 kb. The genome can be classified according to the expression time point of two major overlapping regions; the early E and the late L ORFs. More than 50 distinct human adenovirus serotypes are known, some of which can cause respiratory, intestinal, or eye infections.

Adenoviral vectors have been generated by deleting E1 and E3 ***(17)*** (first generation), E1, E3, and E2/4 ***(18)*** (second generation), and adenoviruses devoid of any viral genes containing only the packaging signal and the viral ITRs (high-capacity, "gutless" vectors) ***(19)***. First-generation vectors are able to host transgenes of up to 8 kb. Following deletion of additional information in second- and third-generation vectors, the packaging capacity has been increased to 35 kb for gutless vectors, which was associated with a significant reduction in immunogenicity.

Several different production systems for first-generation adenoviral vectors are available. Older systems use homologous recombination in HEK293 cells to generate recombinant adenoviral vectors. Homologous recombination between sequences on the left end of the adenovirus genome and sequences on the shuttle plasmid containing the transgene as well as the right end of the adenoviral genome lead to recombinant virus ***(20)***. Newer systems take advantage of recombination-triggering enzymes, such as the phage P1-derived enzyme Cre, to enable efficient recombination of the viral components (1) in vitro ***(18)***, (2) in bacteria ***(21,22)***, or (3) in a Cre-expressing helper cell line ***(23)***.

The following protocol is a standard method for producing first-generation adenoviral vectors by homologous recombination in HEK293 cells. This protocol can be adapted to other production systems by changing helper plasmids and production cell lines.

3.3.1. Production of Adenoviral Particles

3.3.1.1. Production of First-Round Virus

Day 1: Seed 8×10^5 HEK293 (Microbix) cells in a 6-cm culture dish.

Day 2: $CaCl_2$ transfection for virus production:

1. Prepare Eppendorf tube containing 240 μL of HeBS solution.
2. Prepare Eppendorf tube containing 3 μg of pJM17 (genomic plasmid) and 3 μg of shuttle plasmid (e.g., pVN31(9)) encoding the transgene in 180 μL of ddH_2O. Add 60 μL of 1 M $CaCl_2$ solution.
3. Add the DNA mix dropwise to the HeBS solution while bubbling with a pipet, vortex 5 s, and let sit for 2 min. Add the precipitates to the cells and let sit for 6 h.
4. Remove medium and overlay cells with a mix of complete DMEM containing 0.5% agarose (*see* **Note 6**).

Days 5–10: Plaques should appear within 5–10 d:

1. Pick plaque with sterile Pasteur pipet and transfer to 1 mL of PBS containing 10% glycerol (optionally freeze at –80°C until use).

3.3.1.2. Analysis of Viral DNA by Test Digestion

It is recommended that restriction analysis of the backbone is performed prior to amplification of the virus.

Day 1: Seed 8×10^5 HEK293 (Microbix) cells in a 6-cm dish.

Day 2: Transduction:

1. Add 200 µL of virus suspension from **Subheading 3.3.1.1.** to the cells (cells should be at 80% confluence) and adsorb virus for 30 min by shaking plate from time to time to disperse solution over cells.
2. Add 5 mL of DMEM +10% FCS and incubate at 37°C.
3. Cytopathic effect (CPE) should be visible after 3–4 d.

Days 3–4: DNA isolation and restriction digests:

1. Remove medium gently by avoiding the transfer of cells, transfer to Falcon tube, and freeze at -80°C.
2. Remove remaining medium on the cells and add 360 µL of TE buffer containing 0.5% sodium dodecyl sulfate and 0.2 mg/mL of Proteinase K and digest overnight.
3. Transfer lysate to Eppendorf tube and extract one time with saturated phenol. Transfer aqueous phase to new tube.
4. Add 1 mL of ethanol and vortex briefly.
5. Wash twice with ethanol to remove phenol.
6. Dry pellet and resuspend in 50 µL of TE buffer.
7. Digest with appropriate restriction enzyme and run on agarose gel (*see* **Note 7**).

3.3.1.3. Amplification of Adenoviral Particle Stock

After vector verification, the initial stock can be used for further amplification steps.

Day 1: Seed desired numbers of HEK293 cells in 10-cm dishes.

Day 2: When cells reach 80–90% confluence:

1. Remove medium and infect at a multiplicity of infection (MOI) of 1–10 (*see* **Note 8**) per cell.
2. After 30–60 min remove virus solution and refeed with DMEM +10% FCS. Incubate in humidified atmosphere and check daily for CPE.

Days 4–6: When most of the cells are rounded up but not detached:

1. Scrape the cells from the plate with a cell scraper and transfer cells with the medium to a Falcon tube.
2. Centrifuge at 2000g for 15 min, remove supernatant, and resuspend pellet in desired volume of PBS and freeze at –80°.
3. Freeze–thaw three times by shuttling the sample between liquid nitrogen and a 37°C water bath.

4. Centrifuge the cell lysate for 15 min at 2000*g* to pellet cell debris.
5. Transfer supernatant to new tube → crude viral stock (*see* **Note 5**).

3.3.1.4. Purification of Particles by Cesium Chloride (CsCl) Gradient

For certain applications (mostly in vivo experiments), subsequent purification of crude adenoviral preparations is required. Presented below is a standard purification method using CsCl centrifugation:

1. Add 1.5 ml of a 1.45 g/mL CsCl solution and 2.5 mL of a 1.33 g/mL CsCl solution to a 7-mL ultraclear tube (place the least dense layer at the bottom of the tube; float this layer on top of the more dense layer by placing the tip of the syringe at the bottom of the tube).
2. Thaw crude viral stock and layer the virus solution on top of the gradient.
3. Add 2 mm of mineral oil on top of the virus solution and equilibrate tubes.
4. Centrifuge for 2 h at 90,000*g* (at 4°C).
5. Remove the lowest of the three visible white bands by side-wall puncture with a 16G needle (Becton-Dickinson).
6. Dilute the virus fraction with 0.5 volumes of TE buffer pH 7.8.
7. Layer the solution on top of a second gradient prepared as above but this time with 1 mL of 1.45 g/mL and 1.5 mL of 1.33 g/mL CsCl and proceed as in **steps 1**, **2**, and **3**.
8. Recover the virus particle-containing band, transfer to dialysis tubing, seal with clamps, and dialyze for 1 h against 1l of buffer A.
9. Transfer tubing to another beaker containing 1l of buffer B (buffer A + 10% glycerol) and dialyze for another 2 h.
10. Aliquot virus and freeze at –80°C (*see* **Note 5**).

3.4. Adeno-Associated Virus

In recent years AAV vectors have become widely used in clinical initiatives because of their unique safety properties. So far, no disease has been attributed to the wild-type virus, even though a majority of the human population is seropositive for one or more of the different subtypes. For productive infection AAV is dependent on a helper virus—either adenovirus or herpes simplex virus. In the absence of helper functions, the AAV enters a nonproductive life cycle and integrates site specifically into a locus on human chromosome 19 ***(24,25)***. Recombinant AAVs lack this characteristic and remain predominantly episomal (approx 90%), although a small proportion does still randomly integrate into the host chromosome (approx 10%) ***(26)***.

The unique design of recombinant adeno-associated viral vectors (rAAVs), where all viral ORFs have been replaced and only the ITRs are delivered *in cis*, renders this transduction system exceptionally safe. Furthermore, the ability of AAV vectors to transduce nondividing cells and provide sustained expression of the transgene in a wide variety of tissues makes AAVs one of the most promising candidates for clinical applications.

Standard production protocols use transient transfection of plasmids encoding the desired transgene(s) flanked by viral ITRs together with a plasmid providing the AAV Rep and Cap proteins *in trans*. HEK293 production cells already provide the crucial adenoviral functions performed by viral E1A and E1B. In addition, other adenoviral proteins such as E2A, E4orf6, and VA are required for efficient production of AAVs. In older protocols these functions are provided by helper adenovirus infection of producer cells *(27,28)*. A major drawback of such protocols is the contamination of rAAV stocks with wild-type helper adenovirus. State-of-the-art protocols replace helper virus infection completely by providing the essential proteins on engineered helper plasmids.

In this section we provide a production protocol where all adenoviral helper functions (except E1) are combined with Rep/Cap on a single plasmid ***(10)***. Even though titers obtained with such helper-virus-free production protocols are occasionally lower, the advantages in respect of purity and safety outbalance the drawbacks.

Also provided in this section are detailed protocols for the production and purification of crude AAV stocks, as well as purification and concentration strategies for use in animal studies.

3.4.1. Production of Crude Adeno-Associated Viral Vector Preparations

Day 1: Seeding of cells for transfection:

1. Seed 300,000 HEK293-T cells per well of a 6-well plate (or for medium-scale production seed 2.5 Mio cells into a 10-cm culture dish).

Day 2: Calcium phosphate transfection of HEK293-T:

1. Prepare tube containing 80 µL (400 µL) of HeBS solution.
2. Prepare tube containing 1.2 µg (8 µg) plasmid containing transgene of interest flanked by the viral ITRs and 2.4 µg (16 µg) pDG[10] (helper plasmid containing all required adenovirus functions plus AAV rep and cap ORFs) in 60 µL (300 µL) of ddH_2O. Add 20 µL (100 µL) of 1M $CaCl_2$ solution.
3. Add DNA dropwise to HeBS solution while bubbling the solution using a pipet, vortex the mix for 5 s, and let sit for 2 min. Add the precipitates dropwise to the cells and let sit for another 6 h.
4. Replace medium with fresh complete medium and incubate for 60 h.

Day 5: Virus harvest:

1. In the morning, remove medium from cells, wash carefully with 400 µL (2 mL) PBS and add another 500 µL (2 mL) of PBS to the cells.
2. Detach cells by pipetting with a 1000-µL micropipet, transfer cell-suspension to an Eppendorf (15-mL Falcon) tube, and vortex vigorously.
3. Freeze–thaw three times by shuttling the sample between a 37°C water bath and liquid nitrogen (vortex vigorously after each thawing step).

4. Centrifuge for 5 min at 8000g.
5. Transfer supernatant to new tube (→ crude viral stock) and store at –80°C (*see* **Note 5**).

3.4.2. Virus Purification (for Medium-Scale High-Purity Virus Produced in 10-cm Culture Dishes)

1. Add 50 U/mL Benzonase (Sigma) to crude viral stock and incubate for 30 min at 37°C.
2. Dilute crude viral stock in PBS to a final volume of 12 mL.
3. In an ultracentrifuge tube prepare an iodixanol step gradient by pipetting the crude viral preparation onto a solution containing 6 mL of a 15% iodixanol and 1 *M* NaCl in PBS-MK buffer. Continue to sequentially pipet 5 mL of a 25%, 4 mL of a 40% and 4 mL of a 60% iodixanol solution (all in PBS-MK) under the 15% iodixanol solution. For better distinction of the iodixanol layers, add 2.5 μL/mL of a 0.5% Phenol Red stock solution to the 60% and 25% iodixanol layers.
4. Centrifuge for 3.5 h at 150,000*g* (18°C).
5. Harvest the clear 40% iodixanol fraction after puncturing the tube on the side with a 16G needle equipped with a syringe.
6. Equilibrate heparin affinity column with 5–10 bed volumes of binding buffer.
7. Filter gradient-purified virus preparation through a 0.45-μm filter (Schleicher & Schuell), and run it over the column at a flow rate of approx 3 mL/min.
8. Wash with 10 bed volumes of binding buffer.
9. Elute bound virus with 5 bed volumes of elution buffer.
10. Transfer eluate to a vivaspin column and spin for 20 min at 2000*g*, resuspend in 5 mL PBS, and spin another 20 min at 2000*g*.
11. Resuspend in the desired volume of PBS, aliquot, and freeze at –80°C.

3.5. Trigger-Inducible Transgene Transduction Using Engineered AAV

Systems enabling regulated transgene expression are invaluable tools in different aspects of molecular life sciences. Delivery of a desired transgene coupled with the capability to tightly regulate its expression is essential for various applications ***(29–32)***.

Presented above are different methods for efficient delivery of transgenes to target cells capitalizing on viral transduction systems. Here we provide an application example using AAV vectors to transfer and regulate intracellular and secreted reporter transgenes. This protocol may be amenable to other combinations of viral vectors, transgenes, and transgene control systems.

3.5.1. Viral Transduction Using AAV

Day 1: Seeding of cells:

1. Seed 40,000 HT-1080 cells per well of a 24-well plate (9 wells) and 200,000 HT-1080 cells per well of a 6-well plate (2 wells).

Day 2: Transduction:

1. Thaw vector stocks in a 37°C water bath and vortex tubes briefly.
2. Add pDF141-derived AAVs at a multiplicity of infection (MOI) of 5 (*see* **Note 8**) to 2 wells of 6-well plate and pDF77- and pDF51-derived AAVs at identical MOIs to 6 wells of a 24-well plate (*see* **Fig. 2** for plasmid maps).
3. Add erythromycin (EM) at 50 ng/mL to one of the wells transduced with pDF141 and to 3 of the wells containing pDF143.
4. Incubate overnight at 37°C in a humidified atmosphere containing 5% CO_2.

Day 3: Medium change:

1. Change medium the following day and add new DMEM complete, supplemented with 50 ng/mL EM where transgene expression should be repressed.
2. Incubate for another 48 h in a humidified atmosphere containing 5% CO_2.

Day 5: Analysis:

Fluorescence microscopy: analyze samples using a fluorescence microscope (Leica) equipped with appropriate filters (CFP/YFP filter cube, Leica) (*see* **Fig. 3**).

SEAP quantification:

1. Harvest 120 μL of supernatants per well from pDF77/51-transduced cells and 120 μL from nontransduced cells as a negative control.
2. Heat-inactivate supernatants for 20 min at 65°C.
3. Incubate for 5 min on ice and then spin down at top speed for 5 min.
4. Prepare 9 wells of a 96-well assay plate with 100 μL of 2x SEAP assay buffer, prewarm at 37°C.
5. Transfer 80 μL culture supernatant to wells containing 2x SEAP buffer.
6. Add 20 μL of substrate solution to each well.
7. Quantify absorbance at 405 nm for up to 1 h in a microplate reader (Tecan) (*see* **Fig. 3**).

- Subtract values of the blank from sample values.
- Calculate enzymatic activity:

$$\text{EA [U/L]} = \Delta\text{Abs/min}*\nu/(\varepsilon*d)*10^6$$

where dilution factor n (volume measurement/volume sample) = 200/80; absorption factor of ε = 18600/M/cm; and lightpath d = 0.5 cm.

4. Notes

1. To perform transfections we recommend using ethanol or iso-propanol-precipitated DNA for sterility reasons. For constructs used in larger amounts (e.g., helper constructs) we recommend performing midi- or maxi- DNA preparations in order to minimize batch-to-batch variations of virus preparations. For small-scale virus preparations, DNA purification methods based on rapid precipitation-free protocols (e.g., Wizard mini DNA purification system) are sufficient.
2. We observed decreasing efficiency of viral particle production upon transient transfection when using HEK293 cells at higher passage number and recommend therefore working with low-passage cell populations.

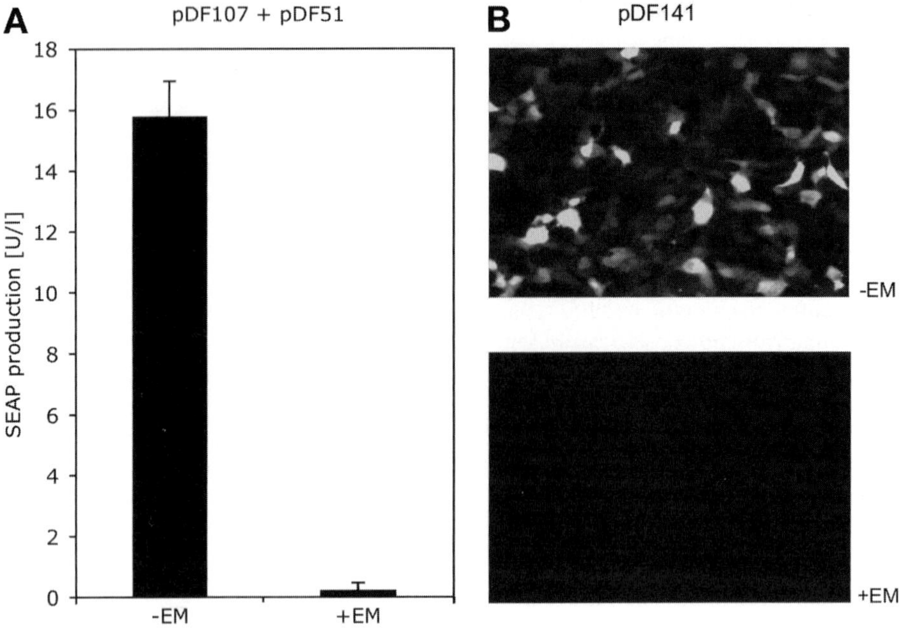

Fig. 3. Macrolide-responsive transgene transduction using engineered adeno-associated viral (AAV) particles. **(A)** SEAP expression levels of HT-1080 cells co-transduced with pDF51- and pDF77-derived AAV particles (*see* **Fig. 2**). **(B)** Fluorescence micrographs of HT-1080 transduced with pDF141-derived AAV particles delivering a self-regulated macrolide-responsive EYFP expression unit and grown in the presence (+EM) and absence of erythromycin (-EM). Abbreviations: EM, erythromycin; SEAP, secreted alkaline phosphatase.

3. Production of lentiviral particle stock is feasible using standard DMEM containing 10% FCS without providing cholesterol, egg lecithin, and chemically defined lipid concentrate, but infectious titers are significantly lower *(33)*.
4. It is crucial to use low-protein-binding filters.
5. Storage at -80° (up to 6 mo) is sufficient. For long-term storage we recommend liquid nitrogen. Oncoretroviral and lentiviral vectors should be aliquotted in small volumes since preparations lose approximately half of their infectivity with every freeze–thaw step.
6. Some protocols forgo this step by directly cultivating the cells in liquid medium and then harvesting the entire supernatant for further procedures. We recommend initial plaque purification and analysis of the first produced viral stocks to ensure consistent starting material for further steps.
7. Upon complete CPE the DNA preparation should be sufficiently pure for restriction analysis. Nevertheless, genomic DNA will be visible as background smear and some enzymes (e.g., *Hind*III) repeatedly cut genomic DNA yielding bands, which are not to be confused with viral DNA.

8. Vectors encoding for fluorescent proteins are titrated by applying serial dilutions of vector preparations to target cells in 96-well plates (at low density) and subsequent counting of transgene-expressing cells by fluorescence microscopy.

References

1. Edelstein, M. L., Abedi, M. R., Wixon, J., and Edelstein, R. M. (2004) Gene therapy clinical trials worldwide 1989-2004-an overview. *J. Gene Med.* **6**, 597–602.
2. Somia, N., and Verma, I. M. (2000) Gene therapy: trials and tribulations. *Nat. Rev. Genet.* **1**, 91–99.
3. Danthinne, X. and Werth, E. (2000) New tools for the generation of E1- and/or E3-substituted adenoviral vectors. *Gene Ther.* **7**, 80–87.
4. Xiao, X., Li, J., and Samulski, R. J. (1998) Production of high-titer recombinant adeno-associated virus vectors in the absence of helper adenovirus. *J. Virol.* **72**, 2224–2232.
5. Morsy, M. A., Gu, M., Motzel, S., et al. (1998) An adenoviral vector deleted for all viral coding sequences results in enhanced safety and extended expression of a leptin transgene. *Proc. Natl. Sci. USA* **95**, 7866–7871.
6. Mitta, B., Rimann, M., Ehrengruber, M. U., Ehrbar, M., Djonov, V., Kelm, J., and Fussenegger, M. (2002) Advanced modular self-inactivating lentiviral expression vectors for multigene interventions in mammalian cells and in vivo transduction. *Nucleic Acids Res.* **30**, e113.
7. Mochizuki, H., Schwartz, J. P., Tanaka, K., Brady, R. O., and Reiser, J. (1998) High-titer human immunodeficiency virus type 1-based vector systems for gene delivery into nondividing cells. *J. Virol.* **72**, 8873–8883.
8. Reiser, J., Harmison, G., Kluepfel-Stahl, S., Brady, R. O., Karlsson, S., and Schubert, M. (1996) Transduction of nondividing cells using pseudotyped defective high-titer HIV type 1 particles. *Proc. Natl. Sci. USA* **93**, 15,266–15,271.
9. Gonzalez-Nicolini, V., and Fussenegger, M. (2005) A novel binary adenovirus-based dual-regulated expression system for independent transcription control of two different transgenes. *J. Gene Med.* **7**, 1573–1585.
10. Grimm, D., Kay, M. A., and Kleinschmidt, J. A. (2003) Helper virus-free, optically controllable, and two-plasmid-based production of adeno-associated virus vectors of serotypes 1 to 6. *Mol. Ther.* **7**, 839–850.
11. Sambrook, J. and Russell, D. W. (2001) *Molecular Cloning A Laboratory Manual*, Cold Spring Harbor Laboratory Press, Cold Spring Harbor, New York.
12. Curran, M. A. and Nolan, G. P. (2002) Nonprimate lentiviral vectors. *Curr. Top. Microbiol. Immunol.* **261**, 75–105.
13. Mann, R., Mulligan, R. C., and Baltimore, D. (1983) Construction of a retrovirus packaging mutant and its use to produce helper-free defective retrovirus. *Cell* **33**, 153–159.
14. Miller, A. D., Miller, D. G., Garcia, J. V., and Lynch, C. M. (1993) Use of retroviral vectors for gene transfer and expression. *Methods Enzymol.* **217**, 581–599.
15. Naldini, L., Blomer, U., Gallay, P., et al. (1996) In vivo gene delivery and stable transduction of nondividing cells by a lentiviral vector. *Science* **272**, 263–267.

16. Bukovsky, A. A., Song, J. P., and Naldini, L. (1999) Interaction of human immunodeficiency virus-derived vectors with wild-type virus in transduced cells. *J. Virol.* **73**, 7087–7089.
17. Danthinne, X. and Imperiale, M. J. (2000) Production of first generation adenovirus vectors: a review. *Gene Ther.* **7**, 1707–1714.
18. Wang, Q., Jia, X. C., and Finer, M. H. (1995) A packaging cell line for propagation of recombinant adenovirus vectors containing two lethal gene-region deletions. *Gene Ther.* **2**, 775–783.
19. Kochanek, S., Schiedner, G., and Volpers, C. (2001) High-capacity 'gutless' adenoviral vectors. *Curr. Opin. Mol. Ther.* **3**, 454–463.
20. Bett, A. J., Haddara, W., Prevec, L., and Graham, F. L. (1994) An efficient and flexible system for construction of adenovirus vectors with insertions or deletions in early regions 1 and 3. *Proc. Natl. Sci. USA* **91**, 8802–8806.
21. Chartier, C., Degryse, E., Gantzer, M., Dieterle, A., Pavirani, A., and Mehtali, M. (1996) Efficient generation of recombinant adenovirus vectors by homologous recombination in *Escherichia coli*. *J. Virol.* **70**, 4805–4810.
22. Aoki, K., Barker, C., Danthinne, X., Imperiale, M. J., and Nabel, G. J. (1999) Efficient generation of recombinant adenoviral vectors by Cre-lox recombination in vitro. *Mol. Med.* **5**, 224–231.
23. Ng, P., Parks, R. J., Cummings, D. T., Evelegh, C. M., Sankar, U., and Graham, F. L. (1999) A high-efficiency Cre/loxP-based system for construction of adenoviral vectors. *Hum. Gene Ther.* **10**, 2667–2672.
24. Kotin, R. M., Siniscalco, M., Samulski, R. J., et al. (1990) Site-specific integration by adeno-associated virus. *Proc. Natl. Sci. USA* **87**, 2211–2225.
25. Samulski, R. J., Zhu, X., Xiao, X., et al. (1991) Targeted integration of adeno-associated virus (AAV) into human chromosome 19. *EMBO J.* **10**, 3941–3950.
26. Nakai, H., Yant, S. R., Storm, T. A., Fuess, S., Meuse, L., and Kay, M. A. (2001) Extrachromosomal recombinant adeno-associated virus vector genomes are primarily responsible for stable liver transduction in vivo. *J. Virol.* **75**, 6969–6976.
27. Samulski, R. J., Chang, L. S., and Shenk, T. (1989) Helper-free stocks of recombinant adeno-associated viruses: normal integration does not require viral gene expression. *J. Virol.* **63**, 3822–3828.
28. Hermonat, P. L. and Muzyczka, N. (1984) Use of adeno-associated virus as a mammalian DNA cloning vector: transduction of neomycin resistance into mammalian tissue culture cells. *Proc. Natl. Sci. USA* **81**, 6466–6470.
29. Clackson, T. (1997) Controlling mammalian gene expression with small molecules. *Curr. Opin. Chem. Biol.* **1**, 210–218.
30. Fussenegger, M. (2001) The impact of mammalian gene regulation concepts on functional genomic research, metabolic engineering, and advanced gene therapies. *Biotechnol. Prog.* **17**, 1–51.
31. Fussenegger, M., Schlatter, S., Datwyler, D., Mazur, X., and Bailey, J. E. (1998) Controlled proliferation by multigene metabolic engineering enhances the productivity of Chinese hamster ovary cells. *Nat. Biotechnol.* **16**, 468–472.

32. Weber, W. and Fussenegger, M. (2002) Artificial mammalian gene regulation networks-novel approaches for gene therapy and bioengineering. *J. Biotechnol.* **98**, 161–187.
33. Mitta, B., Rimann, M., and Fussenegger, M. (2005) Detailed design and comparative analysis of protocols for optimized production of high-performance HIV-1-derived lentiviral particles. *Metab. Eng.* **7**, 426–436.

2

Cells for Gene Therapy and Vector Production

Christophe Delenda, Miguel Chillon, Anne-Marie Douar, and Otto-Wilhelm Merten

Summary

The treatment of genetic diseases and of several acquired diseases can only reasonably be performed by using gene or cell therapy approaches. In principle, practically all viruses can be used as a base for vector systems (→ viral vectors) for the transfer of genetic material/genes into target cells, and in dependence of the viral vector system used the gene is integrated into the cellular genome or stays in an episomal state. The most important viral vectors presently used are retroviral (RV) and adenoviral (AdV) vectors, which are classical vectors, followed by adeno-associated virus (AAV) and lentiviral (LV) vectors, both of which have a very bright future because of their superior characteristics with respect to RV and AdV vectors. Although this chapter does not deal with other vector systems, it should be mentioned that further viral vector systems with other interesting characteristics, such as those vector systems based on herpes simplex virus or α viruses, are in development and will probably be used for future clinical applications. This methods chapter deals with cells and production systems used for producing RV, LV, AdV, and AAV vectors, as well as with a short update on the purification and the titration of the produced viral vectors. With respect to retroviral vectors, the developments leading to optimal and safe (third- and fourth-generation producer cell lines for oncoretro- and lentiviral vector production, respectively) producer cells are presented. Concerning adenoviral vectors, the different vector generations as well as producer cells (from HEK293 to Per.C6) are shown, whereas the various production approaches are briefly described for the AAV vector system (from a transient production system via producer cells to the insect cell/baculovirus production system). Basic titration methods are also presented.

Key Words: Viral vectors; producer cells; insect cell/baculovirus expression system.

1. Introduction

The term gene therapy applies to approaches to disease treatment based upon the transfer of genetic material (DNA, or possibly RNA) into an individual. The concept of gene therapy has been inspired by major discoveries made in basic

From: *Methods in Biotechnology, Vol. 24: Animal Cell Biotechnology: Methods and Protocols, 2nd Ed.* Edited by: R. Pörtner © Humana Press Inc., Totowa, NJ

genetic research since the 1950s, and strategies for gene delivery were matured through the 1980s leading to the first clinical trial in 1990. By the beginning of January 2004 more than 3500 patients had received a gene therapy treatment (source: http://www.wiley.uk.co/genmed). Today's gene therapy research may be seen as pursuing intelligent drug design through a logical extension of results of fundamental biomedical research on the molecular basis of disease.

To allow transfer and proper function in a patient the therapeutic gene must be built into a vector. The most efficient gene delivery systems currently available are based upon the gene transfer machinery used in nature by animal viruses of which retroviral and adenoviral vectors are the most widely used. They will be treated in more detail in this chapter. However, in principle any virus can be modified in a way to eliminate the pathogenic/toxic genes and to maintain its capacity to transfer the gene of interest into the target cell. Since LV and AAV vectors provide several advantages in comparison to retroviral and adenoviral vectors and have thus a rather bright future in the gene therapy field, they will also be treated in more detail here. Other gene transfer systems use naked DNA or DNA coupled to chemicals that may facilitate various steps of entry of DNA into cells. Although of theoretical interest, mainly as a result of the improved safety profile in comparison to viral vectors, they are not really used today because of their inefficiency to transfer genes into the cellular genome; however, the ongoing scientific work will improve this situation in the future.

Gene and cell therapy are very much related because many gene therapy approaches are ex vivo approaches. By the ex vivo gene transfer principle, genetic material is transferred to cells outside the host; following transfer the genetically modified cells are then implanted into the host (= cell therapy). By the in vivo gene transfer principle, genetic material is transferred directly to cells located within the host and no ex vivo cell cultures are necessary for cell amplification and transduction.

This methods paper deals with vector production, the cells used for this purpose (the principle of stable producer cells for RV and LV vector production, transient transfection-based production protocols, principally applicable for the production of all vector types, and the use of the Sf9/baculovirus system for AAV vector production), and other related issues concerning the safe production of viral vectors for gene therapy.

1.1. Production of RV and LV Vectors (see *Notes 1–4*)

1.1.1. Cell Line Designs for the Production of RV and LV Vectors

Many different retroviruses have been modified as nonreplicative vectors for their use in gene transfer applications. These include (1) the spumavirus genus with the foamy virus as the prototype ***(1)***, (2) the gammaretrovirus genus, such as the Moloney leukemia virus (MLV) ***(2)*** or the avian leukemia virus (ALV) ***(3)***, as well as (3) the LV genus that comprises immunodeficiency viruses from human (HIV-1 and HIV-2) ***(4,5)***, simian (SIV) ***(6)***, feline (FIV) ***(7)***, or bovine (BIV) ***(8)*** origins, and other viruses such as the equine infectious anemia virus (EIAV) ***(9)***, the caprine arthritis encephalopathy virus (CAEV) ***(10)***, or the Jembrana disease virus (JDV) ***(11)***.

This chapter will only point out the technical considerations relative to MLV RV-based and LV vectors, although some of these conclusions are also considered for the other types of RV vectors.

Several types of cell lines have been used as reservoirs for the production of these RV vectors. Basically, most of the cell lines permissive for the production of RV vectors are derived from mammalian species with the exception of some viral vectors that are produced in their original species, such as ecotropic, xenotropic, or amphotropic MLV vectors that have been historically produced in murine NIH3T3 cells.

Before treating the exact subject on the cell aspect, a short molecular description is given to understand the evolution of the elements required for RV vector production, i.e., packaging and envelope constructs as well as transfer vectors encoding transgenes of interest. For being able to address and compare the efficacy and potency of different vector production schemes, this chapter highlights results only obtained from transfer vectors expressing reporter or selection genes.

Whether it is derived from LV or RV vector systems, the molecular design of all packaging constructs consists in removing from the viral genome the encapsidation site (ψ) and in replacing the 3′LTR by a heterologous polyadenylation signal. Although additional differences exist for the LV genus because of its more complexed genomic structure, RV vector packaging processes consist in expressing at least the *gag-pol* polyprotein as well as an envelope glycoprotein.

Concerning transfer vector backbones, the major improvement being realized has concerned the removal of the promoting U3 sequence in the 3′LTR, therefore reducing the risk for susceptible oncogenesis after random integration in the host genome of target cells. These so-called self-inactivating (SIN) RV vectors are now the only acceptable vectors for fitting regulatory requirements in human gene transfer applications.

1.1.2. Production by Transient Transfection Processes

The transient transfection protocols for LV vector production are derived from those originally designed for gammaretroviral vectors. The routine procedure involves the co-transfection of different combinations of packaging/envelope/vector into human embryonic kidney (HEK) 293T cells.

Table 1 summarizes the different generations of the packaging elements that have been used for RV vector production assays. Whereas the first designs

Table 1
Different Generations of RV Packaging Constructs Used in Transient Transfection-Based Vector Production Processs

	Packaging genes	Viral origin	Transfer vector and pseudotyping	Cell line	Maximal titer	Ref.
First generation (maintenance of all viral genes)	CMV.*GagPolTatRevNefVifVprVpu*.RRE	HIV-1	Non-SIN, lacZ (VSV/G)		10^5 LFU/mL	***173***
		HIV-2	Non-SIN, lacZ (VSV/G)		8×10^5 LFU/mL	***5***
	CMV.*GagPolTatRevNefVifVpxVpu*.RRE	SIV	SIN and non-SIN, eGFP (SIV & MLV *Env*, VSV/G)	293T	10^6 TU/mL	***174***
	CMV.*GagPolRevVif*Orf2.RRE	FIV	Non-SIN, lacZ (VSV/G)		10^7 LFU/mL	***7***
	CMV.*GagPolTatRev*S2.RRE	EIAV	Non-SIN, lacZ or puro (VSV/G)	293	7×10^4 LFU/mL 2×10^5 CFU/mL	***175***
	CMV.*GagPolTatRevVif*.RRE	JDV	Non-SIN, eGFP or neo (VSV/G)	293T	10^6 TU or CFU/mL	***11***
	CMV.*GagPol*	MLV	Non-SIN, neo (MLV and RSV *Env*)	COS-7	10^5 CFU/mL	***119***
Second generation (removal of the accessory genes)	CMV.*GagPolTatRev*.RRE	HIV-1	Non-SIN, luciferase (VSV/G)	293T	8×10^5 RLU/mL	***176***

		SIV	Non-SIN, eGFP (VSV/G)	293	10^7 TU/mL	177
	LTR.GagPolTatRev.RRE	EIAV	Non-SIN, lacZ (VSV/G, rabies G, MLV Env)		10^6 LFU/mL	9
	CMV.GagPolRev.RRE	FIV	Non-SIN, lacZ (VSV/G)	293T	3×10^6 LFU/mL	178
Third generation (removal of the Tat gene)	CMV.GagPol.RRE + CMV.Rev	HIV-1	Non-SIN, lacZ (VSV/G)		6×10^5 LFU/mL	166
	CMV.GagPol.IRES.Rev.RRE	SIV	Non-SIN, eGFP (VSV/G)	293	2×10^7 TU/mL	177
Fourth generation (removal of the Rev gene)	CMV.GagPol-CO	HIV-1	Non-SIN, lacZ (VSV/G)		10^5 LFU/mL	179
		SIV	SIN, eGFP (VSV/G)		5×10^6 TU/mL	180
	CMV.GagPol.CTE (MPMV)	HIV-1	Non-SIN, lacZ (VSV/G)	293T	5×10^4 LFU/mL	166
	CMV.GagPol.CTE (RSV-1)		Non-SIN, neo (VSV/G)		10^6 CFU/mL	181

TU, transducing unit; CFU, colony-forming unit; LFU, lacZ-forming unit; CMV, cytomegalovirus promoter; RRE, Rev-responsive element; CTE, constitutive transport element from monkey Pfizer Mason virus (MPMV) or virus (SRV-1); VSV/G, vesicular stomatitis virus glycoprotein; MLV, moloney leukemia virus; HIV, human immunodeficiency virus; SIV, simian immunodeficiency virus; FIV, feline immunodeficiency virus; EIAV, equine infection anemia virus; JDV, jembrana disease virus.

maintained the *env* gene in the original genome, further molecular improvements have been realized by splitting the expression of RV genes from heterologous envelope glycoprotein synthesis using a separate plasmid entity. Concerning some of LV vector designs, the *tat*-independent production scheme is only possible if the *tat*-dependent 5'LTR, which originally drives the transcription of viral vector RNAs in producer cells, is replaced by a heterologous promoter sequence such as cytomegalovirus (CMV) or rhinosyncytial virus (RSV). In contrast, *rev*-independent vector particles have been produced by replacing the natural *rev* responsive element (RRE), present in the transfer vector RNA, by additional nuclear export elements that use cellular proteins as *trans*-acting factors.

1.1.3. Production by Stable Packaging Cell Lines

The use of stable producer cell lines eliminates the risk of homologous recombination between the different plasmid constructs, as well as the problem of carrying over plasmid DNA in vector batches. In addition, they are better suited for reproducibility and scalability, where they could be adapted to high-density culture conditions in large bioreactors.

From the early beginning to highly improved recent generations, descriptions of MLV- and LV-based stable producer cells are listed in **Tables 2** and **3**, respectively. Whereas all MLV-based packaging cell lines are based on a constitutive expression system, the cytotoxicity of some viral packaging proteins (*Pro*, *Rev*, or VSV/G), which are involved in the LV vector production process, has made necessary regulatory pathways to switch on or off the relevant packaging genes, such as the tetracycline- or ecdysone-inducible expression systems. In addition to these tight expression controls, one report has also added, during their packaging cell cloning steps, an additional drug specific for the HIV protease, i.e., saquinavir at 1.5 μM (Roche).

Although inducible systems downregulate the expression of toxic genes, the cytotoxic inhibition effect cannot last long after transcription switches on. Therefore, the scalability of such inducible cell lines is less interesting as compared to constitutive-based packaging cell line designs.

1.1.4. Production by Shuttle Viral Vectors

Cell transfection being difficult to scale up, it has been proposed to introduce the different genetic components of the vector system via other viral vector systems. Several groups have developed systems to generate RV packaging cells that take advantage of the broad host range, efficient gene delivery, and expression mediated by viral vectors derived from herpes simplex virus (HSV) *(12,13)*, Semliki Forest virus (SFV) *(14–17)*, and adenovirus *(18–23)*. Concerning the design of LV vector production, only adenoviral vectors have yet been developed *(24,25)*. Vectors derived from these different viruses have been designed to induce expression in infected cells of either or both the vector packageable RNAs and the packaging units encoding *gag-pol* polyprotein and an envelope glycoprotein (*env*, VSV/G). Because of their high infectious capacities, these viral vectors can easily introduce into cultured cells numerous copies of the different transcription units required to generate the production of RV/LV particles.

1.2. Cell Line Designs for the Production of AdV Vectors (see *Notes 5 and 6*)

1.2.1. Biology of AdV/AdV Vectors

Adenovirus is a nonenveloped 60- to 90-nm-diameter virus presenting an icosahedral symetry. Human adenoviruses contain a linear, double-stranded DNA genome, with a terminal protein (TP) attached covalently to the 5′ termini. The DNA, which has a length of 30–38 kb, has inverted terminal repeats of 100–140 bp at both ends, which act as origins of replication. Histone-like viral core proteins are associated with the DNA packaged within the capsid (two TPs and the condensing proteins V and VII). The viral chromosome also contains about 10 copies of the cysteine protease p23 and is linked to the outer capsid by protein VI. The capsid is composed of 240 hexon capsomeres forming the 20 triangular faces of the icosahedron, and 12 penton capsomeres with spike-shaped protrusions located at the 12 ventrices **(Fig. 1)**. In addition to the major capsid protein (hexon), the penton base, and fiber proteins associated to form the penton capsomere, several hexon- and penton-base-associated proteins are stabilizing components of the capsid (hexon-associated and -stabilizing proteins IIIA, VIII, and IX). For a schematic presentation, *see* **Fig. 1**. Altogether the AdV particle has a molecular weight of about 150 MDa *(26,27)*.

There are more than 50 different adenovirus serotypes that can infect humans, which, depending on the GC content, oncogenicity in rodents, and ability to agglutinate red blood cells, are classified into six groups (A–F). All serotypes except those from group B have been shown to use the coxsackie/adenovirus receptor (CAR) as a primary receptor *(28)*. The terminal globular domain ("knob" region) of the homotrimeric protruding fibers of the Ad capsid is responsible for the primary virus attachment to the cellular receptor (CAR) *(29,30)*. For the most commonly used human adenoviruses (type 2 and type 5), after the initial attachment, the interaction between an RGD (arginine-glycine-aspartate) motif exposed in a protruding loop of the penton base protein with a cell surface integrin molecule serving as secondary or internalization receptor triggers the virus uptake by clathrin-dependent receptor-mediated endocytosis *(31,32)*. The endosomal uptake and release into the cytoplasma is accompanied by a stepwise dismantling of the capsid, leading to the microtubule-assisted transport and finally very efficient delivery of the core protein-coated viral genome to the nucleus (details on Ad endocytosis can be found in **ref. *33***).

Table 2
Stable Packaging Cell Lines Dedicated to MLV Vector Production

Expression system	Packaging genes inserted	Pseudotyping	Name of the packaging cell line	Transfer vector	Cell line	Maximal titer	Ref.
	LTR.*GagPolEnv* (TK)	MLV amphotropic *Env*	PA12, PA317	Non-SIN, neo, DHFR or HPRT		10^6–10^7 CFU/mL	***133***
	LTR.*GagPol*.gpt LTR.*Env* (gpt)	MLV ecotropic *Env*	GP+E-86	Non-SIN, neo		10^5–10^6 CFU/mL	***182***
	LTR.*GagPol*.gpt LTR.*Env* (Hygro)	MLV amphotropic *Env*	GP+*env*Am12		NIH3T3	10^3–10^6 CFU/mL	***135***
	LTR.Gag^+ Pot^+ Env^- (Hygro)		ψCRIP	Non-SIN, lacZ or neo		10^6 LFU or CFU/mL	***134***
	LTR.Gag^- Pot^- Env^+ (gpt)	MLV ecotropic *Env*	ψCRE				
	LTR. *GagPol3'-CO* (gpt) LTR.*Env3'-CO* (gpt)		ΩE	Non-SIN, neo		7.5×10^5 CFU/mL	***137***
	LTR.*GagPol* (TK) LTR.*Env* (mtx)	GALV *Env*	PG13	Non-SIN, neo	NIH3T3	10^5–10^6 CFU/mL	***183***
Constitutive	LTR.*GagPol*.bsr LTR(Friend).*Env*.phleo	RD114 *Env*	FLYRD18[a]	Non-SIN, lacZ	HT1080	4×10^6 LFU/mL	***184***
		MLV amphotropic *Env*	FLYA13[a]			2×10^5 LFU/mL	
			CEMFLY	Non-SIN, lacZ	CEM	10^6 CFU/mL	***185***

LTR.*GagPol.* (puro) CMC.*Env* (hygro)	MLV xenotropic *Env*	ProPak-X			>10^6 CFU/mL	***186***
		ProPak-A.52	Non-SIN, neo			
CMV.*GagPol.* (puro) CMV.*Env* (hygro)	MLV amphotropic *Env*	ProPak-A.6		293	2×10^6 CFU/mL	***187***
LTR.*Gag*$^+$ *Pot*$^+$ *Env*$^-$ (gpt) LTR.*Gag*$^-$ *Pot*$^-$ *Env*$^+$ (hygro)		293-SPA	Non-SIN, lacZ		6×10^6 LFU/mL	***188***
RSV.*GagPol*.IRES.CD8 (hygro)	MLV ecotropic *Env*	Phoenix-A	SIN and non-SIN, puro (episomal maintenance)	293T	10^7 CFU/mL	***189***
CMV.*Env* (diphtheria resistance)		Phoenix-E				
CMV.*GagPol* (phleo) CMV.*Env* (mtx)	MLV amphotropic *Env*	DA	Non-SIN, lacZ	D-17	2×10^7 LFU/mL	***190***
		2A		293	10^7 LFU/mL	
	MLV xenotropic *Env*	2X			2×10^6 LFU/mL	
		HX		HT1080	4×10^6 LFU/mL	

TK, thymidine kinase gene from herpes simplex virus (HSV); gpt, hypoxanthine-guanine phosphoribosyltransferase resistance gene; hygro, hygromycin resistance gene; mtx, methotrexate drug for dihydrofolate reductase (DHFR) resistance gene; bsr, blasticidin S deaminase resistance gene; phleo, phleomycin resistance gene; hygro, hygromycin resistance gene; puro, puromycin resistance gene; neo, neomycin resistance gene; IRES, internal ribosome entry site.

[a]Similar packaging cells based on the use of T671 cells have been established by the same authors and have been named TeFLY.

Table 3
Stable Packaging Cell Lines Dedicated to LV Vector Prodution

Expression system	Packaging genes inserted	Pseudotyping	Transfer vector	Cell line	Maximal titer	Ref.
Inducible (tetracycline)	LTR.*GagPolTatNefVifVprVpu*.RRE TRE.*RevEnv*.RRE	HIV-1 *env*	Non-SIN, neo or puro	HeLa.tTA (HtTA-1)	7×10^3 CFU/mL	***139***
	TRE.*GagPolTatNefVifVprVpu*.RRE TRE.*RevEnv*.RRE		Non-SIN, eGFP.IRES.puro		3×10^4 TU/mL	***138***
	TRE.*GagPolTatRevNefVifVprVpu*.RRE TRE.VSV/G		Non-SIN, eGFP	293.tTA (SODk0)	3×10^6 TU/mL	***141***
	TRE.*GagPolTatRevVpu*.RRE TRE.VSV/G		SIN, eGFP (TRE in the 3'LTR)	293T.tTA (SODkl)	$>10^6$ TU/mL	***144***
	TRE.*GagPolTatRev*.RRE		Non-SIN, eGFP	293G (TRE.VSV/G, tTA)	2×10^7 TU/mL	***140***
	TRE.*GagPol*.RRE TRE.*Rev*	VSV/G	Non-SIN, eGFP (CMV in the 3'LTR)		5×10^6 TU/mL	***142***
	TRE.tTA.SV40.Puro.TRE.*RevTat-CO* U3Tar.*GagPol*.SCMV.Neo SCMV.VSV/G.IRES.RRE.TK.Hygro		Non-SIN, eGFP	293	1×10^7 TU/mL	***191***
Inducible (ecdysone)	RSV.RxR.CMV.VgEcR (zeo) Ec.*GagPolRev*.RRE (zeo)		Non-SIN, eGFP.IRES.puro	293T	2×10^5 TU/mL	***145***

	Ec.VSV/G.IRES.eGFP.SV40.Hygro		(CMV in the 5'LTR *tat*-independent)			
	RSV.RxR.CMV.VgEcR Ec.*GagPol*.RRE Ec.*Tat*.IRES.Rev	VSV/G (transient transfection)	SIN, eGFP	293	3×10^5 TU/mL	***146***
	SV40.Zeo.CMV.VgEcR.RSV.RxR Ec.*GagPol-CO*.SV40.Neo Ec.VSV/G.SV40.Hygro	VSV/G	Non-SIN, eGFP	293	2×10^5 TU/mL	***147***
Constitutive	LTR.*GagPol-CO*, LTR.*Tat*, LTR.*Rev* (transduction by non-SIN MLV vectors)	MLV ampho, GALV, and RD114 *Env*	Non-SIN, eGFP	HeLa, 293T HT1080	1×10^7 TU/mL	***163***
			SIN, eGFP		3×10^5 TU/mL	

With the exception of one report that concerns the SIV LV genus ***(147)***, all other publications describe HIV-1-derived packaging cell lines. TRE, Tet-responsive (promoter) element; tTA, doxycycline-dependent transcription factor; RSV, U3-promoting sequence from Rous sarcoma virus; Ec, Ecdysone promoter; RxR & VgEcR, ponasterone-dependent transcription factors; CO, codon-optimized; zeo, zeocin resistance gene.

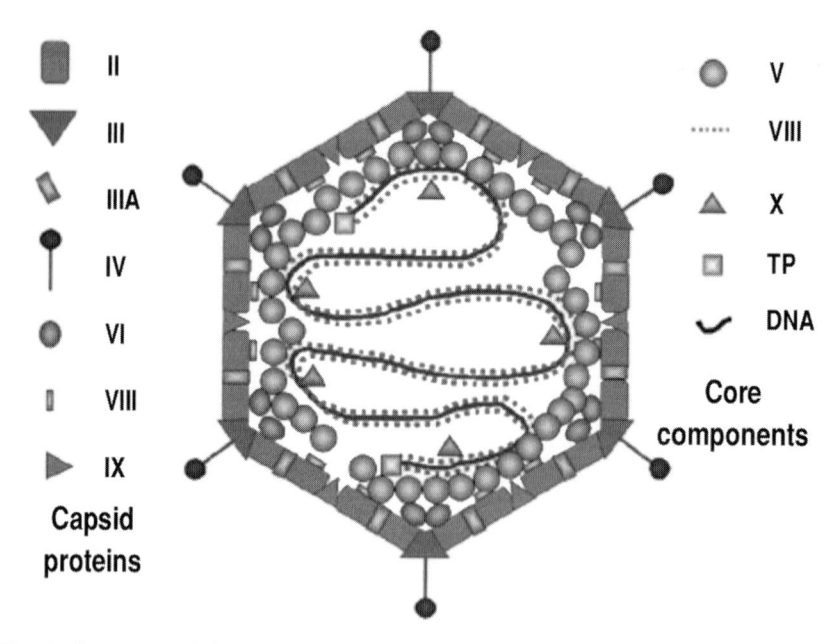

Fig. 1. Structure of the adenovirus particle: the core components are DNA and the core proteins V, VII, Mu (X), and TP. The capsid proteins are hexon (II), penton base (III), fiber (IV), and the hexon-associated proteins IIIa, VI, VIII, and IX. (From http://pathmicro.med.sc.edu/mhunt/adeno-diag.jpg.)

Ad2 and Ad5 genomes consist of a linear 36-kb double-stranded DNA molecule. Both strands are transcribed, and nearly all transcripts are heavily spliced. Viral transcription is conveniently defined as early (encoding E1, E2, E3, and E4), delayed early (encoding transcripts IX and IVa), and late genes (L1–L5, the structural genes), depending on their temporal expression relative to the onset of viral DNA replication *(34)* (for the structural organization, *see* **Fig. 2**). The first Ad gene to be expressed is the immediate early E1A gene encoding a transactivator for the transcription of the early genes E1B, E2A, E2B, E3, and E4, as well as protein functions involved in cellular transformation, together with an E1B protein. The E2 proteins encode for proteins necessary for the DNA replication: DNA-polymerase, DNA-binding protein, and the precursor for the terminal protein. The E3 gene, dispensable for Ad replication, codes for proteins that interferes with the host immune response against virus infection. Finally, the E4 genes are involved in the transition from early to late gene expression, the shut-off of host-cell gene expression, viral replication, and the assembly of the virion *(35)*.

Replication of the viral genome, which depends on the ITRs (inverted terminal repeats) as *cis*-acting elements (origin of replication) and one copy of the TP covalently attached to each 5′ end as initiation primer, starts about 5–6 h after

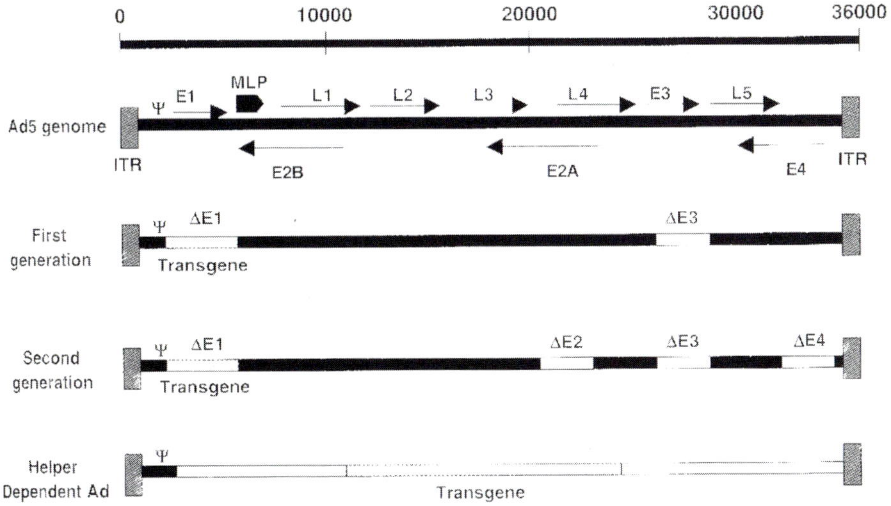

Fig. 2. Map of adenovirus serotype 5 genome and different generations of adenoviral vectors. Early transcripts are represented by E1–E4 regions and late transcripts are represented by L1–L5 regions. MLP: major late promoter; ψ: packaging signal. (From **ref. *107***.)

infection. Core and capsid proteins and the cysteine protease important for proteolytic trimming of TP and other structural proteins are expressed from a common major late promoter after splicing of long precursor transcripts. Virion assembly occuring in the nucleus starts about 8 h after infection after which different proteins link with the packaging signal (ψ) to carry viral genome into the capsid during the packaging, leading to the production of 10^4–10^5 particles per cell. Viral particles can be released after final proteolytic maturation by cell lysis, which occurs 30–40 h postinfection ***(34,36)***.

For a brief overview on adenoviruses, the reader is referred to the following website: http://www-micro.msb.le.ac.uk/3035/Adenoviruses.html.

1.2.2. AdV Design and Vector Production

First-generation recombinant AdV vectors **(Fig. 2)** are usually deleted in their E1A or E1B regions and eventually in their E3 region, providing space for the insertion of a therapeutic gene up to 5 kb ***(37)***. Expression of E1A initiates adenoviral replication and activates adenoviral transcription ***(38)***; therefore, an adenovirus-transformed cell line that provides the missing E1A and E1B genes *in trans* is required to grow the replication-defective virus ***(39)***. However, expression of the therapeutic gene is transient, because as a result of the expression of viral genes/proteins in the transduced cells, adenovirus vectors can either be potentially immunogenic and the transduced cells will be removed by the cellular immune response, or the vectors cause a direct cytopathic effect ***(40)***.

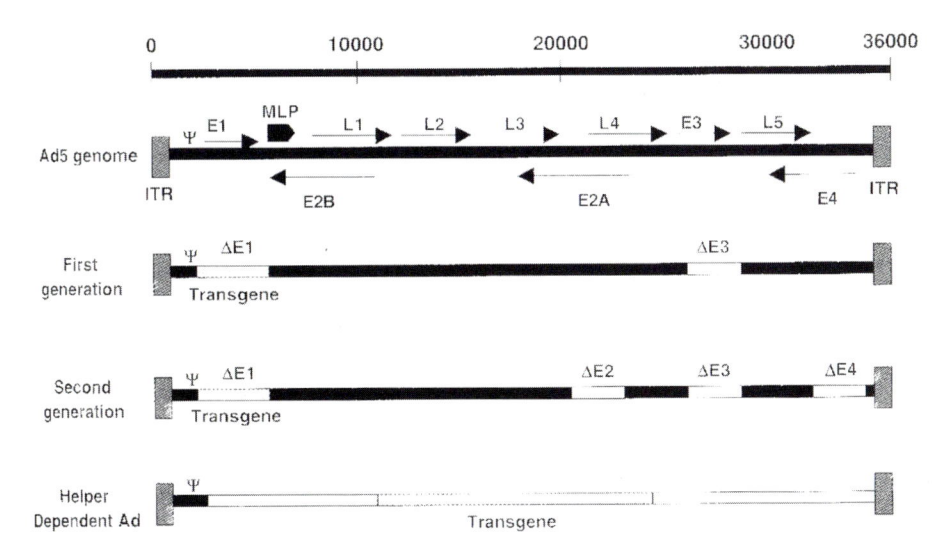

The most common and best documented packaging cell line for adenovirus production is the human embryonic kidney 293 (HEK293) cell line, which contains the E1 region of the adenovirus ***(39)***. Packaging cells based on the HEK293 cell line are versatile and can be grown as adherent cultures or as suspension cultures in serum-containing low-Ca^{2+} media ***(41)*** and serum-free low-Ca^{2+} media ***(42,43)***. Unfortunately, homologous recombination between the left terminus of first-generation Ad vector or helper virus DNA and partially overlapping E1 sequences in the genome of HEK293 cells may lead to the emergence of E1-positive, replication-competent adenoviruses (RCA), a serious safety concern if non-replicating vectors are desired ***(44)***. To overcome this problem, alternative host cell lines have been developed by reducing these overlapping sequences, as for 911 cells (E1 transformed human retinoblast cells) ***(45)***, or by eliminating any overlap, as for N52.E6 ***(46)*** or PER.C6 (E1 transformed human retinoblast cells) cells ***(47)***. The 911 cell line exhibits similar frequencies of homologous recombination to HEK293 cells, although vector yields are three times higher in 911 cells ***(45)***. The PER.C6® cell line, derived from human embryonic retinal cells, has been established for industrial applications (e.g., full tracability available) and is a good vector producer. The cell population is clonal and the cell line was designed to minimize RCA production. Unfortunately, this cell line is only available through licensing. Yields from the PER.C6 cell line are similar to those achieved in HEK293 and 911 cells; however, as overlap between the genome and the E1 sequence has been eliminated, this clone, as well as the N52.E6 cell line, does not generate RCAs ***(46,47)***. Another cell line is A549, a lung carcinoma cell. This cell line has been engineered with low recombination homology and is used to support the replication of E1-deleted adenovirus vectors ***(48)***. A HeLa-derived complementing cell line has also been developed: GH329 ***(49)***.

Immunogenicity and cytotoxicity of these first-generation vectors became apparent from many experiments in animals and from clinical studies in humans, which led to the development of second generation vectors (**Fig. 2**). These vectors are replication-defective Ad, which are further deleted in E2 and/or E4 functions, showing reduced immunogenicity and RCA generation, but engineering of stable cell lines that complement these vectors can be cumbersome and lead to poor cell growth and viral titers ***(50)***. This vector type can accommodate up to 14 kb and hence increases the vector-coding capacity ***(51,52)***.

The third-generation vectors, called gutless or gutted Ad (**Fig. 2**), are devoid of all coding viral regions, which avoid activation of cellular immune response against viral proteins in transduced cells. They are also called helper-dependent adenoviruses because of the need for a helper adenovirus that carries all coding regions and high-capacity adenoviruses because they can accommodate up to 36 kb of DNA. Briefly, the gutless adenoviruses only keep the 5′ and 3′ ITRs as well as ψ from the wild-type adenovirus. More details on gutless adenoviruses can be found in the review by Alba et al. ***(53)***.

The general principles of the different AdV constructs are presented in **Fig. 2**.

1.2.3. Production and Purification Methods

Conventional methods for producing small volumes of viral vectors involve culturing cells in stationary, adherent cultures, such as T-flasks or roller bottles ***(54)***. The principal protocols of this small-scale production method based on the use of HEK293 or 911 cells are presented in **Subheadings 2.2.** and **3.2**. Methods for large-scale production of Ad viral vectors can be found in the literature, but a brief overview is given here:

A general observation is that adherent cell cultures have a higher cell-specific productivity compared to suspension cells ***(55)***; however, scale-up of adherent cell cultures is limited by the available surface area. To date, very few studies have investigated cell growth and virus production using microcarriers as a potential means of scaling-up production. Solid microcarriers have been shown to be superior to macroporous carriers, with Cytodex 3 offering the best results. However, the maximum specific virus titer remains inferior to T-flask cultures ***(54)***. Hydrodynamic shear forces are believed to affect cellular receptor levels and thus hinder virus entry into HEK293 cells at high multiplicity of infection (MOI). Additionally, HEK293 cells have been cited as having a low ability to colonize microcarriers because of poor attachment ***(40)***.

Suspension cultures are more appealing for large-scale production because of the ease of operation and scale-up. HEK293 and Per.C6® cells can be grown in stirred tank bioreactors, and, depending on the mode of operation, cell densities are in excess of 5×10^6 cells/mL at scales from 3 to 20 L ***(42,56,57)***. Typical titers are in the range of $1–6 \times 10^{10}$ virus particles (VPs) per mL. A recent study using the HeLaS3 human tumor cell line for propagating recombinant adenoviruses demonstrated yields of 6×10^{11} VP/mL and showed that this productivity could be maintained at pilot (70 L) scales ***(58)***. Successful large-scale production for all cell lines is dependent on optimization of the culture conditions and the best mode of operation. For more details, *see* Kamen and Henry ***(56)*** and Warnock et al. ***(59)***.

Although out of the scope of this methods chapter, the protocols for purification of AdV vectors have evolved over the last decade. The most classical and easy to acquire for a nonspecialized laboratory remains the ultracentrifugation on a CsCl. This method of purification is limited by the capacity of cell lysate volume that can be processed and by the low purity achieved. However, this method is still widely used and most of the time is sufficient for fundamental studies and early in vivo preclinical evaluation of the vectors, wherefore the CsCl-based ultracentrifugation protocol is also presented here **(Subheading 3.2.4.)**. More complex techniques based on column chromatography and membrane techniques are now well developed for the generation of high-purity-grade and up-scaled production suitable for human clinical applications ***(56,60,61)***.

1.2.4. Titration

The most important endpoint of all vector production protocols is the vector titer and the total vector quantity produced: infectious particle units (ivp) as well as viral particle units (vp) are used. The most common method for the determination of total physical particles relies on absorbance reading at 260 nm, which is converted to the number of total particles using published extinction coefficients ***(62)***. The reproducibility and precision of this quantification method are significantly enhanced by using anion-exchange high-performance liquid chromatography prior to detection by UV spectroscopy ***(63)***.

The determination of infectious particles has more relevance in particular for in vivo applications but also to evaluate vector batch quality. It is based on the direct or indirect observations of modifications of permissive cells following contact with virus particles. The result of the infection process such as cytopathogenicity, expression of a marker gene such as β-galactosidase or the green fluorescent protein (GFP), will then be considered to infer the number of ivp, titer, plaque forming units (PFU), or gene transfer units (GTU). Thus, the precisions of the titers derived from these assays strongly depend on the biological material and operation conditions ***(64)***. Generally, standard plaque assay methods underestimate titers because of diffusion limitations ***(65)***.

Other assays have been developed to measure subcomponents of the virus, such as the DNA-binding protein, an early protein encoded by the vector that is detected by quantitative immunofluorescence ***(66)***, or to detect the major capsid protein hexon by specific antibodies and endpoint dilution assay ***(67)***.

Today, to help worldwide standardization of quantification methods and facilitate interpretation of preclinical and clinical data, a reference material for adenovirus vector has been generated ***(68)***.

1.2.5. AdV Vector Design and Production Systems

For first- and second-generation vectors, design of the adenovirus vector genome is quite similar and it is based on a two plasmid system: (1) one plasmid called shuttle plasmid (*see* phADV in **Fig. 3**) containing the 5′-ITR, the packaging signal a multicloning site and between 3 and 4 kb containing the E2B region; and (2) a second plasmid (*see* pKP1.3 in **Fig. 4**) of about 35 kb, containing the whole adenovirus genome, but the E1 region, which has been substituted by a unique target site for a restriction enzyme, usually SwaI or ClaI. Because of the evident difficulties for a direct cloning on 40-kb-long plasmids, the expression cassette of interest is first cloned into the shuttle plasmid. Then both plasmids,

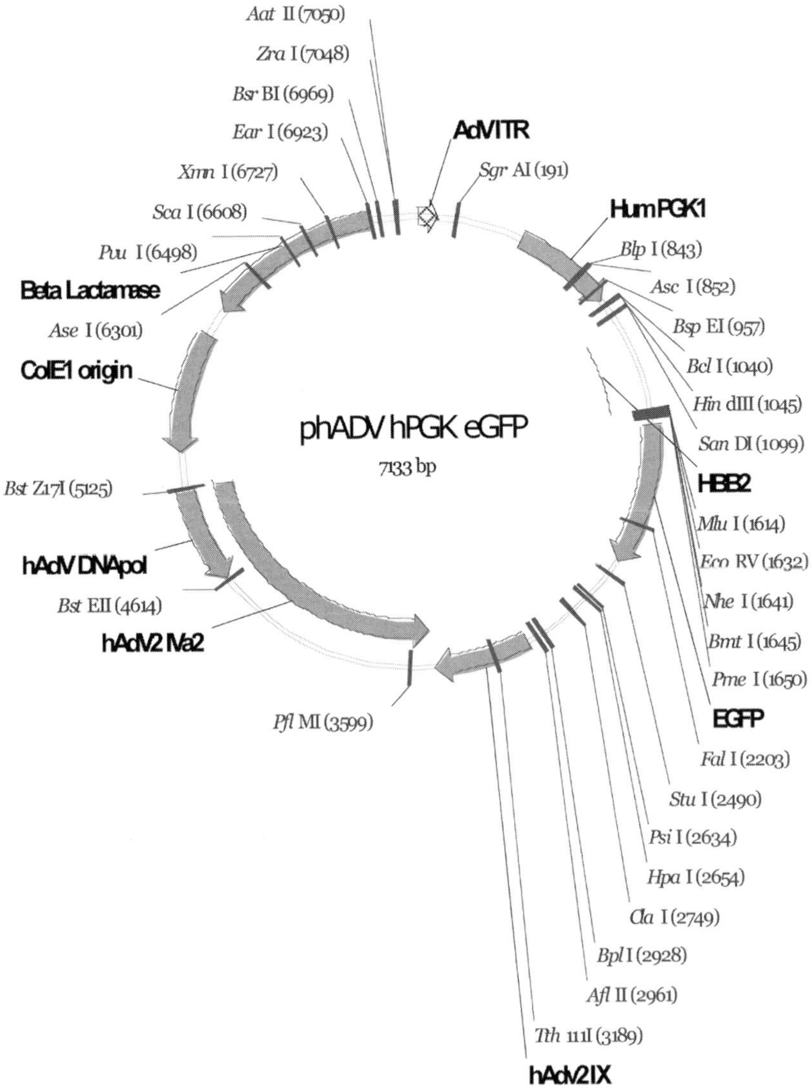

Fig. 3. Shuttle plasmid carrying a green fluorescent protein (GFP) expression cassette contains the left inverted terminal repeat (ITR) (between nt 1 and 103) and the packaging signal (from nt 200 to 400) of the adenovirus. The GFP expression cassette with the eGFP gene is driven by the human PGK1 promoter, followed by almost 3 kb of adenoviral genome and the ColE1 origin and β-lactamase gene.

shuttle and receiving 40-Kb, are linearized (the latter by SwaI or ClaI or the unique site inserted in E1) to allow homologous recombination either in HEK293 cells or in *Escherichia coli* strain BJ5183.

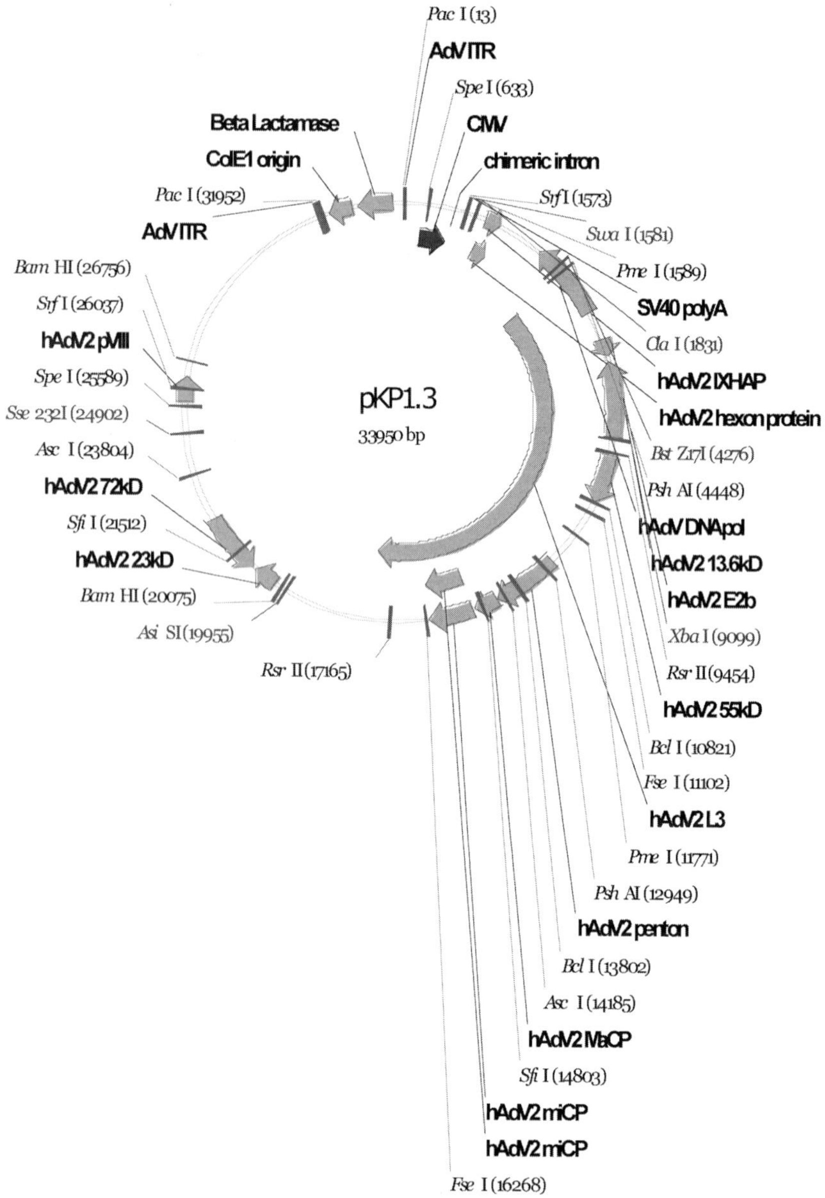

Fig. 4. Genomic organization of the pKP1.3 plasmid containing the adenovirus genome. Nucleotides 457–3328 (from E1 region) and 28450–30757 (from E3 region) have been removed to generate this first-generation vector. The pKP1.3 plasmid also has two adenoviral ITR sequences, both having a PacI site just outside the viral genome.

Prior to 1996, almost all recombinant (E1, E3-deleted) adenovirus vectors were generated by homologous recombination of plasmid and digested adenoviral DNA in HEK293 cells. "Vector clones" were isolated by amplifying individual plaques (3–4 wk) followed by testing for transgene expression (if possible) or restriction digest patterns of purified vector DNA.

This rather time-consuming method was replaced by generating recombinant plasmids that contain the entire vector genome, although homologous recombination is performed in recBC sbcBC *E. coli* strain BJ5183. Because one has a clonal population at the plasmid stage, isolation of plaques is theoretically (when the vector has the appropriate size and the transgene does not code for a cytotoxic protein) not needed. However, it is occasionally performed nonetheless.

1.3. Cell Line Designs for the Production of AAV Vectors (see Notes 7–9)

1.3.1. Biology of AAV/AAV Vectors

AAVs are small nonenveloped single-stranded (ss) DNA viruses with a diameter of 20–25 nm. They belong to the Parvoviridae family and are classified in the *Dependovirus* genus. Eleven strains have been isolated and characterized from humans and primates, and new serotypes are continuously being discovered ***(69–72)***. Phylogenetic and functional analyses revealed that primate AAVs are segregated into six clades, whose members are phylogenetically closely related and share functional and serological similarities ***(73)***. All serotypes share similar structure, genome size, and organization, i.e., structure and location of the open reading frames (ORF), promoters, introns, and polyadenylation site. The most divergent serotype is AAV5, with notable differences at the level of the ITR size (167 nucleotides for AAV5 compared to 143–146 for AAV1 and 4 for AAV6) and function ***(74)***. In addition, at the biological level they are all dependent on the presence of a helper virus for replication and gene expression.

AAVs are frequently found in the human populations, 70–80% of individuals having been exposed to an infectious event ***(75,76)***. No known adverse clinical consequences are associated with AAV infection or latency in humans.

The particle is composed of an icosahedric capsid, consisting of 60 capsomeres, and one single molecule of the viral genome of either polarity, and has a density of 1.41 g/cm ***(72)***. AAVs are very resistant to extreme conditions of pH, detergent, and temperature, making them easy to manipulate. Finally, in the absence of helper virus, wild-type AAV has the unique property of integration at a specific site into the human genome (it [about 70%] integrates in the long arm of chromosome 19, the S1 site), which has been extensively described for AAV2 ***(77–79)***. However, because of viral genome manipulation, this property is not maintained in AAV vectors.

As the first serotype used to generate vectors, AAV type 2 is so far the best characterized, and the majority of gene transfer studies have been based on the use of this serotype ***(77–79)***. Thus it can be considered as a prototype. However, in recent years rAAVs have been actively developed from alternative serotypes. The rationale lies in their very similar structure, but different tropism and immunological properties allow one to design a whole set of vectors ***(73)***. Whereas, for instance, AAV1 leads to 10- to 100-fold higher gene transfer in the muscle, AAV5 can significantly improve AAV-vector-mediated gene transfer to the central nervous system and lung tissue. The physical characteristics of the different serotypes are close enough to handle them under similar conditions as AAV2 for production and purification. Only the purification systems based on fine capsid structure (isoelectric point/range, specific ligands) must be adapted for each serotype. Many groups are actively evaluating the in vivo performances of these serotypes in various animal and disease models ***(73,80)***.

The 4.68-kb genome of AAV2 (**Fig. 5**) contains two ORFs that encode four regulatory proteins, the Rep proteins, and three structural proteins, the Cap proteins ***(81,82)***. The compacted genome of AAV is framed by two ITRs, which are base-paired hairpin structures 145 nucleotides in length. The ITRs contain the only necessary regulatory *cis*-acting sequences required by the virus to complete its life cycle, namely the origin of replication of the genome, the packaging, and the integration signals. The two major Rep proteins, Rep78 and Rep68, are involved in viral genome excision, rescue, replication, and integration ***(83)*** and also regulate gene expression from AAV and heterologous promoters ***(84,85)***. The minor Rep proteins, Rep52 and Rep40, are involved in replicated ssDNA genome accumulation and packaging ***(86)***. The cap ORF is initiated at the p40 promoter and encodes the three structural capsid proteins VP1, VP2, and VP3. Their stoichiometry in the assembled particle is 1:1:10. Finally, all transcripts share the same polyadenylation signal and equal amounts of virions are found containing strands of plus or minus polarity. Details on the encapsidation of AAV genomes were presented in a recent review ***(87)***.

AAVs are naturally replication-defective viruses, which made them dependent on the presence of an auxiliary virus in order to achieve their productive cycle. This feature gave the name to the *Dependovirus* genus. Several viruses can ensure the auxiliary functions: adenovirus, herpes simplex virus, and vaccinia virus ***(88,89)***, as well as cytomegalovirus ***(90)*** and human papillomavirus ***(91)***. More details can be found in a recent review ***(92)***. The infection scheme is fairly simple. Once AAV has entered the cell and has been conveyed to the nucleus, its ss genome is converted into a replicative double-stranded form required for gene expression. In the absence of an auxiliary virus, AAV genome integrates into the host genome and latently persists in a proviral form. In the presence of a helper virus (either by concomitant infection or superinfection), the AAV

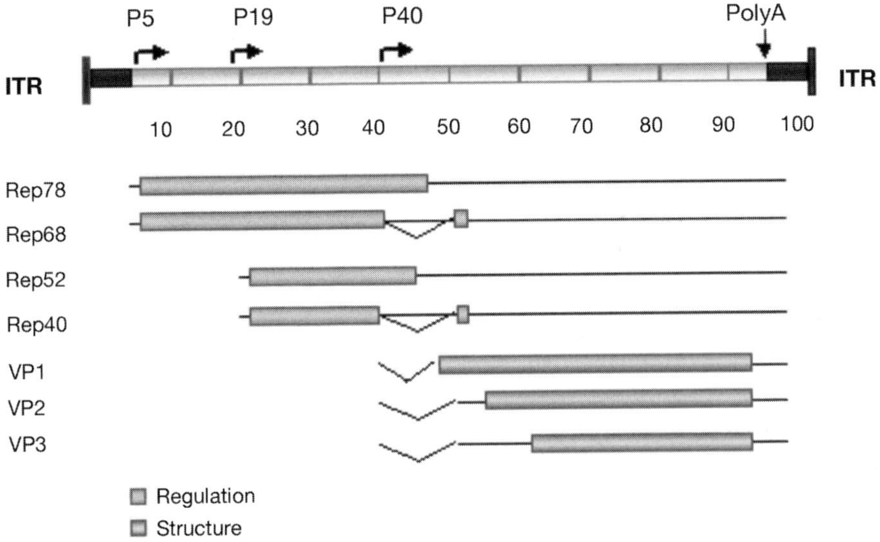

Fig. 5. General organization of the genome and genetic elements of AAV type 2 (a scale of 100 map units is used, with 1 map unit being equivalent to approx 47 nucleotides). The general organization of the other serotypes is similar. T-shaped brown boxes indicate inverted terminal repeats (ITRs). The horizontal arrows indicate the three transcriptional promoters. The solid lines indicate the transcripts; the introns are shown by the broken lines. A polyadenylation signal present at position 96 is common to all transcripts. The first ORF encodes the four regulatory proteins arising from the promoters p5 and p19 and the alternative splicing. The second orf (promoter p40) encodes the three capsid proteins from two transcripts. VP-1 is initiated from the first cap transcript, and VP-2 and VP-3 are initiated at two different codon sites from the second cap transcript. Note that the initiation site of VP2 is an acycloguanosine (ACG).

genome can undergo the process of active replication, during which capsid proteins are synthesized and DNA packaged, these steps taking place inside the nucleus ***(93)***. AAV does not possess a lytic capability by itself, and in the natural infection process the liberation of AAV virions usually relies on the lytic effect of the helper virus.

1.3.2. AAV Vector Design, Vector Production, and Purification

The design of AAV-based vectors is straightforward, the ITRs being retained, and the exogenous sequences (the sequence of the transgene) to be transferred cloned in-between. The *rep* and *cap* functions have therefore to be supplied in *trans*. Similarly, the helper functions from the auxiliary virus have to be provided. The genome size of a recombinant AAV vector should be in the range of 3555–4712 bp in order to obtain efficient encapsidation and vector production ***(87)***.

Several production systems coexist, with their own advantages and drawbacks. In laboratory scale, the mostly utilized production method of viral particles is still based on transient transfection of HEK293 cells with minor variations. The traditional transfection methods employs the use of HEK293, A549, or HeLa cells that have been co-transfected with two plasmids that contain the rAAV vector and the *rep* and *cap* genes, followed by an infection with helper virus (e.g., wtAd) to induce the replication of rAAV. The major drawback is the co-production of rAAV and helper virus. Thus, the helper virus (in most of the cases adenovirus [wtAd]) was replaced by a helper plasmid bringing in the adenoviral functions necessary for the replication of the rAAV (this helper plasmid, e.g., pXX6, provides the following functions: E2A, E4, and VA RNA genes; the other essential adenoviral functions E1A and E1B are provided by the HEK293 cells). In the case of the use of HeLa or A549 cells, the helper plasmid has to provide all essential functions, including the adenoviral E1A and E1B functions. This production method is based on the co-transfection of HEK293 cells (in general) with three plasmids that contain the rAAV vector (pAAV-vector), the *rep* and *cap* genes (pAAV-'Helper'), and the adenovirus helper genes (pAV-'Helper') ***(94,95)*** (*see* **Fig. 6**).

With the discovery and the development of new AAV serotypes, the idea has emerged rapidly to develop vectors from these serotypes ***(70,72,96–98)***. In the vast majority of the studies, a pseudotyping strategy has been adopted for simplicity reason. Basically, all recombinant genomes are based on AAV2 ITRs, and only the capsid is of the serotype of interest. The production strategies for pseudotyped particles are exactly the same as for the classical AAV2 vectors.

This transient production system is largely used for research and developmental purposes because of its great flexibility (easy change of the plasmids, thus the transgene as well as the AAV serotype can rapidly be adapted to specific needs). Because of interest in its R&D purposes, this production protocol will be presented in detail in **Subheadings 2.3.1.** and **3.3.1**.

The main drawbacks are the limitations in scalability—although roller bottle processes (Avigen) and CellCube-based processes ***(99)*** have been established—and the relatively high incidents of recombination events between the plasmids used, leading to rep+ rAAVs and rcAAV production. Thus, for improving scalability many attempts have been made to develop producer cells. The use of producer cell lines has mainly focused on HeLa cells, although some investigators have evaluated the use of HEK293 and A549 cells. These cells contain the rAAV vector and the *rep* and *cap* genes of AAV. For inducing the production of rAAV, the cells have to be infected with the helper virus. Thus, generation of stable cell lines is better suited to large-scale production than transient transfection; nevertheless, generation of such cells can be tedious and time consuming. Furthermore, the highest virus titers for these production methods are typically about 10^7 IP/mL. It has been estimated that 10^{12}–10^{14} rAAV particles are required for clinical

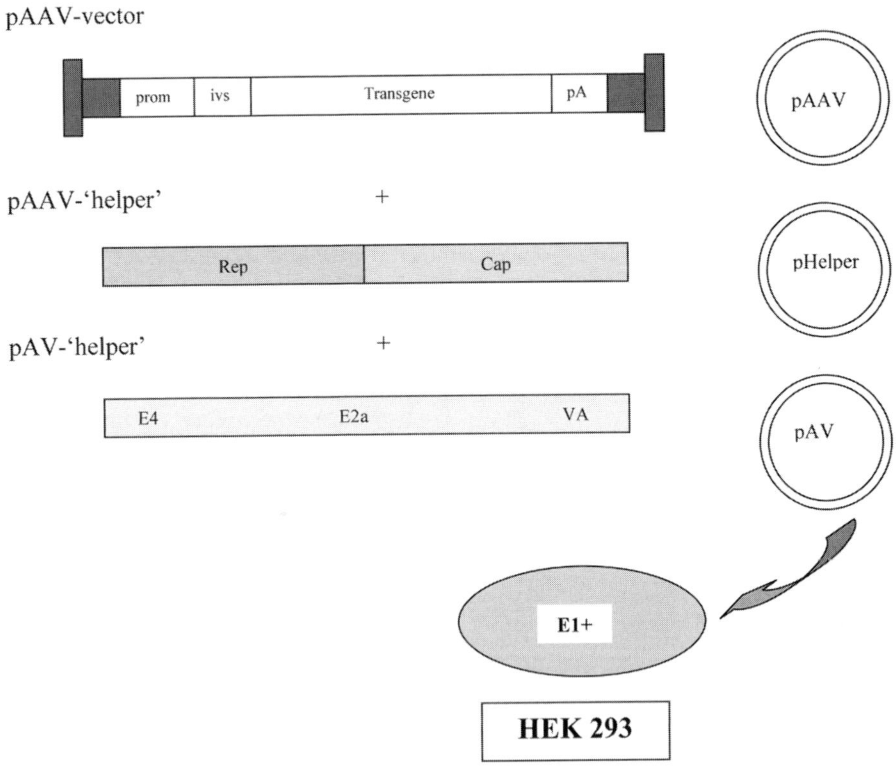

Fig. 6. Classical transfection production principle using HEK293 cells. In a transient production system, the pAAV-'helper,' the pAV-'helper' (a mini AV plasmid, such as pXX6), and the pAAV-vector plasmids will be brought to the producer cells (HEK293) by transfection. This method leads to the generation of about 10^3–10^4 particles per cell. Prom: promoter, ivs: intervening sequence (e.g., intron).

human use ***(100)***. Because to date no really satisfying cell line is available, this approach will not be further developed in these protocols, and the reader is referred to the specific literature for more details ***(101)***.

In order to overcome these limitations, recent studies have focused on producing rAAV vectors in insect cell cultures, using the recombinant baculovirus system derived from the *Autographa californica* nuclear polyhedrosis virus (*Ac*NPV) ***(102)***. Production of rAAV particles is achieved by coinfecting Sf9 cells with three baculovirus vectors: BacRep, BacCap, and Bac-rAAV **(Fig. 7)**. These encode the respective components of the rAAV production machinery. This system lends itself to large-scale production under serum-free conditions, as Sf9 cells grow in suspension. Infection of Sf9 cells at an MOI of 5 and a ratio of 1:1:1 for all three baculoviruses was able to yield a total of 2.2×10^{12} IP in a 3-L bioreactor. The process was scaled-up to 20 L without loss of productivity and demonstrated that

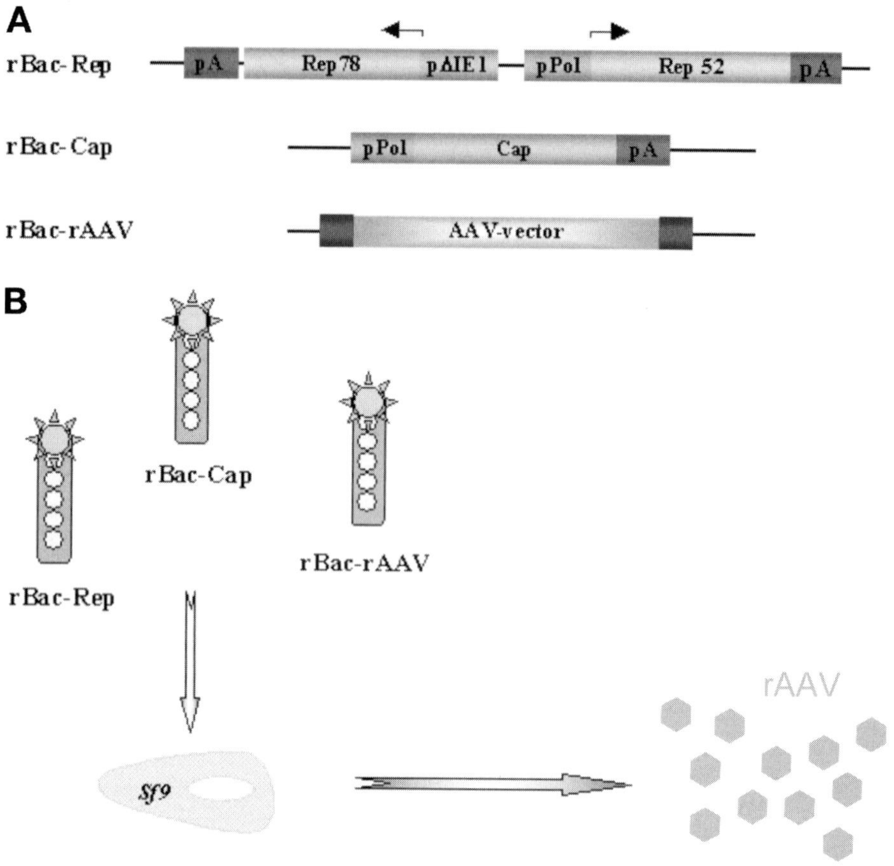

Fig. 7. The baculovirus/insect cell production system for rAAV. **(A)** Genetic constructions of the recombinant baculoviruses. The 2-split *rep orf* are driven by two insect promoters: the polyhedrine promoter (pPol) of AcMNPV and a truncated form of the immediate-early 1 gene promoter (pΔEI1) of *Orgyia pseudotsugata* nuclear polyhedrosis virus. The difference in promoter strength allows high expression of the small Rep and a reduced expression of the toxic large Rep. rBac-Cap express capsid proteins. The three proteins are directly translated from one transcript. rBac-GFP carries a rAAV-GFP genome. The presence of a cytomegalovirus and p10 promoter allows green fluorescent protein (GFP) expression in both mammalian and insect cells. **(B)** Adeno-associated virus (AAV) production is done by a triple infection of Sf9 cells. The three recombinant baculoviruses are used at a 1:1:1 ratio with an multiplicity of infection of 5 per recombinant baculovirus. Three days postinfection rAAVs are harvested from cells and supernatant and are purified.

Table 4
Production Yields of rAAV Using Different Production Systems

Production method	Yield (vg/cell)	Scale-up	Ref.
293, triple transfection	10–10^3	Small scale	***95,115,192–194***
HeLa-based producer cell, rAAV production induced by infection with wild-type Ad5	10^4–10^6	Reactor scale possible	***108,195***
A549 based producer cell, rAAV production induced by infection with a Ad ts	10^5	Reactor scale: 15 L, larger scale possible	***116,196***
Baculovirus system: Sf9 infected with three different recombinant baculoviruses	10^4–10^5	Reactor scale: 20 L, larger scale possible	***102***

quantities sufficient to meet clinical demand could be met ***(103)***. Further optimization of this system is required to improve productivity at high cell densities, but this production system provides the possibility to easily produce large quanitites of rAAVs with the advantages of a rather elevated flexibility to change the transgene (use of another Bac-rAAV construct) or to modify the AAV serotype (use of another BacCap construct). The establishment of a new baculovirus construct (cloning, selection, and amplification included) takes about 6 wk. Today, the Sf9/baculovirus production system has been developed for the production of the following AAV serotypes: AAV1 (R. M. Kotin, personal communication), AAV2 ***(103)***, AAV5 ***(104)***, and AAV8 ***(105)***. These advantages make the Sf9/baculovirus system the most interesting and promising production system for rAAV, and thus protocols are also presented here (*see* **Subheadings 2.3.2.** and **3.3.2.**).

A comparison of the different production systems is presented in **Table 4**. Further details on production issues of rAAV can be found in a review by Merten et al. ***(101)***.

Although out of the scope of this methods chapter, the protocols for purification of rAAV have evolved over the last decade. The most classical and easy to acquire for a nonspecialized laboratory remain the ultracentrifugation on a CsCl ***(106)*** or iodixanol gradient ***(107)***. However, these purification methods are limited by the capacity of cell lysate volume that can be processed and by the low purity achievable. Despite these limitations, these methods are still widely used and most of the time sufficient for fundamental studies and early in vivo preclinical evaluation of the vectors, for which reason the CsCl-based ultracentrifugation protocol is also presented here (*see* **Subheadings 2.3.1.** and **3.3.1.3.**). More complex techniques based on column chromatography and membrane techniques are now well developed for the generation of high-purity-grade and upscaled production suitable for human clinical applications ***(60,61,108–111)***.

1.3.3. Titration

The most important endpoint of all vector production protocols is the vector titer and the total vector quantity produced. A classical method is the dot blot ***(112)***, which serves for the determination of the physical particles by dot blot hybridization (detection of the packaged rAAV genomes using DNA probes specific for the transgene cassette). A positive signal in this assay indicates that rAAV virions were produced, and quantification yields a particle number in virions per milliliter. However, this assay will not indicate if the virus is infectious or if the expression cassette is functional. This test is described in **Subheadings 2.3.1.4.** and **3.3.1.5**. A considerable improvement is the use of quantitative polymerase chain reaction (PCR) developed to precisely determine the physical titers, allowing the titration of vector preparations independently from the transgene and without the presence of a marker gene ***(113,114)***. This method has a much higher reproducibility and improved inter- and intratest variability than the dot blot method.

The determination of infectious particles has more relevance in particular for in vivo applications but also to evaluate vector batch quality. It is based on infection of target cells and detection of the infectious hit by a hybridization-based method ***(115)***. A similar assay based on the use of quantitative PCR detecting the gene sequence of the transgene in the transduced target cells has also been used ***(116)***. For simplicity purposes the transducing titer may be determined, but it is rather imprecise and does not allow comparing productions differing in their expression cassettes ***(117)***. However, the main importance of the transduction assay here is to validate the functionality of the expression cassettes.

To help worldwide standardization of quantification methods and facilitate interpretation of preclinical and clinical data, reference material for AAV vectors will be generated (Richard Snyder, Director of Biotherapeutic Programs, University of Florida, ICBR, Building 62, South Newell Drive, PO Box 110580, Gainesville, FL 32611-0580).

2. Materials (*see* Notes 10 and 11)

2.1. Production of RV and LV Vectors

2.1.1. Transient Production

2.1.1.1. Cell Lines

1. Retroviral MLV-based vectors were originally produced in murine cell lines such as NIH3T3, not particularly well suited because of their low production capabilities and also the presence of endogenous RV (ERV) sequences that

could participate in recombination with murine-derived RV packaging elements and lead to RCR generation *(118)*. Therefore, the next production schemes of MLV-based vectors have used nonmurine cell lines, such as simian COS-7 *(119)* and human HEK293T cells *(120)*.
2. Because of its high permissivity for DNA uptake and expression, the HEK293T cell line has emerged as a standard for RV vector production. The presence of the SV40 large T antigen in these cells results in the amplification of transfected DNA plasmids that contain the SV40 origin of replication. However, this feature is not considered optimal in the context of pharmaceutical manufacturing, since SV40 large T DNA or protein may contaminate the preparations and because SV40 has been arguably linked to human cancers *(121)*.
3. To avoid this discrepancy, the HEK293 cell line has been validated for the production of clinical grade LV vectors *(122)*, even though the presence in these cells of the adenovirus E1 region could also be a matter of safety concern.
4. LV vectors produced from cells expressing galactosyl (α1-3) transferase (αGal) are sensitive to human serum as a result of the activities of anti-αGal antibodies *(123,124)*. For this reason, and especially if the vector is to be used for direct in vivo gene transfer in human subjects, production needs to be performed using cells (old world monkey cells, human, and baboon cells included) that do not contain the galactosyl transferase activity, such as the human TE671 myosarcoma cell line *(125–127)*.

Cell Culture Medium: Dulbecco's modified Eagle's medium (DMEM, Gibco-BRL) or Iscove's modified Eagle's medium (IMDM, Gibco-BRL) supplemented with 10% fetal bovine serum (FBS, HyClone) for the growth of cell lines. To facilitate downstream processing of RV vectors and eliminate the risk of adventitious contaminations, several different serum-free media have been used for RV vector production, such as Ultraculture (BioWhittaker) or Optimem-I (Life Technologies).

Plasmids: Low- or high-copy plasmids are produced and purified from bacteria in sterile endotoxin-free conditions. In terms of purity, the research-grade preparation of plasmid DNA using a final ultracentrifugation in a cesium chloride gradient is certainly less clean than the procedures based on the use of an anion-exchange matrix (Qiagen, Nucleobond), but the production levels obtained with both sources of DNA are identical *(128)*. The majority of plasmids used in the RV vector transfection process contain the SV40 origin of replication in order to replicate and maintain their episomal status in large T-antigen-expressing cells such as COS-7 and HEK293T cell lineages.

Transfection Reagents: The most routinely used transfection method is based on calcium phosphate that precipitates the plasmid mixture. The replacement of the traditional hydroxyethyl piperazine ethane sulfonate (HEPES) buffer by *N,N*-bis(2-hydroxyl)-2-aminoethanesulfonic acid (BES) has made it possible to increase the LV vector production by a factor of 2.5 *(128)*. Commercially available transfection reagents have also been used in the transient transfection process for producing RV vector particles. These include the multicomponent lipid-based Fugen6 (Roche Diagnostics) *(129)*, liposome-based Lipofectamine (Gibco-BRL) *(130)* and DOTAP *(5)*, the cationic polymer PEI *(131)*, and the dendrimer-based Superfect (Qiagen) *(132)*.

2.1.2. Production of RV and LV Vectors Using Stable Packaging Cell Lines

2.1.2.1. CONSTITUTIVE MLV PACKAGING CELL LINES

The first packaging cells for producing MLV RV vectors were based on the genomic integration into the host genome of NIH3T3 murine cells of a single plasmid that contains all required packaging functions (*Gag-Pol-Env*) *(133)*. In order to reduce and/or abrogate susceptible RCR generation, further developments have been performed in NIH3T3 cells, where *Gag-Pol* and *Env* expression were split into two separate constructs *(134–136)*, associated with gene modifications by wobble mutations *(137)* that participate in the overall reduction of susceptible recombination events between constructs.

2.1.2.2. TETRACYCLINE-REGULATED LV PACKAGING CELL LINES

1. The tetracycline-inducible expression system uses the regulatory elements of the Tn10-specific tetracycline resistance operon of *E. coli*. The tetracycline switching-off (Tet-off) system has been privileged because it imposes the absence of the drug during vector collections, a preferred situation for facilitating the downstream processing and for accepting vector transfer for clinical use.
2. In the Tet-off expression system, the tetracycline-dependent transcription factor (tTA) is a fusion protein of the amino-terminal DNA-binding domain of the tetracycline repressor and the carboxy-terminal activation domain of VP-16 from HSV. In the absence of a tetracycline analog (doxycycline), tTA binds to tetracycline responsive elements (TREs) and efficiently activates transcription from downstream minimal promoters.
3. Several different laboratories have used the Tet-off system to generate HIV-1-based LV packaging cell lines with pseudotyping by the natural *Env (138,139)* or by the VSV/G *(140–144)* surface glycoproteins (*see* **Table 3**).

2.1.2.3. ECDYSONE-REGULATED LV PACKAGING CELL LINES

The ecdysone-inducible system (InVitrogen), which is derived from *Drosophila melanogaster*, has been used in the concern to establish inducible packaging cells for the production of HIV-1- *(145,146)* or SIV-based *(147)* LV vectors (*see* **Table 3**). In the presence of an ecdysone analog (ponasterone A, 1–5 µM), the transcription is controlled by a heterodimeric transcription factor consisting of a modified ecdysone receptor fused to the transcriptional activation domain of the HSV protein VP-16 (VgEcR), and RxR, the mammalian homolog of the natural ecdysone receptor-binding partner.

All three reports have demonstrated robust vector production yields, with packaging cell lines maintaining their stability for long periods in culture. Although these reports noticed relevant vector production until day 4 postinduction, no information is given for more prolonged harvests.

2.1.3. Production of RV and LV Vectors Using Shuttle Viral Vectors

2.1.3.1. ADENOVIRUS SHUTTLE EXPRESSION SYSTEM

Production studies using RV packaging cell lines have demonstrated *trans*-complementation by a chimeric adeno-RV transfer vector to produce high-titer retrovirus containing supernatants *(15)*. The broad range of cells transducible by adenovirus, the greater reproducibility of adenoviral transduction vs transfection, and the high-level expression of RV genes from adenoviral vectors can be used to test the relative contribution of individual components in generating viable, high-titer RV supernatants from multiple packaging cell lines. Moreover, the large cloning capacity of adenoviral vectors (up to 35 kb) has allowed the insertion of five gene expression cassettes, which were necessary for the generation of VSV-G pseudotyped LV vectors *(25)*.

The ability of adenoviral vectors that carry the RV *gag-pol* gene and an envelope glycoprotein gene—amphotropic *env* for MLV *(20)* and VSV/G for HIV-1 *(25)*—to *trans*-complement efficiently may offer new possibilities for corrective gene therapy via the generation of in vivo RV producer cells *in situ*.

2.1.3.2. ALPHAVIRUS SHUTTLE EXPRESSION SYSTEM

The major advantages of alphavirus (SFV and Sindbis virus) expression systems are their ease of use and the high levels of RNAs and proteins produced in transfected cells. These systems monopolize on the ability of the alphavirus RNA genome to self-replicate very efficiently in the cell cytoplasm. Li and Garroff have shown that the cytoplasmic RNA-based SFV expression system can be used for the production of MLV-based particles at high titers *(14,15)*. This supports the idea that all retrovirus assembly events, including genome encapsidation, can take place in the cytoplasm of infected cells.

Delivering the RV vector RNA outside the nucleus has permitted to maintain the integrity of introns in encapsidated RV RNAs in situations where introns and flanking regions are involved in the stability of some transcripts such as human clotting factors *(17)*.

Two methods for producing RV vectors from the SFV-based expression system have been described in which the whole process is carried out in the cytoplasm of either RV packaging cells (Method 1 *[17]*) or hamster BHK-21 cells (Method 2 *[65]*) (*see* **Subheading 3.1.3.3.**).

2.1.3.3. Herpes Shuttle Expression System

In contrast to the two above-described shuttle expression systems, the duration of RV gene expression and therefore vector production is prolonged when using modified amplicon-based HSV-derived vectors. HSV amplicons are plasmid-based vectors that, in addition to the transgene of interest and corresponding expression elements, need only two noncoding HSV sequences, an origin of DNA replication (*ori-S*) and a DNA cleavage-packaging signal (pac), to be packaged in HSV virions in the presence of helper functions ***(148)***. These virions can infect a wide range of dividing and nondividing cells and can package about 150 kb of DNA. The amplicon DNA is packaged as a concatemer as a result of a rolling circle mode of viral DNA replication that makes it attractive for generating multiple copies of transgenes. One major limitation of these vectors has been the loss of amplicon DNA from the host cell nucleus over time and therefore of gene expression, especially in dividing cells ***(149)***. A hybrid amplicon system has been developed to increase retention of the amplicon DNA therefore extending transgene expression. Insertion in HSV amplicons of the replication origin (*ori-P*) associated with its *trans*-acting protein (EBNA-1) from Epstein-Barr virus has been shown to support nuclear replication of the HSV amplicon DNA in dividing cells ***(150)*** and to highly produce MLV-based RV vector particles ***(13)***.

2.2. Production of AdV Vectors

(Note: For cell culture work, all solutions must be sterile and made with tissue-culture-quality reagents.)

2.2.1. Generation of the Recombinant Vector Genome in a Plasmid

- Restriction enzymes: e.g., Swa I, Eco RI, ClaI, PvuI, and Xmn I.
- pKP1.3: plasmid containing AdV genome **(Fig. 4)**.
- GFP shuttle vector (shuttle vector DNA from Quantum Biotechnologies or Stratagen) **(Fig. 3)**.
- Competent *E. coli* strain BJ5183 (rec+) (Quantum Biotechnologies or Stratagen).
- Miniprep kits (Promega & solutions 1, 2, 3).
- 1% agarose gel in TAE 1X (Tris-acetate 40 m*M* pH 8.0 EDTA 1 m*M*).
- Electrophoresis system.
- LB/amp$^+$ plates.

2.2.2. Transfection of Recombinant Genome in Trans-Complementing Cells

- HEK293 cells and 911 cells.
- Infection medium: DMEM (Gibco ref. 31960-021), supplemented with 2% FBS, penicillin/streptomycin (PS) (100X, Gibco ref. 15140-122), and 2 m*M* glutamine (glutamine 100X, Gibco ref. 25030-024).
- Production medium: DMEM, supplemented with 10% FBS, PS, and 2 m*M* glutamine.

- For calcium phosphate transfection: calcium phosphate transfection kits (i.e., Stratagene ref. 200388).
- For PEI transfection: PEI 25KD (Aldrich ref. 40,872-7). PEI 10X: dilute 4.5 mg of pure PEI in 10 mL of water, adjust to pH 7.0 with HCl.
- Pac I.
- Cold absolute ethanol, 70% ethanol.
- 150 m*M* NaCl, 3 *M* sodium acetate pH 5.2, TE (10 m*M* Tris/1 m*M* EDTA) pH 8.0.
- Dry ice, 1.5-mL Eppendorf tubes, 15-mL tubes.
- Electrophoresis system.
- 0.7% agarose gel, TBE 1X (TrisHCl, 90 m*M*; boric acid, 72 m*M*; EDTA, 2.4 m*M*, adjust to pH = 8.3 with acetic acid).

2.2.3. Preparation of an E1-Deleted Vector Stock

- E1-deleted vector stock (purified or crude cell lysate, e.g., AdGFP).
- Twenty 10-cm plates of 911 (or HEK293) cells, 80–90% confluent.
- DMEM, supplemented with 10% FCS, PS, and nonessential amino acids (NEAA; Gibco ref. 11140-035).
- Tabletop centrifuge.

2.2.4. Purification of Vector Stock by Banding on CsCl

- Ultracentrifuge: Beckman Coulter Optima L90K o L100XP.
- Rotor: Beckman Coulter SW40Ti.
- CsCl solutions: 1.4 g/mL, 1.34 g/mL, and 1.25 g/mL in phosphate-buffered saline (PBS) 1X.
- Polyallomer centrifuge tubes for SW40 rotor (Beckman Coulter ref. 331374).
- 18-gage needles.
- 2-mL syringes, pipet-aid, 5-mL pipets.
- Amersham/Pharmacia PD-10 columns Sephadex G-25 (ref. 17-0851–01).
- PBS 1X, 0.5-mL tubes, 5-mL clear collection tubes, mineral oil, Saran wrap, empty 500-mL PBS or medium bottle.
- PBS 1X Ca^{2+}/Mg^{2+} (Gibco ref. 14080–048).
- Glycerol, anhydride (Fluka ref. 49769).

2.2.5. Titration of an E1-Deleted Vector Stock

(Note: All solutions must be sterile.)

- E1-deleted vector stock (purified or crude cell lysate).
- 6-well plates of 911 (or HEK293) cells, 80–90% confluent.
- 12-well plates of 911 (or HEK293) cells, 80–90% confluent.
- 2X MEM without phenol red (100 mL).
- FBS, glutamine, PS.
- 2% LMT agarose in water (autoclaved).
- 10% SDS in PBS (5 mL).
- Adeno-XTM Rapid Titer Kit (BD Clontech ref. 631028).

2.3. Production of rAAV Vectors

2.3.1. Production of Adenovirus-Free rAAV by Transient Transfection of 293 Cells

(Note: All solutions are prepared by using double distilled water or equivalent.)

2.3.1.1. rAAV Production by Transient Transduction of 293 Cells

- pAAV Vector cloning: a helper-free system can be purchased from Stratagene, or the psub201 plasmid (for cloning the transgene between the AAV termini) is available (ATCC ref. 68065), as is a model AAV2 vector plasmid (pAAV-CMV(nls)lacZ can be obtained from A. Salvetti, Laboratoire de Thérapie Génique, CHU Hôtel-DIEU, Nantes, France).
- pXX6 plasmid: the adenoviral helper plasmid (can be obtained from the UNC Vector Core Facility (Gene Therapy Center, Division of Pharmaceutics, University of North Carolina at Chapel Hill, NC 27599) ***(95)***.
- pRepCap4 plasmid: the AAV2 helper plasmid can be obtained from Généthon or A. Salvetti from the Laboratoire de Thérapie Génique, CHU Hôtel-DIEU, Nantes, France ***(115)***.
- HEK293 tissue culture cell line (ATCC ref. CRL 1573), derived from a controlled frozen cell stock.
- Cell culture media:
  DMEM (4.5 g/L glucose) (Invitrogen ref. 41966), supplemented with 10% fetal calf serum (FCS).
  DMEM (4.5 g/L glucose) (Invitrogen ref. 41966), supplemented with 1% FCS.
  DMEM (1 g/L glucose) (Invitrogen ref. 31885), supplemented with 10% FCS.

(Note: The media can be stored at 4°C in the dark for up to 4 wk; cell culture media are not stable because of the inactivation of glutamine (if ala-gln or gly-gln are not used), some other amino acids and some vitamins [e.g., B-vitamins]).

- Trypsin/EDTA (0.05%/0.2 g/L) (Invitrogen ref. 25300).
- T-flask (175 cm^2, Corning).
- Trypan blue solution.
- Disposable 15- or 50-mL polystyrene and polypropylene centrifuge tubes (Falcon).
- Polyethyleneimine (PEI, 25 kDa) (10 m*M*) (Aldrich ref. 40872-7). Preparation: dissolve 4.5 mg of pure PEI in 8 mL of deionized water (mix well), neutralize with HCl (pH 6.5–7.5), adjust the volume to 10 mL, sterilize by filtration through a 0.22-μm filter). The solution is equivalent to 10 m*M* expressed in nitrogen.
- NaCl (150 m*M*) (Sigma).
- Tabletop centrifuge (e.g., Jouan).

2.3.1.2. Viral Vector Harvest and Purification Using Cesium Chloride Gradient Ultracentrifugation

- Cell scrapers (Corning ref. 3010 or ref. 3011).
- PBS (Ca^{2+}, Mg^{2+}) (Invitrogen ref. 14040).

- Lysis buffer (50 m*M* HEPES, 150 m*M* NaCl, 1 m*M* $MgCl_2$, 1 m*M* $CaCl_2$) (reagents from Sigma).
- Disposable 50-mL polystyrene and polypropylene centrifuge tubes (Falcon),
- Tabletop centrifuge (e.g., Jouan).
- Dry ice/ethanol bath.
- Water bath (37°C).
- Benzonase (250 U/µL) (Merck ref. 101694).
- Saturated ammonium sulfate $(NH_4)_2SO_4$ (Merck), pH 7.0, 4°C: add 450 g of $(NH_4)_2SO_4$ to 500 mL of water. Heat on a stir plate until $(NH_4)_2SO_4$ dissolves completely. Filter through Whatman paper while still warm and allow to cool (upon cooling, crystals will form, which should not be removed). Adjust the pH to 7.0 with ammonium hydroxide. Store up to 1 yr at 4°C.
- Beckman high-speed centrifuge and JA17 rotor or equivalent.
- CsCl (Fluka ref. 20966) gradient solutions:
  1.35 g/mL density CsCl (47.25 g CsCl filled up to 100 mL with PBS).
  1.5 g/mL density CsCl (67.5 g CsCl filled up to 100 mL with PBS).
  1.4 g/mL CsCl density CsCl (54.5 g CsCl filled up to 100 mL with PBS).

(Note: For all CsCl gradient solutions, check the density of each solution by weighing 1 mL. Then filter the solutions through 0.22-µm filters. Store the gradient solutions up to 1 yr at room temperature.)

- Beckman ultracentrifuge with 90Ti rotor.
- 8.9-mL Beckman Optiseal tubes for the 90Ti.
- Beckman Fraction Recovery System.
- Refractometer (Abbé A.B986).
- Pierce Slide-A-Lyzer dialysis cassettes (MWCO 10000; PIERCE).
- 21-gage needles.

2.3.1.3. Titration: Determination of Physical Particles by Dot Blot Hybridization

- DMEM (Invitrogen ref. 31966).
- 2x Proteinase K buffer (0.5% SDS, 10 m*M* EDTA; both from Sigma).
- Proteinase K (10 mg/mL) (Merck ref. 1245680100).
- DNAse I (Invitrogen ref. 18047-019).
- 25:24:1 (v/v/v) Phenol/Chloroform/Isoamyl alcohol.
- Sodium acetate 3 *M*.
- Glycogen (Boehringer M901393) (20 mg/mL).
- 100% and 70% ethanol.
- 0.4 *N* NaOH/10 m*M* EDTA.
- 2x SSC (Ameresco 0804-41; 0.3 *M* NaCl/0.03 *M* Na-citrate).
- rAAV plasmid used to make recombinant viral vector (*see* **Subheading 2.3.1.1.**).
- Water bath (37°C).
- 100°C heating block.
- 0.45-µm N^+ nylon membrane (Hybond RPN119B).

- Dot-blot device (BioRad 170 3938).
- Nonradioactive labeling system (Alk Phos Direct, Amersham; *see* manufacturer's instructions).
- ECF substrate detection (Amersham; *see* manufacturer's instructions).

2.3.1.4. Solutions for the Supplementary Protocols for Dot Blot Hybridization

1. **Hybridization buffer:** Add NaCl to the hybridization buffer to give a concentration of 0.5 *M*. Add blocking reagent to a final concentration of 4%. Immediately, mix thoroughly, to get the blocking reagent into a fine suspension. Continue mixing at room temperature for 1–2 h on a magnetic stirrer. This buffer can be used immediately or stored in suitable aliquots at –5 to –30°C.

2. **Primary wash buffer (200 mL):**

Urea	24 g	2 *M*
SDS (0.2 g/mL)	1 mL	0.1%
0.5 *M* sodium phosphate pH 7.0	20 mL	50 m*M*
NaCl	1.74 g	150 m*M*
1.0 *M* $MgCl_2$	2 mL	10 m*M*
Blocking reagent (Amersham RPN3680)	0.4 g	0.2%

The primary wash buffer can be kept for up to 1 wk at 2–8°C.

3. **Secondary wash buffer—20X stock:**

Tris base	121 g	1 *M*
NaCl	112 g	2 *M*

Adjust pH to 10.0. Make up to 1 L with water. This buffer can be kept for up to 4 mo at 2–8°C.

4. **Secondary wash buffer—working dilution:**
Dilute stock 1:20 and add 2 mL/L of 1 *M* $MgCl_2$ to give a final concentration of 2 m*M* Mg^{2+} in the buffer. This buffer should not be stored.

2.3.2. Materials for the rAAV Production by Infection of Insect Cells With Recombinant Baculoviruses

(Note: All solutions are prepared by using double distilled water or equivalent.)

2.3.2.1. rAAV Production by Infection of SF9 Insect Cells With Recombinant Baculoviruses

- Baculovirus stocks:
  Bac-AAV1VP (cap proteins).
  Bac-LSR (rep proteins).
  Bac-ITReGFP-ITR.

- Sf9 cell line (InVitrogen ref. 11496-015).
- Sterile Bellco spinner flasks (100 mL, 500 mL).
- Agitator plate from Bellco, rpm = 150.

- SFM900 II (InVitrogen ref 10902-153).
- SFM900II 1.3X (InVitrogen ref. 10967-032).
- 4% Agarose gel (InVitrogen ref. 18300-012).
- 6-Well tissue-culture plates (Corning).
- Waters baths at 40 and 70°C.
- Bac-to-Bac® Baculovirus Expression System Version C (InVitrogen).

The baculovirus stocks are produced with the Bac-to-Bac Baculovirus Expression SystemVersion C (InVitrogen ref. 10584-027) using standard procedures.

The baculovirus plasmids were described by Urabe et al. ***(103)*** and can be obtained from R. Kotin (Laboratory of Biochemical Genetics, National Heart, Lung, and Blood Institute, National Institutes of Health, Bethesda, MD 20892).

2.3.2.2. $CaCl_2$-Based Purification Protocol

- Lysis buffer (50 m*M* HEPES, 150 m*M* NaCl, 1 m*M* $MgCl_2$, 1 m*M* $CaCl_2$).
- 250-mL disposable (polypropylene) centrifuge tubes (Corning).
- Tablctop centrifuge.
- Waterbath (60°C).

3. Methods

3.1. Production of RV and LV Vectors (see Notes 12–15)

3.1.1. Transient Production

1. In small-scale designs, approx 1×10^6 cells permissive for the production of RV vectors (*see* **Subheading 2.1.1.**) are seeded in 10-cm-diameter dishes the day before transfection. Whereas initial production processes were described by using classical cell culture plates (10- or 15-cm-diameter Petri dishes) or flasks (T75, T150), current efforts aim at scaling up the method using multifloored cell factories ***(131,151,152)***.
2. The day after, cells are in their exponential growing phase with a confluency estimated at approx 60–80%. A few hours before the transfection starts, the culture medium is removed and replaced by fresh medium. At that time some transfection agents still require the addition of serum for reaching optimal vector production yields.
3. The time to generate precipitates, liposomes, or other associated chemicals is dependent on the product by itself, but it has been variable according to external parameters such as temperature, pH, or agitation parameters.
4. The amount of each plasmid DNA needs to be addressed as soon as further modifications have been added. The amount of total plasmid DNA (packaging/ envelope/ vector) is a parameter that is also variable according to the different combinations. Briefly, between 5 and 30 μg of each plasmid are required to fulfill production criteria.
5. The time required for the adsorption of the plasmid transfection mixture to cells is dependent on the type of chemical agent that has been used. Around 3–12 h are

normally sufficient. At that time, supernatants that still contain unwanted DNA mixtures are replaced by fresh medium.

6. Two collections of supernatants are generally addressed 24 and 48 h after medium exchange. Two successive collects in the same day with a minimum of 8 h of delay can also be performed in order to enrich the infectivity of RV vectors. Interestingly, the use of some specific media such as the serum-free ultraculture medium has allowed a more prolonged production phase ***(153)***.
7. The collected supernatants stored at 4°C are then afterfiltered through a 0.45-μm pore-size filter and concentrated by ultracentrifugation at high (50000*g* for 2 h) and low speeds (26000*g* for 12 h) or by ultrafiltration using dead-end filtration cartridges ***(154)***. Further purification can also be performed. The most routinely used technique to purify viral particles is based on ion-exchange chromatography using columns or membrane cartridges, such as the Mustang Q capsules (Pall).
8. Several different analytical methods have been designed for the titration of RV or LV vector preparations. The quantities of total particles can be evaluated by measuring the amount of capsid protein, the reverse transcriptase (RT), activity or the amount of RNA genomes in vector preparations. These methods will, however, mostly detect defective particles, which commonly make up 99% of the viral material isolated from producer cells ***(155)***. Classical RT assays measure the incorporation of radiolabeled deoxyribonucleoside triphosphate into DNA using synthetic RNA/DNA matrices ***(156)***. Nonradioisotopic assays based on the incorporation of digoxigenin-labeled dUTP are also available ***(157)***. The sensitivity of these assays is enhanced when using product-enhanced RT (PERT), in which neosynthesized cDNA is further amplified by standard ***(158)*** or quantitative ***(159)*** PCR. A standardized assay is available (Perkin Elmer) for quantifying the amount of capsid protein (p24) in HIV-derived vector preparations. Vector RNA is measured using crude dot blot assays or more sensitive and quantitative reverse transcriptase–polymerase chain reaction (RT-PCR) assays ***(160–162)***.
9. The final read-out in vector titration assays consists of measuring the amount of transduction-competent particles in vector preparations. The dedicated analytical methods are then based on the measurement of the transgene signal in cells serially transduced by different dilutions of the RV vector stock. Different types of transgene signal have already been quantified. The first titration method consists of determining the number of transduced cells that express the transgenic protein. FACS analysis is often used for measuring the percentage of cells expressing living color reporter genes such as the green fluorescent protein (eGFP, titer being termed as transducing unit TU/mL). Other protein-based approaches have been considered by counting the number of βGal-expressing cells (*lacZ*-forming unit [LFU]/mL), the survival of cells after expression of antibiotic resistance genes (colony-forming unit [CFU]/mL), or by quantifying the level of luciferase protein expression (relative luciferase activity [RLU]/mL). The second type of transgene signal being evaluated in transduced cells is the DNA-based provirus genome. Quantitative PCR (qPCR) analysis is often used to quantitate different proviral forms (total, integrated, or 2LTR-based intermediate circular forms). Normalization with parallel amplification of a cellular gene has allowed us to exactly measure the number of proviral copies per transduced

Table 5
Selection Marker Genes and Corresponding Drug(s) Commonly Used to Establish Resistant Stable Cell Lines

Selection genes	Corresponding drug(s)	Concentration
Hypoxanthine-guanine	Hypoxanthine	15 µg/mL
phosphoribosyltransferase (gpt)	Xanthine	250 µg/mL
HXM-selective medium	Mycophenolic acid	25 µg/mL
Thymidine kinase from	Hypoxanthine	1×10^{-4} *M*
herpes simplex virus	Aminopterin	2×10^{-6} *M*
HAT-selective medium	Thymidine	1.6×10^{-5} *M*
Blasticidin S deaminase gene (bsr)	Blasticidin S	10 µg/mL
Hygromycin phosphotransferase	Hygromycin B	200 µg/mL
Neomycin	Geneticin (G418)	800 µg/mL
Phleomycin/Zeocin	Phleomycin	400 µg/mL
	Zeocin	200 µg/mL
Puromycin	Puromycin	1 µg/mL
Dihydrofolate reductase	Methotrexate (mtx)	100 n*M*

cell and therefore determine indirectly, after having taken into consideration the dilution factor, the number of viral genomes (VGs) per milliliter. Last, by using a RT-qPCR approach, some laboratories have also standardized the vector titration method by evaluating the amount in transduced cells of transgenic mRNAs.

3.1.2. Production of RV and LV Vectors Using Stable Packaging Cell Lines

3.1.2.1. Integration in the Host Cell Genome by Stable Transfection

Plasmids encoding packaging and envelope genes as well as, in some cases, transfer vector RNAs have been stably introduced into the genome of permissive cell lines by the use of marker selection genes. **Table 5** summarizes the different antibiotic genes already being used for the isolation of selectants in transfected eukaryotic cells.

1. The genes that confer resistance to specific antibiotic drugs are present either on the same plasmids that express packaging/envelope functions or on a separate plasmid construct. For the latter version, a 10:1 ratio between the plasmid that encodes packaging or envelope genes and the one that encodes for the marker resistance gene is generally used to enlarge the number of double transfectant/selectants.
2. Concerning the integration in the host cell genome of transfer vectors, the stable transfection process has only been developed for SIN-derived materials. Since non-SIN transfer vectors still contain an active promoter element (original U3 sequence or a heterologous promoter inserted in the deleted U3 region of the 3′LTR) to drive the expression of primary transcripts, viral transduction has been preferred for their stable integration in the host cell genome.

3.1.2.2. INTEGRATION IN THE HOST CELL GENOME BY RV TRANSDUCTION

The isolation of transfected cell clones expressing the packaging functions is a low-yield process, with hundreds of clones having to be screened in order to get a high-performance clone. Surprisingly, Ikeda *(163)* reported that when LV packaging functions were introduced into the cell using an MLV-based vector, a much greater proportion of high-performance packaging clones was obtained *(163)*. This allows for the rapid generation of different packaging lines with different designs and therefore simplifies early development steps.

1. Compared to the classical stable transfection process, the number of packaging cell selectants obtained after RV-mediated transduction was significantly higher for their performance to transcribe the transgene signal.
2. Although it is generally admitted that the stable transfection process could generate concatemerized plasmid integration, the intrinsic nature of RV vectors to integrate in the heterochromatin should help the transgene to be better transcribed.

3.1.3. Production of RV and LV Vectors Using Shuttle Viral Vectors

3.1.3.1. ADENOVIRAL SHUTTLE EXPRESSION SYSTEM: METHOD 1 FOR RV VECTOR PRODUCTION *(164)*

1. Three different E1-deleted replication-defective adenoviral vectors expressing either RV *gag-pol* core particle proteins, gibbon ape leukemia virus (GALV) envelope glycoprotein, or a RV vector genome are linearized, and 10 μg of each plasmid are co-transfected in independent assays with 2 μg of helper adenovirus by the calcium phosphate method into HEK293 cells. Each recombinant adenovirus is purified through four rounds of plaque purification. Replication-competent helper-free viruses are finally propagated on HEK293 cells and purified twice by cesium chloride gradient centrifugation. Adenoviral vector stocks are plaque-titered by infecting HEK293 cells with serial dilutions in DMEM supplemented with 2% FCS for 2 h. Media are then discarded before addition of agar (agar overlay). The number of PFU is counted 7 d later ($\rightarrow$ PFU/mL).
2. TE671 cells are seeded at 4×10^5 cells/well in 6-well plates. The next day they are infected with all three recombinant adenoviruses at various MOIs, from 5 to 500, and incubated for 2 h in 1 mL of DMEM supplemented with 2% FCS. The cells are then washed once with PBS and left in 2 mL of DMEM plus 10% FCS for 24 h, at which time the supernatant is discarded and replaced by 2 mL of fresh 10% FCS-containing medium. Twelve hours later, RV vector-containing supernatants are then collected, filtered through a 0.45-μm pore-size filter, and finally stored at −80°C.

3.1.3.2. ADENOVIRAL SHUTTLE EXPRESSION SYSTEM: METHOD 2 FOR LV VECTOR PRODUCTION *(24)*

1. E1-deleted adenoviral vectors expressing VSV/G, codon-optimized HIV-1 (or SIV) *gag-pol* proteins under the control of a tetracycline-regulatable promoter are

produced in T-Rex-293 cells, which constitutively express the tetracycline repressor (Invitrogen). Production and purification protocols are roughly equivalent to those described in Method 1 with the exception of a further downstream step, which involves the dialysis of CsCl-purified materials against 1 L of PBS containing 25% glycerol and 10 mM of MgCl$_2$ using a 10,000 MW cutoff Slide-A-Lyser dialysis cassette (Pierce). The concentration of the adenoviral vector particles in the vector stocks is determined by measuring the optical density at 260 nm. For the determination of 50% tissue culture infectious dose (TCID$_{50}$) of the adenoviral vector stocks, T-Rex-293 cells are plated at a density of 10^4 cells/well in 96-well plates and infected 1 d later with 100 mL of vector stocks serially diluted with DMEM 2% FCS. Ten days after infection, the number of wells showing cytopathic effects is depicted for each dilution and the adenoviral vector titer thereafter determined.
2. Cos-7 cells, which are permissive for *vif*-deleted lentiviruses and do not support the replication of E1-deleted adenoviruses, are stably transfected with a Zeocin-expressing LV vector genome. Zeocin-resistant cells are thereafter plated in 25-cm^2 flasks and co-infected 1 d later at different MOIs with both VSV/G- and *gag-pol*-expressing adenoviral vectors. Four hours after infection, cells are washed three times with PBS and fed with fresh DMEM. The conditioned media are harvested at different time points after infection, passed through a 0.45-µm filter, and stored at −80°C.

3.1.3.3. ALPHAVIRUS SHUTTLE EXPRESSION SYSTEM: METHOD 1 (*17*)

1. Cytoplasmic chimeric SFV/RV vectors are produced as follow. In brief, 20 µg of helper 2 SFV RNA (Life Technologies) along with 40–50 µg of hybrid SFV/RV vector RNA are electroporated into BHK-21 cells, which are then incubated at 37°C for 24 h.
2. Supernatants are collected and activated by adding chymotrypsin for 15 min at room temperature. The protease is inactivated by adding a one-fourth volume of aprotinin protease inhibitor (500 µg/mL, Sigma). Activated viruses are stored on ice before being added to RV packaging cells.
3. For retrovirus production, Phoenix amphotropic or ecotropic RV packaging cells are transduced at a MOI of 100 with activated supernatants.
4. SFV-transduced cells are washed twice with PBS and incubated in their normal growth medium at 32°C for 16 h. RV vector particles are collected and stored at −80°C.

3.1.3.4. METHOD 2 (*65*)

1. RV *gag-pol* and *env* genes, as well as a recombinant RV genome (LTR-ψ+-neo^R-LTR), are inserted into SFV expression plasmids that also encode for helper SFV proteins.
2. Run-off transcripts are produced in vitro from these linearized plasmids using SP6 RNA polymerase. Replication-competent SFV-based RNAs (20 µL of each) are transfected into 8×10^6 BHK-21 cells by electroporation.
3. The transfected BHK-21 cells are then diluted with 9 mL of complete BHK medium, and 6 mL of the cell suspension (containing $3–4 \times 10^6$ living cells) are plated into a 60-mm culture dish.

4. Cells are incubated at 37°C, and the medium is harvested at 5-h intervals and replaced by 2-mL aliquots of fresh medium. Collected supernatants are passed through a 0.45-μm filter and stored at –80°C.

3.1.3.5. METHOD 3 (***12***)

1. The RV packaging transcription unit *gag-pol-env* is cloned under the control of HSV regulatory sequences into a HSV amplicon plasmid. Five micrograms of this plasmid are transfected into M64A cells that express the HSV IE3 gene. Transfected cells are then infected the next day by an E3-deleted HSV helper virus at 1 PFU per cell. When total cytopathic effect is observed, cells are frozen and thawed three times to release infectious viruses. Titers of the helper virus in viral stocks are determined by plaque assays on M64A cells. The relative amount of amplicon vector DNA is determined by blot hybridization, in comparison with the helper virus DNA, by using a ^{32}P-labeled probe corresponding to cloned *ori-S* sequences, which recognize both helper and amplicon vector DNAs.
2. The vector population (5 μL from passage P5, corresponding to an MOI of 0.1 for the helper particles and of about 0.02 for the amplicon particles) is then used to infect human TE671 cells (approx 10^6 cells) harboring a *lacZ* RV provirus. Cell culture supernatants are collected 48 h later, filtered with a 0.2-μm pore size filter, and centrifuged at 1000 rpm to eliminate cell debris. RV vector stocks are kept at 4°C or stored at –80°C.

3.2. Production of AdV Vectors (see *Notes 16–18*)

3.2.1. Generation of the Recombinant Vector Genome in a Plasmid

1. Linearize 1 μg of plasmid pKP1.3 with 10 units of Swa I at 25°C for 2 h.
2. Digest 1 μg of the shuttle plasmid (containing the expressing cassette) with two different enzymes leaving at least 1 kb per each side flanking the cassette. Digest at 37°C for 2 h and use 10 units of each restriction enzyme (e.g., Xmn I and Pvu I).
3. Mix 100 ng of Swa I-linearized pKP1.3 with 30 ng of shuttle plasmid and 100 ng of Swa I-linearized pKP1.3 with 100 ng of shuttle plasmid. Use two different molar ratios by varying the amount of shuttle plasmid.
4. Transform competent BJ5183 cells by heat shock (42°C for 45 s) with each ratio. Controls: Swa I-linearized pKP1.3 and digested shuttle plasmid alone. Incubate overnight at 37°C. (Control plates should have at least 5–10 times fewer colonies.)
5. The next day, pick 10 smaller colonies immediately upon arrival. Grow in LB/amp for 7–8 h. Do minipreps using Promega kit, and assay by digesting plasmids with one or two restriction enzymes to detect correct recombination events (**Fig. 8A**). (Another option is to do PCR with specific primers of the transgene of interest in the colonies.)
6. Select two to three recombinant plasmids and digest them with several restriction enzymes (at least four or five) to confirm correct insertion of the expression cassette into the adenovirus genome, as well as to discard unwanted recombination events involving other sequences of the adenovirus genome (**Fig. 8B**).

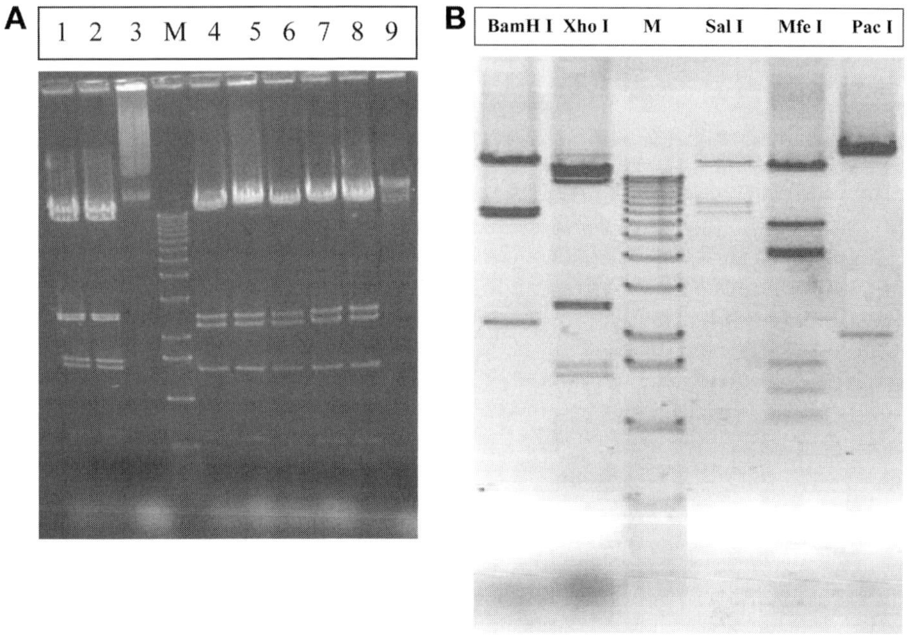

Fig. 8. Screening digestion with restriction enzymes to detect recombinant pKP plasmids containing the transgene of interest. **(A)** Rapid screening to discard non-recombinant plasmids. Lines 1–9: DNA from BJ5183 colonies were digested with EcoRI. All have incorporated the transgene except colonies 3 and 9. M is 1-kb marker. **(B)** Exhaustive analysis by digestion with restriction enzymes to confirm correct cloning of the transgene as well as its genomic stability. One positive clone for homologous recombination was digested by five different informative restriction enzymes overnight. Since all restriction patterns are correct, this clone was selected to generate viral stock.

7. Grow one good colony in 100-mL LB/amp overnight (orbital shaker, 37°C). Do the maxiprep with Qiagen kit (HiSpeed Maxiprep kit) to get a large amount of recombinant plasmid.

3.2.2. Transfection of Recombinant Genome in Trans-*Complementing Cells*

3.2.2.1. Plasmid Preparation

1. Digest 20 μg of recombinant plasmid with Pac I (final volume of 200 μL) for 2 h at 37°C. Pac I removes the adenovirus genome from the plasmid backbone at the inverted terminal repeats. (Make sure that there is not a Pac I site present in the expression cassette.)
2. Check digest on agarose gel: note the presence of 2-kb plasmid backbone band and a high molecular weight band of 30–35 kb (viral genome).

3.2.2.2. DNA Precipitation to Remove Salt (Under Sterile Conditions)

1. Add to the digestion 20 µL of 3 *M* sodium acetate. Mix gently.
2. Add 550 µL of cold absolute ethanol. Mix gently.
3. Incubate 30 min at –80°C.
4. Centrifuge 15 min at 14,000 rpm/4°C and discard supernatant.
5. Wash pellet with 70% ethanol.
6. Centrifuge 5 min at 14,000 rpm and 4°C and discard supernatant.
7. Dry pellet for 15–20 min in the cell culture hood.
8. Resuspend DNA in 100 µl of TE. Incubate 1 h at 37°C.
9. Use immediately to transfect 911 and HEK293 cells, or store DNA at –20°C.

3.2.2.3. Transfection Protocol 1: Calcium Phosphate Transfection

1. Plate HEK293 (or 911) cells in 6-well plates at 80,000 c/cm^2 1 d before transfection.
2. Use two wells from 6-well plates 70–80% confluent 911 and HEK293 cells. Change medium at least 2 h before transfection.
3. Prepare calcium phosphate/DNA complexes in 2-mL Eppendorf tubes following manufacture's instructions. Briefly:
   a. In a tube labeled A: add 5–10 mg DNA and 25 mL 2.5 *M* $CaCl_2$ in a final volume of 250 mL sterile H_2O. Mix well.
   b. Using a Pasteur pipet, slowly add 250 µL of BBS 2X solution (from the Stratagene's kit), dropwise to solution A.
   c. Incubate for 20–30 min at room temperature.
4. Add calcium phosphate/DNA complexes dropwise to the cells.
5. Incubate at 37°C/5% CO_2.
6. Seventy to 80 h later (two virus cycles), cells should be rounded, refractory, with a large nucleus, and floating. Take cells plus medium and put them in a 15-mL tube.
7. Freeze–thaw three times to lyse cells and liberate the virus. Freeze in dry ice/ethanol; thaw in 37°C water bath.
8. Centrifuge 5 min at 4000 rpm. Use supernatant to infect a plate containing 911 or HEK293 cells, and check for cytopathic effect 36 h post-infection.

3.2.2.4. Transfection Protocol 2: PEI Transfection

1. Plate HEK293 (or 911) cells in 6-well plates at 80,000 c/cm^2 1 d before transfection.
2. Use two wells from 6-well plates 70–80% confluent 911 and HEK293 cells. Change medium at least 2 h before transfection.
3. Prepare PEI/DNA complexes in 2-mL Eppendorf tubes:
   a. In a tube labeled A: put 6 µg DNA and 150 µL of sterile 150 m*M* NaCl. Mix well.
   b. In a tube labeled B: put 1.35 µL of PEI 10X and 150 µL of sterile 150 m*M* NaCl. Mix well.
   c. Using a Pasteur pipet, slowly add solution B dropwise to solution A.
   d. Incubate for 30 min at room temperature.
4. Add 700 µL of DMEM (1% FBS) to PEI/DNA complexes and mix.

5. Aspirate medium from cells and add PEI/DNA complexes dropwise to the cells.
6. Incubate at 37°C/5% CO_2. Change for fresh medium (with 10% FBS) 4–6 h later.
7. Seventy to 80 h later (two virus cycles), cells should be rounded, refractory, with a large nucleus, and floating. Take cells plus medium and put them in a 15-mL tube.

3.2.3. Preparation of an E1-Deleted Vector Stock

During the amplification of Ad5 vectors, a reasonable estimate is that each infected cell produces ~10^4 physical particles. Optimal amplification varies between different vectors, but a general rule of thumb is that 50–100 input particles/cell are needed to infect >80% of the cells and maximise the input:output ratio.

3.2.3.1. Phase 1

1. Plate two 10-cm plates with 911 or HEK293 cells at 10^6 cells/plate. (Note: Cells that have been confluent for too long often detach.)
2. Mix the vector with DMEM/10% FCS/PS/nonessential amino acids. Use one-third of the vector produced by transfection (*see* **Subheading 3.2.2.**) per plate. Add 2 mL of complete medium.
3. Gently agitate the plates every 15–30 min to augment vector cell contact. Then add 7 mL of complete medium. (Note: Certain cell lines attach tightly and can be rocked continuously.)
4. Incubate for 34–36 h at 37°C/5% CO_2. (Note: The medium should turn acidic, i.e., red to yellow, demonstrating signs of glycolysis and active propagation of the vector after ~30 h. HEK293 and 911 cells round up and may detach.)
5. Collect the cells by scraping with rubber policeman or tapping on the plate. Put cells and supernatants in 50-mL tubes for tabletop low-speed centrifuge. Spin for 5 min at 1800 rpm. Keep cell pellet with about 20–22 mL of medium.
6. Begin three freeze–thaw cycles in dry-ice-ethanol bath and water bath at 37°C. Vortex between thaw–freeze cycles. (Cover tubes with Parafilm to avoid ethanol leaking into tubes.)
7. Centrifuge for 5 min at 4000 rpm in tabletop centrifuge.
8. Keep cleared vector supernatant at –80°C, or continue with amplification.

3.2.3.2. Phase 2

9. Plate 20 10-cm plates with 911 or HEK293 cells at 10^6 cells/plate.
10. Add 1 mL of vector from point 8 per plate. If amplifying from a titered stock, use 50–100 particles per cell.
11. Gently agitate the plates every 15–30 min to augment vector cell contact. Then add 9 mL of complete medium.
12. Incubate for 30–36 h at 37°C/5% CO_2. (Note: The medium should turn acidic after ~30 h. HEK293 and 911 cells round up and may detach.)
13. Collect the cells by scraping with rubber policeman or tapping on the plate. Put cells and supernatant in 50-mL tubes for tabletop low-speed centrifuge. Spin for 5 min at 1800 rpm. Keep cell pellet with about 1 mL of medium per tube and pool in one 50-mL tube.

14. Begin three freeze–thaw cycles in dry-ice-ethanol bath and water bath at 37°C. Vortex between thaw–freeze cycles.
15. Centrifuge for 5 min at 4000 rpm in tabletop centrifuge.
16. Keep cleared vector supernatant at –80°C, or continue with purification.

3.2.4. Purification of Vector Stock by Banding on CsCl

3.2.4.1. Initial Step Gradient

1. In an SW40 polyallomer centrifuge tube, add 2.5 mL of 1.4 g/mL of CsCl.
2. Add 2.5 mL of 1.25 g/mL of CsCl by placing tip of a 5-mL pipet, **slowly** dispensing solution to make two phases.
3. Gently add ~7 mL of cleared vector supernatant on top of 1.25 g/mL CsCl. Leave ~0.5 cm at the top of the tube.
4. Add ~0.5 mL of mineral oil on top of cleared vector supernatant.
5. Balance tubes against closest sample and load in rotor.
6. Centrifuge 90 min at 35,000 rpm at 18°C.
7. Remove tubes from rotor with forceps.
8. Vector appears as opaque band at interface of 1.25 g/mL and 1.4 g/mL CsCl **(Fig. 9A)**. Remove band by piercing tube ~1 cm below vector with 2 mL syringe loaded with 18-gage needle. (Note: Carry pierced tube to plastic bottle using loaded syringe, withdraw needle, and let drain into 500-mL plastic bottle—an old medium bottle works well.)
9. Combine aliquots of vectors in a new polyallomer centrifuge tubes. (Option: Some labs do a second isopycnic spin.)

3.2.4.2. Second Isopycnic Gradient

1. Add 5–6 mL of 1.34 g/mL of CsCl to the polyallomer centrifuge tube from previous step to fill the tube. Cover with ~0.5 mL of mineral oil.
2. Equilibrate tubes and centrifuge 18 h at 35,000 rpm, 18°C.
3. The following day prepare hood, PD-10 column (by rinsing with PBS), collection tubes, and disposable bottle.
4. Remove tube from rotor and place in black safety tube holder. Vector appears as opaque band near the centre of the tube **(Fig. 9B)**. Remove band as above. Collect vector by keeping volume to a minimum.

3.2.4.3. Desalting Column and Storage

1. Desalt prep over column. Load ≤2 mL of vector on column.
2. Collect by adding 0.5 mL of PBS 1X Ca^{+2}/Mg^{+2}. Repeat step 9–10 times. Label 0.5-mL tubes.
3. The vector is clearly visible as opaque elute in a final volume of ~2.0–3.5 mL (from aliquots 4–7), depending on initial volume size.
4. Combine the most opaque tubes (excluding the extremities, which can be used to isolate DNA for further analysis) and add glycerol to a final concentration of 10%. (Note: Another option is to dialyze the vector in several changes of PBS at 4°C.)

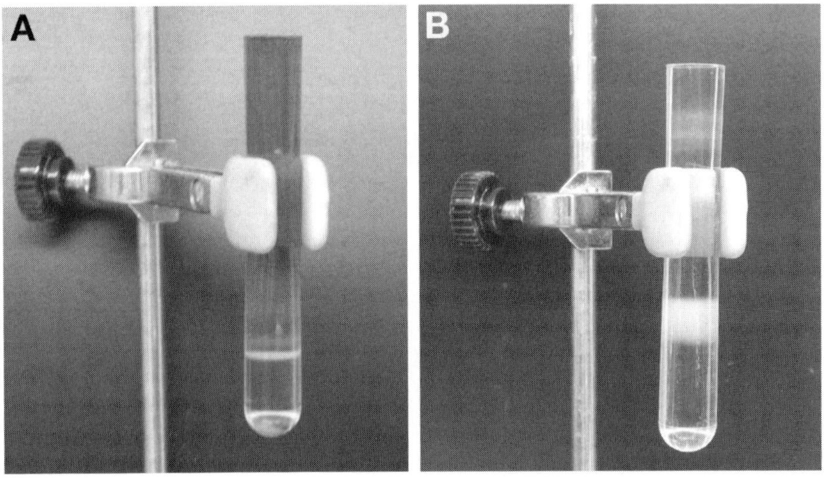

Fig. 9. Adenovirus purification by ultracentrifugation. **(A)** First cesium chloride: step gradient. Crude lysate from a virus preparation of 20 15-cm plates was distributed in six polyallomer tubes containing a cesium chloride step gradient (1.4–1.25 g/mL). After 90 min at 35,000 rpm in a SW40 rotor, virus was clearly visible in the cesium chloride phase. Sometimes other minor bands containing nonfully mature or empty capsids can be seen. **(B)** Second cesium chloride: isopycnic gradient. Recovered bands from polyallomer tubes after cesium chloride step gradient were combined and added to 1.34 g/mL solution of cesium chloride to perform an isopycnic gradient for 18 h at 35,000 rpm in a SW40 rotor. A big and intense virus-containing band can be seen after centrifugation. Virions at intermediate steps of the maturation process can be seen at the top of the tube.

5. Aliquot (in 10, 50, and 100 µL) in 0.5-mL tubes and store at –80°C as quickly as possible.

3.2.5. Titration of an E1-Deleted Vector Stock

3.2.5.1. Titration of Physical Particles (by OD_{260})

1. Make a 1/10 and 1/20 dilution of the purified vector stock in 0.1% SDS (minimum final volume 100 µL—depends on the size of the quartz cuvette); heat to 56°C for 10 min to free the DNA from the capsid.
2. Briefly (30 s) centrifuge (10,000 rpm), transfer the supernatant to a quartz cuvette, and measure OD_{260}. 1 OD is equal to 1.1×10^{12} particle/mL. Do assay at least in duplicate.

3.2.5.2. Titration of PFUs by Plaque Assay

1. Make serial 5- or 10-fold dilutions of vector in DMEM/10% FBS/PS. Use a final volume of 1 mL for each dilution (dilution range between 10^{-7} and 10^{-11}; use the minimum volume to increase the vector-to-cell contact).

2. Remove medium from 6-well plates containing 1 × 10^6 911 or HEK293 cells (plated the day before) and add dilutions gently (Note: 911 cells are less fragile—they don't detach as readily—and larger than HEK293 cells. The size allows a rapid identification of plaques.)
3. Incubate plates at 37°C/5% CO_2 for predetermined time (e.g., 2–4 h). Rock gently every 15–30 min to increase vector to cell contact. (Note: The longer the vector is in contact with the cells, the more accurate is the titration.)
4. Melt 2% agarose in microwave oven. Store at 37°C in the incubator. Prepare 2X MEM/4% FCS/PS/glutamine solution and store at 37°C. Allow 15–30 min to come to 37°C. Mix agarose and MEM extemporaneously before use.
5. Aspirate the medium and replace with premixed MEM/agarose solution at 37°C. Add 4 mL/plate. Let the dish sit at room temperature for 15–30 min. Return to 37°C/5% CO_2 in "Ad-infected" incubator. (Note: Be careful not to disturb the monolayer; it can be fragile depending on the passage number and health of the cells.)
6. Add 4 mL of melted agarose/MEM at 37°C every 3–4 d and begin checking for plaques after 5 d. (Note: The plates/cells are not overly fragile at this point, and the agarose mixture can be added rapidly.)
7. Count plaques at day 9 (if possible) by holding plates against the light (looking through the bottom of course). Determine titer (PFU/mL). (Some labs add neutral red to the agarose in order to detect the plaques.)

3.2.5.3. Titration of Infectious Units (by Microscopy)

Serial Endpoint Dilutions: (Note: This is a very useful protocol, because the vector can be titered whether there is a marker gene or not.)

1. Viral suspension to be titered should be thawed on ice and kept on ice at all times. Use 24-well plates with cells at 80% confluency. Work in triplicate per dilution per virus (from 10^{-7} to 10^{-11}).
2. When serial dilutions are ready, remove media and add, in duplicate, 300 μL of the viral solutions per well.
3. Incubate for 48 h. Aspirate medium and fix cells by very gently adding 300 μL of ice-cold 100% methanol to each well. After 10 min at –20°C, aspirate methanol. Gently rinse wells three times with 300 μL of PBS + 1% BSA.
4. Use Mouse Anti-Hexon Antibody from kit following manufacture's instructions (Adeno-XTM Rapid Titer kit from BD Clontech). Incubate 1 h at 37°C and aspirate Anti-Hexon Antibody. Rinse wells three times with 300 μL of PBS + 1% BSA.
5. Add Rat Anti-Mouse Antibody. Incubate 1 h at 37°C and aspirate Anti-Hexon Antibody. Rinse wells three times with 300 μL of PBS + 1% BSA.
6. After removing the final PBS + 1% BSA rinse, add DAB working solution to each well. Incubate at room temperature for 10 min (5 min may be enough, watch for color development against a white background).
7. Aspirate DAB and add 300 μL of PBS to each well. (Note: Background develops with time. More extensive washes may avoid the problem, but it is recommended that the results be observed soon after completing the color development.)

8. Calculate titer by taking into account the well with stained cells at the highest dilution. (Note: It is recommended to use a standard if possible.)

Serial Dilutions (for eGFP-Expressing Vectors): Viral suspension to be titered should be thawed on ice and kept on ice at all times. Use cells at 80% confluent. Work in duplicate per dilution per virus (from 10^{-7} to 10^{-11}).

1. When serial dilutions are ready, remove media and add, in duplicate, 500 µL of the viral solutions per well. Incubate cells for 2–4 h then add 1.5 mL of complete medium.
2. After the plates have been incubated for 2–3 d, read fluorescence and calculate titer (based on standard if possible).
3. Read fluorescence by noting which wells have fluorescent cells. Be careful not to misinterpret labeled artefacts for labeled cells. Average and calculate the titer depending on which dilution was the last to show fluorescence (prepare at least two out three wells per condition).

Infectious Units Assay by FACS (for GFP-Expressing Vectors):

1. Using the serial dilution from above (from 10^{-7} to 10^{-11}), infect 911 or HEK293 cells in a 6-well plate with 1 mL. Do in triplicate. Incubate at 37°C/5% CO_2 for 2–4 h, then add 4 mL of complete medium.
2. Check infection efficiency 24 h postinfection with fluorescent microscope for GFP expression. Trypsinize the cells, centrifuge, and suspend them in 150 m*M* NaCl. Count the number of cells per well. Do a FACS analysis of the remaining cells.
3. Estimate the titer. Calculate the total number of GFP positive cells/well using the total number of cells and % GFP positive.
4. Using the known physical particle titer, calculate the number of input particles vs infectious units.

3.3. Production of rAAV Vectors

3.3.1. rAAV Production by Transient Transfection of HEK293 Cells

(Note: All cell culture work should be performed according to Good Cell Culture Practice ***[165]***.)

(Note: The equipment and reagents dedicated to tissue culture cells and vector purification should be sterile. All tissue culture incubations are performed in a humidified 37°C, 5% CO_2 incubator unless otherwise specified.)

(Note: A schematic of the transient production protocol is presented in **Fig. 10.**)

3.3.1.1. Thawing and Preparation of HEK293 Cells

1. Thawing of an ampule from the cell stock prepared in DMEM supplemented with 10% FCS and 10% DMSO (5×10^6 c/ampule): After rapid thawing of the cells (incubation of the cryovial in a water batch (37°C), the content (1 mL) is

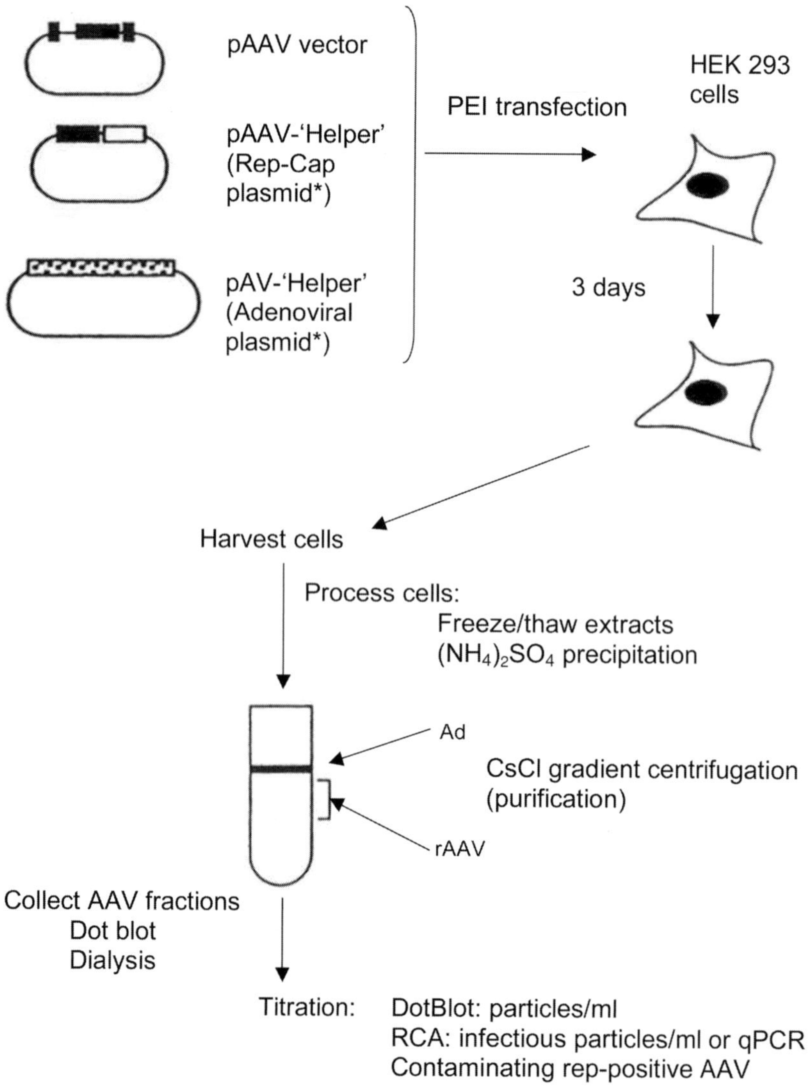

Fig. 10. rAAV is produced using the triple transfection protocol: HEK293 cells are transfected with the rAAV vector (pAAV-vector), the rep-cap plasmid (pAAV-'Helper'), and the adenoviral plasmid (pAV-'Helper') (1:1:2 ratio), fresh medium is added 5–7 h after transfection, and the cells are collected 3 d later and processed (freezing–thawing, $(NH_4)_2SO_4$ precipitation). After CsCl gradient centrifugation, rAAV-containing fractions are pooled, dialyzed, and titered by dot blot, RCA ***(168)***, or qPCR ***(166,167)***.

transferred to a 15- or 50-mL Falcon tube and 9 mL of fresh medium (DMEM + 10% FCS) (preheated to 37°C) is added dropwise.

2. Centrifuge at 140*g* at 20°C for 5 min (Jouan) and eliminate the supernatant.
3. The cell pellet is taken up in 30 mL of DMEM + 10% FCS and plated in a T-flask (175 cm^2).
4. Incubate the T-flask in a CO_2 incubator (37°C, 5% CO_2). The medium is eventually changed 3 d postinoculation.
5. At confluence the medium is eliminated and the cells are once washed with 10 mL of PBS (without Ca^{2+}/Mg^{2+}).
6. The cells are trypsinized by adding 4 mL of trypsin/EDTA; 0.5–5 min later, the T-flask is agitated to detach all cells and 16 mL of DMEM + 10% FCS is added.
7. The cell suspension is transferred to a 50-mL tube (Falcon) and centrifuged at 120*g* (Jouan) for 5 min.
8. The cells are taken up in 10 mL of fresh medium, the suspension is homogenized, and a sample is counted (use of Trypan blue).
9. The subcultures are started with about 30,000 c/cm^2, use of 30 mL of DMEM + 10% FCS per T-flask (175 cm^2).

3.3.1.2. Transfection of HEK293 Cells in View of rAAV Production

(Note: The production procedure described in this protocol is for 20 15-cm dishes but can be scaled up four times.)

1. Seed 3×10^7 cells in 15-cm dishes and maintain the cells in complete DMEM (4.5 g/L glucose) + 10% FBS medium. Incubate overnight in a humidified 37°C, 5% CO_2 incubator. (Note: Alternatively, cells can be seeded 2 d prior transfection at 1.2×10^7 cells in 15-cm dishes.)
2. The cell culture should be at 70–80% of confluence for optimal transfection. Change media 1 h before transfection. (Note: This step can be omitted if the cells appear too fragile, which often occurs with HEK293 cells.)
3. For transfection of five 15-cm dishes, combine first the following in a disposable 50-mL polystyrene tube:

   125 μg pXX6 helper plasmid (adenoviral helper genes).
   62.5 μg rAAV vector plasmid.
   62.5 μg pRepCap4 helper plasmid (AAV helper genes).
   Fill up to 5 mL with a NaCl solution (150 m*M*).

   (This is referred to as the DNA solution. Total DNA is equal to 50 μg per plate.)
4. Perform a set of four tubes for 20 15-cm dishes.
   (Note: The transfection is performed at an optimized ratio of PEI 25 kDa to DNA [R = 2.25 = volume (μL) PEI/weight (μg) DNA])
5. For transfection of five 15-cm dishes, combine the following in a disposable 50-mL polystyrene tube:

   562.5 μL PEI 25 kDa (10 m*M*).
   Fill up to 5 mL with a NaCl solution (150 m*M*).

   (This is referred to as the PEI solution.)

6. Perform a set of four tubes for 20 15-cm dishes.
7. Add dropwise the PEI solution into the DNA solution. Incubate 15–20 min at room temperature to allow the formation of the PEI/DNA complexes.
8. Optional: Meanwhile, rinse the cells with 10 mL of DMEM 4.5 g/L glucose + 1% FBS.
9. Add 12 mL of DMEM 4.5 g/L glucose + 1% FBS.
10. Add 2 mL of the transfection complex drop wise to the medium in each of four 15-cm plates of cells (from **step 5**). Swirl the plates to disperse homogeneously.
11. Repeat **steps 6–8** to transfect each set of five dishes at a time until all 20 dishes have been transfected.
12. Incubate cells for 5–7 h.
13. Add 12 mL of medium (DMEM [1 g/L glucose] + 10% FBS) to each 15-cm plate. (Note: The final glucose concentration will be about 2.75 g/L, allowing cells to divide at a slow rate, thereby producing more viruses.)
14. Incubate the cells until 72 h posttransfection.

3.3.1.3. Viral Vector Harvest and Purification Using Cesium Chloride Gradient Ultracentrifugation

In this procedure, ammonium sulfate precipitation removes most of the cellular debris from the virions before the CsCl gradient is performed. This allows a greater number of cells to be processed and results in more concentrated stocks of rAAV following the CsCl gradient. This recipe is developed for 20–40 15-cm plates, but can be scaled up two to four times.

1. Collect the cells and the supernatant from the tissue culture plates by scraping the cells with a cell scraper to collect cells and medium. Transfer the cell suspension to 50-mL polypropylene centrifuge tubes (10 50-mL tubes will be necessary for a 20-dish preparation).
2. Centrifuge the tubes for 10 min at 700*g* (1500 rpm, Jouan CL412), 8°C. Decant the supernatant. (Note: The pellet can be stored at this stage at –20°C.)
3. Dissolve the pellet with 2 mL of lysis buffer per 50-mL tube (equivalent to 1 mL per initial 15-cm plate) and vortex vigorously. Pool the pellets in one 50-mL tube.
4. Freeze and thaw the cell suspension four times by transferring the tubes between a dry ice/ethanol bath and a 37°C water bath. Vortex vigorously between each cycle. (Note: Freezing and thawing liberates most of the virus particles from the cells. This suspension can be stored at –20°C for up to 6 mo.)
5. Centrifuge the tubes for 15 min at 1500*g* (2600 rpm, Jouan CL412), 4°C. Decant the supernatant into fresh 50-mL polypropylene centrifuge tubes.
6. Treat the lysate with Benzonase (250 U/μL) at a final concentration of 25 U/mL (0.1 μL per mL of lysate) and incubate for 15 min at 37°C in a water bath.
7. Centrifuge the tubes for 20 min at 10000*g* (8500 rpm, Beckman High Speed JA17), 4°C. Decant the supernatant into fresh 50-mL polypropylene centrifuge tubes.
8. Add 1 volume of ice-cold saturated ammonium sulfate to the supernatant. Mix thoroughly and precipitate on ice for 1 h. (Note: athe ammonium sulfate will precipitate the virus in the supernatant.)

9. Centrifuge the tubes 30 min at 12000*g* (10000 rpm, Beckman High Speed JA17 rotor) at 4°C.
10. Slowly decant the supernatant into a container, avoiding touching the pellet. (Note: Do not use bleach to decontaminate ammonium sulfate supernatants because a noxious odor will be produced.)

3.3.1.4. CsCl Gradient Purification

11. Dissolve pellet in 2.5 mL of PBS (Ca^{2+}, Mg^{2+}).
12. Add 3 mL of a 1.5 g/mL CsCl solution to 8.9-mL Beckman Optiseal centrifuge tubes. Gently overlay with 3 mL of a 1.35 g/mL CsCl solution. Then overlay with the 2.5 mL-sample. Fill up the tube with PBS (Ca^{2+}, Mg^{2+}) (about 0.4 mL).
13. Centrifuge the samples at 67,000 rpm for 6 h in a Beckman 90Ti rotor, 8°C. (Note: It is convenient at this stage to perform the ultracentrifugation step overnight.)
14. Decelerate with brake to 1000 rpm, then turn the brake off. (Note: AAV should form a diffuse band in the middle of the tube; however, most of the time the AAV band is not visible.)
15. Using a Beckman Fraction Recovery System, collect 10 fractions of 500 µL (equivalent to 30 drops) from the bottom of the tube.
16. By using a refractometer, analyze the fractions by measuring the refraction index (RI). Pool the positive fractions. (Note: RI for positive fractions containing rAAV usually range between 1.376 and 1.368. Alternatively, positive fraction can be determined by dot blot hybridization using a rAAV-specific probe.)
17. Add 1.40 g/mL CsCl solution to the pooled fractions to attain a final volume of 8.9 mL and transfer the virus solution to an Optiseal tube to re-band the virus. Re-band, drip, and assay the gradient as in **steps 12–15**. (Note: Re-banding the fractions will increase the purity and concentration of the rAAV, but some loss will occur with these supplementary manipulations.)
18. Dialyze the rAAV in MWCO 10000 Slide-A-Lyzer dialysis cassettes (Pierce) against three 500-mL changes of sterile 1x PBS (Ca^{2+}, Mg^{2+}) for at least 3 h each at 4°C. For in vivo application, five changes of sterile PBS are recommended for the dialysis. (Note: Overnight dialysis will not result in any loss of titer.)
19. Transfer the virus suspension into convenient aliquots (typically 100–200 µL) to avoid repeated freezing and thawing. (Note: The virus can be stored at –20 or –80°C for more than 1 yr.)

3.3.1.5. Titration: Determination of Physical Particles by Dot Blot Hybridization

Digest Virus Particles to Release DNA:

1. Place samples (usually 2 and 10 µL) of the viral solution in microtubes. Complete to 200 µL with DMEM.
2. Add 10 U of DNAse I and incubate for 30 min at 37°C in a water bath. (Note: This step digests any DNA that may be present and that has not been packaged into virions.)
3. Release viral DNA by adding 200 µL of 2x Proteinase K buffer and 10 µL of 10 mg/mL Proteinase K (0.25 mg/mL final concentration). Incubate for 60 min at 37°C in a water bath.

4. Extract with 1 volume of 25:24:1 (v/v/v) phenol/chloroform/isoamyl alcohol.
5. Precipitate the viral DNA by adding 40 μg of glycogen as a carrier, 0.1 volume of 3 *M* sodium acetate, and 2.5 volumes of 100% ethanol. Incubate for 30 min at –80°C.
6. Centrifuge at 10000*g* for 20 min, 4°C. Discard the supernatant and wash the pellet with 70% ethanol, then centrifuge at 10000*g* for 10 min, 4°C. Air-dry the pellet for 10 min and resuspend it in 400 μL of 0.4 *N* NaOH/10 m*M* EDTA.
7. Prepare a twofold serial dilution of the pAAV plasmid used to generate the rAAV stock in 20 μl. Add 400 μl of 0.4 *N* NaOH/10 m*M* EDTA. (Note: The dilution range should be 40–0.3125 ng of plasmid.)
8. Denature samples by heating at 100°C for 5 min, then chill on ice.

Dot Blot:

9. Set up the dot blot device with a prewetted 0.45-μm nylon membrane.
10. Wash the wells with 400 μ: of deionized water. Apply vacuum to empty the wells.
11. Add the viral DNA and the plasmid range samples without vacuum. Apply the vacuum until each well is empty.
12. Rinse wells with 0.4 *N* NaOH/10 m*M* EDTA; apply vacuum to dry the membrane.
13. Disassemble the device and rinse the membrane in 2x SSC.

Hybridization:

14. Probe the membrane with a nonradioactive labeled probe (*see* next section).
15. Expose the membrane following the manufacturer instruction and quantify the signal using a PhosphorImager associated with the appropriate software.

3.3.2. Supplementary Protocols Needed for Dot Blot Hybridization

3.3.2.1. Nonradioactive Labeling of DNA Probe (Alk Phos Direct, RPN 3680, Amersham)

(Note: The kit from Amersham is mentioned here, but any comparable kit can be used.)

Preparation of Labeled Probe:

1. Dilute 2 μL of the cross-linker solution with 8 μL of the water supplied to give the working concentration.
2. Dilute DNA to be labeled to a concentration of 10 ng/μL using the water supplied.
3. Place 10 μL of the diluted DNA sample in a micro-centrifuge tube and denature by heating for 5 min in a vigorously boiling water bath.
4. Immediately cool the DNA on ice for 5 min. Spin briefly in a micro-centrifuge to collect the contents at the bottom of the tube.
5. Add 10 μL of reaction buffer to the cooled DNA. Mix thoroughly but gently.
6. Add 2 μL of labeling reagent. Mix thoroughly but gently.
7. Add 10 μL of the cross-linker working solution. Mix thoroughly. Spin briefly in a micro-centrifuge to collect the contents at the bottom of the tube.

8. Incubate the reaction for 30 min at 37°C.
9. The probe can be used immediately or kept on ice for up to 2 h. For long-term storage, labeled probes may be stored in 50% (v/v) glycerol at –15 to –30°C for up to 6 mo.

Hybridization:

10. Preheat the required volume of prepared Alk Phos Direct hybridization buffer to 55°C. The volume of the buffer should be equivalent to 9 mL for blot hybridized in bottles.
11. Place the blot into the hybridization buffer and prehybridize for at least 15 min at 55°C in a hybridization oven.
12. Add the labeled probe to the buffer used for the prehybridization step. Typically use 100 ng per 9 mL of buffer.
13. Hybridize at 55°C overnight in a hybridization oven.

Posthybridization Stringency Washes:

14. Preheat the primary wash buffer to 55°C. This is used in excess at a volume of 100 mL for a blot.
15. Carefully transfer the blot to this solution and wash for 10 min at 55°C in a hybridization oven.
16. Perform a further wash in fresh, primary wash buffer at 55°C for 10 min.
17. Place the blot in a clean container and add an excess of secondary wash buffer (150 mL). Wash under gentle agitation for 5 min at room temperature.
18. Perform a further wash in fresh, secondary wash buffer at room temperature for 5 min.

Detection With ECF Substrate:

19. Pipet 2 mL of ECF substrate onto the blot and incubate for 1 min. Transfer the blot directly to a fresh detection bag. Fold the plastic over the top of the blot and immediately spread the reagent evenly over the blot.
20. Incubate at room temperature in the dark for the required length of time.
21. Scan the blot.

3.3.3. rAAV Production by Infection of SF9 Insect Cells With Recombinant Baculoviruses

3.3.3.1. Preparation of Sf9 Cells

1. Thaw an ampule from the cell stock prepared in SFM900 II medium (5–10 × 10^6 c/ampule) and transfer these cells into a 100-mL spinner containing 50 mL of medium (SFM900 II). Incubate at 27°C on a Bellco agitator, rpm = 150.
2. Four to 5 d later: transfer the whole culture to two 100 mL spinners, dilute the culture with fresh medium (SFM900 II) to a starting cell density of 3 × 10^5 c/mL.
3. Four to 5 d later: transfer the cultures to 400 mL spinners, dilute the culture with fresh medium (SFM900 II) to a starting cell density of 3 × 10^5 c/mL.
4. Four to 5 d later the cells can be used for virus production.

3.3.3.2. Amplification of the Recombinant Baculoviruses on Sf9 Cells

1. Seed an appropriate spinner Bellco (depending on the amount available of the recombinant baculovirus stock to amplify and the needs) with a cell suspension adjusted to 10^6 cells/mL in an appropriate volume (80% of indicated maximum volume on the spinner).
2. Proceed directly to the infection with the recombinant baculovirus to amplify at a multiplicity of infection of 0.1.
3. Using the following relation, calculate the volume (V) to take from the baculovirus stock:

$$V\ (\text{mL}) = (10^6 \times \text{medium volume in mL} \times 0.1)/\text{baculovirus stock titer (pfu/mL)}$$

4. Incubate the cells on an agitator for 3 d at 27°C, rpm = 150.
5. Harvest the supernatant and centrifuge for 15 min at 500*g*.
6. Harvest the supernatant (= virus stock) and eliminate the pellet.
7. Stock the amplified baculovirus stock at +4°C or –80°C in the dark.

3.3.3.3. Production of rAAV Using the Sf9/Baculovirus System

1. Seed a 500-mL Bellco spinner with a cell suspension adjusted to 10^6 cells/mL in a total volume of 400 mL.
2. Proceed directly to the infection with the three baculoviruses at a 1:1:1 ratio and a multiplicity of infection of 5 for each baculovirus:
   a. With the following relation, calculate the volume to take from each viral stock (Bac-AAV1VP, Bac-LSR, Bac-ITReGFP-ITR):

$$V\ (\text{mL}) = (10^6 \times 400 \times 5)/\text{baculovirus stock titer (pfu/mL)}$$

   b. Add the three previously calculated volumes to the spinner flask.
   c. Incubate the cells on an agitator for 3 d, at 27°C, rpm = 150.

3.3.4. Viral Plaque Assay (Titration of Recombinant Baculoviruses)

(*See* also guide of the bac-to-bac system.)

3.3.4.1. Preparing the Plaquing Medium (Sf-900 Plaquing Medium)

1. Combine 30 mL of SFM900II 1.3X with 10 mL of the melted 4% agarose gel (by placing the bottle in a 70°C water bath for 20 min) and mix gently.
2. Return the bottle of plaquing medium to the 40°C water bath until use.

3.3.4.2. Plaque Assay Procedure

(Note: Amounts are suitable to titer one baculovirus stock.)

1. Seed two 6-well tissue-culture plates with 1×10^6 cells/well in 2 mL of SFM900II. Incubate cells for 1 h at room temperature.
2. Prepared an 8-log serial dilution (10^{-1} to 10^{-8}) of the baculovirus stock in SFM900II.
3. Remove the medium from each well and replace with 1 mL of the virus dilution. Incubate cells with virus for 1 h at room temperature.

4. Remove the medium containing virus from the wells and replace with 2 mL of plaquing medium. Incubate the cells in a 27°C humidified incubator for 7–10 d until plaques are visible and ready to count.
5. Count the number of plaques present in each dilution, then use the following formula to calculate the titer (plaque forming unit (pfu/mL) of the viral stock:

$$\text{Titer (pfu/mL)} = \text{number of plaques} \times \text{dilution factor}$$

3.3.4.3. CsCl-Based Purification Protocol

1. Three days postinfection, transfer the cell suspension into two 250-mL polypropylene centrifuge tubes (Corning).
2. Centrifuge the tubes for 5 min at 1300*g* (2500 rpm in a Sigma 4K15 rotor) to collect the cells. Decant the supernatant. (Note: rAAV is frequently present in both the cells and the supernatant (ratio 1/1). For easier handling, only the cells will be processed here.)
3. Dissolve the pellet with 10 mL of lysis buffer per 250-mL tube and pool both pellets together in a 50-mL polypropylene centrifuge tube.
4. Proceed as described for rAAV produced by triple-transfection (*see* **Subheading 3.3.1.3.**) for the lysis.
5. Inactivate baculoviruses in the supernatant from the crude lysate by heating at 60°C for 30 min.
6. Proceed as described for rAAV produced by triple transfection (*see* **Subheading 3.3.1.3**) to the end of the protocol.

4. Notes

1. The scalability for the transient transfection process is limited since huge variations are observed as far as the production volume increases. As an example, the simple rule of three to calculate the concentration of DNA and transfecting agent mixtures is no longer predictive for generating efficient complexes or precipitates.
2. Interestingly, sodium butyrate has been shown to stimulate the activity of a number of promoters and to enhance RV vector production when added to transiently transfected cells ***(151,166–168)***.
3. Differences observed in the lipid composition from cell culture media could also contribute to variable cell membrane rigidity, a factor that has been shown to affect HIV virion stability and infectivity ***(169)***. As an example, the addition of cholesterol during the production process has increased the stability of MLV vectors ***(170)*** and the infectivity of HIV-1-based vectors ***(129)***.
4. Particular attention should be paid to the ability to pellet serum-free supernatants by ultracentrifugation since less than 10% of vector particles could be pelleted without the addition of serum ***(131)***. Alternative solutions have been given by adding poly-L-lysine ***(171)*** or calcium phosphate ***(172)*** to crude extracts.
5. As observed for the transient transfection process, the vector recovery yield obtained from packaging cells has been enhanced by adding sodium butyrate during the production process ***(141)***.

6. The number of supernatant harvests from stable packaging cell lines is dependent on the type of production system. Vector recovery is generally undertaken early, either when the expression system is inducible or when the transfer vector is given transiently. For producer cell lines that are based on a constitutive expression system, vector particle harvests will be limited as far as cells reach their confluency.
7. The methodologies to produce RV vectors from these different viral shuttle expression systems are surely not exhaustive. Many other methods have been described to asset the production of RV vectors by this type of helper system. As an example, another alphavirus system has been used for the production of RV vectors in COS-7 cells, i.e., the simian virus 40-derived DNA vectors ***(119)***.
8. It is also important to notice that two types of HSV-based vectors have already been developed: recombinant and amplicon viral vectors. In recombinant vectors, the transgene of interest is inserted directly into the viral genome. Amplicon vectors that consist of plasmid units (*see* **Subheading 2.1.3.3.**) can only be amplified in the presence of *trans*-acting factors provided by helper virus.
9. Low MOIs of helper viral vectors are privileged to drive the expression of RV elements since higher MOIs have been shown to alter the efficiency of vector production. In the case of amplicon-based HSV vectors, in which low MOIs (between 1 to 5) have been preferred, strong expression of helper proteins resulting from high MOIs has likely impaired retrovirus assembly and protein expression from the amplicon genome ***(13)***. In the same extent, in vitro adenoviral tolerance for RV vector production can be achieved only at optimal MOIs (between 50 and 100) and not at higher MOIs ***(164)***.
10. The key step is to generate adenoviral recombinants in BJ5183 cells. Once the recombinants are obtained, generation of adenoviruses in HEK293 or 911 cells is practically guaranteed. Thus, provided that homologous recombination in bacterial cells works, the overall success rate for generating adenoviruses is high. Depending on the transfection efficiency of HEK293 or 911 cells, one can expect a virus titer of the initial virus lysate from one 25-cm^2 flask ranging from 10^8 to 10^{10} PFU/mL.
11. If every step works well, one can expect to get the initial virus production in 2–3 weeks and the subsequent large-scale purification in an additional 1–2 wk.
12. In these protocols, 20 15-cm tissue culture plates should yield 10^{12}–10^{13} particles as assayed by dot blot. This would translate into 10^9–10^{10} rAAV particles that are capable of delivering and expressing the transgene (~1000 particles per cell). The typical ratio of particles to transducing units is 100–1000:1 when assayed on a cell type that transduces well, such as 293 or HeLa cells. Cells that transduce less well will give a higher ratio. The particle preparations should be free of contaminating cellular and adenovirus proteins and free of infectious adenovirus.
13. Transfection of the cells requires 5 d. On day 1 the cells are split. On day 2 the cells are transfected. On day 5 the cells are harvested.
14. The purification requires 3–4 d: 1 d to perform cell lysis and fractionation leading to the overnight CsCl gradient centrifugation, 1 d to locate the rAAV by reading the refraction index and overnight rebanding of the virus, and 1 d for dialysis (the last round can be performed overnight).
15. The dot blot assay requires 2 d: 1 d to process the samples, bind to the filter, and hybridize, and another day to wash the filter and expose.

16. In this protocol, the yield of the vector is about 5×10^4 particles per SF9 cell as assayed by dot blot.
17. Infection of SF9 cells requires 5 d. On day 1 the cells are seeded and infected. On day 5 the cells are harvested.
18. The time considerations concerning the other protocols are similar those used for the production of rAAV using the transfection of HEK293 cells.

Acknowledgments

This chapter is based on the theoretical and practical courses of the 1st (HCPV-2002-00388) and the 3rd EuroLabCourse (FP6-2002-503219), which were organized in Evry/F in 2002 and 2004. In addition to the authors, all persons having been active in these events are thanked for their involvement.

References

1. Russell, D. W. and Miller, A. D. (1996) Foamy virus vectors. *J. Virol.* **70**, 217–222.
2. Perkins, A. S., Kirschmeier, P. T., Gattoni-Celli, S., and Weinstein, I. B. (1983) Design of a retrovirus-derived vector for expression and transduction of exogenous genes in mammalian cells. *Mol. Cell. Biol.* **3**, 1123–1132.
3. Girod, A., Cosset, F. L., Verdier, G., and Ronfort, C. (1995) Analysis of ALV-based packaging cell lines for production of contaminant defective viruses. *Virology* **209**, 671–675.
4. Naldini, L., Blomer, U., Gage, F. H., Trono, D., and Verma, I. M. (1996) Efficient transfer, integration, and sustained longterm expression of the transgene in adult rat brains injected with a lentiviral vector. *Proc. Natl. Acad. Sci. USA* **93**, 11,382–11,388.
5. Poeschla, E., Gilbert, J., Li, X., Huang, S., Ho, A., and Wong-Staal, F. (1998) Identification of a human immunodeficiency virus type 2 (HIV-2) encapsidation determinant and transduction of nondividing human cells by HIV-2-based lentivirus vectors. *J. Virol.* **72**, 6527–6536.
6. Mangeot, P. E., Negre, D., Dubois, B., et al. (2000) Development of minimal lentivirus vectors derived from simian immunodeficiency virus (SIVmac251) and their use for gene transfer into human dendritic cells. *J. Virol.* **74**, 8307–8315.
7. Poeschla, E. M., Wong-Staal, F., and Looney, D. J. (1998) Efficient transduction of nondividing human cells by feline immunodeficiency virus lentiviral vectors. *Nat. Med.* **4**, 354–357.
8. Matukonis, M., Li, M., Molina, R. P., Paszkiet, B., Kaleko, M., and Luo, T. (2002) Development of second- and third-generation bovine immunodeficiency virus-based gene transfer systems. *Hum. Gene Ther.* **13**, 1293–1303.
9. Mitrophanous, K., Yoon, S., Rohll, J., et al. (1999) Stable gene transfer to the nervous system using a non-primate lentiviral vector. *Gene Ther.* **6**, 1808–1818.
10. Mselli-Lakhal, L., Favier, C., Da Silva Teixeira, M. F., et al. (1998) Defective RNA packaging is responsible for low transduction efficiency of CAEV-based vectors. *Arch. Virol.* **143**, 681–695.
11. Metharom, P., Takyar, S., Xia, H. H., et al. (2000) Novel bovine lentiviral vectors based on Jembrana disease virus. *J. Gene Med.* **2**, 176–185.

12. Savard, N., Cosset, F. L., and Epstein, A. L. (1997) Defective herpes simplex virus type 1 vectors harboring gag, pol, and env genes can be used to rescue defective retrovirus vectors. *J. Virol.* **71**, 4111–4117.
13. Sena-Esteves, M., Saeki, Y., Camp, S. M., Chiocca, E. A., and Breakefield, X. O. (1999) Single-step conversion of cells to retrovirus vector producers with herpes simplex virus-Epstein-Barr virus hybrid amplicons. *J. Virol.* **73**, 10,426–10,439.
14. Li, K. J. and Garoff, H. (1996) Production of infectious recombinant Moloney murine leukemia virus particles in BHK cells using Semliki Forest virus-derived RNA expression vectors. *Proc. Natl. Acad. Sci. USA* **93**, 11,658–11,663.
15. Li, K. J. and Garoff, H. (1998) Packaging of intron-containing genes into retrovirus vectors by alphavirus vectors. *Proc. Natl. Acad. Sci. USA* **95**, 3650–3654.
16. Wahlfors, J. J. and Morgan, R. A. (1999) Production of minigene-containing retroviral vectors using an alphavirus/retrovirus hybrid vector system. *Hum. Gene Ther.* **10**, 1197–1206.
17. Wahlfors, J. J., Xanthopoulos, K. G., and Morgan, R. A. (1997) Semliki Forest virus-mediated production of retroviral vector RNA in retroviral packaging cells. *Hum. Gene Ther.* **8**, 2031–2041.
18. Bilbao, G., Feng, M., Rancourt, C., Jackson, W. H., Jr., and Curiel, D. T. (1997) Adenoviral/retroviral vector chimeras: a novel strategy to achieve high-efficiency stable transduction in vivo. *FASEB J.* **11**, 624–634.
19. Caplen, N. J., Higginbotham, J. N., Scheel, J. R., et al. (1999) Adeno-retroviral chimeric viruses as in vivo transducing agents. *Gene Ther.* **6**, 454–459.
20. Feng, M., Jackson, W. H., Jr., Goldman, C. K., et al. (1997) Stable in vivo gene transduction via a novel adenoviral/retroviral chimeric vector. *Nat. Biotechnol.* **15**, 866–870.
21. Lin, X. (1998) Construction of new retroviral producer cells from adenoviral and retroviral vectors. *Gene Ther.* **5**, 1251–1258.
22. Ramsey, W. J., Caplen, N. J., Li, Q., Higginbotham, J. N., Shah, M., and Blaese, R. M. (1998) Adenovirus vectors as transcomplementing templates for the production of replication defective retroviral vectors. *Biochem. Biophys. Res. Commun.* **246**, 912–919.
23. Yoshida, Y., Emi, N., and Hamada, H. (1997) VSV-G-pseudotyped retroviral packaging through adenovirus-mediated inducible gene expression. *Biochem. Biophys. Res. Commun.* **232**, 379–382.
24. Kuate, S., Stefanou, D., Hoffmann, D., Wildner, O., and Uberla, K. (2004) Production of lentiviral vectors by transient expression of minimal packaging genes from recombinant adenoviruses. *J. Gene Med.* **6**, 1197–1205.
25. Kubo, S. and Mitani, K. (2003) A new hybrid system capable of efficient lentiviral vector production and stable gene transfer mediated by a single helper-dependent adenoviral vector. *J. Virol.* **77**, 2964–2971.
26. Doerfler, W. and Boehm, P. (1995) *The Molecular Repertoire of Adenoviruses*, Springer, Berlin.
27. Stewart, P. L. (2002) *Adenovirus Structure*, Academic Press, San Diego, CA.

28. Roelvink, P. W., Lizonova, A., Lee, J. G., et al. (1998) The coxsackievirus–adenovirus receptor protein can function as a cellular attachment protein for adenovirus serotypes from subgroups A, C, D, E, and F. *J. Virol.* **72**, 7909–7915.
29. Bergelson, J. M., Cunningham, J. A., Droguett, G., et al. (1997) Isolation of a common receptor for coxsackie B viruses and adenoviruses 2 and 5. *Science* **275**, 1320–1323.
30. Tomko, R. P., Xu, R., and Philipson, L. (1997) HCAR and MCAR: the human and mouse cellular receptors for subgroup C adenoviruses and group B coxsackieviruses. *Proc. Natl. Acad. Sci. USA* **94**, 3352–3356.
31. Wickham, T. J., Mathias, P., Cheresh, D. A., and Nemerow, G. R. (1993) Integrins alpha v beta 3 and alpha v beta 5 promote adenovirus internalization but not virus attachment. *Cell* **73**, 309–319.
32. Li, E., Brown, S. L., Stupack, D. G., Puente, X. S., Cheresh, D. A., and Nemerow, G. R. (2001) Integrin v 1 is an adenovirus coreceptor. *J. Virol.* **75**, 5405–5409.
33. Meier, O. and Greber, U. F. (2004) Adenovirus endocytosis. *J. Gene Med.* **6 (Suppl 1)**, S152–S163.
34. Shenk, T. (1996) *The Viruses and Their Replication*, Lippincott, Philadelphia: Lippincott.
35. Imler, J. L., Bout, A., Dreyer, D., et al. (1995) Trans-complementation of E1-deleted adenovirus: a new vector to reduce the possibility of codissemination of wild-type and recombinant adenoviruses. *Hum. Gene Ther.* **6**, 711–721.
36. Doerfler, W. and Boehm, P. (1995) *The Molecular Repertoire of Adenoviruses*, Berlin, Springer.
37. Breyer, B., Jiang, W., Cheng, H., et al. (2001) Adenoviral vector-mediated gene transfer for human gene therapy. *Curr. Gene Ther.* **1**, 149–162.
38. Chuah, M. K., Collen, D., and VandenDriessche, T. (2003) Biosafety of adenoviral vectors. *Curr. Gene Ther.* **3**, 527–543.
39. Graham, F. L., Smiley, J., Russell, W. C., and Nairn, R. (1977) Characteristics of a human cell line transformed by DNA from human adenovirus type 5. *J. Gen. Virol.* **36**, 59–74.
40. Nadeau, I. and Kamen, A. (2003) Production of adenovirus vector for gene therapy. *Biotechnol. Adv.* **20**, 475–489.
41. Garnier, A., Cote, J., Nadeau, I., Kamen, A., and Massie, B. (1994) Scale-up of the adenovirus expression system for the production of recombinant protein in human 293S cells. *Cytotechnology* **15**, 145–155.
42. Cote, J., Garnier, A., Massie, B., and Kamen, A. (1998) Serum-free production of recombinant proteins and adenoviral vectors by 293SF-3F6 cells. *Biotechnol. Bioeng.* **59**, 567–575.
43. Peshwa, M. V., Kyung, Y. S., McClure, D. B., and Hu, W. S. (1993) Cultivation of mammalian cells as aggregates in bioreactors: effect of calcium concentration of spatial distribution of viability. *Biotechnol. Bioeng.* **41**, 179–187.
44. Lochmuller, H., Jani, A., Huard, J., et al. (1994) Emergence of early region 1-containing replication-competent adenovirus in stocks of replication-defective

adenovirus recombinants (delta E1 + delta E3) during multiple passages in 293 cells. *Hum. Gene Ther.* **5**, 1485–1491.

45. Fallaux, F. J., Kranenburg, O., Cramer, S. J., et al. (1996) Characterization of 911: a new helper cell line for the titration and propagation of early region 1-deleted adenoviral vectors. *Hum. Gene Ther.* **7**, 215–222.
46. Schiedner, G., Hertel, S., and Kochanek, S. (2000) Efficient transformation of primary human amniocytes by E1 functions of Ad5: generation of new cell lines for adenoviral vector production. *Hum. Gene Ther.* **11**, 2105–2116.
47. Fallaux, F. J., Bout, A., van der Velde, I., et al. (1998) New helper cells and matched early region 1-deleted adenovirus vectors prevent generation of replication-competent adenoviruses. *Hum. Gene Ther.* **9**, 1909–1917.
48. Imler, J. L., Chartier, C., Dreyer, D., et al. (1996) Novel complementation cell lines derived from human lung carcinoma A549 cells support the growth of E1-deleted adenovirus vectors. *Gene Ther.* **3**, 75–84.
49. Gao, G., Qu, G., Burnham, M. S., et al. (2000) Purification of recombinant adeno-associated virus vectors by column chromatography and its performance in vivo. *Hum. Gene Ther.* **11**, 2079–2091.
50. Volpers, C. and Kochanek, S. (2004) Adenoviral vectors for gene transfer and therapy. *J. Gene Med.* **6 (Suppl 1)**, S164–S171.
51. Amalfitano, A., Hauser, M. A., Hu, H., Serra, D., Begy, C. R., and Chamberlain, J. S. (1998) Production and characterization of improved adenovirus vectors with the E1, E2b, and E3 genes deleted. *J. Virol.* **72**, 926–933.
52. Armentano, D., Zabner, J., Sacks, C., et al. (1997) Effect of the E4 region on the persistence of transgene expression from adenovirus vectors. *J. Virol.* **71**, 2408–2416.
53. Alba, R., Bosch, A., and Chillon, M. (2005) Gutless adenovirus: last-generation adenovirus for gene therapy. *Gene Ther.* **12 (Suppl 1)**, S18–S27.
54. Wu, S. C., Huang, G. Y., and Liu, J. H. (2002) Production of retrovirus and adenovirus vectors for gene therapy: a comparative study using microcarrier and stationary cell culture. *Biotechnol. Prog.* **18**, 617–622.
55. Iyer, P., Ostrove, J. M., and Vacante, D. (1999) Comparison of manufacturing techniques for adenovirus production. *Cytotechnology* **30**, 169–172.
56. Kamen, A. and Henry, O. (2004) Development and optimization of an adenovirus production process. *J. Gene Med.* **6 (Suppl 1)**, S184–S192.
57. Nadeau, I., Garnier, A., Côté, J., Massie, B., Chavarie, C., and Kamen, A. (1996) Improvement of recombinant protein production with the human adenovirus/ 293S expression system using fed-batch strategies. *Biotechnol. Bioeng.* **51**, 613–623.
58. Yuk, I. H., Olsen, M. M., Geyer, S., and Forestell, S. P. (2004) Perfusion cultures of human tumor cells: a scalable production platform for oncolytic adenoviral vectors. *Biotechnol. Bioeng.* **86**, 637–642.
59. Warnock, J. N., Merten, O. W., and Al-Rubeai, M. (2006) Cell culture processes for the production of viral vectors for gene therapy purposes. *Cytotechnology* **50**, 141–162.
60. Burova, E. and Ioffe, E. (2005) Chromatographic purification of recombinant adenoviral and adeno-associated viral vectors: methods and implications. *Gene Ther.* **12 (Suppl 1)**, S5–S17.

61. Duffy, A. M., O'Doherty, A. M., O'Brien, T., and Strappe, P. M. (2005) Purification of adenovirus and adeno-associated virus: comparison of novel membrane-based technology to conventional techniques. *Gene Ther.* **12 (Suppl 1)**, S62–S72.
62. Maizel, J. V., Jr., White, D. O., and Scharff, M. D. (1968) The polypeptides of adenovirus. I. Evidence for multiple protein components in the virion and a comparison of types 2, 7A, and 12. *Virology* **36**, 115–125.
63. Shabram, P. W., Giroux, D. D., Goudreau, A. M., et al. (1997) Analytical anion-exchange HPLC of recombinant type-5 adenoviral particles. *Hum. Gene Ther.* **8**, 453–465.
64. Nyberg-Hoffman, C., Shabram, P., Li, W., Giroux, D., and Aguilar-Cordova, E. (1997) Sensitivity and reproducibility in adenoviral infectious titer determination. *Nat. Med.* **3**, 808–811.
65. Mittereder, N., March, K. L., and Trapnell, B. C. (1996) Evaluation of the concentration and bioactivity of adenovirus vectors for gene therapy. *J. Virol.* **70**, 7498–7509.
66. Lusky, M., Christ, M., Rittner, K., et al. (1998) In vitro and in vivo biology of recombinant adenovirus vectors with E1, E1/E2A, or E1/E4 deleted. *J. Virol.* **72**, 2022–2032.
67. Zabner, J., Chillon, M., Grunst, T., et al. (1999) A chimeric type 2 adenovirus vector with a type 17 fiber enhances gene transfer to human airway epithelia. *J. Virol.* **73**, 8689–8695.
68. Hutchins, B. (2002) Development of a reference material for characterization adenovirus vector. *Bioproc. J.* **1**, 25–28.
69. Gao, F. G., Jeevarajan, A. S., and Anderson, M. M. (2004) Long-term continuous monitoring of dissolved oxygen in cell culture medium for perfused bioreactors using optical oxygen sensors. *Biotechnol. Bioeng.* **86**, 425–433.
70. Gao, G. P., Alvira, M. R., Wang, L., Calcedo, R., Johnston, J., and Wilson, J. M. (2002) Novel adeno-associated viruses from rhesus monkeys as vectors for human gene therapy. *Proc. Natl. Acad. Sci. USA* **99**, 11,854–11,859.
71. Mori, S., Wang, L., Takeuchi, T., and Kanda, T. (2004) Two novel adeno-associated viruses from cynomolgus monkey: pseudotyping characterization of capsid protein. *Virology* **330**, 375–383.
72. Rutledge, E. A., Halbert, C. L., and Russell, D. W. (1998) Infectious clones and vectors derived from adeno-associated virus (AAV) serotypes other than AAV type 2. *J. Virol.* **72**, 309–319.
73. Gao, G., Vandenberghe, L. H., and Wilson, J. M. (2005) New recombinant serotypes of AAV vectors. *Curr. Gene Ther.* **5**, 285–297.
74. Qiu, J. and Pintel, D. J. (2004) Alternative polyadenylation of adeno-associated virus type 5 RNA within an internal intron is governed by the distance between the promoter and the intron and is inhibited by U1 small nuclear RNP binding to the intervening donor. *J. Biol. Chem.* **279**, 14,889–14,898.
75. Erles, K., Sebokova, P., and Schlehofer, J. R. (1999) Update on the prevalence of serum antibodies (IgG and IgM) to adeno-associated virus (AAV). *J. Med. Virol.* **59**, 406–411.

76. Tobiasch, E., Burguete, T., Klein-Bauernschmitt, P., Heilbronn, R., and Schlehofer, J. R. (1998) Discrimination between different types of human adeno-associated viruses in clinical samples by PCR. *J. Virol. Methods* **71**, 17–25.
77. Kotin, R. M., Siniscalco, M., Samulski, R. J., et al. (1990) Site-specific integration by adeno-associated virus. *Proc. Natl. Acad. Sci. USA* **87**, 2211–2215.
78. Samulski, R. J., Zhu, X., Xiao, X., et al. (1991) Targeted integration of adeno-associated virus (AAV) into human chromosome 19. *EMBO J.* **10**, 3941–3950.
79. Shelling, A. N. and Smith, M. G. (1994) Targeted integration of transfected and infected adeno-associated virus vectors containing the neomycin resistance gene. *Gene Ther.* **1**, 165–169.
80. Grimm, D. and Kay, M. A. (2003) From virus evolution to vector revolution: use of naturally occurring serotypes of adeno-associated virus (AAV) as novel vectors for human gene therapy. *Curr. Gene Ther.* **3**, 281–304.
81. Srivastava, A., Lusby, E. W., and Berns, K. I. (1983) Nucleotide sequence and organization of the adeno-associated virus 2 genome. *J. Virol.* **45**, 555–564.
82. Tratschin, J. D., Miller, I. L., and Carter, B. J. (1984) Genetic analysis of adeno-associated virus: properties of deletion mutants constructed in vitro and evidence for an adeno-associated virus replication function. *J. Virol.* **51**, 611–619.
83. Weitzman, M. D., Kyostio, S. R., Kotin, R. M., and Owens, R. A. (1994) Adeno-associated virus (AAV) Rep proteins mediate complex formation between AAV DNA and its integration site in human DNA. *Proc. Natl. Acad. Sci. USA* **91**, 5808–5812.
84. Horer, M., Weger, S., Butz, K., Hoppe-Seyler, F., Geisen, C., and Kleinschmidt, J. A. (1995) Mutational analysis of adeno-associated virus Rep protein-mediated inhibition of heterologous and homologous promoters. *J. Virol.* **69**, 5485–5496.
85. Pereira, D. J., McCarty, D. M., and Muzyczka, N. (1997) The adeno-associated virus (AAV) Rep protein acts as both a repressor and an activator to regulate AAV transcription during a productive infection. *J. Virol.* **71**, 1079–1088.
86. King, J. A., Dubielzig, R., Grimm, D., and Kleinschmidt, J. A. (2001) DNA helicase-mediated packaging of adeno-associated virus type 2 genomes into preformed capsids. *EMBO J.* **20**, 3282–3291.
87. Timpe, J., Bevington, J., Casper, J., Dignam, J. D., and Trempe, J. P. (2005) Mechanisms of adeno-associated virus genome encapsidation. *Curr. Gene Ther.* **5**, 273–284.
88. Schlehofer, J. R., Ehrbar, M., and zur Hausen, H. (1986) Vaccinia virus, herpes simplex virus, and carcinogens induce DNA amplification in a human cell line and support replication of a helpervirus dependent parvovirus. *Virology* **152**, 110–117.
89. Weindler, F. W. and Heilbronn, R. (1991) A subset of herpes simplex virus replication genes provides helper functions for productive adeno-associated virus replication. *J. Virol.* **65**, 2476–2483.
90. McPherson, R. A., Rosenthal, L. J., and Rose, J. A. (1985) Human cytomegalovirus completely helps adeno-associated virus replication. *Virology* **147**, 217–222.

91. Chon, S. K., Rim, B. M., and Im, D. S. (1999) Adeno-associated virus Rep78 binds to E2-responsive element 1 of bovine papillomavirus type 1. *IUBMB Life* **48**, 397–404.
92. Geoffroy, M. C. and Salvetti, A. (2005) Helper functions required for wild type and recombinant adeno-associated virus growth. *Curr. Gene Ther.* **5**, 265–271.
93. Wistuba, A., Kern, A., Weger, S., Grimm, D., and Kleinschmidt, J. A. (1997) Subcellular compartmentalization of adeno-associated virus type 2 assembly. *J. Virol.* **71**, 1341–1352.
94. Grimm, D., Kern, A., Rittner, K., and Kleinschmidt, J. A. (1998) Novel tools for production and purification of recombinant adenoassociated virus vectors. *Hum. Gene Ther.* **9**, 2745–2760.
95. Xiao, X., Li, J., and Samulski, R. J. (1998) Production of high-titer recombinant adeno-associated virus vectors in the absence of helper adenovirus. *J. Virol.* **72**, 2224–2232.
96. Chiorini, J. A., Kim, F., Yang, L., and Kotin, R. M. (1999) Cloning and characterization of adeno-associated virus type 5. *J. Virol.* **73**, 1309–1319.
97. Chiorini, J. A., Yang, L., Liu, Y., Safer, B., and Kotin, R. M. (1997) Cloning of adeno-associated virus type 4 (AAV4) and generation of recombinant AAV4 particles. *J. Virol.* **71**, 6823–6833.
98. Xiao, W., Chirmule, N., Berta, S. C., McCullough, B., Gao, G., and Wilson, J. M. (1999) Gene therapy vectors based on adeno-associated virus type 1. *J. Virol.* **73**, 3994–4003.
99. Brown, P., Barret, S., Godwin, S., et al. (1998) Optimization of production of adeno-associated virus (AAV) for use in gene therapy. Presented at Cell Culture Engineering VI, San Diego, CA.
100. Clark, K. R. (2002) Recent advances in recombinant adeno-associated virus vector production. *Kidney Int.* **61 (Suppl 1)**, S9–S15.
101. Merten, O. W., Geny-Fiamma, C., and Douar, A. M. (2005) Current issues in adeno-associated viral vector production. *Gene Ther.* **12 (Suppl 1)**, S51–S61.
102. Urabe, M., Ding, C., and Kotin, R. M. (2002) Insect cells as a factory to produce adeno-associated virus type 2 vectors. *Hum. Gene Ther.* **13**, 1935–1943.
103. Meghrous, J., Aucoin, M. G., Jacob, D., Chahal, P. S., Arcand, N., and Kamen, A. A. (2005) Production of recombinant adeno-associated viral vectors using a baculovirus/insect cell suspension culture system: from shake flasks to a 20-L bioreactor. *Biotechnol. Prog.* **21**, 154–160.
104. Urabe, M., Nakakura, T., Xin, K. Q., et al. (2006) Scalable generation of high-titer recombinant adeno-associated virus type 5 in insect cells. *J. Virol.* **80**, 1874–1885.
105. Kohlbrenner, E., Aslanidi, G., Nash, K., et al. (2005) Successful production of pseudotyped rAAV vectors using a modified baculovirus expression system. *Mol. Ther.* **12**, 1217–1225.
106. Tenenbaum, L., Hamdane, M., Pouzet, M., et al. (1999) Cellular contaminants of adeno-associated virus vector stocks can enhance transduction. *Gene Ther.* **6**, 1045–1053.

107. Zolotukhin, S., Byrne, B. J., Mason, E., et al. (1999) Recombinant adeno-associated virus purification using novel methods improves infectious titer and yield. *Gene Ther.* **6**, 973–985.
108. Blouin, V., Brument, N., Toublanc, E., Raimbaud, I., Moullier, P., and Salvetti, A. (2004) Improving rAAV production and purification: towards the definition of a scaleable process. *J. Gene Med.* **6 (Suppl 1)**, S223–S228.
109. Brument, N., Morenweiser, R., Blouin, V., et al. (2002) A versatile and scalable two-step ion-exchange chromatography process for the purification of recombinant adeno-associated virus serotypes-2 and -5. *Mol. Ther.* **6**, 678–686.
110. Kaludov, N., Handelman, B., and Chiorini, J. A. (2002) Scalable purification of adeno-associated virus type 2, 4, or 5 using ion-exchange chromatography. *Hum. Gene Ther.* **13**, 1235–1243.
111. Zolotukhin, S., Potter, M., Zolotukhin, I., et al. (2002) Production and purification of serotype 1, 2, and 5 recombinant adeno-associated viral vectors. *Methods* **28**, 158–167.
112. Flotte, T. R., Solow, R., Owens, R. A., Afione, S., Zeitlin, P. L., and Carter, B. J. (1992) Gene expression from adeno-associated virus vectors in airway epithelial cells. *Am. J. Respir. Cell Mol. Biol.* **7**, 349–356.
113. Rohr, U. P., Wulf, M. A., Stahn, S., Steidl, U., Haas, R., and Kronenwett, R. (2002) Fast and reliable titration of recombinant adeno-associated virus type-2 using quantitative real-time PCR. *J. Virol. Methods* **106**, 81–88.
114. Veldwijk, M. R., Topaly, J., Laufs, S., et al. (2002) Development and optimization of a real-time quantitative PCR-based method for the titration of AAV-2 vector stocks. *Mol. Ther.* **6**, 272.
115. Salvetti, A., Oreve, S., Chadeuf, G., et al. (1998) Factors influencing recombinant adeno-associated virus production. *Hum. Gene Ther.* **9**, 695–706.
116. Farson, D., Harding, T. C., Tao, L., et al. (2004) Development and characterization of a cell line for large-scale, serum-free production of recombinant adeno-associated viral vectors. *J. Gene Med.* **6**, 1369–1381.
117. Rohr, U. P., Heyd, F., Neukirchen, J., et al. (2005) Quantitative real-time PCR for titration of infectious recombinant AAV-2 particles. *J. Virol. Methods* **127**, 40–45.
118. Otto, E., Jones-Trower, A., Vanin, E. F., et al. (1994) Characterization of a replication-competent retrovirus resulting from recombination of packaging and vector sequences. *Hum. Gene Ther.* **5**, 567–575.
119. Landau, N. R., and Littman, D. R. (1992) Packaging system for rapid production of murine leukemia virus vectors with variable tropism. *J. Virol.* **66**, 5110–5113.
120. Yang, S., Delgado, R., King, S. R., et al. (1999) Generation of retroviral vector for clinical studies using transient transfection. *Hum. Gene Ther.* **10**, 123–132.
121. Barbanti-Brodano, G., Sabbioni, S., Martini, F., Negrini, M., Corallini, A., and Tognon, M. (2004) Simian virus 40 infection in humans and association with human diseases: results and hypotheses. *Virology* **318**, 1–9.
122. Manilla, P., Rebello, T., Afable, C., et al. (2005) Regulatory considerations for novel gene therapy products: a review of the process leading to the first clinical lentiviral vector. *Hum. Gene Ther.* **16**, 17–25.

123. Takeuchi, Y., Porter, C. D., Strahan, K. M., et al. (1996) Sensitization of cells and retroviruses to human serum by (alpha 1-3) galactosyltransferase. *Nature* **379**, 85–88.
124. Welsh, R. M., O'Donnell, C. L., Reed, D. J., and Rother, R. P. (1998) Evaluation of the Galalpha1-3Gal epitope as a host modification factor eliciting natural humoral immunity to enveloped viruses. *J. Virol.* **72**, 4650–4656.
125. Chang, L. J., Urlacher, V., Iwakuma, T., Cui, Y., and Zucali, J. (1999) Efficacy and safety analyses of a recombinant human immunodeficiency virus type 1 derived vector system. *Gene Ther.* **6**, 715–728.
126. Higashikawa, F., and Chang, L. (2001) Kinetic analyses of stability of simple and complex retroviral vectors. *Virology* **280**, 124–131.
127. Iwakuma, T., Cui, Y., and Chang, L. J. (1999) Self-inactivating lentiviral vectors with U3 and U5 modifications. *Virology* **261**, 120–132.
128. Karolewski, B. A., Watson, D. J., Parente, M. K., and Wolfe, J. H. (2003) Comparison of transfection conditions for a lentivirus vector produced in large volumes. *Hum. Gene Ther.* **14**, 1287–1296.
129. Mitta, B., Rimann, M., and Fussenegger, M. (2005) Detailed design and comparative analysis of protocols for optimized production of high-performance HIV-1-derived lentiviral particles. *Metab. Eng.* **7**, 426–436.
130. Nakajima, T., Nakamaru, K., Ido, E., Terao, K., Hayami, M., and Hasegawa, M. (2000) Development of novel simian immunodeficiency virus vectors carrying a dual gene expression system. *Hum. Gene Ther.* **11**, 1863–1874.
131. Geraerts, M., Michiels, M., Baekelandt, V., Debyser, Z., and Gijsbers, R. (2005) Upscaling of lentiviral vector production by tangential flow filtration. *J. Gene Med.* **7**, 1299–1310.
132. Coleman, J. E., Huentelman, M. J., Kasparov, S., et al. (2003) Efficient large-scale production and concentration of HIV-1-based lentiviral vectors for use in vivo. *Physiol. Genomics* **12**, 221–228.
133. Miller, A. D. and Buttimore, C. (1986) Redesign of retrovirus packaging cell lines to avoid recombination leading to helper virus production. *Mol. Cell. Biol.* **6**, 2895–2902.
134. Danos, O. and Mulligan, R. C. (1988) Safe and efficient generation of recombinant retroviruses with amphotropic and ecotropic host ranges. *Proc. Natl. Acad. Sci. USA* **85**, 6460–6464.
135. Markowitz, D., Goff, S., and Bank, A. (1988) Construction and use of a safe and efficient amphotropic packaging cell line. *Virology* **167**, 400–406.
136. Markowitz, D., Hesdorffer, C., Ward, M., Goff, S., and Bank, A. (1990) Retroviral gene transfer using safe and efficient packaging cell lines. *Ann. NY Acad. Sci.* **612**, 407–414.
137. Morgenstern, J. P. and Land, H. (1990) Advanced mammalian gene transfer: high titer retroviral vectors with multiple drug selection markers and a complementary helper-free packaging cell line. *Nucleic Acids Res.* **18**, 3587–3596.
138. Kaul, M., Yu, H., Ron, Y., and Dougherty, J. P. (1998) Regulated lentiviral packaging cell line devoid of most viral cis-acting sequences. *Virology* **249**, 167–174.

139. Yu, H., Rabson, A. B., Kaul, M., Ron, Y., and Dougherty, J. P. (1996) Inducible human immunodeficiency virus type 1 packaging cell lines. *J. Virol.* **70**, 4530–4537.
140. Farson, D., Witt, R., McGuinness, R., et al. (2001) A new-generation stable inducible packaging cell line for lentiviral vectors. *Hum. Gene Ther.* **12**, 981–997.
141. Kafri, T., van Praag, H., Ouyang, L., Gage, F. H., and Verma, I. M. (1999) A packaging cell line for lentivirus vectors. *J. Virol.* **73**, 576–584.
142. Klages, N., Zufferey, R., and Trono, D. (2000) A stable system for the high-titer production of multiply attenuated lentiviral vectors. *Mol. Ther.* **2**, 170–176.
143. Scherr, M., Battmer, K., Schultheis, B., Ganser, A., and Eder, M. (2005) Stable RNA interference (RNAi) as an option for anti-bcr-abl therapy. *Gene Ther.* **12**, 12–21.
144. Xu, K., Ma, H., McCown, T. J., Verma, I. M., and Kafri, T. (2001) Generation of a stable cell line producing high-titer self-inactivating lentiviral vectors. *Mol. Ther.* **3**, 97–104.
145. Pacchia, A. L., Adelson, M. E., Kaul, M., Ron, Y., and Dougherty, J. P. (2001) An inducible packaging cell system for safe, efficient lentiviral vector production in the absence of HIV-1 accessory proteins. *Virology* **282**, 77–86.
146. Sparacio, S., Pfeiffer, T., Schaal, H., and Bosch, V. (2001) Generation of a flexible cell line with regulatable, high-level expression of HIV Gag/Pol particles capable of packaging HIV-derived vectors. *Mol. Ther.* **3**, 602–612.
147. Kuate, S., Wagner, R., and Uberla, K. (2002) Development and characterization of a minimal inducible packaging cell line for simian immunodeficiency virus-based lentiviral vectors. *J. Gene Med.* **4**, 347–355.
148. Spaete, R. R. and Frenkel, N. (1982) The herpes simplex virus amplicon: a new eucaryotic defective-virus cloning-amplifying vector. *Cell* **30**, 295–304.
149. Johnston, K. M., Jacoby, D., Pechan, P. A., et al. (1997) HSV/AAV hybrid amplicon vectors extend transgene expression in human glioma cells. *Hum. Gene Ther.* **8**, 359–370.
150. Wang, S. and Vos, J. M. (1996) A hybrid herpesvirus infectious vector based on Epstein-Barr virus and herpes simplex virus type 1 for gene transfer into human cells in vitro and in vivo. *J. Virol.* **70**, 8422–8430.
151. Lu, X., Humeau, L., Slepushkin, V., et al. (2004) Safe two-plasmid production for the first clinical lentivirus vector that achieves >99% transduction in primary cells using a one-step protocol. *J. Gene Med.* **6**, 963–973.
152. Przybylowski, M., Hakakha, A., Stefanski, J., Hodges, J., Sadelain, M., and Riviere, I. (2005) Production scale-up and validation of packaging cell clearance of clinical-grade retroviral vector stocks produced in cell factories. *Gene Ther.* **13**, 95–100.
153. Reiser, J. (2000) Production and concentration of pseudotyped HIV-1-based gene transfer vectors. *Gene Ther.* **7**, 910–913.
154. Steffens, S., Tebbets, J., Kramm, C. M., Lindemann, D., Flake, A., and Sena-Esteves, M. (2004) Transduction of human glial and neuronal tumor cells with different lentivirus vector pseudotypes. *J. Neurooncol.* **70**, 281–288.

155. Martin-Rendon, E., White, L. J., Olsen, A., Mitrophanous, K. A., and Mazarakis, N. D. (2002) New methods to titrate EIAV-based lentiviral vectors. *Mol. Ther.* **5**, 566–570.
156. Goff, S., Traktman, P., and Baltimore, D. (1981) Isolation and properties of Moloney murine leukemia virus mutants: use of a rapid assay for release of virion reverse transcriptase. *J. Virol.* **38**, 239–248.
157. Eberle, J. and Seibl, R. (1992) A new method for measuring reverse transcriptase activity by ELISA. *J. Virol. Methods* **40**, 347–356.
158. Pyra, H., Boni, J., and Schupbach, J. (1994) Ultrasensitive retrovirus detection by a reverse transcriptase assay based on product enhancement. *Proc. Natl. Acad. Sci. USA* **91**, 1544–1548.
159. Arnold, B. A., Hepler, R. W., and Keller, P. M. (1998) One-step fluorescent probe product-enhanced reverse transcriptase assay. *Biotechniques* **25**, 98–106.
160. Ikeda, Y., Collins, M. K., Radcliffe, P. A., Mitrophanous, K. A., and Takeuchi, Y. (2002) Gene transduction efficiency in cells of different species by HIV and EIAV vectors. *Gene Ther.* **9**, 932–938.
161. Lizee, G., Aerts, J. L., Gonzales, M. I., Chinnasamy, N., Morgan, R. A., and Topalian, S. L. (2003) Real-time quantitative reverse transcriptase-polymerase chain reaction as a method for determining lentiviral vector titers and measuring transgene expression. *Hum. Gene Ther.* **14**, 497–507.
162. Sastry, L., Johnson, T., Hobson, M. J., Smucker, B., and Cornetta, K. (2002) Titering lentiviral vectors: comparison of DNA, RNA and marker expression methods. *Gene Ther.* **9**, 1155–1162.
163. Ikeda, Y., Takeuchi, Y., Martin, F., Cosset, F. L., Mitrophanous, K., and Collins, M. (2003) Continuous high-titer HIV-1 vector production. *Nat. Biotechnol.* **21**, 569–572.
164. Duisit, G., Salvetti, A., Moullier, P., and Cosset, F. L. (1999) Functional characterization of adenoviral/retroviral chimeric vectors and their use for efficient screening of retroviral producer cell lines. *Hum. Gene Ther.* **10**, 189–200.
165. Coecke, S., Balls, M., Bowe, G., et al. (2005) Guidance on good cell culture practice. A report of the second ECVAM task force on good cell culture practice. *Altern. Lab. Anim.* **33**, 261–287.
166. Gasmi, M., Glynn, J., Jin, M. J., Jolly, D. J., Yee, J. K., and Chen, S. T. (1999) Requirements for efficient production and transduction of human immunodeficiency virus type 1-based vectors. *J. Virol.* **73**, 1828–1834.
167. Sakoda, T., Kasahara, N., Hamamori, Y., and Kedes, L. (1999) A high-titer lentiviral production system mediates efficient transduction of differentiated cells including beating cardiac myocytes. *J. Mol. Cell Cardiol.* **31**, 2037–2047.
168. Sena-Esteves, M., Tebbets, J. C., Steffens, S., Crombleholme, T., and Flake, A. W. (2004) Optimized large-scale production of high titer lentivirus vector pseudotypes. *J. Virol. Methods* **122**, 131–139.
169. Aloia, R. C., Jensen, F. C., Curtain, C. C., Mobley, P. W., and Gordon, L. M. (1988) Lipid composition and fluidity of the human immunodeficiency virus. *Proc. Natl. Acad. Sci. USA* **85**, 900–904.

170. Beer, C., Meyer, A., Muller, K., and Wirth, M. (2003) The temperature stability of mouse retroviruses depends on the cholesterol levels of viral lipid shell and cellular plasma membrane. *Virology* **308**, 137–146.
171. Zhang, B., Xia, H. Q., Cleghorn, G., Gobe, G., West, M., and Wei, M. Q. (2001) A highly efficient and consistent method for harvesting large volumes of high-titer lentiviral vectors. *Gene Ther.* **8**, 1745–1751.
172. Pham, L., Ye, H., Cosset, F. L., Russell, S. J., and Peng, K. W. (2001) Concentration of viral vectors by co-precipitation with calcium phosphate. *J. Gene Med.* **3**, 188–194.
173. Naldini, L., Blomer, U., Gallay, P., et al. (1996) In vivo gene delivery and stable transduction of nondividing cells by a lentiviral vector [*see* comments]. *Science* **272**, 263–267.
174. Schnell, T., Foley, P., Wirth, M., Munch, J., and Uberla, K. (2000) Development of a self-inactivating, minimal lentivirus vector based on simian immunodeficiency virus. *Hum. Gene Ther.* **11**, 439–447.
175. Olsen, J. C. (1998) Gene transfer vectors derived from equine infectious anemia virus. *Gene Ther.* **5**, 1481–1487.
176. Zufferey, R., Nagy, D., Mandel, R. J., Naldini, L., and Trono, D. (1997) Multiply attenuated lentiviral vector achieves efficient gene delivery in vivo. *Nat. Biotechnol.* **15**, 871–875.
177. Negre, D., Mangeot, P. E., Duisit, G., et al. (2000) Characterization of novel safe lentiviral vectors derived from simian immunodeficiency virus (SIVmac251) that efficiently transduce mature human dendritic cells. *Gene Ther.* **7**, 1613–1623.
178. Johnston, J. and Power, C. (1999) Productive infection of human peripheral blood mononuclear cells by feline immunodeficiency virus: implications for vector development. *J. Virol.* **73**, 2491–2498.
179. Kotsopoulou, E., Kim, V. N., Kingsman, A. J., Kingsman, S. M., and Mitrophanous, K. A. (2000) A Rev-independent human immunodeficiency virus type 1 (HIV-1)-based vector that exploits a codon-optimized HIV-1 gag-pol gene. *J. Virol.* **74**, 4839–4852.
180. Wagner, R., Graf, M., Bieler, K., et al. (2000) Rev-independent expression of synthetic gag-pol genes of human immunodeficiency virus type 1 and simian immunodeficiency virus: implications for the safety of lentiviral vectors [In Process Citation]. *Hum. Gene Ther.* **11**, 2403–2413.
181. Mautino, M. R., Keiser, N., and Morgan, R. A. (2000) Improved titers of HIV-based lentiviral vectors using the SRV-1 constitutive transport element. *Gene Ther.* **7**, 1421–1424.
182. Markowitz, D., Goff, S., and Bank, A. (1988) A safe packaging line for gene transfer: separating viral genes on two different plasmids. *J. Virol.* **62**, 1120–1124.
183. Miller, A. D., Garcia, J. V., von Suhr, N., Lynch, C. M., Wilson, C., and Eiden, M. V. (1991) Construction and properties of retrovirus packaging cells based on gibbon ape leukemia virus. *J. Virol.* **65**, 2220–2224.
184. Cosset, F. L., Takeuchi, Y., Battini, J. L., Weiss, R. A., and Collins, M. K. (1995) High-titer packaging cells producing recombinant retroviruses resistant to human serum. *J. Virol.* **69**, 7430–7436.

185. Pizzato, M., Merten, O. W., Blair, E. D., and Takeuchi, Y. (2001) Development of a suspension packaging cell line for production of high titer, serum-resistant murine leukemia virus vectors. *Gene Ther.* **8**, 737–745.
186. Forestell, S. P., Dando, J. S., Chen, J., de Vries, P., Bohnlein, E., and Rigg, R. J. (1997) Novel retroviral packaging cell lines: complementary tropisms and improved vector production for efficient gene transfer. *Gene Ther.* **4**, 600–610.
187. Rigg, R. J., Chen, J., Dando, J. S., Forestell, S. P., Plavec, I., and Bohnlein, E. (1996) A novel human amphotropic packaging cell line: high titer, complement resistance, and improved safety. *Virology* **218**, 290–295.
188. Davis, J. L., Witt, R. M., Gross, P. R., et al. (1997) Retroviral particles produced from a stable human-derived packaging cell line transduce target cells with very high efficiencies. *Hum. Gene Ther.* **8**, 1459–1467.
189. Swift, S., Lorens, J., Achacoso, P., and Nolan, G. P. (1999) Rapid production of retroviruses for efficient gene delivery to mammalian cells using 293T cell-based systems. *Curr. Protocols Immunol.* **10**, 14–29.
190. Sheridan, P. L., Bodner, M., Lynn, A., et al. (2000) Generation of retroviral packaging and producer cell lines for large-scale vector production and clinical application: improved safety and high titer. *Mol. Ther.* **2**, 262–275.
191. Ni, Y., Sun, S., Oparaocha, I., et al. (2005) Generation of a packaging cell line for prolonged large-scale production of high-titer HIV-1-based lentiviral vector. *J. Gene Med.* **7**, 818–834.
192. Matsushita, T., Elliger, S., Elliger, C., et al. (1998) Adeno-associated virus vectors can be efficiently produced without helper virus. *Gene Ther.* **5**, 938–945.
193. Grimm, D., Kern, A., Pawlita, M., Ferrari, F., Samulski, R., and Kleinschmidt, J. (1999) Titration of AAV-2 particles via a novel capsid ELISA: packaging of genomes can limit production of recombinant AAV-2. *Gene Ther.* **6**, 1322–1330.
194. Collaco, R. F., Cao, X., and Trempe, J. P. (1999) A helper virus-free packaging system for recombinant adeno-associated virus vectors. *Gene* **238**, 397–405.
195. Jenny, C., Toublancs, E., Danos, O., and Merten, O. W. (2006) Serum-free production of rAAV-2 using HeLa derived producer cells. *Cytotechnology* **49**, 11–23.
196. Gao, G. P., Lu, F., Sanmiguel, J. C., et al. (2002) Rep/Cap gene amplification and high-yield production of AAV in an A549 cell line expressing Rep/Cap. *Mol. Ther.* **5**, 644–649.

3

Technology and Production of Murine Monoclonal and Recombinant Antibodies and Antibody Fragments

Alexandra Dorn-Beineke, Stefanie Nittka, and Michael Neumaier

Summary

MAbs (mAbs) are unique and versatile molecules that have found applications in research, in vitro and in vivo diagnostics, and the treatment of several diseases. Antibody technology has been revolutionized by the hybridoma technology for mAb production described by Köhler and Milstein in 1975 *(1)*. In the last 30 yr a number of genetically engineered antibody constructions have emerged, including chimeric and human-like antibodies as well as different antibody fragments. Nowadays, 18 mAb products are currently on the market and more than 100 are in use in clinical trials *(2)*. The total market for biopharmaceuticals like therapeutic mAbs is a multibillion dollar opportunity, with cancer and arthritis as the major therapeutic applications, with estimated total markets of \$15 and \$25 billion in 2010, respectively *(3)*.

In this chapter we give an overview of the state of the art of antibody engineering and production methods and describe standard protocols for producing a hybridoma cell line, the building up of transfectomas, and a suitable bacterial expression system for the production of antibodies and antibody fragments established in our laboratory.

Key Words: Murine antibodies; human-like antibodies; chimeric antibodies; humanized antibodies; human antibodies; antibody fragments; hybridomas; transfectomas; phage libraries; transgenic mice.

1. Introduction

Since the beginning of the mAb technology, murine monoclonal antibodies (mAbs) have been used most exclusively for in vitro diagnostics ***(4)***, for in vivo diagnostics of tumor progression and metastasis ***(5)***, or for therapeutic purposes ***(6–8)*** (*see* **Tables 1** and **2**).

While being highly antigen-specific molecules, murine mAbs suffer from a number of drawbacks. In therapeutic regimens, murine mAbs do not efficiently trigger human biological effector functions. Moreover, repeated in vivo

From: *Methods in Biotechnology, Vol. 24: Animal Cell Biotechnology: Methods and Protocols, 2nd Ed.* Edited by: R. Pörtner © Humana Press Inc., Totowa, NJ

Table 1
Examples of In Vitro and In Vivo Applications of Monoclonal Antibodies

In vitro application	In vivo appliation
Diagnostic and scientific purposes	**Diagnostic**
Gel precipitation methods	In vivo imaging
• Immunofixation	• Radiolabeled mAbs
• RID (Manchini technique)	• SPECT ("SPECT mAb")
• Rocket immunodiffusion (Laurell technique)	• PET
Comparative and immunometric assay systems	
• ELISA, RIA, LIA, FIA, IRMA, ILMA, CLIA	**Therapeutic**
Nephelometry/turbidimetry	Immunotherapy
Agglutination assays	mAb or Ab fragments "naked" or labeled/fused to
Immunoblotting	• Radionuclides (radioimmunotherapy, RIT)
• Western blot	• Immunotoxins
• Dot blot	• Cytokines
Flow cytometry	• Growth factors
Microscopy	• Costimulatory signals
• IHH	• Drugs
• ICC	• Enzymes (ADEPT: antibody-directed enzyme prodrug therapy)
• Immunofluorescence	• Retroviruses (gene therapy)
Multiplex analyses by fluorescent beads	Vaccination
Immunoaffinity chromatography	Intracellular protein function
Biosensors	knockout by intrabodies

RID, radial immunodiffusion; ELISA, enzyme-linked immuosorbant assay; RIA, radioimmunoassay; LIA, luminescence immunoassay; FIA, fluorescence immunoassay; IRMA, immunometric radioimmunoassay; ILMA, immunometric luminescence immunoassay; CLIA, chemiluminescence immunoassay; IHH, immunohistochemistry; ICC, immunocytochemistry; SPECT, single-photon emission computerized tomography; PET, positron emission tomography.

administration provokes the development of a human anti-murine antibody (HAMA) response that not only will lead to a faster clearance of the complexed murine mAb from the circulation, but also may cause severe adverse side effects, which are sometimes fatal ***(9–11)***. Finally, immunological in vitro diagnostic tests, e.g., to detect tumor markers in patient sera, can be compromised by HAMAs of patients previously administered with a murine mAbs, thereby simulating false-positive or false-negative results that can lead to misdiagnosis and inadequate disease management ***(12)***.

The reduction of immunogenicity of xenogenic antibodies has therefore been a major goal of recombinant antibody technology during the last two decades.

Table 2
Examples of Therapeutic and Diagnostic Antibodies and Antibody Fragments in Clinical or Preclinical Trials

Generic name	Brand name	Type	Target antigen	Indication
Oncology, solid tumors				
Arcitumomab[a]	CEA-scan	Fab/murine, ^{99m}Tc-labeled	CEA	CRC imaging
Bevacizumab[a]	Avastin®	Humanized	VEGF	CRC, breast cancer, NSCLC
Cetuximab[a]	Erbitux®	Humanized	EGFR	CRC
Oregovomab[b]	OvaRex®	Murine	CA125	Ovarian cancer
Panitumimab[b]	ABX-EGF	Human	EGFR	EGFR-receptor-positive tumors, CRC
SGN-17[c]		scFv fused to β-lactamase human	P97 antigen	Melanoma ADEPT prodrug activation
T84.66[c]		Diabody human, ^{131}I-labeled	CEA	CRC imaging
cT84.66[b]		Chimeric minibody, ^{123}I-labeled	CEA	CRC imaging
Trastuzumab[a]	Herceptin®	Humanized	HER2/ neu-receptor	Breast cancer
Oncology, hematological malignancies, hematology				
Alemtuzumab[a]	Mab-Campath®	Humanized	CD52 antigen	PNH
Eculizumab[b]	Alexion®	Humanized	C5 complement	NHL, AID, ALL
Epratuzumab[b]	LymphoCide®	Humanized	CD22 antigen	
Gemtuzumab[a]	Mylotarg®	Humanized, linked to Calicheamicin	CD33	AML
Ibritumomab[a]	Zevalin®	Murine, ^{90}Y-labeled	CD20 antigen	NHL
Tituximab[a]	MabThera®	Chimeric	CD20 antigen	NHL
Tositumomab[a]	Bexxar®	Murine, ^{131}I-labeled	CD20	NHL
Autoimmunity, Transplant rejection, cardiovascular disease				
Abciximab[a]	ReoPro®	Fab/chimeric	GPIIb/IIIa	Prophylaxis of vascular occlusion after PTCA

(*Continued*)

Table 2 (*Continued*)

Generic name	Brand name	Type	Target antigen	Indication
Adalimumab[a]	Humira®	Human	TNF-α	RA, psoriasis arthritis
Basiliximab[a]	Simulect	Chimeric	CD25 antigen	Prophylaxis of acute rejection in renal transplantation
Daclizumab[a]	Zenapax®	Humanized	CD25 antigen	Prophylaxis of acute rejection in renal transplantation
Infliximab[a]	Remicade®	Chimeric	TNF-α	Crohn's disease, RA, Bechterew's disease, ulcerative colitis
Muronomab[a]	Orthoclone OKT3®	Murine	CD3 antigen	Acute rejection in heart- lung and renal transplantation

[a]FDA approved. [b]In clinical trials. [c]Preclinical trials.

CEA, carcinoembryonic antigen; Fab, monovalent antibody fragment; CRC, colorectal cancer; VEGF, vascular endothelial growth factor; NSCLC, non-small-cell lung cancer; EGFR, epidermal growth factor receptor; scFv, single-chain variable fragment; ADEPT, antibody-directed enzyme prodrug therapy; CLL, chronic lymphocytic leukemia; ALL, acute lymphoblastic leukemia; PNH, paroxysmal nocturnal hemoglobinuria; NHL, non-Hodgkin lymphoma; AID, Autoimmune disease; AML, acute myeloid leukemia; TNF, tumor necrosis factor; RA, rheumatoid arthritis.

Classically, the elimination of antigenic structures has been carried out by genetic engineering to yield chimeric human/mouse mAb ***(13,14)***.

This strategy by which the Ig-constant region domains of the murine antibody are being replaced by the respective human homologs has also been used by distributors of tumor marker assays in order to enhance system robustness ***(12,15)***. A more thorough design is the production of humanized mAb by grafting of the murine complementary-determining regions (CDR) onto the backbone of a human antibody ***(16)*** (**Fig. 1**). Recently developed techniques (phage display and transgenic mice) enable the selection and identification of fully human mAbs originated from immunizations of human or the genetic transfer of human immunoglobulin gene loci into animals, respectively. Such approaches are expected to most efficiently remove allogenic and anti-idiotypic antigenic determinants and also allow for the improvement of antibody affinity ***(17)***. However, with respect to the effort necessary to generate these modifications, data reported by Hwang et al. ***(18)*** have clearly shown that the largest step in

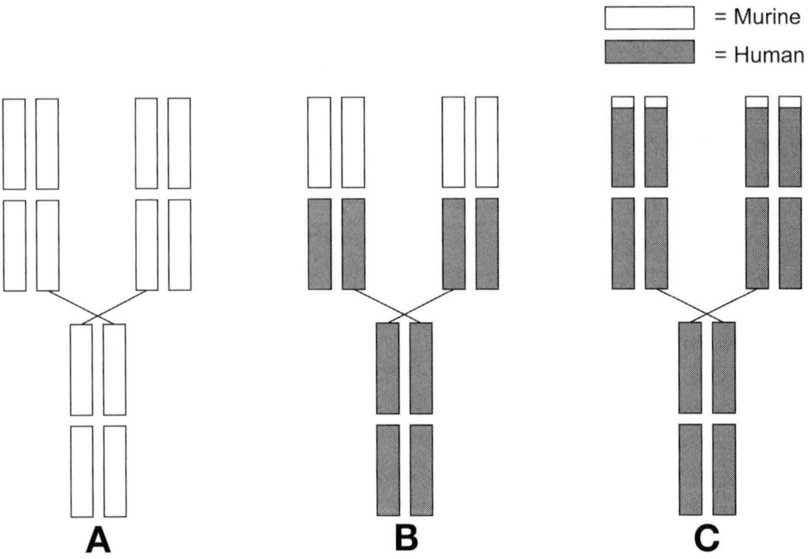

Fig. 1. Chimeric and humanized whole antibodies. **(A)** Murine monoclonal antibody (mAb). **(B)** Chimeric murine/human mAb. Chimeric mAbs are 60–70% human and are obtained by exchanging the Fvs of the human antibody heavy- and light-chain genes for those derived from the rodent monoclonal. They are constructed by joining the DNA segments encoding the mouse variable regions, specific for a certain antigen, to segments of DNA encoding human constant regions. **(C)** Humanized mAb. The epitope is confined to the complementary-determining regions (CDRs), and each domain (Vh and VL) possesses three of these regions, which are flanked by framework regions. Mouse CDRs selected for a specific antigen are grafted onto a human framework, resulting in a humanized antibody (90–95% human), which retains the affinity and specificity characteristics of the original murine mAb for the target antigen. (Modified from **ref. *50***.)

reducing immunogenicity of therapeutic mAbs comes from the replacement of murine constant regions with human homologs. Humanization of variable regions appears to decrease immunogenicity further **(Table 3)**. Data from a fully human anti-tumor necrosis factor (TNF)-α mAb, "Humira," derived from phage-display human libraries, show that, after repetitive in vivo use of Humira, 12% of the patients developed anti-antibody response (AAR), a proportion quite high for a fully human antibody.

Antibody fragments designed as less immunogenic in vivo biopharmaceuticals possess superior biodistribution and blood clearing properties ***(19)*** **(Fig. 2)**. In detail, variable fragments (Fv) consisting of the heavy- and light-chain variable regions of an immunoglobulin molecule are the smallest fragments still retaining the antigen-binding specificity and affinity of the whole antibody. Antigen binding is in most cases a cooperative function of both variable domains that are not

Table 3
Anti-Antibody Reactions

			Incidence		
Antibody construct	*n*	AAR	Marked	Tolerable	Negligible
Murine mAb	44	HAMA	84%	7%	9%
Mouse/Human mAb	15	HACA	40%	27%	33%
Humanized mAb	22	HAHA	9%	36%	55%

Anti-antibody reactions (AAR) were investigated by a Medline search from January 1984 to December 2003. The incidence were grouped into three categories: marked (AAR reported in >15% of patients), tolerable (AAR reported in 2–15% of patients), negligible (AAR reported in <2% of patients). HAMA: human anti-mouse antibodies; HACA: human anti-mouse-human chimeric antibodies; HAHA: human anti-humanized antibodies. Results were reported as percentage of antibodies investigated.

Modified from ref. ***18***.

covalently linked, but associate through hydrophobic interaction. The association between these domains during biotechnological production and their stabilization has been improved by introduction of disulfide bridges or short peptide linkers, the latter resulting in so-called single-chain Fv (scFv). In some cases antibody V_H domains have been isolated from lymphocytes of immunized mice that were shown to be fully functional. However, the hydrophobic surface for the side-to-side adhesion to the V_L domains adversely affects their solubility and stability. Because of their smaller size they are used to affect potentially cryptic epitopes ***(20)***. Selection of these fragments by phage display may be advantageous provided these difficulties can be solved ***(21)***. A further improvement can be achieved by mimicking camelid light chain sequences ***(22)***.

When the genes coding for V_L and V_H domains are joined in-frame by a sequence coding for a flexible hydrophilic polypeptide linker (e.g., $(Gly_4Ser)_n$ and are translated, a single-chain fragment (scFv) results that possesses antigen-binding properties. The length of the connecting linker influences the aggregation of the scFv molecules. The optimal linker length to form scFv monomers is 15–25 amino acids (aa) ***(23)***. Shorter linkers as well as the stability of V_H and V_L associations may force the noncovalent association of the scFv to dimers (diabodies) or trimers (triabodies). ScFv molecules are half the size of Fabs and thus have lower retention times in nontarget tissues, more rapid blood clearance, and better tumor penetration. They are also less immunogenic and are versatile building blocks for the construction of fusion proteins with specific antigen-binding properties. Potential applications are tissue imaging and the target delivery of drugs, toxins, or radionuclides to a tumor side. Multivalent antibody fragments (diabodies, triabodies, tetrabodies) can improve the functional affinity by introducing avidity into the antigen–antibody interaction not available

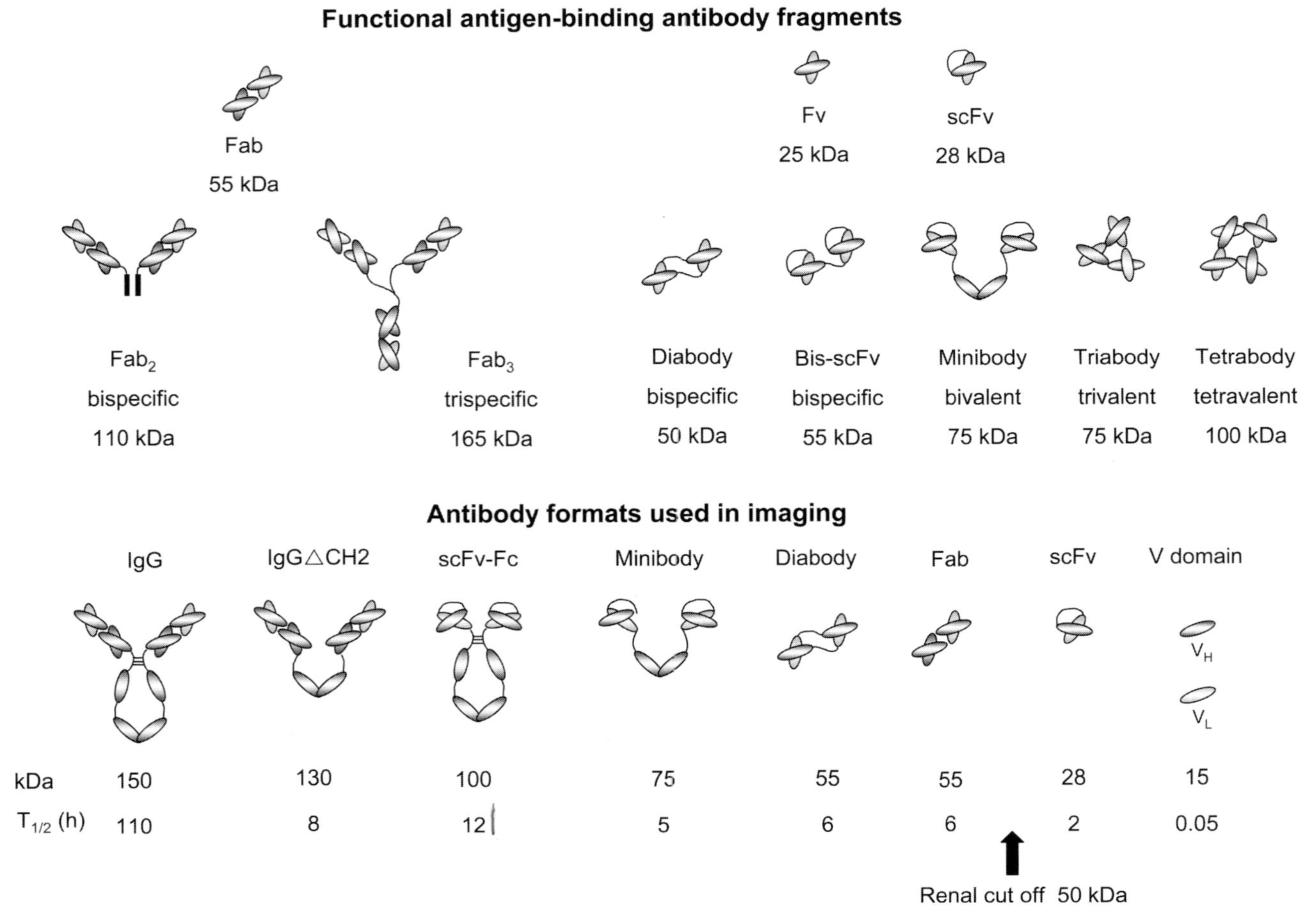
Functional antigen-binding antibody fragments
Fab
55 kDa
Fv
25 kDa
scFv
28 kDa
Fab$_2$
bispecific
110 kDa
Fab$_3$
trispecific
165 kDa
Diabody
bispecific
50 kDa
Bis-scFv
bispecific
55 kDa
Minibody
bivalent
75 kDa
Triabody
trivalent
75 kDa
Tetrabody
tetravalent
100 kDa
Antibody formats used in imaging
IgG
IgG△CH2
scFv-Fc
Minibody
Diabody
Fab
scFv
V domain
V_H
V_L
kDa 150 130 100 75 55 55 28 15
$T_{1/2}$ (h) 110 8 12 5 6 6 2 0.05
Renal cut off 50 kDa

Fig. 2.

to the monomeric scFv. Monomeric Fab fragments have been chemically crosslinked to yield di- and trivalent multimers, which had improved retention and internalization properties as compared to the parent full divalent IgG *(24,25)*. Bispecific mAbs (bis-Abs), most commonly constructed as bisepcific diabodies, comprise two different binding specificities in a single molecule and allow for binding of two different antigens. Bispecificity is advantageous for numerous applications. For an overview of recombinant antibody derivative molecules, *see* **Fig. 2**.

The development of antibody technology for diagnostic and therapeutic applications through the years can be exemplifed by mAb T84.66 (k, IgG1), a murine anti-CEA antibody with high specificity and affinity **(Table 4)**. T84.66 has been developed from a research reagent for enzyme-linked immunosorbent assay (ELISA) and immunohistochemistry *(26,27)* to a chimeric and fully engineered fragment and fusion protein used for experimental and clinical in vitro and in vivo applications. T84.66 has also been instrumental for the generation of anti-idiotypic antitumor vaccination *(28,29)*.

Of the biopharmaceuticals that are being evaluated in clinical trials today, more than 30% are accounted for by antibody molecules *(30)*. The worldwide production of therapeutic antibodies in 2002 was reported to be 1000 kg. It is estimated that this number will increase severalfold in the near future and will be in the vicinity of 20,000 kg for a single therapeutic antibody *(31)*. Obviously, the requirement for mass antibody production is changing. On the other hand, depending on the structure and application of the antibody to be developed, different possibilities with respect to the expression system can be used. Indeed,

---

Fig. 2. Antibody engineering: overview of different antibody formats and functional antibody fragments. **(A)** Functional antigen-binding antibody fragments. Fab_2 and Fab fragments are obtained by proteolysis of the IgG molecule with pepsin and papain. ScFv (single-chain varible fragments): the linker length has an influence of the folding and aggregation. Peptide linker (>12 amino acids) allow the V_L and V_H domains to form monomeric scFv. Shorter linkers (3–12 amino acids) force the domains to associate with domains from another molecule forming scFv dimers (diabodies) *(52)*. Linkers of less than three residues can result in the formation of scFv trimers (triabodies) *(53)* or scFv tetramers (tetrabodies) *(54)*. Multimeric antibody fragments (e.g., diabodies, triabodies, tetrabodies) are represented as multivalent structures, although they can also be engineered to be multispecific. Bis-scFv: two different scFv are joined in tandem. Miniantibodies are scFv dimers that are linked to the C_H3 fragment with highly flexible protein helixes, allowing extensive adaptation to the antigen. **(B)** Antibody formats used for in vivo imaging. Depicted are the respective molecular weights (kDa) and the serum half-lifes (β-phase). Sizes given in kDa are approximate. (Modified from **refs.** *2* and *51*.)

Table 4
Development of T84.66 in Preclinical and Clinical Studies

Construct	Use/Purpose	Ref.
Murine mAb T84.66	Engineering, routine clinical ELISAs	***26,27***
^{111}In (Indacea)-labeled murine mAb T84.66	RIS[a]	***5***
Mouse/Human chimeric (chT84.66)	Engineering	***13***
^{111}In (Indacea)-labeled DTPA-chT84.66	Imaging[b]	***14***
^{90}Y-labeled DTPA-chT84.66	RIT[c]	***45***
^{99m}Tc-labeled Fab/murine T84.66	Imaging[d]	***46***
^{123}I-labeled chT84.66 minibody	Imaging[e]	***47***
131I-labeled T84.66 diabody	Imaging[f]	***48***
^{123}I- and ^{131}I-labeled chimeric scFv-Fc T84.66	Engineering; Imaging[g]	***49***
^{124}I-labeled chimeric scFv-Fc T84.66	Imaging,[g] PET	

ELISA, enzyme-linked immunosorbent assay; DTPA: diethylenetriaminepentaacetic acid; PET: positron emission tomography; RIT: radioimmunotherapy; RIS, radiology information system.

[a]Detection of primary tumors and metastases done with sensitivities from 45 to 80%.

[b]Imaging of at least one known tumor site was observed in 93.0% of patients. The sensitivity rate was 45.1% and the positive predictive value 94.1% in analyzing 47 lesions. $t_{1/2}$: 90.9 h (β-phase).

[c]In 5 of 22 patients (22.7%) antitumor effects (41–100%) were observed.

[d]78% per-lesion concordance with computed tomography, 73% sensitivity and 94% specificity in nine confirmed surgically.

[e]Tumor imaging was observed in seven of eight not pretreated patients (87.5%).

[f]Preclinical, $t_{1/2}$ in mice: 2.68 h (β-phase).

[g]Preclinical.

numerous expression systems including bacteria, yeast, plants, insect cells, and mammalian cell lines have been used. Recently, the production has been demonstrated in transgenic animals.

The starting point for recombinant antibody engineering is classically DNA or RNA from a hybridoma producing the suitable antibody. Antibody-producing hybridomas are most commonly obtained by fusion of spleen cells of immunized mice with myeloma cells using the classic Köhler and Milstein procedure. Several nonsecretor myeloma cell lines that are deficient for the enzyme hypoxanthine guanine phosphoribosyl transferase (HGPRT) have been established as fusion partners with murine spleen lymphocytes, e.g., SP2/0-Ag/4, P3-X63-Ag8.653, and FO ***(32–34)***. Cell fusion is facilitated by polyethylene glykol (PEG), which promotes cell adherence, membrane fusion, and the exchange of nuclei. The culture is grown in hypoxanthine–aminopterin–thymine (HAT) medium, which can sustain only the HGPRT$^+$ hybrid cells. After testing for specificity by ELISA, appropriate cultures are cloned and selected by limiting dilution. Hybridoma cells can be stored in liquid nitrogen for many years.

Attempts to produce human mAbs with the same technology have been hampered because of inefficient immortalization procedures. Satisfactory results were obtained by fusion of an Epstein-Barr virus-transformed human B-lymphocyte line with a mouse myeloma cell line ***(35)*** and by using heteromyelomas (mouse × human hybrids) or triomas (mouse myeloma × human myeloma × human lymphocyte) as fusion partners ***(36)***. New approaches are molecular biology techniques involving transgenic mice, (semi)-synthetic or natural V-region libraries, and phage display technology with subsequent expression of cloned and characterized antibody candidates in either mammalian cells or bacteria.

Transgenic mice have been produced that lack the native immune repertoire and instead harbor most of the human antibody gene repertoire in germline configuration. Immunization of these mice leads to the generation of human antibodies that have undergone mouse somatic hypermutation and selection to relatively high affinity ***(37)***. After immunization of these transgenic mice, cloning and production of the fully human mAbs (HuMab) can be carried out using the hybridoma technology. The XenoMouse® and HuMAb® mice were the first engineered mice carrying a majority of human V_H and V_L (κ) repertoire, and in 2002 five fully human mAbs generated from the XenoMouse® were introduced into clinical trials ***(38)***.

Phage display is a well-established technique and currently the most widely used strategy for antibody display and library screening. The development of this technique in 1990 by McCafferty et al. ***(39)*** led to large repertoires of V_H and V_L regions being amplified by polymerase chain reaction. The products are introduced into a phagemid (hybrid DNA vectors between viral phages and plasmids) vector backbone, because phagemids can be packaged as stable virus particles that allow (1) high multiplicity of infection of bacteria and (2) high-throughput library screening of very large ($>10^{10}$ clones) antibody repertoires that are displayed as fusion proteins on the surface of filamentous bacteriophages (e.g., M13, fl, fd). This method allows phages expressing scFvs fusion proteins with a desired specificity on their virus coat to be panned/retrieved from antigen-coated plates, thereby also delivering the genetic sequences of the antibody fragment responsible for the antigen-binding. These combined features greatly facilitate downstream genetic engineering and expression strategies. Antibody repertoires (or libraries) are classified according to the source of the antibody fragments in (1) immunized libraries, (2) semi-synthetic libraries, and (3) naïve libraries. Both transgenic mice and display libraries typically produce human antibodies with moderate to strong binding affinity constants ranging from 10^{-7} to 10^{-9} *M*.

Prokaryotic cells are favored for expression of small, nonglycosylated Fab and scFv fragments, diabodies, and V-domain fragments. Bacterial expression, i.e., almost exclusively *E. coli* expression, has been the method of choice for expression of these molecules. Two basic strategies have been applied to express various antibody fragments in *E. coli.* The two approaches involve directing the antibody product to either the reducing environment of the cytoplasm or the oxidizing environment of the periplasmic compartment, a space between the cytoplasmic and outer membranes. The intracellular expression approach benefits from a high expression level of antibodies using a strong promoter (e.g., T7 promoter vectors, like pET vectors). The protein expressed accumulates within the cytoplasm as insoluble aggregates ("inclusion bodies") because of the high expression yield and the lack of disulfide bond formation within the reducing environment of the bacterial cytoplasm in *E. coli.* The amount of protein expression is up to 0.5 g/L in shaker flask cultures or up to 3.0 g/L in fermentors. To reconstitute immunologically active antibody fragments, the inclusion bodies first have to be washed and solubilized under strong denaturing conditions, e.g., in 8 *M* urea or 6 *M* guanidine hydrochloride. Subsequently, thc antibodies can be refolded *in vitro* by dialysis of the denaturing agent in the presence of a redox pair, like reduced and oxidized glutathione. Moreover, this approach is useful for producing antibody-based fusion proteins, such as immunotoxins that might be toxic for bacterial cells or mAbs that are unstable due to intracellular degradation when expressed in soluble or secreted form. The periplasmatic strategy imitates the natural folding and secretion process in eukaryotic expression systems by using N-terminal fusion of signal peptide-coding sequences to the scFv gene. These leader sequences originate from genes like pelB, phoA, and ompA, which are naturally targeted to the periplasmic space of *E. coli* bacteria. The periplasmic space is a more oxidizing environment than the cytoplasm and is equipped with a number of proteins important for folding and assembly, such as those catalyzing disulfide bond formation and rearrangement (DsbA, PDI, and DsbC) or functioning as chaperone-like structures (SKp or FkpA). Periplasmic expression has been reported to yield antibody fragments in concentrations of 0.1–100 mg/L in shaker flasks or 1.0–2.0 g/L in fermentors. Extraction of proteins from the periplasm can be performed by osmotic shock. The purification of these proteins from the periplasmatic extract is less problematic than that from cell lysates, because there are fewer contaminating bacterial proteins. A disadvantage is that protein export to the periplasm is rate-limiting and high protein expression often results in the accumulation of insoluble product in the periplasm. An alternative, the usage of *E. coli* strains, which promote the correct folding and oxidation of recombinant antibodies in vivo in the cytoplasm, has been promoted. Functional and soluble antibody fragments have been produced in *E. coli* strains carrying mutations in the genes coding for thioreduxin reductase (trxB) and glutathione oxidoreductase (gor). *E. coli* trxB/gor mutant strains (e.g., Rosetta-gami™ strains of *E. coli*) have an oxidizing cytoplasm and are capable of forming disulfide bonds. Alternatively to secretion into the periplasm, Fernandez et al. ***(40)*** have described an expression strategy in which antibody fragments were secreted into the culture supernatants using an α-hemolysin (HlyA) system in *E. coli.* The monomeric toxin HlyA is secreted directly from the cytoplasm to the extracellular medium. Antibody fragments fused to the C-terminal domain of HlyA are secreted in the culture supernatant by *E. coli* cells expressing TolC/HlyB/HlyD, a three-compartment protein channel. The yields of secreted antibody fragments are similar to those obtained with periplasmatic expression.

The initial need for human viral vaccines in the 1950s accelerated the evolution of large biotechnological processes for mammalian cells. Mammalian cell hosts usually fold recombinant proteins correctly and assemble and glycosylate mAbs in order to enable biological activities such as antibody-dependent cell cytotoxicity (ADCC) and complement-mediated lysis in vivo. Recombinant antibodies can be expressed by mammalian cells either transiently or stably. For transient expression, suitable cells are COS-1 and COS-7 (derived from the African green monkey kidney). MAbs can be harvested from the culture supernatant for 48–72 h. Large-scale production processes (e.g., stirred bank bioreactors) need stably transfected systems employing Chinese hamster ovary (CHO) cells, myeloma cell lines (Ag8.653, SP/0, NSO), baby hamster kidney (BHK) cells, human embryonic kidney cells (HEK-293), or human-retina-derived cells (PER-C6) as production vehicles. Cotransfection of cells with the gene of interest along with an amplifiable gene like dihydrofolate reductase (DHFR) or glutamine synthetase (GS) is the preferred method. The target gene is delivered to the cells along with the DHFR marker gene, usually on the same plasmid vector ***(41)***. The expression vector normally contains a strong viral promotor to drive transcription of the recombinant gene. For transfection, numerous methods including viral transfer or nonviral transfer techniques like calcium phosphate, electroporation, lipofection, and polymer-mediated gene transfer have been successfully used. One system routinely used in DHFR-negative CHO cells uses selective pressure of increasing amounts of methotrexate to the culture medium to boost production. The GS-expression system is an alternative that works as a dominant selectable marker and does not require the use of specific mutant cells ***(42)***. For these techniques, suitable vectors are required, e.g., DHFR vectors containing DHFR coding sequences (pSVM.dhfr and pSV2.dhfr) or GS expression vectors (e.g., pEE14). The expression levels of the two systems are comparable but generally lower than yield from prokaryotic expression systems. For example, stable clone

Table 5
Aspects of Antibody Production in Different Host Systems

	Transgenic mice	Mammalian cells	Insect cells	Bacterial cells
Time to develop the production system	Long	Medium	Medium	Short
Feasibility of success	High	Limited	Medium	Limited
Reproducibility of production	++	+++	+++	+++
Stability	Limited	Limited	Limited	Limited
Time for scale-up	Fast	Medium	Medium	Medium
Costs for scale-up	Medium	High	High	High
Volumimetric product yield	Unlimited	Limited	Limited	Limited
Biological compatibility	Very good	Good	Good	Limited
Contamination with human pathogens	Yes	Yes	No	Yes
Contamination with toxins	No	No	No	Yes

usually produce 10–20 mg/L in roller bottles. In summary, mammalian or plant cells are favored hosts for high-yield expression of larger intact antibodies and minibodies, whereas bacterial and yeast systems are most useful for mAb fragments and V domains. An overview of the aspects of the antibody production in different host systems as discussed above is given in **Table 5**.

2. Materials

2.1. Hybridoma Technology and Transfection and Selection of Mammalian Cells

2.1.1. Hybridoma Culture

1. Dulbeccos modified Eagle's medium (DMEM) supplemented with 10% (v/v) fetal bovine serum Gold and penicillin 100 U/streptomycin 100 µg/mL (all PAA, Pasching, Austria).
2. Hybridoma Express™ (PAA), supplemented with 10% fetal bovine serum Gold.
3. Dulbecco's phosphate-buffered saline (PBS) without calcium and magnesium (PAA).

2.1.2. Identification of Binding Specificity of Hybridoma Cultures

1. 3% (w/v) Bovine serum albumin (BSA) resuspended in Dulbecco's PBS (Sigma, Taufkrichen, Germany).
2. Goat anti-mouse IgG secondary antibody (H+L), horseradish peroxidase conjugated (Dianova, Hamburg, Germany).
3. 3,3′,5,5′-Tetramethylbenzidine (TMB) peroxidase substrate (Perbio Science Germany GmbH, Bonn, Germany).

4. Murine IgG (Sigma) and antigen (purified protein or 0.4 µm filtered crude supernatant).
5. ELISA plate (Microlon 600, flat bottom, Greiner Bio-One, Solingen-Wald, Germany).

2.1.3. Calcium Phosphate Transfection of HEK293 Cells

1. 2 M Calcium chloride solution, sterile filtered and stored at 4°C.
2. 2x concentrated HBS: 400 mM NaCl, 1.5 mM Na$_2$HPO$_4$, 55 mM HEPES, pH 7.0, stored in aliquots at –20°C; discard aliquot after use.
3. Purified, endotoxin-free plasmid (2 µg/ well to be transfected).
4. Water for embryo cultures (Sigma).
5. pEGFP-C (Clontech, BD Biosciences, Heidelberg, Germany).

2.1.4. Generation of Stable HEK293 Transfectants: Limited Dilution

1. DMEM supplemented with 10% (v/v) fetal bovine serum. Gold and penicillin 100 U/streptomycin 100 µg/mL (all PAA).
2. Dulbecco's PBS without calcium and magnesium (PAA).
3. Trypsin/EDTA (1:250) prewarmed to 37°C (PAA).
4. Selection antibiotic G-418 sulfate (50 mg/mL) (PAA) in aliquots stored at –20°C.

2.2. Bacterial Expression of Recombinant scFv 6G6.C4/Granulocyte Macrophage Colony-Stimulating Factor (GM-CSF) Fusion Protein

###

Murine Antibodies 107

(50 μg/mL) or kanamycin (50 μg/mL) or ampicillin (100 μg/mL for control reaction). Pour into Petri dishes (25 mL/100-mm plate).
2. LB medium: reconstitute LB broth powder according to the manufactor's instructions with aqua dest and autoclave. Cool down and add chloramphenicol (50 μg/mL) and kanaycin (50 μg/mL).
3. 10x TE buffer: 10 mM ethylene diamine tetraacetic acid (EDTA; Sigma-Aldrich), 100 mM Tris-HCl (Trizma® base, Sigma-Aldrich), pH 8.0.
4. SOC medium: 0.5% yeast extract, 2.0% tryptone, 10 mM NaCl, 2.5 mM KCl, 10 mM MgCl$_2$, 20 mM MgSO$_4$, 20 mM glucose (Biosource International, Camarillo, CA).
5. SDS–polyacrylamide gel electrophoresis (PAGE) buffer (4x): 4.6 mL glycerol (100%), 2.5 mL 1 M Tris-HCl (pH 6.8), 0.8 g SDS, 1.0 mL bromphenol blue (1%), 1.0 mL 2-ME ad 10 mL aqua dest, store in 1-mL aliquots at –20°C.

2.2.2. Preparation and Refolding of Insoluble Inclusion Bodies

2.2.2.1. CHEMICALS, STOCK SOLUTIONS, AND BUFFERS

1. Lysozyme (Sigma-Aldrich, Steinheim, Germany) is dissolved at 50 mg/mL; store at –20°C.
2. DNase (Sigma-Aldrich) is dissolved at 1 mg/mL in 50% glycerol/75 mM NaCl (store at –20°C). Prepare a diluent for the DNase by mixing 100% glycerol and 150 mM NaCl 1+1.
3. 0.5 M MgCl$_2$.
4. 2.5 M Tris-HCl, pH 8.0.
5. 0.5 M Na-EDTA in 50 mM Tris-HCl, pH 8.0, autoclave.
6. 10% Na-deoxycholate (w/v).
7. 10% Triton X-100 (v/v).
8. 5 M NaCl.
9. 1 M Dithiothreonine (DTT).
10. Phenylmethylsulfonylfluoride (PMSF, Roche Applied Science, Mannheim, Germany): prepare a 100-mM solution in 2-propanol (Merck, Darmstadt, Germany).
11. 8 M Urea in 20 mM Tris-HCl, pH 8.0.
12. 2-Mercaptoethanol.
13. 100-mM Reduced gluthathione (GSH; Sigma-Aldrich): prepare in 8 M urea/20 mM Tris-HCl, pH 8.0.
14. 50 mM Oxidized gluthathione (GSSG; Sigma-Aldrich): prepare in 8 M urea/20 mM Tris-HCl, pH 8.0.
15. 480 mM L-Arginine (Merck): prepare in 8 M urea/20 mM Tris-HCl, pH 8.0, check pH.
16. Solution buffer, pH 8.0: 50 mM Tris-HCl (260 μL stock), 25% sucrose (w/v) (Merck), 1 mM Na-EDTA (26 μL stock), 10 mM DTT (130 μL stock), 1 mM PMSF (130 μL stock), ad 13 mL aqua dest.
17. Lysis buffer, pH 8.0: 50 mM Tris-HCl (250 μL stock), 1% (v/v) Triton X-100 (1.25 mL 10% stock), 1% (v/v) Na-deoxycholate (1.25 mL stock), 10 mM DTT (125 μL stock), ad 12.5 mL aqua dest.

18. Washing buffer with Triton X-100, pH 8.0: 50 m*M* Tris-HCl (200 μL stock), 0.5% (v/v) Triton X-100 (0.5 mL 10% stock), 100 m*M* NaCl (200 μL stock), 1 m*M* NaEDTA (20 μL stock), 1 m*M* DTT (10 μL stock), ad 10 mL aqua dest.
19. Washing buffer without Triton X-100, pH 8.0: 50 m*M* Tris-HCl (200 μL stock), 100 m*M* NaCl (200 μL stock), 1 m*M* NaEDTA (20 μL stock), 1 m*M* DTT (10 μL stock), ad 10 mL aqua dest.
20. Folding buffer, pH 8.0: 10 m*M* GSH (100 μL stock), 1 m*M* GSSG (20 μL stock), 400 m*M* L-arginine (830 μL stock), 0.05 *M* DTT (50 μL stock).
21. Tris-HCl buffer, 20 m*M*, pH 7.6: prepare from 2.5 *M* stock.

2.2.3. Purification of scFv6G6.C4/GM-CSF by Affinity Chromatography

1. HisTrap HP Kit (Amersham Biosciences).
2. 8x Phosphate buffer (160 m*M* phosphate, 4 *M* NaCl): 1.42 g $Na_2HPO_4 \times 2H_2O$, 1.11 g $NaH_2PO_4 \times H_2O$, 23.38 g NaCl ad 90 mL aqua dest. Dissolve completely. Adjust pH to 7.4. Add aqua dest to 100 mL and filter through a 0.45-μm filter.
3. 2 *M* Imidazole, pH 7.4: 13.61 g imidazole ad 90 mL aqua dest, dissolve completely, adjust pH to 7.4. Add aqua dest to 100 mL and filter through a 0.45-μm filter.
4. Binding buffer: 20 m*M* phosphate, 0.5 *M* NaCl, 20 m*M* imidazole. Mix 3 mL phosphate buffer 8x stock solution with 0.24 mL 2 *M* imidazole, add aqua dest ad 24 mL. Adjust pH to 7.6 if necessary.
5. Elution buffer: 20 m*M* phosphate, 0.5 *M* NaCl, 500 m*M* imidazole. Mix 1 mL phosphate buffer 8x stock solution with 2 mL 2 *M* imidazole and add aqua dest ad 8 mL. Adjust pH to 7.6 if necessary.

3. Methods

3.1. Hybridoma Technology and Transfection and Selection of Mammalian Cells

Essentially, the culture of hybridoma cells involves the same techniques as any cell line growing in suspension, although hybridoma cells tend to growth semi-adherent. Subculturing of hybridoma usually does not require dissociation by trypsin-EDTA or other detachment reagents. They can be detached from the plastic surface of the culture flask or plate by repeated rinsing of the surface with culture medium, using a standard sterile serological pipet.

The amount of antibody production depends on the hypridoma, but careful choice of the culture medium can help to improve antibody production. A good monitoring method for antibody production involves screening of cell culture supernatants in a direct ELISA approach.

In contrast to hybridoma cells, transfected mammalian cells were cultured under constant selection corresponding to the resistance transferred with the plasmid used for transfection.

3.1.1. Hybridoma Culture

1. Quick-thaw frozen hybridoma T84.66 cells (10^7 cells/mL) at 37°C in a prewarmed water bath.
2. Resuspend cells in 5 mL prewarmed DMEM culture media.
3. Sediment the cells by centrifugation at 300*g* and aspirate supernatant.
4. Resuspend the cell pellet in 10 mL prewarmed DMEM and seed the carefully resuspended cells into an 80-cm^2 culture flask or 25-cm culture dish. Culture in a humidified atmosphere at 37°C and 5% CO_2.
5. For every subsequent medium change, keep the old medium as it contains the produced antibody.
6. First control for antibody production can be done after 3–7 d of culture. Store the spent culture medium at 4°C until testing.
7. Subculture of hybridoma cultures is necessary every 3–4 d, depending on the cell density and cell division rate. A split ratio of 1:5 resulted usually in a sufficient reduction of the cell density.
8. As soon as the cells show stable growth, e.g., no occurrence of excess cell death, it is possible to adapt the culture to a specialized hybridoma medium for improved antibody production rates.
9. The cells are harvested by centrifugation (300*g*), and the resulting pellet is resuspended in medium composed of 70% DMEM and 30% Hybridoma Express™ medium.
10. The cells are subsequent cultured for 2 d before the spent medium is replaced by a mixture of 50% DMEM and 50% Hybridoma Express.
11. At 70–80% confluency, the cells are split at a ratio of 1:5 and seeded in 100% Hybridoma Express (15 mL for 80-cm^2 flasks).
12. Growth of the hybridoma cells in Hybridoma Express is somewhat slower than in DMEM, but the amount of produced antibody increases up to 10-fold.

See also **Notes 1–4**.

3.1.2. Screening for Antibody Production

1. Supernatants from new generated or thawed hybridoma cultures are tested in a direct ELISA for production of anti-murine antibodies.
2. ELISA plates are coated with murine IgG, diluted in cell culture medium. The standard curve, ranging from 1 to 100 μg/mL, should be included in every assay.
3. 50 μL of culture supernatants or standards are coated at 4°C overnight.
4. Wells are rinsed 3 × 5 min with 200 μL PBS at room temperature.
5. Additional protein-binding capacities are blocked by incubation with 3% BSA/PBS.
6. Wells are rinsed 3 × 5 min with 200 μL PBS at room temperature.
7. Detection of murine antibodies with goat anti-murine horseradish peroxidase conjugate (1:10.000), diluted in 1% BSA/PBS, 100 μL/well.
8. Incubate covered plate for 1 h at 37°C.

9. Rinse wells as described before.
10. Apply 100 µL TMB substrate and allow color to develop at room temperature until samples show at least slight blue color.
11. Stop color development by addition of 25 µL 1 *N* sulfuric acid.
12. Read OD at 450 nm in an ELISA plate reader.

3.1.3. Calcium Phosphate Transfection of HEK293 Cells

1. The day before the transfection, split cells to 50% confluency and culture cells as usual at 37°C and 5% CO_2.
2. Detach HEK293 cells using trypsin-EDTA and seed 3×10^5 cells/well in 6-well plates and allow settling back for at least 4 h. If the purpose is stable transfection, include one additional well of untransfected cells as selection control for later determination of the selection endpoint. Include a positive control, e.g., constitutive expressing vector pEGFP, into the transfection scheme.
3. Thaw sufficient aliquots of 2x HBS at room temperature. Bring $CaCl_2$ and plasmid DNA to room temperature.
4. To a volume containing 2 µg plasmid DNA add 7.47 µL 2 *M* $CaCl_2$ and bring to 60 µL using endotoxin-free water.
5. Slowly add 60 µL 2x HBS and incubate at room temperature for 15 min.
6. Add mixture in drops to the cells; do not mix after addition, as it will interfere with the precipitate formation.
7. Incubate cells for 48 h at 37°C without change of the culture medium.
8. For transient transfection, harvest the cells and proceed as required for further analysis. For stable transfection begin with selection.

3.1.4. Generation of Stable HEK293 Transfectants: Limited Dilution

1. Two days after transfection add appropiate antibiotics to the culture medium. Good starting concentrations for HEK293 cells are 1 mg/mL of G-418 or 400 µg/mL of hygromycin, respectively.
2. Replace medium and antibiotics every 2 d. Primary selection should be finished after 7–9 d, estimated by complete cell death in wells of untransfected cells containing the antibiotic in equal concentrations.
3. Allow cells to grow for another week with further selection.
4. Detach cells, determine cell number and dilute the cell suspension with culture medium until it contains about 30 cells/10 mL.
5. Seed 100 µL/well in a 96-well flat-bottom cell culture plate and incubate at 37°C and 5% CO_2 for 6–12 h.
6. Screen every well for cell content. Mark wells containing only one single well and proceed with culture.
7. After 7–10 d, wells containing viable cells should reach 40–70% confluency. Detach cells from 10 wells with subclones showing good and uniform growth and seed them into a 6-well plate, each clone in a separate well.
8. Further expand cultures according to their growth.

9. At this point, screening for plasmid-transferred characteristics disposal of cryo backups should be done. Select two good expressing clones and transfer them to the subsequent subcloning round.
10. To ensure monoclonality, the subcloning procedure described in **steps 4–9** should be repeated two times.

See also **Notes 5–7**.

3.2. Bacterial Expression of Recombinant scFv 6G6.C4/GM-CSF Fusion Protein

This section describes the production of the anti-idiotype antibody scFv6G6.C4/GM-CSF in bacteria. MAb 6G6.C4 mimics an epitope specific for human CEA and generates a CEA-specific response (Ab3) in various experimental animals *(28,43)*. In recent studies we showed that the fusion protein binds specifically to T84.66 and is biologically active. Mice immunized with scFv6G6.C4/GM-CSF exhibit a higher specific Ab3 response compared to scFv6G6.C4 *(29)*.

The protocol starts with the preparation of plasmid DNA. The gene of interest has been cloned into the pET27b(+) vector (Novagene, Madison, WI) (*see* **Note 8**) and was subsequently transformed into the BL21(DE3) *E. coli* cells. In this protocol, we intended to use the BL21-Codon Plus® (DE3)-RIL cells for protein expression. This strain is engineered to contain extra copies of argU, ileY, and leuW tRNA genes, which encodes for tRNA recognizing the arginine codons AGA and AGG, the isoleucin codon AUA, and the leucin codon CUA. These tRNAs allow high-level expression of many heterologous recombinant genes in *E. coli*. Both expression systems are T7 polymerase-based and inducible with IPTG (isopropyl-1-thio-β-D-galactopyranoside). BL21-Codon Plus (DE3)-RIL cells can also be used for expression with vectors driven by non-T7 promotors. In our protocol, the fusion protein accumulates in the cytoplasm as unsoluble inclusion bodies. The pelB leader signal sequence of the pET27b(+) is not translated because of cloning strategies to extend the yield of protein produced. Therefore, denaturation and an in vitro refolding procedure is needed. For purification purposes, two C-terminal vector-encoded affinity tags, HisTag and HSVTag, are available.

3.2.1. Expression of 6G6.C4/GM-CSF in E. coli

3.2.1.1. Preparing Plasmid-DNA from Glycerol Stocks

1. Prepare LB agar plates containing 50 µg/mL kanamycin.
2. Scratch with a loop some material from the glycerol stock—pET27b(+) with scFv6G6.C4/GM-CSF gene insert, transformed in BL21(DE3)—and spread the bacteria on an appropriate number of agar plates and incubate the plates overnight at 37°C.

3. Prepare tubes with 4 mL LB medium containing 50 µg/mL kanamycin and pick some clones, incubate the cultures for 8 h at 37°C while shaking (225 rpm).
4. Check optical density (OD_{600}) of the cultures; optimal result is OD_{600} 0.6–0.8.
5. Pipet 100 µL from the cultures in 50 mL of LB medium containing kanamycin (50 µg/mL), incubate the cultures at 37°C overnight while shaking (225 rpm).
6. Check optical density (OD_{600}) of the cultures.
7. Spin down the cultures (10.000*g*, 15 min, 4°C).
8. Resuspend the bacteria pellets and purify plasmid DNA with the QIAprep® Spin Miniprep kit according the manufacturer's instructions (*see* **Note 9**).
9. Assess photometrically the purity of plasmid-DNA by the $A_{260/280}$ ratio (optimal 1.8–2.0) and check photometrically the concentration of the plasmid DNA by the A_{260}.
10. Store the purified plasmid DNA in 10x TE-buffer at –80°C in 10-µL aliquots.

3.2.1.2. Transformation of Competent BL21-Codon Plus® (DE3)-RIL *E. coli* Strains With pET 27b(+)

1. Thaw the competent cells on ice (*see* **Note 10**).
2. Gently mix the competent cells, aliquot 50 µL into prechilled 14-mL round-bottom polypropylene tubes. Prepare one tube for transformation control.
3. Add 0.85 µL 2-ME into each tube (hood!).
4. Mix the tubes gently for every 2 min while incubating the cells on ice for 10 min.
5. Add 30–50 ng of pET27b(+) with scFv6G6.C4/GM-CSF gene insert and add 1 µL of pUC18 control plasmid into the respective tubes, mix gently, incubate for 30 min on ice.
6. Preheat SOC medium at 42°C (water bath).
7. Heat-shock the cells for 20 s in the water bath at 42°C.
8. Place tubes immediately on ice for 2 min.
9. Add 450 µL SOC medium to each transformation reaction, incubate the tubes at 37°C for 60 min while shaking (225 rpm).
10. Spread 200 µL of the transformed cells and 200 µL of the control reaction on preheated LB agar plates and incubate overnight at 37°C. LB agar plates contain either chloramphenicol (34 µg/mL) for maintaining the pACYC plasmid in the BL21-Codon Plus strains and kanamycin (50 µg/mL) for maintenance of the expression plasmid) or ampicillin (100 µg/mL) for the control reaction.
11. Pick overnight colonies (without satellites). Prepare a patch plate and incubate at 37°C overnight (*see* **Note 11**).
12. Pick colonies (without satellites) from the patch plate and incubate each colony in 5 mL of LB medium containing chloramphenicol (50 µg/mL) and kanamycin (50 µg/mL) at 37°C while shaking (225 rpm) and incubate overnight. Check the optical density (OD_{600}) of the overnight cultures and place the patch plate at 4°C.
13. Prepare glycerol stocks from 5 mL cultures: pipet into a kryotubes 700 µL of overnight culture and 300 µL glycerol 50%, mix and keep in liquid nitrogen until the sample is frozen, then store at –80°C (freezer).

3.2.1.3. IPTG-Induction Protocol

1. Prepare for each clone a 14-mL polypropylene tube with 1 mL of LB medium containing chloramphenicol (50 μg/mL) and kanamycin (50 μg/mL).
2. Pick the clones from the patch plate and incubate the subcultures overnight at 37°C while shaking at 225 rpm.
3. Check the optical density of the overnight cultures (OD_{600}).
4. Take 50 μL from each culture into a fresh tube with 1 mL LB medium without antibiotics and shake for 2 h at 37°C (225 rpm).
5. Take 100 μL from each culture and place on ice (= noninduced samples) and measure optical density (OD_{600}); optimal OD_{600} 0.6–0.8.
6. IPTG induction of T7 polymerase from lacUV5 promotor: add 100 m*M* IPTG stock at a final concentration of 1 m*M* to the cultures and incubate for 2–3 h at 37°C (225 rpm) (*see* **Note 12**).
7. Place all induced and noninduced samples on ice.
8. Pipet 30 μL of each induced and noninduced culture and 10 μL SDS-PAGE buffer (4x) into a fresh Eppendorf cup and check protein induction by SDS-PAGE and Western blot (*see* **Note 13**).
9. Spin ITPG-induced cell cultures down (10,000*g*, 20 min, room temperature) and freeze, if necessary, the bacteria pellets at –80°C (*see* **Note 14**).

3.2.2. Preparation and Refolding of Insoluble Protein Fractions

3.2.2.1. Protocol Adjusted to Bacteria Pellets of 100–200 mL

1. Thaw bacteria pellets (–80°C) and resuspend on ice in 2.6 mL solution buffer.
2. Ultrasound: 30 strokes on ice.
3. Add 20 μL lysozyme, 50 μL DNase (stock), and 10 μL $MgCl_2$ (stock) and vortex.
4. Add 2.5 mL lysis buffer, vortex, and incubate at room temperature for 30– 60 min.
5. Add 70 μL Na-EDTA (stock) and keep vial in liquid nitrogen until no more bubbles float up.
6. Thaw for 30 min at 37°C and add 40 μL $MgCl_2$ (stock).
7. Wait for 30–60 min, until viscosity decreases, add 70 μL NaEDTA (stock); if the viscosity is still high, repeat **step 3** without lysozyme at 37°C.
8. Spin down pellet (11,000*g*, 20 min, 4°C) and discard supernatant.
9. Resuspend pellet in 2.0 mL washing buffer (with Triton X-100) on ice.
10. Ultrasound: 30 strokes on ice.
11. Spin down pellet (11,000*g*, 20 min, 4°C) and discard supernatant.
12. Resuspend pellet in 2.0 mL washing buffer (without Triton X-100) on ice.
13. Ultrasound: 30 strokes on ice.
14. Spin down pellet (11,000*g*, 20 min, 4°C) and discard supernatant.

3.2.2.2. Denaturation of scFv6G6.C4/GM-CSF in 8 *M* Urea

1. Dissolve the pellet in 2 mL 8 *M* urea, pH 8.0 and add DTT to a final concentration of 4 m*M* or add 2-ME to a final concentration of 20 m*M*.

2. Shake gently at room temperature until the pellet is solubilized.
3. Aliquot in 1-mL fractions and store at –80°C or keep at 4°C when refolding directly.

3.2.2.3. Refolding of scFv6G6.C4/GM-CSF

1. Thaw protein aliquots in 8 *M* urea and spin down precipitates (10,000*g*, 20 min, 4°C).
2. Mix the supernatant 1:2 with folding buffer, pH 8.0.
3. Shake overnight in an overhead shaker and dilute the denatured proteins 1:5 (v/v with 8 *M* urea and incubate for 60 min while shaking.
4. Dialyze against 20 m*M* TRIS-HCl, pH 7.6, at 4°C with frequent change of dialysis buffer and centrifuge (12,000*g*, 20 min, 4°C). The supernatant contains the soluble, refolded protein.

3.2.3. Purification of scFv6G6.C4/GM-CSF by Affinity Chromatography

HisTrap HP columns are precharged with Ni^{2+} ions. Using this procedure, His-tagged proteins can be purified directly from pretreated bacterial lysates. $(His)_6$ increases the affinity for and generally makes the His-tagged protein the strongest binder among other proteins.

1. Prepare binding buffer and elution buffer (*see* **Note 15**).
2. Equilibrate the HisTrap HP columns with 10 mL binding buffer.
3. Apply the sample with the syringe, collect the flow-through fraction.
4. Wash with 10 mL binding buffer, collect the wash fraction.
5. Eluate with 5 mL elution buffer; collect the eluates in 1-mL fractions.
6. Check the purification with on SDS-PAGE or better on silver-stained gel (*see* **Note 16**).
7. Regenerate the columns by washing with binding buffer (*see* **Notes 17** and **18**).

4. Notes

1. Supernatants containing the antibody should be free from cells or cell debris. Centrifugation at 3000–5000*g* is usually enough to eliminate these contaminants and facilitate the subsequent screening.
2. If outright testing of the supernatants is impossible, add 0.05% (w/v) sodium azide to the medium to prevent microbial growth that otherwise could deteriorate your antibody.
3. For long-term storage of supernatants at –20°C, filtration through a 0.22-µm membrane enhances their stability, although some antibody may be lost by adsorption to the membrane.
4. Reducing the serum content in the culture medium facilitates the subsequent antibody-purification procedure. It is often possible the reduce the serum content in the culture medium gradually over 2–3 wk, if specialized hybridoma medium is used.
5. For stable transfection of other cell lines, a kill curve to determine the optimal antibiotic concentration should be performed before starting the first transfection. The optimal concentration should lead to 100% cell death after 6–7 d in non-resistant cells.

6. After generation of stable transfectants, it is often possible to reduce the antibiotic concentration at 25–50% without loss of expression. This allows faster cell growth and reduces expenses for the culture of stable transfected cells.
7. Problems with transfection procedures often arise from impurities within the plasmid preparation or poor transfection efficiencies. Use of a control plasmid (e.g., pEGFP-C, Clontech) helps to determine the type of problem. If cells are 50–80% EGFP positive but enter cell death a few days after transfection although they still are EGFP positive, it is likely that impurities within the plasmid preparation are the reason.
8. After cloning a gene of interest into a vector, you should always check the orientation of your insert by restriction enzymes. It is useful to choose an enzyme that cleaves the insert asymmetrically. In our protocol we checked the orientation of scFv6G6.C4/GM-CSF gene with the restriction enzyme Xba I (Fermentas GmbH, Leon-Rot, Germany). Add 1 μg DNA to 1 μL enzyme and 2.5 μL Y(Yellow) Tango buffer, add aqua dest ad 25 μL total volume. Incubate for 2 h at 37°C, followed by 20 min at 60°C. Check the sizes of your fragments by agarose gel electrophoresis (0.9% gel).
9. The principle of these kits is the alkaline lysis of bacterial cells followed by adsorption of DNA onto silica in the presence of high salt concentrations. In our hands, the concentration of the plasmid -DNA from a 50-mL culture was 162 ng/μL. The pellet wet weight was determined with 1 g. For purification of endotoxin-free plasmid DNA use other approaches, e.g., the EndoFree plasmid purification kits (QIAGEN). Endotoxin-free DNA will improve transfection into sensitive eukaryotic cells and is essential for gene therapy research. You should use in these cases endotoxin-free plasticware, glassware, and buffers.
10. For checking the cells for competence, use a control vector as described in our protocol. The manufacturer gives transformation summaries for each plating quantity (μL of transformation reaction) plated on the agar plates with the expected colony number in colony forming units (CFU) and the efficiency (CFU/μg of control plasmid DNA). If the efficiency is too low, use the calcium chloride procedure to make cells chemically competent ***(44)***.
11. A patch plate is a suitable tool to go back (e.g., to scale up or to make glycerol stocks) to the original clone after checking the protein induction by SDS-PAGE or Western blot. The patch plate contains the same antibiotics as the LB broth of the cultures.
12. The optimal conditions for the transformation have to be evaluated. We performed IPTG induction in cultures with different OD_{600} (0.5–0.9). The time of IPTG induction varied from 2 to 8 h. The expression results from the different cultures were examined by SDS-PAGE on a 9% polyacrylamide gel. The conditions with the best expression yields of protein were chosen for scale-up experiments.
13. The protein expression yield was checked by SDS-PAGE and Western blot under standard conditions **(Fig. 3)**. The antibody used to detect His-tagged fusion protein was anti-his_6-peroxidase in a concentration of 0.2 μg/mL (Roche Applied Science, Mannheim, Germany).

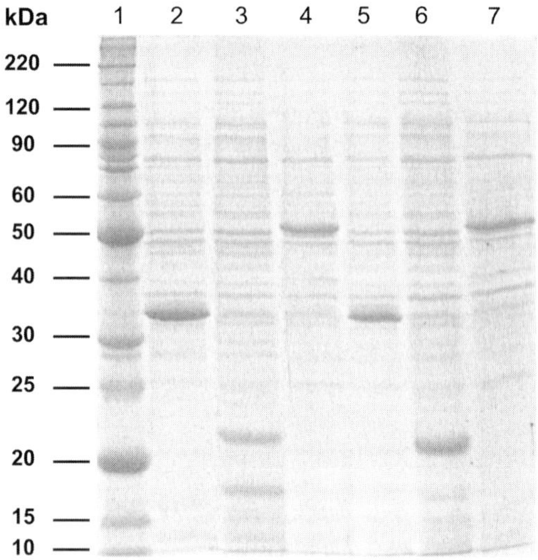

Fig. 3. SDS-PAGE analysis of recombinant proteins: unpurified protein fractions achieved after culture of transfected BL21(DE3) codon plus *E. coli* cells separated in a 9% polyacrylamide gel and stained with Coomassie brilliant blue G250 stain. Lane 1: Bench Mark Protein Ladder (Invitrogen); lane 2: scFv6G6.C4 with an apparent mass of ~35 kDa (1:10 diluted, stored at 4°C); lane 3: mGM-CSF (~25 kDa, 1:10 dilute, stored at 4°C); lane 4: scFv6G6.C4/GM-CSF (~50 kDa, 1:10 diluted, stored at 4°C); lane 5: scFv6G6.C4 (1:10 diluted, stored at 37°C); lane 6: mGM-CSF (1:10 dilute, stored at 37°C); lane 7: scFv6G6.C4/GM-CSF (1:10 diluted, stored at 37°C).

14. The given protocol is for analytical purposes. We scaled up the yield of protein expression by modifying the IPTG-induction protocol: using 50 mL overnight cultures **(step 1)** we came up with 1-L LB cultures **(step 4)** and induced with 1 mL 1 *M* IPTG stock **(step 6)**. Ten bacteria pellets from 1 L LB medium were stored at –80°C. The yield of total protein from 1 L culture was determined to be 362 mg/L by the Biorad protein assay based on the method of Bradford (Biorad, München, Germany).
15. Two questions must be clarified in this protocol: which conditions of binding and elution are optimal and whether the protein should be denatured or renatured during affinity chromatography. If the protein is difficult to dissolve or precipitates occur during the purification, several detergents, urea, and guanidine hydrochloride can be used. Check the concentrations of these substances according the manufacturer's instructions. If the protein is unproblematic, the refolded, renatured protein can run the column. Then check the conditions of optimal binding and elution of your protein by an optimization protocol with different imidazole concentrations (20–500 m*M*) in the binding and elution buffer.
16. After each purification step of your protocol, check the purity of your sample. SDS-PAGE may be not sensitive enough to detect contamination of your desired

protein fraction. By preparing a silver gel, the sensitivity can be below 1 ng per band for most proteins. Coomassie staining shows a 50-fold decreased sensitivity compared to silver staining.

17. When affinity chromatography with HisTrap HP columns is finished, the protein sample must be desalted from the high imidazole concentrations. For this purpose we use PD-10 desalting columns (Amersham). These columns contain Sephadex™ G-25 medium to separate high (M_r >5000) from low (M_r <1000) molecular weight substances. The contamination with low molecular weight substances should be less than 4%. MicroBeads-based arrays for direct immunoprecipitation of His-tagged proteins from cell lysates may also used (e.g., Miltenyi Biotech, Bergisch Gladbach, Germany). Further approaches to purify the fusion protein have been gel filtration with a Superdex™ 200 column on a FPLC system (both Amersham Pharmacia). Therefore, the volume of the sample should be reduced to 1–2 mL (e.g., Centricon Filter YM 30, Millipore, Eschborn, Germany). The total protein content should be 1–15 mg. It is noteworthy that the protein must be refolded when running gel filtration.
18. Following the purification steps, the functional properties of the protein must be tested. In the case of scFv6G6.C4/GM-CSF, we tested the immunologically capability of the fusion protein to bind to its idiotype mAb chT84.66 by ELISA. The GM-CSF-bioactivity of the fusion protein was shown using GM-CSF and interleukin-3-dependent FDP-P1-cells *(29)*. For in vivo approaches, there is a need for endotoxin removal, for which Polymyxin B agarose (Sigma-Aldrich) is suitable. Endotoxin testing of the so purified protein should result in less than 1 EU/mg protein.

References

1. Kohler, G. and Milstein, C. (1975) Continuous cultures of fused cells secreting antibody of predefined specificity. *Nature* **256**, 495–497.
2. Holliger, P. and Hudson, P. J. (2005) Engineered antibody fragments and the rise of single domains. *Nat. Biotechnol.* **23**, 1126–1136.
3. Monzavi-Karbassi, B. and Kieber-Emmons, T. (2001) Current concepts in cancer vaccine strategies. *Biotechniques* **30**, 170–172, 174, 176 passim.
4. Yu, K. Y., Noh, Y., Chung, M., et al. (2004) Use of mAbs that recognize p60 for identification of Listeria monocytogenes. *Clin. Diagn. Lab. Immunol.* **11**, 446–451.
5. Beatty, J. D., Williams, L. E., Yamauchi, D., et al. (1990) Presurgical imaging with indium-labeled anti-carcinoembryonic antigen for colon cancer staging. *Cancer Res.* **50**, 922s–926s.
6. Watanabe, T., Terui, S., Itoh, K., et al. (2005) Phase I study of radioimmunotherapy with an anti-CD20 murine radioimmunoconjugate (Y-ibritumomab tiuxetan) in relapsed or refractory indolent B-cell lymphoma. *Cancer Sci.* **96**, 903–910.
7. Sevmis, S., Emiroglu, R., Karakayali, F., et al. (2005) OKT3 treatment for steroid-resistant acute rejection in kidney transplantation. *Transplant Proc.* **37**, 3016–3018.
8. Behr, T. M., Memtsoudis, S., Vougioukas, V., et al. (1999) Radioimmunotherapy of colorectal cancer in small volume disease and in an adjuvant setting: preclinical

evaluation in comparison to equitoxic chemotherapy and initial results of an ongoing phase-I/II clinical trial. *Anticancer Res.* **19**, 2427–2432.
9. Levy, R. and Miller, R. A. (1983) Biological and clinical implications of lymphocyte hybridomas: tumor therapy with mAbs. *Annu. Rev. Med.* **34**, 107–116.
10. Houghton, A. N., Mintzer, D., Cordon-Cardo, C., et al. (1985) Mouse monoclonal IgG3 antibody detecting GD3 ganglioside: a phase I trial in patients with malignant melanoma. *Proc. Natl. Acad. Sci. USA* **82**, 1242–1246.
11. Sears, H. F., Herlyn, D., Steplewski, Z., and Koprowski, H. (1984) Effects of mAb immunotherapy on patients with gastrointestinal adenocarcinoma. *J. Biol. Response Mod.* **3**, 138–150.
12. Nussbaum, S. and Roth, H. J. (2000) Human anti-mouse antibodies: pitfalls in tumor marker measurement and strategies for enhanced assay robustness; including results with Elecsys CEA. *Anticancer Res.* **20**, 5249–5252.
13. Neumaier, M., Shively, L., Chen, F. S., et al. (1990) Cloning of the genes for T84.66, an antibody that has a high specificity and affinity for carcinoembryonic antigen, and expression of chimeric human/mouse T84.66 genes in myeloma and Chinese hamster ovary cells. *Cancer Res.* **50**, 2128–2134.
14. Wong, J. Y., Thomas, G. E., Yamauchi, D., et al. (1997) Clinical evaluation of indium-111-labeled chimeric anti-CEA mAb. *J. Nucl. Med.* **38**,1951–1959.
15. Kuroki, M., Yamanaka, T., Matsuo, Y., Oikawa, S., Nakazato, H., and Matsuoka, Y. (1995) Immunochemical analysis of carcinoembryonic antigen (CEA)-related antigens differentially localized in intracellular granules of human neutrophils. *Immunol. Invest.* **24**, 829–843.
16. Jones, P. T., Dear, P. H., Foote, J., Neuberger, M. S., and Winter, G. (1986) Replacing the complementarity-determining regions in a human antibody with those from a mouse. *Nature.* **321**, 522–525.
17. Aujame, L., Geoffroy, F., and Sodoyer, R. (1997) High affinity human antibodies by phage display. *Hum. Antibodies* **8**, 155–168.
18. Hwang, W. Y. and Foote, J. (2005) Immunogenicity of engineered antibodies. *Methods* **36**, 3–10.
19. Wu, A. M., Chen, W., Raubitschek, A., et al. (1996) Tumor localization of anti-CEA single-chain Fvs: improved targeting by non-covalent dimers. *Immunotechnology* **2**, 21–36.
20. Ward, E. S., Güssow, D., Griffiths, A. D., Jones, P. T., and Winter, G. (1989) Binding activities of a repertoire of single immunoglobulin variable domains secreted from Escherichia coli. *Nature* **341**, 544–546.
21. Jespers, L., Schon, O., James, L. C., Veprintsev, D., and Winter, G. (2004) Crystal structure of HEL4, a soluble, refoldable human V(H) single domain with a germ-line scaffold. *J. Mol. Biol.* **337**, 893–903.
22. Hamers-Casterman, C., Atarhouch, T., Muyldermans, S., et al. (1993) Naturally occurring antibodies devoid of light chains. *Nature.* **363**, 446–448.
23. Plückthun, A., Krebber, A., Krebber, C., et al. (1996) Producing antibodies in Escherichia coli: from PCR to fermentation. In *Antibody engineering: a practical approach* (McCafferty, J., Hoogenboom, H., Chiswell, D., ed.), Oxford University Press, pp. 203–252.

24. Casey, J .L., Napier, M. P., King, D. J., et al. (2002) Tumor targeting of humanised cross-linked divalent-Fab' antibody fragments: a clinical phase I/II study. *Br. J. Cancer* **86**, 1401–1410.
25. Weir, A. N., Nesbitt, A., Chapman, A. P., Popplewell, A. G., Antoniw, P., and Lawson, A. D. (2002) Formatting antibody fragments to mediate specific therapeutic functions. *Biochem. Soc. Trans.* **30**, 512–516.
26. Wagener, C., Clark, B. R., Rickard, K. J., and Shively, J. E. (1983) MAbs for carcinoembryonic antigen and related antigens as a model system: determination of affinities and specificities of mAbs by using biotin-labeled antibodies and avidin as precipitating agent in a solution phase immunoassay. *J. Immunol.* **130**, 2302–2307.
27. Neumaier, M., Fenger, U., and Wagener, C. (1985) MAbs for carcinoembryonic antigen (CEA) as a model system: identification of two novel CEA-related antigens in meconium and colorectal carcinoma tissue by Western blots and differential immunoaffinity chromatography. *J. Immunol.* **135**, 3604–3609.
28. Gaida, F.-J., Fenger, U., Wagener, C., and Neumaier, M. (1992) A monoclonal anti-idiotypic antibody bearing the image of an epitope specific to the human carcinoembryonic antigen. *Int. J. Cancer* **51**, 459–465.
29. Schwegler, C., Dorn-Beineke, A., Nittka, S., Stocking, C., and Neumaier, M. (2005) Monoclonal anti-idiotype antibody 6G6.C4 fused to GM-CSF is capable of breaking tolerance to carcinoembryonic antigen (CEA) in CEA-transgenic mice. *Cancer Res.* **65**, 1925–1933.
30. Hudson, P. J. and Souriau, C. (2003) Engineered antibodies. *Nat. Med.* **9**, 129–134.
31. Franklin, S. E. and Mayfield, S. P. (2005) Recent developments in the production of human therapeutic proteins in eukaryotic algae. *Expert Opin. Biol. Ther.* **5**, 225–235.
32. Shulman, M., Wilde, C. D., and Köhler, G. (1978) A better cell line for making hybridomas secreting specific antibodies. *Nature* **276**, 269–270.
33. Kearney, J. F., Radbruch, A., Liesegang, B., and Rajewsky, K. (1979) A new mouse myeloma cell line that has lost immunoglobulin expression but permits the construction of antibody-secreting hybrid cell lines. *J Immunol.* **123**, 1548–1550.
34. de St Groth, S. F. (1980) MAbs and how to make them. *Transplant. Proc.* **12**, 447–450.
35. Thompson, K. M., Hough, D. W., Maddison, P. J., Melamed, M. D., and Hughes-Jones, N. (1986) The efficient production of stable, human mAb-secreting hybridomas from EBV-transformed lymphocytes using the mouse myeloma X63-Ag8.653 as a fusion partner. *J. Immunol. Methods* **94**, 7–12.
36. Kalantarov, G. F., Rudchenko, S. A., Lobel, L., and Trakht, I. (2002) Development of a fusion partner cell line for efficient production of human mAbs from peripheral blood lymphocytes. *Hum. Antibodies* **11**, 85–96.
37. He, Y., Honnen, W. J., Krachmarov, C. P., et al. (2002) Efficient isolation of novel human mAbs with neutralizing activity against HIV-1 from transgenic mice expressing human Ig loci. *J. Immunol.* **169**, 595–605.
38. Kellermann, S. A. and Green, L. L. (2002) Antibody discovery: the use of transgenic mice to generate human mAbs for therapeutics. *Curr. Opin. Biotechnol.* **13**, 593–597.

39. McCafferty, J., Griffiths, A. D., Winter, G., and Chiswell, D. J. (1990) Phage antibodies: filamentous phage displaying antibody variable domains. *Nature* **348**, 552–554.
40. Fernandez, L. A., Sola, I., Enjuanes, L., and de Lorenzo, V. (2000) Specific secretion of active single-chain Fv antibodies into the supernatants of Escherichia coli cultures by use of the hemolysin system. *Appl. Environ. Microbiol.* **66**, 5024–5029.
41. Lucas, B. K., Giere, L. M., DeMarco, R. A., Shen, A., Chisholm, V., and Crowley, C. W. (1996) High-level production of recombinant proteins in CHO cells using a dicistronic DHFR intron expression vector. *Nucleic Acids Res.* **24**, 1774–1779.
42. Bebbington, C. R., Renner, G., Thomson, S., King, D., Abrams, D., and Yarranton, G. T. (1992) High-level expression of a recombinant antibody from myeloma cells using a glutamine synthetase gene as an amplifiable selectable marker. *Biotechnology (NY).* **10**, 169–175.
43. Gaida, F.-J., Pieper, D., Roder, U. W., Shively, J. E., Wagener, C., and Neumaier, M. (1993) Molecular characterization of a cloned idiotypic cascade containing a network antigenic determinant specific for the human carcinoembryonic antigen. *J. Biol. Chem.* **268**, 14,138–14,145.
44. Mandel, M. and Higa, A. (1970) Calcium-dependent bacteriophage DNA infection. *J. Mol. Biol.* **53**, 159–162.
45. Wong, J. Y. C., Chu, D. Z., Yamauchi, D. M., et al. (2000) A phase I radio-immunotherapy trial evaluating 90yttrium-labeled anti-carcinoembryonic antigen (CEA) chimeric T84.66 in patients with metastatic CEA-producing malignancies. *Clin. Cancer Res.* **6**, 3855–3863.
46. Wegener, W. A., Petrelli, N., Serafini, A., and Goldenberg, D. M. (2000) Safety and efficacy of arcitumomab imaging in colorectal cancer after repeated administration. *J. Nucl. Med.* **41**, 1016–1020.
47. Wong, J. Y., Chu, D. Z., Williams, L. E., et al. (2004) Pilot trial evaluating an 123I-labeled 80-kilodalton engineered anticarcinoembryonic antigen antibody fragment (cT84.66 minibody) in patients with colorectal cancer. *Clin. Cancer. Res.* **10**, 5014–5021.
48. Olafsen, T., Cheung, C. W., Yazaki, P. J., et al. (2004) Covalent disulfide-linked anti-CEA diabody allows site-specific conjugation and radiolabeling for tumor targeting applications. *Protein Eng. Des. Sel.* **17**, 21–27.
49. Kenanova, V., Olafsen, T., Crow, D. M., et al. (2005) Tailoring the pharmacokinetics and positron emission tomography imaging properties of anti-carcinoembryonic antigen single-chain Fv-Fc antibody fragments. *Cancer Res.* **65**, 622–631.
50. Thiel, M. A., Pilkington, G. J., and Zola, H. (2000) Antibody engineering, in MAbs. *The Basics—From Background to Bench* (Zola, H., ed.), BIOS Scientific Publishers Limited, pp. 45–79.
51. Sanz, L., Cuesta, A. M., Compte, M., and Alvarez-Vallina, L. (2005) Antibody engineering: facing new challenges in cancer therapy. *Acta Pharmacol. Sin.* **26**, 641–648.

52. Holliger, P., Prospero, T., and Winter, G. (1993) "Diabodies": small bivalent and bispecific antibody. *Proc. Natl. Acad. Sci. USA* **90**, 6444–6448.
53. Kortt, A. A., Lah, M., Oddie, G. W., et al. (1997) Single-chain Fv fragments of anti-neuraminidase antibody NC10 containing five- and ten-residue linkers form dimers and with zero-residue linker a trimer. *Protein Eng.* **10**, 423–433.
54. Le Gall, F., Kipriyanov, S. M., Moldenhauer, G., and Little, M. (1999) Di-, tri- and tetrameric single chain Fv antibody fragments against human CD19: effect of valency on cell binding. *FEBS Lett.* **453**, 164–168.

4

DNA Fingerprinting and Characterization of Animal Cell Lines

Glyn N. Stacey, Ed Byrne, and J. Ross Hawkins

Summary

The history of the culture of animal cell lines is littered with published and much unpublished experience with cell lines that have become switched, misslabeled, or cross-contaminated during laboratory handling. To deliver valid and good quality research and to avoid waste of time and resources on such rogue lines, it is vital to perform some kind of qualification for the provenance of cell lines used in research and particularly in the development of biomedical products. DNA fingerprinting provides a valuable tool to compare different sources of the same cells and, where original material or tissue is available, to confirm the correct identity of a cell line. This chapter provides a review of some of the most useful techniques to test the identity of cells in the cell culture laboratory and gives methods that have been used in the authentication of cell lines.

Key Words: DNA fingerprint; cell lines; cross-contamination.

1. Introduction

Cell line characterization is obviously of great importance to ensure that the cells have the correct origin (identity) and phenotype. This is critical to carrying out reliable research and to ensure valid data from cell-based assays and successful manufacturing processes using cell lines. However, the history of in vitro animal cell culture is littered with cases where diligent scientists have identified "cross-contamination" of cultures that in many cases probably represented simple mislabeling or switching of cultures. More than 20 yr ago Nelson-Rees et al. published a critical review of this problem which focused scientific attention on the hazards of "cross-contaminated" cell lines *(1)*. Their results were based on painstaking studies involving karyology and isoenzyme analysis, which controversially revealed that a large number of independently derived cell lines from different individuals were in fact all the HeLa cell line.

From: *Methods in Biotechnology, Vol. 24: Animal Cell Biotechnology: Methods and Protocols, 2nd Ed.* Edited by: R. Pörtner © Humana Press Inc., Totowa, NJ

In subsequent years there have been numerous reports of cross-contaminated cell lines and reviews of this problem *(2,3)*, which is still prevalent today. Thus, it is important that all workers using cell lines should be aware of this problem and should be able to take steps to avoid the use of bogus cell lines. This chapter is intended to give the reader an overview of some of the more popular DNA identification tests available, to put them in context with other more traditional tests, and to give some guidance on selection of the most appropriate identity test, which should reflect the type of cell lines in use and the facilities available.

1.1. The Development of Cell Identification Techniques

Early descriptions of cell lines were heavily dependent on the microscopic appearance and morphology of individual cells until the development of a reliable karyotyping technique for cells by Hsu *(4)* and Tjio and Leven *(5)*. Karyology visualizes the entire genome in the form of premitotic condensed chromosomes and enables rapid identification of the species of origin and can also reveal unique genetic markers for an individual cell line. Subsequently, a wide variety of characterization techniques were developed for the identification of cells, including species-specific antigen immunofluorescence *(6)* and isoenzyme analysis *(7)*. Of these, isoenzyme analysis has achieved the widest use, and this has been promoted by the availability of a standardized "kit" (Authentikit™). This technique is based on visualization of certain enzyme activities, which exhibit interspecies polymorphism in their electrophoretic mobility, thus showing characteristic banding patterns for cells from the same species. Different isoenzyme profiles represent the products of different gene alleles, which may also be influenced by posttranslational modification or hybridization in hybrid cells.

Since the 1970s the rapid development of molecular biology delivered a host of new methods for cell identification based on restriction fragment length polymorphisms (RFLPs). A particularly useful technique for discriminating human cells was human leukocyte antigen (HLA) typing *(8)*, which has also been used to resolve cases of cell line cross-contamination *(9)*. However, none of these molecular approaches provides a single method for cell identification that works well over a broad range of species and, at the same time, provides adequate discrimination between human individuals and the cell lines commonly used in industry. In 1985 Jeffreys et al. first described the technique of DNA fingerprinting, which was unique in its capability to differentiate between closely related human individuals *(10)*. This technique was based on RFLPs derived from hypervariable minisatellite sequences called variable number tandem repeats (VNTRs) and revealed patterns of hybridization on Southern blots of genomic DNA representing alleles that are inherited in a Mendelian fashion. Furthermore, the only case in which two individuals could show an identical profile is identical twins. Since the publication of this method a range of polymerase chain reaction (PCR)-based techniques have been developed for DNA typing based on hypervariable DNA sequences and single nucleotide polymorphisms (SNPs) and other RFLP-like approaches. However, some of the methods for PCR of conserved intron sequences, as included in the methods given below, have shown some potential in reproducible identification of cell lines for a wide range of species, although such genes do not generally have sufficient polymorphism to differentiate cells from different individuals of the same species ***(11)***. An additional disadvantage with PCR methods, which is often overlooked, is that different sources of template DNA will compete to react with the same primers resulting in poor sensitivity for detecting mixed cells. Thus, in cases of cross-contamination of cell lines the cell line DNA template of lower concentration may go undetected when using PCR.

The following pages will describe the basis of the DNA fingerprinting and profiling techniques as well as other less well-known techniques and provide examples of protocols validated for cell identification. Approaches to selecting an appropriate identification system to meet the reader's own purposes will also be discussed.

1.2. Methods to Identify Species

1.2.1. Karyology

Karyology is the study of chromosomes, and each species of plant and animal has a specific number of chromosomes. This is referred to as the 2*N*, or diploid, number. For example, humans (*Homo sapiens*) have 2*N* = 46, mouse (*Mus musculus*) 2*N* = 40, and dog (*Canis familiaris*) 2*N* = 78.

The chromosomes are commonly viewed by arresting the cells at the metaphase stage of mitosis in which the chromosomes are highly condensed and can be visualized by staining to reveal banding patterns along the chromosomes.

The morphology and number of chromosomes can be used to determine the species of cells in culture. However, in cell lines the chromosome content can be unstable, and a modal number of chromosomes may need to be determined by counting a large number (i.e., more than 100) of mitoses. In addition to identifying species, the technique may also identify particular cell lines carrying a signature chromosome content, e.g., HeLa.

1.2.2. Isoenzymes

Many enzymes produced by a cell are not a single protein but a group of proteins, all of which have the same activity. These enzymes, called isoenzymes (or isozymes), differ in their amino acid composition and can be separated electrophoretically by their mass:charge ratio, producing a pattern often characteristic for particular species. Isoenzymes that will discriminate between species include glucose-6-phosphate dehydrogenase (G6PD), lactate dehydrogenase (LDH), malate dehydrogenase (MDH), and nucleoside phosphorylase (NP). While isoenzyme patterns are usually helpful for confirming the species of origin of a cell line, the patterns for certain enzymes may be different in cells of embryonic origin, and genetic changes arising during in vitro hybridization of cells (e.g., hybridomas) may also alter the isoenzyme profile.

1.2.3. DNA Barcoding

Using PCR amplification and DNA sequencing technology it is possible to directly visualize DNA polymorphism. Until recently it has been necessary to perform this type of work in a species-specific manner. However, the characterization of the mitochondrial cytochrome *c* oxidase I (*COI*) gene in many species has revealed a rapidly evolving region of DNA flanked by highly conserved regions. This had enabled the design of primer pairs, which should allow the PCR amplification of *COI* in most, if not all, plant and animal species ***(12)***. The PCR product (648 bp in most species) is then sequenced to yield a barcode specific to the species being tested. The method has been widely suggested as a panacea for molecular taxonomy ***(11)***. It is therefore likely that the method will become widely used to identify the species of cell cultures. A database of DNA barcodes will be produced and maintained by the US National Center for Biotechnology Information (http://www.ncbi.nih.gov).

Primer sequences for DNA barcoding are

Primer LCO1490: 5′-GGTCAACAAATCATAAAGATATTGG-3′
Primer HCO2198: 5′-TAAACTTCAGGGTGACCAAAAAATCA-3′

The DNA barcode of the cytochrome *c* oxidase gene may be represented as a four-color pattern as illustrated in **Fig. 1**.

1.3. Methods to Identify Individuals in a Range of Species

1.3.1. Variation on Numbers of Tandem Repeats

Repetitive DNA comprises a large proportion of the genome of many higher organisms. While the function of many types of repetitive DNA has yet to be elucidated, some have proven highly useful for identification of individuals and the cell lines derived from them.

Two groups of repetitive DNA called minisatellites (repeat units of 10–100 bp) and microsatellites (repeat units of usually 1–4 bp) have proven particularly useful for cell identification. While minisatellite loci are often found towards chromosome ends (telomeres), microsatellites appear to be randomly distributed throughout the genome.

Fig. 1. The DNA barcodes of the cytochrome c oxidase gene of human and mouse genomes represented as a four-color pattern. Numbers indicate numbers of base pairs across the amplified sequence.

1.3.2. Multilocus DNA Fingerprinting

Certain VNTR sequences such as the Jeffreys probes for the human myoglobin locus, 33.15 and 33.6 ***(13)***, and the microsatellite sequence from the M13 phage protein III gene ***(14)*** will cross-hybridize (under the appropriate experimental conditions) with a wide range of families of VNTR sequence. These probes have proven especially successful in delivering specific identification methods for a wide range of species including plants and animals (for reviews, *see* **ref. *15***). Multilocus DNA fingerprinting benefits from the presence of related satellite sequences present in the genomes of a wide spectrum of eukaryotic organisms. These sequences are revealed in Southern blot hybridization in which the stringency of hybridization is set to permit visualization of DNA–DNA hybrids that have some homology but are not entirely complementary (i.e.,derived from related satellite families) and reveal alleles from a range of genetic loci widely distributed in the genome (**Fig. 2**).

Multilocus DNA fingerprinting has been applied to cell culture in research ***(16,17)***, culture collections ***(18,19)***, and in the manufacture of biological products from animal cells ***(20)***. This technique is now providing a valuable tool for assuring the quality of cell culture processes by excluding cross-contamination of cell lines between master and working banks and screening for common contaminants such as HeLa ***(20,21)*** and cell line genetic stability ***(22,23)***.

1.3.3. Amplified Fragment Length Polymorphisms

Amplified fragment length polymorphism (AFLP) analysis is a PCR-based version of RFLP analysis (*see* next section). In this method adaptors are ligated to digested DNA. The PCR primers used are complementary in sequence to the adaptors, but also have an additional short (e.g., 3 bp) 3′ sequence. This enables the specific amplification of fragments carrying this short terminal sequence, allowing co-amplification of a large number of fragments without knowledge of the DNA sequence ***(24)***. The approach will differentiate between species and is also capable, but with lesser power and robustness, to differentiate between individuals.

1.3.4. Randomly Amplified Polymorphic DNA

In ramdomly amplified polymorphic DNA (RAPD) analysis, instead of using two primers that are designed based on pre-existing knowledge of the target sequence, RAPDs are produced from short primers (typically 10 mers) of random sequence that, under low-stringency conditions, produce PCR products ***(25)***. The number of products or bands on the gel depends on the number of appropriately oriented and spaced target sites present in DNA in that species or individual. A significant disadvantage of the RAPD approach has been that, without careful

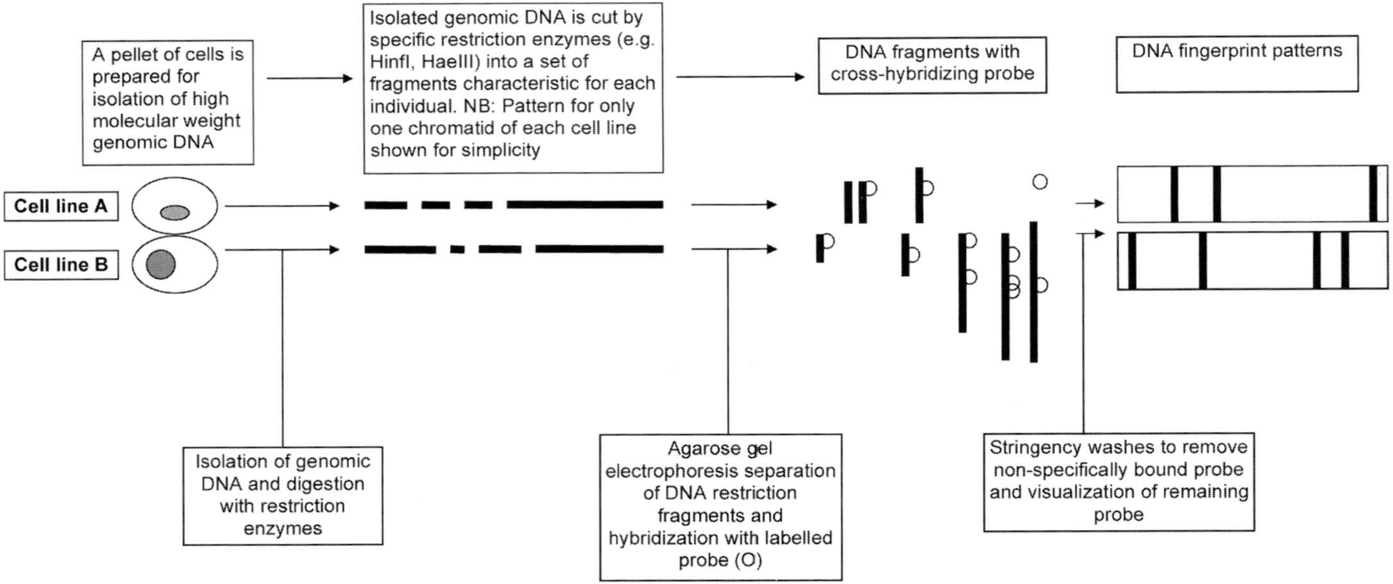

Fig. 2. DNA multilocus fingerprinting process.

standardization, reproducibility is often poor. Both the RAPD and AFLP approach may also be used as a method of species identification.

1.3.5. Aldolase Intron G PCR

PCR primers have been designed that enable amplification of a region of the Aldolase gene intron G in a wide variety of animal species ***(26)***. The PCR products vary in size and number between different species because of differences in the aldolase gene and thus provide a simple mechanism for species identification.

1.4. Methods to Identify Individuals Within a Species

Many cell identification needs are limited to a single species, often human. For these applications it is not necessary for the assay to work on a range of species, and simpler assays specific to a particular species may be performed.

1.4.1. Single Locus DNA Fingerprinting

A variety of minisatellite single locus probes are available for Southern blot-based analysis of specific VNTR loci ***(27)***. If desired, cocktails of probes can be used, increasing the information derived, giving very specific identification for human and other cells. Although the result appears similar to multilocus DNA fingerprinting, it is significantly easier to perform and standardize. The method has, however, been largely replaced by PCR-based methods of microsatellite typing.

1.4.2. RFLPs

RFLP detection of polymorphism is a Southern-blot based method in which (usually) a single polymorphic site causing the creation or destruction of a restriction endonuclease cleavage site is detected on the basis of the length of the DNA fragment visualized. The method is cumbersome and provides limited polymorphism information. It has therefore been superseded by other (PCR-based) methods.

1.4.3. Microsatellite-Based DNA Profiling

PCR-based DNA short tandem repeat (STR) profiling takes advantage of the high number of STRs that occur frequently within the eukaryotic genome. STR DNA profiling was originally reported by Tautz ***(28)*** and Litt and Luty ***(29)***. In this technique template DNA is amplified using a multiplex PCR reaction, with each pair of primers amplifying a different STR allele. When the method was first developed ***(28,29)***, the amplified samples had to be manually run on a gel. Recent developments have replaced the manual agarose gel with a capillary system where the PCR products are drawn through a capillary by electrophoresis. The samples pass a laser where a fluorescent label on the primers is excited and

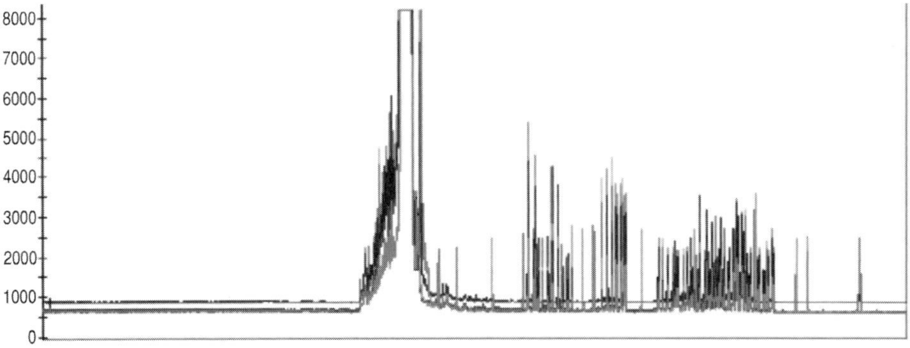

Fig. 3. Raw capillary electrophoresis data showing relative florescence units against time.

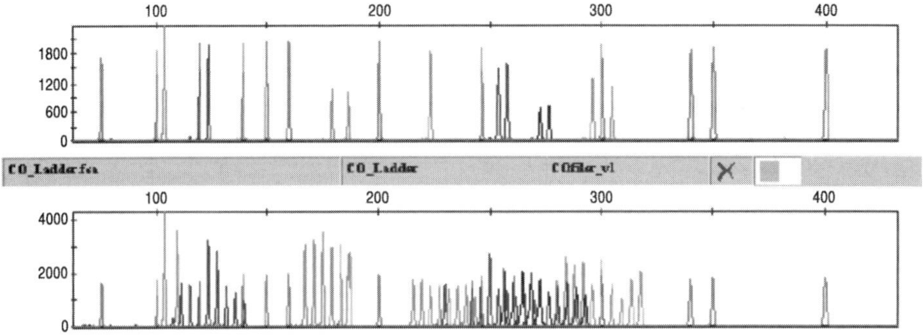

Fig. 4. DNA profiling data for a human diploid fibroblast cell culture. Processed sample data (above) aligned with the allelic ladder, which shows all possible allele locations (below).

the relative florescence is detected using a CCD camera. The raw data are delivered as an electrophoretogram that appears as a series of peaks of fluorescence along a time line **(Fig. 3)** and is resolved using specialist software to align allele peaks with reference alleles in a reference allele ladder **(Fig. 4)**.

Despite the popularity of the STR technique, promoted by the availability of reagent kits (e.g., Promega, ABI), it requires very careful scientific validation for reliability and reproducibility before it can be applied in a routine setting. However this technique has been successfully applied in the authentication of cell lines ***(30,31)***, detection of cross-contamination among human tumor cell lines ***(32–34)*** and have potential in monitoring genetic stability (e.g., **ref. *35***). Having been developed for forensic work, commercially available kits for fluorescent multiplexed STR analysis are designed for use with human material and also to avoid cross-reaction with nonhuman material. Accordingly, their value for use with other species can be expected to be neglegible.

1.4.4. SNP Analysis

Single nucleotide polymorphisms (SNPs) also provide a useful method for the identification of cell lines. This is a much more common form of polymorphism, occurring roughly once every 1000 bp of DNA, but as there are only two alternative forms for each SNP (rather than the many forms seen for each VNTR locus), this form of analysis is much less informative at individual loci. This lack of information content can, however, be countered by highly multiplexed forms of SNP analysis. The most common forms of highly multiplexed analysis are those that anchor DNA molecules to defined solid-surface locations (microarrays) or attach DNA molecules to beads either anchored or detected by laser in liquid ***(36)***. Commercial systems are available for these fomats (e.g., Affymetrix™, Illumina™, Luminex™) for several species, but are expensive both in terms of equipment and reagents. The throughput needs to be very high to offset this expense. Simpler systems of SNP analysis using less highly multiplex assays or multiple simplex assays may allow the development cheap and simple SNP-based cell line identification.

1.5. Methods to Monitor Stability of Cell Lines

Both isoenzyme analysis and karyology, in addition to identifying cell line species, may be used to detect cell line change. Alteration of isoenzyme patterns may occur as cells transform, hybridize, differentiate, or lose isoenzyme gene loci on long-term passaging. Alteration in the karyotype may also be an indicator of permanent changes. This may present as a change in the chromosomal modal number or as a number of changes to chromosome structure including deletions and translocations.

Microarrays may be used to monitor cell change both in terms of gene expression and DNA content ***(37,38)***. SNP microarrays can detect loss of heterozygosity, which may occur through partial chromosome loss or through mitotic recombination. Microarrays specifically produced to detect chromosome rearrangments, known as comparative genome hybridization (CGH) arrays, are a new phenomenon ***(39,40)***. They have the potential to detect submicroscopic duplications and deletions and are thus likely to become a reference method for cell line quality control.

2. Materials

2.1. Multilocus DNA Fingerprinting

2.1.1. Equipment

1. 1.5-mL microtubes.
2. Microfuge.

3. 60°C water bath.
4. Ice bucket.
5. Spectrophotometer (260 nm) or fluorimeter for DNA determinations.
6. Submarine gel electrophoresis equipment.
7. Plastic or PVC gel wash trays approx 5–10 cm larger than the dimensions of the analytical agarose gel.
8. Paper hand towels (e.g., Kirby Clark).
9. Nylon membrane (20 cm × 20 cm, Hybond N, Genetic Research Instruments Ltd, or equivalent).
10. Transilluminator (315 nm).
11. Transparent polyester sheet (greater than 20 cm × 20 cm).
12. X-ray development cassette.
13. X-ray film (e.g., Fuji-RX).

2.1.2. Materials

1. Phosphate-buffered saline (PBS) pH 7.6 in a sterile container.
2. Ice-cold sterile distilled water.
3. Sucrose/detergent buffer (0.32 *M* sucrose, 10 m*M* Trizma base pH 7.5, 5 m*M* $MgCl_2 \cdot 6H_2O$, 1% (w/v) Triton X-100).
4. Lysis buffer (0.075 *M* NaCl, 0.024 *M* Na_2-ethylene diamine tetraacetic acid [EDTA]).
5. Proteinase K (20 mg/mL, molecular grade).
6. RNAse A (10 mg/mL, molecular grade).
7. 10% Sodium dodecyl sulfate (SDS) (w/v) in distilled water.
8. 5 *M* NaCl in distilled water.
9. Absolute ethanol.
10. 80% (v/v) ethanol in distilled water.
11. TE buffer (10 m*M* Trizma base pH 7.5, 1 m*M* Na_2-EDTA).
12. HinfI (or HaeIII) restriction enzymes.
13. Nucleic acid grade agarose (e.g., Type 1A, Sigma).
14. TBE eletrophoresis buffer (20 m*M* Tris-HCl, 20 m*M* Na borate, 2 m*M* EDTA, 0.25% SDS, 0.5 mg/L ethidium bromide).
15. Loading buffer (1/6 volume of 15% Ficoll 400, 0.25% bromophenol blue in TBE).
16. Lambda phage Hind III digest molecular weight markers.
17. Depurination solution (0.25 *M* HCl, 15 min treatment).
18. Dehybridization solution (1.5 *M* NaCl, 0.5 *M* NaOH, 30-min treatment).
19. Neutralization buffer (3 *M* NaCl, 0.5 *M* Tris-HCl pH 7.5, 30-min treatment).
20. 20X SSC (175 g/L NaCl, 88.2 g/L trisodium citrate, in distilled water at a final pH of 7.4).
21. 1X SSC (1:20 20X SSC in distilled water).
22. Prehybridization solution (990 mL/L 0.5 *M* Na_2HPO_4, in distilled water at a final pH of 7.2 plus 10 mL/L 10% SDS).
23. Hybridization buffer (900 mL/L prehybridization buffer, 100 mL/L of 100 g/L casein Hammarsten in stringency wash solution 2).
24. Wash solution 2 (13.8 g/L maleic acid and 8.7 g/L NaCl in distilled water at pH 7.5).

25. NICE™ probe 33.15 or 33.6 (CellMark Diagnostics, UK).
26. Stringency wash solution 1 (160 mL/L 0.5 *M* Na_2HPO_4 in distilled water at pH 7.2 plus 10 mL/L 10% SDS).
27. Lumiphos 530 luminescent reagent.
28. 0.1% SDS in distilled water.

2.2. Cell Typing by PCR of Intron G of the Aldolase Gene

2.2.1. Equipment

1. PCR cycler 2.
2. Submarine minigel equipment.

2.2.2. Materials

1. Primers Ald1 (TGTGCCCAGTATAAGAAGGATGG) and Ald2 (CCCATCAGGGA-GAATTTCAGGCTCCACAA).
2. Ethidium-bromide-stained submarine minigel (1.4% agarose in 1x TBE).
3. Molecular weight marker (e.g., marker VI, Boehringer-Mannheim).

2.3. DNA STR Profiling Equipment and Materials

2.3.1. Equipment

1. Automated genetic analyzer (e.g., 3130 × l, ABI).
2. Suitable micropipets with sterile tips.
3. PCR thermocycler.
4. Aluminum foil.

2.3.2. Materials

1. ABI AmpF*l*STR COFiler kit.
2. Ultra-pure water.
3. Amplitaq DNA polymerase.
4. Formamide.

3. Methods

The following protocols give an example of a validated multilocus DNA fingerprinting method (see **refs.** ***14*** and ***18***), Aldolase Intron G PCR ***(41)*** for wide-ranging species, and microsatellite profiling of human DNA for human cell identification.

3.1. Multilocus Fingerprinting Protocol (see Note 1)

3.1.1. Preparation of Southern Blots of Genomic DNA

For DNA fingerprinting it is important to obtain undegraded high molecular weight genomic DNA. This may be obtained using one of the many commercially available kits for DNA extraction, but the quality should be checked by confirming the absence of significant amounts of degradation. A cheap and safe alternative minipreparation method (*see* **ref. *22***) that avoids the use of phenol is given here:

1. Resuspend a pellet of at least 5 × 106 cells in 1 mL PBS pH 7.6 in a sterile 1.5-mL microtube, microfuge (9000*g* for 1 min) to pellet the cells, and aspirate the PBS.
2. Resuspend the pellet in 1 mL ice-cold sterile distilled water, microfuge as before for 2 min, and aspirate the supernatant.
3. Resuspend the pellet in ice-cold sucrose/detergent buffer, microfuge at 12,000*g* for 5 min, and again aspirate the supernatant.
4. Resuspend the pellet in 1 mL lysis buffer and, after mixing by inversion, add 20 µL Proteinase K (20 mg/mL), 20 µl RNAse A (10 mg/mL), and 10 µL 10% SDS.
5. Remix by inversion and incubate at 60°C for 1 h or 37°C overnight.
6. After incubation cool the tube on ice and add 400 µL 5 *M* NaCl and repeatedly invert vigorosly to yeild a white precipitate.
7. Microfuge at 12,000*g* for 5–10 min and transfer the clear supernatant to a fresh sterile microtube.
8. Add two volumes of cold ethanol and mix by inversion at least 10 times to precipitate the DNA before microfuging at 12,000*g* for 5 min and aspirating the ethanol.
9. Wash the pellet in 80% ethanol and microfuge at 12,000*g* for 2 min before partially drying the pellet at 37°C.
10. Suspend the pellet in 30 µL TE buffer (10 m*M* Trizma base pH 7.5, 1 m*M* Na_2-EDTA) and dissolve the DNA at 37–40°C with occasional mixing.
11. The DNA is then quantified by spectrophotometer (260 nm) or fluorimeter and 10 g is digested with HinfI (or HaeIII) restriction enzymes according to the enzyme manufacturer's instructions. (Note: For each set of digests, standard DNA samples [e.g., HeLa, K562] should be run in parallel so that the DNA fingerprints on each blot can be checked for consistency.)
12. Run an agarose minigel to check the high molecular weight of 1 uL uncut genomic DNA in parallel with 1 uL restriction digest to confirm satisfactory digestion (i.e., only low molecular weight DNA below 5 kb is visible).

3.1.2. Southern Blot

1. Quantify the DNA in each digest and run an analytical agarose gel of at least 20 cm and electrophorese 5 µg digested DNA with loading buffer. Allow the 2.3-kb fragment of a Lambda/HinDIII molecular weight marker to run the full length of the gel.
2. Gently agitate the gel consecutively in the following solutions: depurination solution (15 min), dehybridization solution (30 min), and neutralization buffer (30 min). Each of these is used in the given order (for the given times) with brief intermediate washes in distilled water.

3. Transfer the gel onto a supported paper (two sheets Whatman 3MM) drawing from a reservoir of 20X SSC.
4. Place a nylon membrane over the top surface of the gel and then cover with four similar sized sheets of Whatman 3MM soaked in the 20X SSC transfer buffer. Place two piles of absorbent paper hand towels (e.g., Kirby Clark) over the soaked sheets and apply a 1-kg weight evenly on top of the towels.
5. Transfer of DNA fragments should be complete after blotting overnight, when the nylon membrane can be removed, dried, and fixed over a UV transilluminator (315 nm for 5 min).

3.1.3. Visualization of DNA Fingerprints by Chemiluminescence With the Multilocus Probe 33.15

1. Wet Southern membranes (up to 10 per hybridization) in 1X SSC (1:20 20X SSC in distilled water) and place into 500 mL sterile prehybridization solution and agitate at 50°C for 20 min.
2. Transfer the membranes individually to 160 mL hybridization buffer, add the contents of one NICE^{TM} probe vial, and gently agitate at 50°C for 20 min.
3. Transfer the membranes individually to 500 mL prewarmed stringency wash solution 1 and gently agitate at 50°C for 10 min.
4. Repeat **step 3**.
5. Transfer each membrane to 500 mL stringency wash solution 2 (*see* above) and gently agitate at room temperature for 10 min.
6. Repeat **step 5**.
7. Drain each membrane and place DNA side up on a transparent polyester sheet, apply 3 mL Lumiphos530 luminescent reagent (Note: Use a fume cabinet), place another polyester sheet on top of the gel, squeezing out excess Lumiphos, and seal the membrane between the sheets with tape. The sealed gels are then fixed into an X-ray development casette.
8. Fix two sheets of X-ray film (e.g., Fuji-RX) over the sealed membranes and incubate at 37°C for 3–5 h.
9. Develop the top X-ray film according to the manufacturer's instructions and check that all the expected bands in the standard HeLa DNA fingerprint are present and clear.
10. If increased exposure is required, reincubate the second X-ray film as before Additional films can be exposed for up to 3 d after the Lumiphos has been added. (Note: Membranes can be stripped with 0.1% SDS at 80°C and rinsed in 1X SSC for reprobing.)

3.2. Cell Typing by PCR of Intron G of the Aldolase Gene (see Notes 2 and 3)

1. Set up PCR reactions as described by Stacey ***(41)*** using the primers Ald1 (TGTGC-CCAGTATAAGAAGGATGG) and Ald2 (CCCATCAGGGAGAATTTCAGGCTC-CACAA) with negative controls (e.g., no primers, no template, unrelated plasmid template) and a standard genomic DNA of known profile.

2. Perform 35 cycles of PCR as follows: 94°C for 1 min (denaturation), 57°C for 45 s (annealing), 72°C for 2 min (extension). An initial denaturation of 94°C for 3 min and a final extension of 72°C for 5 min are also carried out.
3. PCR products are visualized in an ethidium-bromide-stained minigel (1.4% agarose) following electrophoresis at 75 V for 2 h in parallel with a Boehringer-Mannheim molecular weight marker VI.
4. Typical PCR products are:

Human and primate	0.50 kb, 0.38 kb, 0.30 kb, 0.19 kb
Mouse and rat	0.49 kb, 0.38 kb, 0.31 kb, 0.19 kb
Rabbit	0.48 kb, 0.30 kb, 0.19 kb
Dog	0.55 kb, 0.31 kb, 0.19 kb
Cat	0.56 kb, 0.32 kb
Frog	0.29 kb
Mosquito	0.99 kb, 0.85 kb, 0.60 kb

5. Further results showing the value of this technique for testing cell lines from an even wider range of species are given in **ref. *41***.

3.3. DNA STR Profiling

3.3.1. Preparation of Samples for Multiplex Fluorescent STR PCR

Several types of automated genetic analyzer currently exist, one of the most popular of which is the ABI series. Although the versions vary, the basic principles are the same. The protocol below is for the ABI 3130 × l, a 16-capillary system able to analyze fragment analysis and sequencing in the same run using the same polymer.

Several types of fluorescent STR kits are available off the shelf; the type of STR kit used will be determined by several factors, including the required power of resolution, the instrument available, and the financial resources. For the purposes of this protocol, the ABI AmpF*l*STR COFiler kit is used; this kit contains primers for seven loci, each linked to one of four dyes: FAM, JOE, NED, and ROX (**Fig. 5**).

The ABI 3130 × l works in groups of 16 samples; 1 of the 16 must be an allelic ladder and another might be a control sample.

1. Ensure DNA sample is mixed well by pipetting; accurately quantify DNA samples. Adjust to 1–2 ng/μL using ultra-pure water; too much template DNA as well as salts can interfere with subsequent fluorescence detection.
2. Set up the PCR reaction. It may be necessary to make a different volume of master mix specifically for the positive control DNA supplied depending on the concentration supplied. Use 1 μL of nuclease-free water in place of DNA as a negative control.

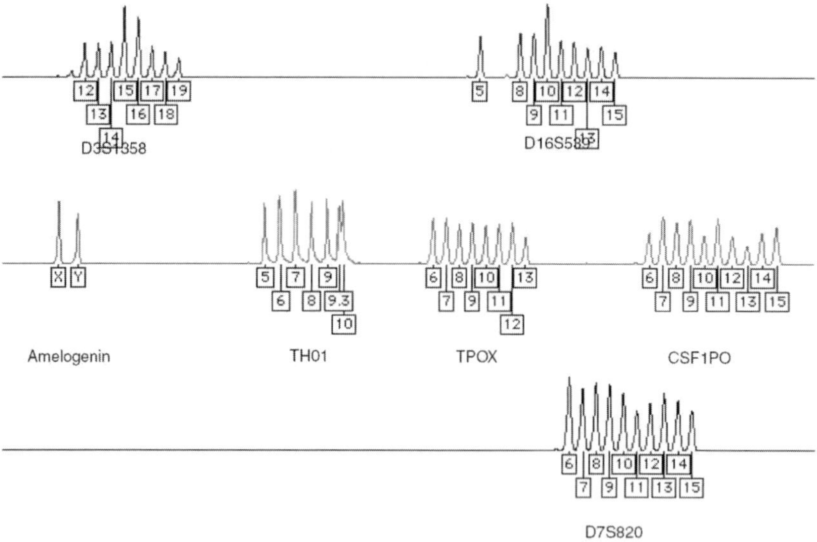

Fig. 5. Electrogram plot of ampF*l*STR COfiler allelic ladder showing the designation of each allele. (Reproduced by permission of ABI.)

	×1 (μL)	Positive control
Reaction mix	6	20
AmpliTaq	0.3	1
Primer	3	10
DNA (1 ng/μL)	1	20

3. Run the reaction in a thermo-cycler; if using a plate, avoid the use of well along the edge of the plate. Particular thermo-cyclers may have been specified by the manufacturer, and use of different models may require optimization.

Step	Temperature	Time	No. of cycles
Incubation	95°C	11 min	1
Melting	94°C	1 min	
Annealing	59°C	1 min	28
Extension	72°C	1 min	
Final extension	60°C	45 min	1
Final hold	25°C	Forever	1

4. Protect the PCR samples from light using aluminum foil if they are not for immediate use. Prepare a formamide solution by mixing 19 parts HiDi formamide to 1 part

ROX size marker. Allow 10 μL of formamide solution per sample, remembering to allow one extra for allelic ladder plus one or two extra.

5. Place 10 μL formamide solution into each well of a 3130 × 1 96-well plate (or tubes or strips) and add 1 μL of the PCR product. The samples must be loaded into plates/tubes specified by the genetic analyzer manufacture; any deviation may result in damaged injection needles. Once this is done, add 1 μL of allelic ladder to the appropriate wells. Ensure all samples/controls are in groups of 16; use blank wells containing formamide if there are any empty wells.
6. Seal the tubes/plates and denature by heating to 95°C for 3 min and immediately place on ice.

3.3.2. Running Samples on the Genetic Analyzer

7. Load the samples into the plate assembly and place this in the genetic analyzer. Ensure the genetic analyzer has been prepared for the run for example, add polymer, and fill the buffer reservoir.
8. Open up collection software and create a plate record for the samples. It is necessary to calibrate the machine and set up the appropriate protocols prior to initiating a run. An internal lane standard is run within each capillary with the sample allowing each sample to be aligned and standardized. This, coupled with the use of fluorescent probes, has facilitated the automation of DNA fingerprinting. Specialist software has been designed to analyze this raw data and convert it to a more user-friendly format using complex algorithms to comparing the raw data to the internal size standard and allelic ladder **(Fig. 2)**.
9. Run the instrument. During the run, the instrument view and capillary view can be observed to ensure the run is proceeding correctly. Once the samples have run, they can be imported and analyzed according to the manufacturer's instructions.

4. Notes

1. Multilocus DNA fingerprinting of minisatellite or microsatellite DNA offers a major advantage in that a single method such as that given above will give specific banding patterns for a diverse range of species ***(18,22)***. However, microsatellite probes (e.g., $[GATA]_4$, $[GACA]_4$, $[GTG]_3$) often show a restricted species range for which they will yield useful intraspecies differentiation. The "Jeffreys" probes 33.15 and 33.6 and the M13 phage protein III gene microsatellite probe ***(14,15)*** have proven valuable in identifying individuals for a remarkably wide range of species in ecological studies and are excellent candidates for a general DNA fingerprinting method where cell lines from many species are handled. However, as for the other VNTR-based methods, although the multilocus DNA fingerprint of a cell line is extremely valuable for identity testing and to some extent for determining genetic stability ***(22,35)***, it cannot be used to identify cross-contaminated cell lines without appropriate samples from the individual of original or standard profiles from contaminating cells such as HeLa.
2. Analysis of minisatellite or microsatellite DNA loci using multiplexed PCR or multiple probes in Southern blot hybridization has provided highly specific

identification of human cell lines ***(13–19,22,30,31)***. Some commercially available STR kits analyze nine or more different human loci simultaneously (e.g., AmpFLSTR Profiler Plus Amplification Kit, ABI). It is important to be aware that the specificity of identification by such methods will depend on matching the origin of the cells tested with the human (or wild) populations of origin for which the methods were validated. Furthermore, while cross-hybridization will occur for some species, the sequences analyzed in probe sets for human analysis are unlikely to be useful for cell lines from other species. Even when using a test fully validated for the species and population of origin, unusual results may occur since cell lines may have genetic defects (e.g., deletions, triploidy), which may confuse these methods of identification (e.g., microsatellite instability caused by defects in DNA repair enzymes). Nevertheless, use of multiple probes for specific VNTR loci provides clearly interpretable data given that there will be a defined set of expected allele sizes for each VNTR locus.

3. STR analysis provides a rapid and reliable method of cell typing now used widely. However, despite continued improvements, the ABI Genetic Analyser instruments are complex machines, requiring careful set up and maintenance prior to running samples. The analysis software is also complex, requiring a high degree of computer literacy and understanding in order to obtain satisfactory results. While setting up a run, it is important to incorporate the appropriate control including the allelic ladder and positive controls (**Fig. 4**).

For authentication of human cell lines there is an ever-increasing number of companies that provide STR and other identity testing services; a list of examples is given in **Table 1**. The authors cannot vouch for the quality of service provided by any of these organizations, and it is important when approaching such companies to ask certain questions to assure yourself that they will provide the service you need. These questions might include:

- Do they perform the testing themselves, or do they outsource testing services?
- What is the specificity of the methods used for individual identification?
- Are the genetic markers used linked with those used by professional bodies or other expert centers?
- Do they have experience in interpreting cell line data?
- Do they have in-house expertise in the methods to assist in interpretation of results?
- Do they have accreditation by an appropriate professional or government body, or do they have formal affiliation with an expert group or organization?

It should be remembered that public service collections such as ATCC (http://www.atcc.org), ECACC (www.hpa.org), DSMZ (http://www.dsmz.de), and JCRB (http://cellbank.nihs.go.jp) will also be able to advise on such testing and may provide it as a service.

5. Conclusion

Identification techniques generally fall into one or more of three categories: (1) those that identify species, (2) those that identify individuals and are not

Table 1
Service Providers for DNA Identity Testing

Region	Country	Identity testing company
Europe	Germany	ID-Labor GmbH
		IMGM Laboratories
		Labtest
		Papacheck GmbH
		Bj-diagnostik GmbH
	Netherlands	Verilabs Europe
	Sweden	DNA-Test Sweden
	UK	Anglia DNA Bioservices Ltd
		Geneservice (GSL)
		LGC
Australasia	Australia	DNA-ID Labs/Identigene
		DNA Solutions (agents in many countries)
North America	Canada	DNA solutions (agents in many countries)
		Genetrack Biolabs
	United States	More than 40 websites advertise identity-testing services. It is advisable to look for AAB accreditation or equivalent when selecting a service

species-specific, and (3) those that identify individuals, but are species-specific. Very often the molecular techniques give highly specific identification but only within a limited group of species (e.g., microsatellites, single locus minisatellite probes) or, while useful over a wide range of species, do not in general determine the species of origin (e.g., multilocus DNA fingerprinting with the Jeffreys probes and the M13 probe). Thus, it is often necessary to decide whether to set up a new system for each species, such as microsatellite profiling, or to run two techniques in parallel: one giving a specific profile (such as multilocus fingerprinting) and the other identifying the species of origin (e.g., species-specific PCR, karyology, isoenzyme analysis). The latter course is the one often selected in culture collections where there is the extreme combination of many species and many cell lines from each species.

When setting up an identification system the older methods of isoenzyme analysis and karyology should not be forgotten: they provide rapid and reliable identification of species of origin and, of particular interest in respect to cells used in manufacturing processes, are identified in the guidelines from regulatory bodies ***(42,43)***. Karyology may also reveal unique marker chromosomes for identification and has been used as an indicator of genetic stability. The main restriction on the use of karyology is the need for experience and expertise if dealing with cell lines from a range of species. Even so, the rapidly developing area of molecular cytogenetics including techniques such as CGH and its microarray versions may provide valuable new techniques for the future ***(39,40)***. Isoenzyme analysis is an extremely reliable and straightforward technique, which can also provide confirmation of the species composition of hybrid cells. The major disadvantages of this technique, however, are the improbability of achieving specific identification of an individual cell line and that, in the absence of only one chromosome, should it carry a key isoenzyme, identification, even to the level of species, may not be possible.

Several important questions will be critical in the selection of an identification system:

- What are the species of origin of the cell lines in use in the laboratory?
- Is there a need for information on characteristics other than identification, such as genomic stability?
- What resources, technology and reagents are available to perform identity tests?

The ideal identification system would be capable of identifying the species or strain of origin of each cell line while also providing a unique marker, code, or profile for each individual cell line. Systems approaching this ideal would be expected to be developed in reference facilities such as culture collections (e.g., ATCC, USA; ECACC, UK; DSMZ, Germany; JCRB, Japan). The specific requirements for identification methods will vary between different laboratories depending on the type of work performed and the available resources. For example, a laboratory dedicated to working with a large number of cell lines from human tumors will have a specific set of demands which are very different from those in a laboratory where cell lines from many different species are in use. While meeting the special needs for identification within a particular laboratory it is vital that the identification system in use will also enable identification of cultures that have been cross-contaminated or switched with another.

References

1. Nelson-Rees, W. A., Daniels, D. W., and Flandermeyer, R. R. (1989) Cross-contamination of cells in culture. *Science* **212**, 446–452.
2. Stacey, G., Masters, J. R. W., Hay, R. J., Drexler, H. G., MacLeod, R. A. F., and Freshney, I. R. (2000) Cell contamination leads to inaccurate data: we must take action now. *Nature* **403**, 356.
3. Stacey, G. N. (2002) Standardization of cell lines. *Dev. Biologicals* **111**, 259–272.
4. Hsu, T. C. (1952) Mammalian chromosomes in vitro the karyotype of man. *J. Hered.* **43**, 167–172.

5. Tjio, J. H. and Leven, A. (1956) The chromosome number of men. *Hereditas* **42**, 1–6.
6. Simpson, W. F., Stulberg, C. S., and Petersen, W. D. (1978) Monitoring species of cells in culture by immunofluorescence. *Tissue Culture Association Manual* **4**, 771–774.
7. O'Brien, S. J., Kleiner, G., Olson, R., and Shannon, J. E. (1977) Enzyme polymorphisms as genetic signatures in human cell cultures. *Science* **195**, 1345–1348.
8. Ferrone, S., Pellegrino, M. A., and Reisfeld, R. A. A. (1971) A rapid method for direct HLA typing of cultured lymphoid cells. *J. Immunol.* **107**, 613–615.
9. Christensen, B., Hansen, C., Debiek-Rychter, M., Kieler, J., Ottensen, S., and Schmidt, J. (1993) Identity of tumorigenic uroepithelial cell lines and 'spontaneously' transformed sublines. *Br. J. Can.* **68**, 879–884.
10. Jeffreys, A. J., Wilson, V., and Thein, S-L. (1985) Individual specific fingerprints of human DNA. *Nature* **316**, 76–79.
11. Hebert, P. D., Cywinska, A., Ball, S. L., and deWaard, J. R. (2003) Biological identifications through DNA barcodes. *Proc. Biol. Sci.* **270**, 313–321.
12. Folmer, O., Black, M., Hoeh, W., Lutz, R., and Vrijenhoek, R. (1994) DNA primers for amplification of mitochondrial cytochrome *c* oxidase subunit I from diverse metazoan invertebrates. *Mol. Mar. Biol. Biotechnol.* **3**, 294–299.
13. Jeffreys, A. J., Wilson, V., and Thein, S-L. (1985) Hypervariable minisatellite regions in human DNA. *Nature* **314**, 67–73.
14. Vassart, G., Georges, M., Monsieur, R., Brocas, H., Lequarre, A. S., and Christophe, D. (1987) A sequence in M13 phage detects hypervariable minisatellites in human and animal DNA. *Science* **235**, 683–684.
15. Burke, T., Dolf, G., Jeffreys, A. J., and Wolf, R. (eds.) (1990) *DNA Fingerprinting: Approaches and Applications*, Birkhauser, Basel.
16. Thacker, J., Webb, M. B. T., and Debenham, P. G. (1988) Fingerprinting cell lines: use of human hypervariable DNA probes to characterize mammalian cell cultures. *Som. Cell Mol. Genet.* **14**, 519–525.
17. Van Helden, P. D., Wiid, I. J. F., Albrecht, C. F., Theron, E., Thornley, A. L., and Hoal-van Helden, E. G. (1988) Cross-contamination of human oesophageal squamous carcinoma cell lines detected by DNA fingerprint analysis. *Can. Res.* **48**, 5660–5662.
18. Stacey, G. N., Bolton, B. J., Doyle, A., and Griffiths, J. B. (1992) DNA fingerprinting: a valuable new technique for the characterization of animal cell lines. *Cytotechnology* **9**, 211–216.
19. Gilbert, D. A., Reid, Y. A., Gail, M. H., et al. (1990) Application of DNA fingerprints for cell line individualization. *Am. J. Hum. Gen.* **47**, 499–517.
20. Doherty, I., Smith, K. T., and Lees, G. M. (1994) DNA fingerprinting as a quality control marker for the genetic stability of production cells, in *Animal Cell Technology: Products for Today Prospects for Tomorrow* (Spier, R. J., Griffiths, J. B., and Berthold, W., eds.), Butterworth-Heinemann Ltd., Oxford, pp. 76–79.
21. Stacey, G. N. (2002) Standardization of Cell Lines. *Dev. Biologicals* **111**, 259–272.
22. Stacey, G. N., Bolton, B. J., Morgan, D., Clark, S. A., and Doyle, A. (1992) Multi-locus DNA fingerprint analysis of cell banks: stability studies and culture identification in human B-lymphoblastoid and mammalian cell stocks. *Cytotechnology* **8**, 13–20.

23. Racher, A. J., Stacey, G. N., Bolton, B. J., Doyle, A., and Griffiths, J. B. (1994) Genetic and biochemical analysis of a murine hybridoma in long term continuous culture, in *Animal Cell Technology: Products for Today Prospects for Tomorrow* (Spier, R. J., Griffiths, J. B., and Berthold, W., eds.), Butterworth-Heinemann Ltd., Oxford, pp. 69–75
24. Vos, P., Hogers, R., Bleeker, M., et al. (1995) AFLP: a new technique for DNA fingerprinting. *Nucleic Acids Res.* **23**, 4407–4414.
25. Williams, J. G. K., Kubelik, A. R., Livak, K. L., Rafalski, J. A., and Tingey, S.V. (1990) DNA polymorphisms amplified by arbitrary primers are useful as genetic markers. *Nucleic Acids Res.* **18**, 6531–6535.
26. Slade, R. W., Moritz, C., Heideman, A., and Hale, P. T. (1993) Rapid assessment of single-copy nuclear DNA variation in diverse species. *Mol. Ecol.* **2**, 359–373.
27. Wong, Z., Wilson, V., Patel, I., Povey, S., and Jeffreys, A. J. (1987) Characterization of a panel of highly variable minisatellites cloned from human DNA. *Ann. Hum. Genet.* **51**, 269–288.
28. Tautz, D. (1989) Hypervariability of simple sequences as a general source of polymorphic DNA markers. *Nucleic Acids Res.* **17**, 6463–6471.
29. Litt, M. and Luty, J. A. (1989) A hypervariable microsatellite revealed by in-vitro amplification of a dinucleotide repeat within the cardiac muscle actin gene. *Am. J. Genet.* **44**, 397–401.
30. Masters, J. R., Thomson, J. A., Daly-Burns, B., et al. (2001) Short tandem repeat profiling provides an international reference standard for human cell lines. *Proc. Natl. Acad. Sci. USA* **98**, 8012–8017.
31. Dirks, W. G., Faehnrich, S., Estella, I. A., Drexler, H. G. (1995) Short tandem repeat DNA typing provides an international reference standard for authentication of human cell lines. *ALTEX* **22**, 103–109.
32. MacLeod, R. A., Dirks, W. G., Matsuo, Y., Kaufmann, M., Milch, H., Drexler, H. G. (1999) Widespread intra-species cross-contamination of human tumor cell lines arising at source. *Int. J. Cancer* **12**, 555–563.
33. Thompson, E. W., Waltham, M., Ramus, S. J., et al. (2004) LCC15-MB cells are MDA-MB-435: a review of misidentified breast and prostate cell lines. *Clin. Exp. Metastasis* **21**, 535–541.
34. Milanesi, E., Ajmone-Marsan, P., Bignotti, E., et al. (2003) Molecular detection of cell line cross-contaminations using amplified fragment length polymorphism DNA fingerprinting technology. *In Vitro Cell Dev. Biol. Anim.* **39**, 124–130.
35. Hussein, M. R., Haemel, A. K., Sudilovsky, O., and Wood, G. S. (2005) Genomic instability in radial growth phase melanoma cell lines after ultraviolet irradiation. *J. Clin. Pathol.* **58**, 389–396.
36. Syvanen, A. C. (2001) Accessing genetic variation: genotyping single nucleotide polymorphisms. *Nat. Rev. Genet.* **2**, 930–942.
37. Pollack, J. R., Perou, C. M., Alizadeh, A. A., et al. (1999) Genome-wide analysis of DNA copy-number changes using cDNA microarrays. *Nat. Genet.* **23**, 41–46.
38. Murray, J. I., Whitfield, M. L., Trinklein, N. D., Myers, R. M., Brown, P. O., and Botstein, D. (2004) Diverse and specific gene expression responses to stresses in cultured human cells. *Mol. Biol. Cell.* **15**, 2361–2374.

39. Ishkanian, A. S., Malloff, C. A., Watson, S. K., et al. (2004) A tiling resolution DNA microarray with complete coverage of the human genome. *Nat. Genet.* **36**, 299–303.
40. Speicher, M. R. and Carter, N. (2005) The New cytogenetics: blurring the boundaries with molecular biology. *Nat. Rev.* **6**, 782–792.
41. Stacey, G. N., Hoelzl, H., Stephenson, J. R., and Doyle, A. (1997) Authentication of animal cell culture by direct visualization of repetitive DNA, aldolase gene PCR and isoenzyme analysis. *Biologicals* **25**, 75–83.
42. CBER (1993) Points to consider in the characterization of cell lines used to produce biologicals. Center for Biologics Evaluation and Research, National Institutes of Health, Bethesda, MD.
43. World Health Organization Expert Committee on Biological Standardization and Executive Board (1997) Requirements for the use of animal cells as in vitro substrates for the production of biologicals. World Health Organization, Geneva.

II

Basic Cultivation Techniques

5

Microcarrier Cell Culture Technology

Gerald Blüml

Summary

Cell culture techniques have become vital to the study of animal cell structure, function, and differentiation and for the production of many important biological materials such as vaccines, enzymes, hormones, antibodies, interferons, and nucleic acids. Microcarrier culture introduces new possibilities and, for the first time, allows practical high-yield culture of anchorage-dependent cells *(1)*. In microcarrier culture, cells grow as monolayers on the surface of small spheres or as multilayers in the pores of macroporous structures that are usually suspended in culture medium by gentle stirring. By using microcarriers in simple suspension culture, fluidized or packed bed systems, yields of up to 200 million cells per milliliter are possible.

The following microcarrier examples described in the procedures represent the major types: (1) Cytodex^{TM} (GE Healthcare) represents spherical microcarriers for anchorage-dependent cells, mainly for vaccine production; (2) Cytopore^{TM} (GE Healthcare) represents macroporous microcarriers for anchorage and suspension cells in stirred tank reactors, mainly for r-protein production and monoclonal antobodies; (3) Cytoline^{TM} (GE Healthcare) represents weighed macroporous microcarriers for fluidized bed applications.

Key Words: Microcarrier; macroporous microcarrier; fluidized bed; perfusion; high-density culture; vaccine production; mAb production; r-protein production; cell retention.

1. Introduction

The idea of culturing anchorage-dependent animal cells on small spheres (microcarriers) kept in suspension by stirring was first conceived by van Wezel *(1)*. In the first experiments, van Wezel used beaded ion exchanger *N,N*-diethyl-aminoethylamine (DEAE) Sephadex^{TM} (GE Healthcare) A-50 as a microcarrier. After optimization of the surface charge, Cytodex 1 was developed for a wide variety of cells. In the meantime, many different spherical microcarriers are now on the market, and in the vaccine industry, microcarrier technology is used up to a scale of 6000 l (Influenza production with Vero cells).

From: *Methods in Biotechnology, Vol. 2: Animal Cell Biotechnology: Methods and Protocols, 2nd Ed.* Edited by: R. Pörtner © Humana Press Inc., Totowa, NJ

The next major step forward was the development of macroporous microcarriers, which allowed growth inside the beads, thereby increasing cell density and protecting the cells *(2)*. Young and Dean *(3)* then described the use of macroporous microcarriers for animal cells in fluidized beds. These developments allowed the immobilization of both anchorage and suspension cells in high-cell-density production systems. Perfusion technology with macroporous microcarrier is used to produce r-proteins *(4)*, monoclonal antibodies *(5)*, or nonlytic virus *(6)*.

2. Materials

2.1. Microcarriers

2.1.1. Cytodex

Cytodex 1 microcarriers are based on a crosslinked dextran matrix that is substituted with positively charged *N,N*-diethylaminoethyl (DEAE) groups. The charged groups are found throughout the entire matrix of the microcarrier:

- Density 1.03.
- Size approx 180 µm.
- Area 4400 cm^2/g dry weight.
- Swelling factor 18 mL/g dry weight.

Cytodex 3 microcarriers consist of a surface layer of denatured pig skin collagen type I covalently bound to a matrix of crosslinked dextran:

- Density 1.04.
- Size approx 175 µm.
- Area 2700 cm^2/g dry weight.
- Swelling factor 14 mL/g dry weight.

2.1.2. Cytopore

Cytopore is a macroporous microcarrier that increases cell density (and therefore yield) and allows suspension cell cultivation and perfusion. Crosslinked cellulose with DEAE groups:

- Pore size 30 µm.
- Mean particle size 235 µm.
- Density 1.03.
- Dry weight 1.1 m^2/g.
- Swelling factor 40 mL/g dry weight.
- Charge density: Cytopore 1, 1.1 meq/g; Cytopore 2, 1.8/g.

2.1.3. Cytoline

Cytoline macroporous microcarriers is composed of high-density polyethylene weighted with silica. They can be used for fluidized bed applications and carry a slight negative charge.

1. Cytoline 1: sedimentation rate 120–220 cm/min; length 1.7–2.5; thickness 0.4–1.1; density 1.3 g/mL; do not swell.
2. Cytoline 2: sedimentation rate 40–90 cm/min; length 1.7– .5; thickness 0.4–1.1; density 1.03 g/mL; do not swell. Can be used for stirred tank applications as well.

2.2. Cell Quantification

1. Trypsin ethylene diamine tetraacetic acid (EDTA) solution: the solution can be prepared in Ca^{2+},Mg^{2+}-free phosphate-buffered saline (PBS), but the following solution is preferred for retention of maximum cell viability: 122 m*M* NaCl, 3.3 m*M* phenol red, 3.0 m*M* KCI, EDTA 0.02% (w/v), 1 m*M* Na_2HPO_4, Tris (hydroxymethyl) aminomethane 2% (w/v), 4.0 m*M* glucose, pH 7.8–8.0. Trypsin is added to this solution at the usual concentration for a given type of cell. For most cells, 100 mg trypsin/mL is sufficient. Strongly adhering cells, such as FS-4, may require 500 mg trypsin/mL. Trypsin solutions can be sterilized by filtration though a 0.2-mm sterile filter. Since trypsin solutions are subject to self-digestion, it is important to divide freshly prepared solutions into small aliquots and store frozen until required. Crude trypsin solutions have a high content of DNA and RNA *(7)*, and therefore pure, recrystallized enzyme is preferred for many biochemical and somatic cell genetics studies. Crude trypsin also shows large batch-to-batch variation in toxicity, and difficulties with cell growth can often be traced to a specific batch. When possible, test new batches of trypsin for toxicity.
2. Collagen solution: prepare the solution in PBS or Krebs II buffer *(8)* and sterilize by filtration through a 0.2-mm sterile filter. Collagenase is usually used at a concentration of 100–500 mg/mL.

3. Methods

3.1. Concentration of Microcarriers

The yield of cells from microcarrier culture is directly related to the surface area for growth and hence the concentration of microcarriers. In most situations, Cytodex microcarriers are used in stirred cultures at a concentration of 0.5–5.0 mg/mL final volume. If an adequate supply of medium is provided and gas tension and pH are controlled, it becomes possible to work with cultures containing more than 5 mg Cytodex/mL, and in some cases $5–10 \times 10^6$ cells/mL can be achieved.

Provided a correct inoculation density is used, a concentration of 3 mg Cytodex/mL final volume is usually optimal for general microcarrier culture. This will result in the greatest proportion of microcarriers bearing cells.

As an approximate guide to expected cell yields, assume that the culture has a density of 10^5 cells/cm^2 at confluence. This corresponds to 6×10^5 cells/mg Cytodex 1 and 4.6×10^5 cells/mg Cytodex 3. The exact yield will depend on the characteristic saturation density of the cell type and on the supply of medium.

For Cytopore, the recommended carrier concentration is 2–3 g/L. Due to the high swelling factor of 40 mL/g, the concentration must be lower than with Cytodex. The maximum concentration is 5 g/L.

For Cytoline 2 in stirred tank applications, the concentration is between 100 and 200 mL/L of culture volume. Higher concentrations are only possible for very robust cells. For Cytoline in fluidized beds, carrier concentration depends only on the degree of expansion in the reactor (cell growth is independent of carrier load inside the fermenter). The rate of microcarrier bed expansion depends on the attachment properties of the cells. The higher their adherence to the beads, the greater the expansion rate of the carrier bed inside the reactor and the lower the filling of the reactor. The maximum carrier load in a fluidized bed is half the working volume.

3.2. Preparing Microcarriers for Culture

3.2.1. Preparing for Sterilization

3.2.1.1. Cytodex

1. Dry Cytodex microcarriers (*see* **Note 1**) are added to a suitably siliconized glass bottle and swollen in Ca^{2+}, Mg^{2+}-free PBS (50–100 mL/g Cytodex) for at least 3 h at room temperature with occasional gentle agitation. The hydration process can be accelerated by using a higher temperature, e.g., 37°C.
2. The supernatant is decanted and the microcarriers are washed once with gentle agitation for a few minutes in fresh Ca^{2+}, Mg^{2+}-free PBS (30–50 mL/g Cytodex).
3. The PBS is discarded and replaced with fresh Ca^{2+}, Mg^{2+}-free PBS (30–50 mL/g Cytodex) and the microcarriers are sterilized by autoclaving with steam from purified water (*see* **Note 2**).
4. Prior to use, the sterilized microcarriers are allowed to settle, the supernatant decanted, and the microcarriers rinsed in warm culture medium (20–50 mL/g Cytodex). This rinse reduces dilution of the culture medium by PBS trapped between and within the microcarriers (a step of particular importance when using small culture volumes or cells with low plating efficiencies). The microcarriers are then allowed to settle, the supernatant removed, and the microcarriers resuspended in a small volume of culture medium and transferred to the culture vessel (*see* **Note 3**).

3.2.1.2. Cytopore

1. Hydrate and swell the dry microcarriers in PBS (50–100 mL/g of Cytopore) in a siliconized bottle. Some of the carriers might float at first, but autoclaving will expel any air trapped in the carriers and allow them to settle.
2. Gently agitate the solution occasionally for approx 10 minutes at room temperature.
3. Adjust pH to 7.0–7.5 and then autoclave the microcarrier at 121°C for 20 min.
4. After autoclaving, allow the sterilized microcarriers to settle, remove the supernatant, and add fresh PBS. A second wash with PBS is recommended. (*see* **Note 4**).

3.2.1.3. Cytoline

1. Add twice the amount of distilled water to the desired volume of Cytoline (1 or 2) and autoclave the mixture for 10 min at 121°C (1 bar). This procedure degasses the dry microcarrier. 121°C is the maximum temperature for this microcarrier.
2. Remove the water supernatant by sieving or, with smaller amounts, by sucking with a pipet.
3. Then add twice the amount of fresh distilled water. Stir the microcarriers for 10 min.
4. Remove the water supernatant and add 0.1 *M* NaOH.
5. Incubate the mixture for 3 h minimum or preferably overnight.
6. Wash with distilled water until alkali is removed (check pH of the wash water). Autoclave the microcarrier in distilled water or PBS at 121°C (1 bar) for 30 min (*see* **Note 5**).

3.2.2. Equilibration Before Inoculation

1. Adjust the culture temperature to a level optimal for cell attachment. In practice, this temperature is usually the same as that of the growth stage of culture (normally 35–37°C) (*see* **Note 6**).
2. After a few minutes the culture will be ready for inoculation. Stirring can hasten the process of gas exchange. Equilibration of cultures with very large volumes may take 2–3 h. Overnight equilibration is recommended. This long equilibration time can also be used as a short sterility test. Always note the exact procedure used for equilibration if you want to obtain reproducible results.

3.3. Inoculum

3.3.1. Inoculation Density

On average, the inoculation density for Cytodex is around 2×10^5/mL of culture volume. For macroporous beads like Cytopore and Cytoline, the concentration should be around 2×10^6 per/mL of settled beads. A lower concentration might be deduced from plating efficiency or small-scale inoculation density tests (*see* **Note 7**).

It is essential to equilibrate and stabilize all culture parameters before adding the inoculum. Temperature and pH are especially important. Also avoid exposing the inoculum to sudden changes in temperature, pH, or osmolarity.

3.3.2. Inoculum Condition

1. When inoculating, use actively growing cells in logarithmic growth. Avoid cells in stationary phase. Stationary phase cells can lose 1 d of process time at each subculture step.
2. The cells used for inoculation should be single cells and definitely not large aggregates, which will otherwise cause heterogeneity.
3. Surface carriers are normally inoculated with $0.5–2 \times 10^5$ cells/mL or $0.5–2 \times 10^4$ cells/cm^2. Macroporous carriers are normally inoculated with about the same cell concentrations. Inoculate the cells in one-third to one-half of the final volume.

4. Stir immediately at the lowest speed to keep the microcarriers in homogeneous suspension.
5. After attachment, add media to the final volume. If the vessel is very well suited for microcarriers, it may be possible to start directly with the final volume.

In some processes, large amounts of cells are generated on microcarriers and then harvested and frozen. These cells are later thawed and used as inoculum straightaway. The process is then scaled up two more steps before final production. This makes production more flexible.

Fluidized beds are inoculated with 1–2 × 10^6 cells/mL of packed bed carrier volume. The inoculation density is highly dependent on both cell line characteristics and media composition. In general, the higher the inoculation density, the earlier the production cell density is reached. Stirring rate is also related to the shear sensitivity of the cell line (*see* **Note 8**).

3.3.3. Initial Stirring and Intitial Culture Volume

The key to achieving maximum yields from microcarrier cultures is to ensure that all microcarriers are inoculated with cells from the very beginning of the culture. Transfer of cells from one microcarrier to another occurs only infrequently during the culture cycle, and it is therefore important to ensure that the maximum possible number of cells from the inoculum attach to the microcarriers.

- One way of initiating a microcarrier culture is to inoculate the cells into the final volume of medium containing microcarriers and immediately begin stirring.
- The rate and proportion of cells attaching to the microcarriers can be increased if the culture remains static with gentle intermittent stirring during the early attachment stage.
- If the cell–microcarrier mixture is contained in a reduced volume (e.g., in one-third of the final volume) at intermittent stirring, cells have a greater chance of coming into contact with a microcarrier and the conditioning effects on the medium are also much greater.
- The increased efficiency of attachment that results from initiating the culture in a reduced volume and with intermittent stirring produces an increase in cell yield at the plateau stage of culture (*see* **Note 9**).
  1. When starting most cultures, cells and microcarriers are incubated in one-third of the final volume. The culture is stirred for 2 min every 30 min at the speed used during the growth phase of the culture.
  2. After 3–6 h, continuous stirring is commenced at a speed just sufficient to keep the microcarriers in suspension and the volume of the culture is increased with prewarmed (37°C) culture medium (*see* **Note 10**).

Slow continuous stirring during the attachment stage is necessary for cell types that tend to clump during settlement (e.g., primary chicken embryo fibroblasts). In such cases, the initial stirring speed only needs to be about 25% of that normally used for the growth phase. Griffiths et al. ***(9)*** have observed that the modified initial culture procedure was essential for good results when growing human fibroblasts.

For suspension cells inoculated on macroporous microcarriers, a reduced inoculation volume is essential if the cells are to have a chance of being entrapped inside the pores. In fluidized beds, reduce the initial stirring so that the carrier bed is just expanded and not fluidized. This expanded bed acts like a dead-end filter unit where all the cells are filtered out of the medium.

3.3.4. Maintaining a Microcarrier Culture

Once a microcarrier culture has been initiated, certain procedures and precautions are required to maintain cell proliferation and achieve maximum yield. If the culture is essentially nonproliferating, as in the case of some primary cultures, (e.g., hepatocytes), conditions must maintain the function and survival of the cells for as long as possible. Although the following comments will be restricted to proliferating cultures, exactly the same principles should be considered when maintaining nonproliferating cultures.

During the culture cycle, changes in the density of the cell population are usually greater than 10-fold. Such growth conditions the medium and thereby encourages cell growth until a saturation density is reached and density-dependent inhibition of proliferation occurs (note, however, that many established and transformed cell lines do not show such inhibition). At the same time, oxygen and medium components are utilized and toxic products of metabolism accumulate. These changes must be taken into account when maintaining a microcarrier culture. In addition, keep in mind when optimizing culture procedure that conditions optimal for cell growth at low density (e.g., gas tension, pH) are not necessarily optimal for later stages of culture. Always remember that the most important aspect of maintaining a microcarrier culture and obtaining the best results is to anticipate changes in the culture.

For example, from the first few cultures with a particular type of cell, it is possible to observe if or when pH changes occur, when oxygen supply or medium components are depleted, and if aggregation of the microcarriers occurs. Once the stage at which these changes occur is known, corrective measures can be taken beforehand. It is often difficult to return to optimal conditions and obtain good results once large deviations have taken place and, in such cases, irreparable damage to the culture has occurred.

3.3.4.1. Stirring Speed

Although other culture systems can be used, the most suitable method for maintaining microcarrier cultures (Cytodex, Cytopore) is stirred suspension.

Stirring (1) ensures that the entire surface of the microcarrier is available for cell growth, (2) creates a homogeneous culture environment, (3) avoids aggregation of microcarriers by cell overgrowth, and (4) facilitates exchange of gases between the culture headspace and the medium or oxygen transfer during sparging or via other oxygenators.

In principle, stirring speed should be just sufficient to keep all the microcarriers in suspension. After the initial culture period, during which there is normally intermittent stirring or a static attachment period, the culture should be stirred continuously (*see* **Note 11**).

- The optimal stirring speed when using traditional magnetic spinner vessels is usually 50–70 rpm.
- 15–30 rpm are used with the modified spinner vessels or cultures stirred with bulb-shaped rods.
- Higher speeds are often required when using the same design of stirrer with culture volumes larger than 500 mL.
- The stirring speed when working with large-scale culture volumes also depends on the design of the stirrer. Progressive increases from 50 to 100–150 rpm during the culture cycle are frequently used for cultures of 100 L or more ***(10,11)*** (*see* **Note 12**). In the Cytopilot reactor (internal loop fluidized bed) the stirring speed is between 100 and 300 rpm. Stirring speed is initially low and is increased as the cell concentration increases. During inoculation, the fluidized bed is run as a packed bed. After a maximum of 5 h, the bed can be fluidized.

When microcarrier cultures are used for production of viruses or cell products, the stirring speed should be reduced during the production phase. In most cases, a reduction to half of the speed used during the growth phase is optimal. Excessive stirring or shear forces result in decreased yields of viruses or cell products.

3.3.4.2. Replenishment of Culture Medium

Careful replenishment of medium during the culture cycle is an important aspect of maintaining microcarrier cultures. There are three reasons for replenishing the medium: (1) replacing essential nutrients depleted by cell growth, (2) removing products of metabolism that inhibit growth or survival, and (3) assisting in control of pH.

The frequency and extent of medium replenishment depends on cell type, culture density, culture medium, and gas tension. Rapidly dividing cells and cultures at high densities require more frequent replenishment than low-density or slowly dividing cultures. Rapid cell division and high cell densities deplete medium components and decrease culture pH. At the same time, metabolites such as lactate, ammonia, and even specific growth inhibitors accumulate. The ideal replenishment scheme is the one that results in the smallest fluctuation of nutrient concentrations and pH during the culture cycle. For this reason, a continuous flow of medium is the preferred method for culture maintenance. However, for small-scale cultures or experiments with cell densities up to 3–5 × 10^6 cells/mL batch, medium replenishment is more convenient.

- The usual procedure is to start with replenishing 50% of the medium volume every 3 d. If necessary, a modified scheme can then be developed to get the best yield from the culture.
- It is common practice to observe the culture every day and determine its density. When samples are taken, 10–20% of the medium volume can be replaced with fresh medium.
- To take advantage of conditioning effects, replenishment should not take place within the first 2 d of culture.
- Another approach to medium replenishment, especially during the later stages of exponential growth, is to feed the culture with a modified medium. During this stage, nutrient and growth factor requirements are not the same as at the beginning of the culture. Many medium components used when initiating the culture, e.g., nonessential amino acids, nucleosides, etc., can be omitted and greater economy achieved by reducing the concentration of the serum supplement (*see* **Note 13**).
- Closed culture systems are frequently used for microcarrier culture at laboratory scale. With such systems, the vessels are sealed and the supply of gas is only renewed when the culture is opened for sampling or replenishment of the medium. One important aspect to consider when working with such culture systems is the ratio of the culture volume to the total internal volume of the vessel. The ratio of working volume to total volume influences greatly the maximum yield of cells per milliliter from the culture. Therefore, for reproducible results, closed culture vessels should always be filled to the same extent (and not more than half full). The reduction in yield in closed culture vessels that are more than half full is probably a result of a decreased supply of oxygen and reduced headspace volume for buffering the usual CO_2–bicarbonate system. This phenomenon is not encountered with open culture systems having a continuous gas supply.

During maintenance of large-scale cultures, nutrients like amino acids may be utilized differently in different culture systems ***(12)***. Check the amino acid concentration in your culture systems at different glucose levels and add the highly utilized ones.

3.3.4.3. Maintaining Cultures at Confluence

The following general points should always be considered when maintaining microcarrier cultures at confluence (they are not as critical for macroporous microcarriers as for surface microcarriers because the cell multilayers are protected inside the pores).

- pH: It is very important to maintain optimum culture conditions, especially pH. Once a drift in pH occurs (usually a decrease), cells tend to detach, even after pH has been returned to the optimal level. The most common cause of a decrease in pH at the later stage of the culture cycle is accumulation of lactate.

- Osmolarity: When pH is being controlled by addition of acid/base or buffers, it is important to avoid changes in osmolarity.
- Serum concentration: The most common way of maintaining cultures at confluence is to reduce the concentration of the serum supplement. A reduction from the usual 5–10% (v/v) supplement to 2–5% (v/v) is required for cells that are contact-inhibited for proliferation. For cells that continue to divide after confluence, consider lower concentrations of serum (down to 0.5% v/v).
- Medium replenishment: The concentration of nutrients should be kept as constant as possible, and toxic products of metabolism should not be allowed to accumulate.
- Antibiotics: When possible, decrease the concentration of antibiotics in the culture medium for long-term maintenance of confluent cultures.

3.4. Cell Quantification

Cell numbers can either be determined directly or indirectly.

1. Direct methods include cell enumeration by counting whole cells (attached cells trypsinized) or crystal violet stained nuclei ***(13)***. Others are determining total cell mass via protein or dry weight measurements. Cell viability is more difficult to quantify, but is often done via trypan blue (0.4%) or erythrosine B (0.4%) staining. An alternative is the MTT colorimetric method ***(14)***. Some cell culture systems cannot be sampled, and therefore numbers cannot be determined directly nor morphology studied via hematoxylin staining/fixation.
2. Indirect methods measure metabolic activities such as glucose or oxygen consumption, lactic or pyruvic acid production, CO_2 production, and increase in product concentration. These are quite useful during logarithmic growth but can be misleading later. Another possibility is to measure enzyme concentrations in the culture. One example is lactate dehydrogenase, which is not as dependent on the different phases of growth. A relatively new technique is to measure capacitance (Aber instruments). The advantage of this technology is that it measures cell mass even within macroporous carriers, so it continuously monitors cell growth within the bioreactor ***(4)***.

All these cell counts give a good picture of cell growth if you always use the same detection method. It is nearly impossible to compare different culture runs (performed with different cell lines) measured with different cell-counting methods. For example, Capiaumont ***(15)*** found an eightfold difference between the various cell-counting methods.

3.4.1. Direct Observation by Microscopy

Examining cells by microscopy is a vital part of microcarrier culture technique. For routine observation, the growth and condition of the cells can be assessed simply with phase contrast optics.

A small sample of evenly suspended culture is placed on a microscope slide and a coverslip is gently lowered over the sample. To avoid crushing the microcarriers, the coverslip should come to rest slightly above the slide; this can be accomplished by placing small pieces of broken coverslip between the slide and the coverslip. Haloes occasionally form with phase optics, but can be avoided by increasing the refractive index of the medium, for example, by the addition of serum or Ficoll™ 400 as an isotonic 30% (w/v) stock.

3.4.2. Counting Cells Released After Trypsinization

1. A 1-mL sample of evenly suspended culture is placed in a test tube.
2. After the microcarriers have settled, the supernatant is removed and the microcarriers briefly washed in 2 mL Ca^{2+}, Mg^{2+}-free PBS containing 0.02% (w/v) EDTA, pH 7.6.
3. When the microcarriers have settled, this solution is decanted and replaced by 1 mL of a 1:1 mixture of 0.25% (w/v) trypsin in Ca^{2+}, Mg^{2+}-free PBS and EDTA (0.02%, w/v) in Ca^{2+}, Mg^{2+}-free PBS.
4. The tube is incubated at 37°C for 15 min with occasional agitation.
5. The microcarriers are allowed to settle and the supernatant is transferred to another test tube.
6. The microcarriers are washed with 2 mL culture medium containing serum (5–10%, v/v), and the supernatant is pooled with the first supernatant.
7. The cell suspension is centrifuged (300 RCF, 5 min, 4°C).
8. The supernatant is discarded and the pellet resuspended in 2 mL Ca^{2+}, Mg^{2+}-free PBS containing 0.05% (w/v) trypan blue.
9. The concentration of cells in the suspension can be counted in a hemocytometer or electronic counter, and the concentration of the cells in the culture can be expressed per mL or per cm^2 of microcarrier surface area.
10. Including trypan blue in the resuspension solution allows cell viability to be estimated at the same time. A similar method can be used when using collagenase in combination with Cytodex 3 microcarriers.

3.4.3. Counting Released Nuclei

A simpler way of monitoring cell growth is to count released nuclei as described by Sanford et al. ***(16)***. In this method, modified by van Wezel ***(17)***, cells growing on the microcarriers are incubated in a hypotonic solution and nuclei released by lysis are stained by a dye in this solution.

1. A 1-mL sample of evenly suspended culture is centrifuged (22*gav*, 5 min) and the supernatant discarded.
2. The pelted microcarriers are resuspended in 1 mL 0.1 *M* citric acid containing 0.1% (w/v) crystal violet.
3. The contents of the tube are mixed well (e.g., with a Whirlimixer or by several traverses over a corrugated surface).

4. Incubate for 1 h at 37°C. Avoid evaporating the tube contents by using either a humidified incubator or by sealing the tube with plastic film.
5. After incubation, the contents of the tube are mixed as above and the released stained nuclei are counted with a hemocytometer.

Microcarriers in the sample do not interfere with the counting, and samples can be stored for up to one week at 4°C. This method of determining the number of cells in the culture is most accurate when cultures are evenly suspended and when culture conditions have avoided aggregation of microcarriers and cells.

3.4.4. Measurement of Cell Viability

Exclusion of dyes provides a convenient measure of cell viability ***(18)***. Trypan blue is the dye most commonly used since it can be used with both living material and material fixed with glutaraldehyde. Trypan blue is the only dye to give reproducible results both before and after fixation. The solution is prepared in PBS (4 mg/mL). Approximately 0.9 mL of diluted cell suspension is mixed with 0.1 mL of trypan blue. After 5 min at room temperature, the viable (unstained) and nonviable (stained) cells are counted in a hemocytometer. Counting can be done in conjunction with determination of cell concentration. Staining tests should be performed at pH 7.0.

3.4.5. Fixing Cells

When fixation and staining are necessary, e.g., for preservation of samples, cytochemistry, electron microscopy, etc., any of the usual cell culture fixation and staining procedures can be used with Cytodex microcarriers.

Methanol and ethanol are the most common fixatives.

1. Rinse the cells and microcarriers with warm PBS.
2. Prefix in 50% (v/v) alcohol in PBS.
3. After 10 min, replace the 50% alcohol by two 10-min changes of cool 70% (v/v) alcohol in PBS.
4. Finally, replace this fixative by 70% (v/v) alcohol in water. Alternatively, use a modified Carnoy's fixative (three parts methanol, one part glacial acetic acid and containing 2% (v/v) chloroform) after the cells and microcarriers have been rinsed in PBS (*see* **Note 14**).

Further processing of the fixed cells attached to the microcarriers depends on the purpose of the study. For example, using standard procedures, the material can be dehydrated in a graded series of alcohol solutions, cleared in xylene, and embedded in paraffin ***(19)***. For electron microscopy, cells growing on the microcarriers can be fixed, embedded, sectioned, and stained by the usual procedures.

Dehydration of the microcarriers in acetone instead of alcohol avoids the use propylene oxide, which has been reported to alter the surface of microcarriers

(20). When sections through cells attached to a solid surface are required, cells growing on Cytodex are easier and more convenient to process than cells growing on the surface of Petri dishes or coverslips. When processing microcarriers with cells attached for microscopy, remember that the time taken for each step should allow for penetration of the microcarrier matrix by the solute or embedding agent, which doubles the usual process times for embedding.

Pawlowski et al. ***(21)*** used the following procedures for preparing the cells in plate:

1. Cytodex with cells attached was allowed to settle onto coverslips coated with gelatin (1%) and fixed in half-strength Karnovsky's fixative (4% glutaraldehyde, 1% paraformaldehyde, 0.1 *M* sodium cacodylate buffer, pH 7.8, and 12 mg $CaCl_2$/100 mL) for 30 min.
2. The microcarriers were then washed briefly in 0.1 *M* cacodylate buffer (pH 7.8) and then postfixed in 1% osmium tetroxide for 1 h.
3. Samples were dehydrated in a graded ethanol series and critical point dried with carbon dioxide.
4. The coverslips were spatter-coated (gold coating device) with gold for 1.5 min at 25 mA and 1.5 kV ***(21)***.

3.4.6. Staining Cells

The most suitable routine procedures for staining cells growing on microcarriers use either Geimsa stain or Harris' hematoxylin. The latter stain can be used when better nuclear detail is required.

3.4.6.1. Staining With Giemsa Stain

1. Microcarriers with cells attached are rinsed in a small volume of warm PBS.
2. Fixation in 50% (v/v) methanol in PBS for 10 min is followed by dehydratration in a cool graded methanol series (70, 90, 95% v/v solutions in PBS) to absolute methanol, with about 5 min at each concentration.
3. The material is stained for 5 min in May-Grünwald's stain (alternatively Jenner's or Wright's stain) and for a further 10 min in dilute Giemsa (1:10 volumes in distilled water).
4. A brief rinse in water will reduce staining to the required intensity.

3.4.6.2. Staining with Hematoxylin

1. Cytodex with cells attached is rinsed in a small volume of warm PBS.
2. After fixation in 50% (v/v) methanol PBS for 10 min, the microcarriers are fixed for a further 10 min in cool 70% (v/v) methanol in PBS.
3. The fixative is removed, and 10 mL distilled water containing two to three drops of hematoxylin are added.
4. The material is left overnight at room temperature.
5. Material is rinsed in tap water for 20 min.

6. If desired, the cells can be counterstained in an aqueous solution of Eosin-Y for 30 s.
7. For permanent storage, the material can be dehydrated in a series of alcohol solutions (50, 70, 90, 95% v/v solutions in distilled water) and two changes of absolute alcohol.
8. The material can then be cleared in xylene and mounted.

3.5. Scale-up

An essential part of scale-up is the cellular multiplication at each scale-up step. If it is possible to inoculate at a low cell density, i.e., 10^4/mL and the final yield is 10^6/mL, the multiplication factor is 100-fold. The larger the multiplication factor at each step of scale-up to final production volume, the fewer steps are needed to achieve it. Maximizing the multiplication factor minimizes the number of steps/operations needed! It also affects the investments required. If the above applies, it is possible to inoculate a 100-L fermenter from a 1-L spinner. In the case of colonization, a 1-L spinner could inoculate a 1000-L fermenter.

Scaling up microcarrier cultures can be done by increasing the size of the vessel or by increasing the microcarrier concentration. Production units up to 6000 L have been achieved by increasing size. Factors that influence this strategy include reactor configuration and the power supplied by stirring. The height-to-diameter ratio is one important factor. For surface aeration, the surface area:height ratio should be 1:1. Variables that affect impeller function include shape, ratio of impeller to vessel diameter, and impeller tip speed. Larger impellers at lower speeds generate less shear forces. Marine impellers have been found to be more effective for cells. Pneumatic energy supplied via air bubbles or hydraulic energy in perfusion can both be scaled up without increasing power input.

Desirable features of a stirred microcarrier reactor include no baffles, curved bottom for better mixing, double jacket for heating and cooling, top-driven stirrer, and a smooth surface finish (electropolished). Scale-up by increasing the microcarrier concentration requires perfusion, which makes it necessary to have a separation device to keep the carriers in the reactor. A settling zone and a spin filter are both suitable for this task.

The limiting factor for higher cell concentrations is usually oxygen supply. It is difficult to use direct sparging of small bubbles with normal microcarriers in stirred tanks because the carriers may accumulate and float in the foam created. This can, however, be achieved by using large bubbles or by sparging inside a filter compartment. Other alternatives that increase oxygen supply include increasing surface aeration, perfusion via an external loop, and oxygenation devices/vessels *(22)*.

It is easier to increase oxygen supply via sparging in packed or fluidized beds as the cells are protected inside the macroporous carriers. Scale-up of fluidized beds is linear, i.e., the diameter to height ratio is the same, which keeps the fluidization velocity constant. As oxygen is supplied via direct sparging with microbubbles (fluidized bed with internal loop), there is a continuous gas-to-liquid transfer throughout the bed. This makes the technology extremely scalable. Airlifts have been scaled up in this way to 10,000-L scale!

3.5.1. Harvesting Cells

Removing cells from microcarriers is usually required when subculturing and scaling up as well as when large numbers of cells are required for biochemical analyses. Various methods can be used to remove cells from microcarriers, and it is important to choose procedures that minimize damage to cells (*see* **Note 15**).

3.5.1.1. Chelating Agents

Chelating agents such as EDTA can be used to remove certain epithelial and transformed cells from Cytodex.

1. After removing and discarding the medium, wash the microcarriers twice in Ca^{2+}, Mg^{2+}-free PBS containing 0.05% (w/v) EDTA (100 mL/g Cytodex).
2. Incubate microcarriers at 37°C with fresh Ca^{2+}, Mg^{2+}-free PBS containing 0.05% (w/v) EDTA (approx 50 mL/g Cytodex).
3. Stir the mixture continuously in the culture vessel at approx 60 rpm for at least 10 min.
4. Stirring speed may need to be increased or aspiration with a pipet may be required for some types of cell.
5. When the cells have detached from the microcarriers, neutralize the EDTA by adding culture medium (100 mL/g Cytodex).

In general, chelating agents alone are not sufficient to remove most cell types and are therefore usually used in combination with protoelytic enzymes. Long periods of exposure to EDTA may be harmful to some fibroblast strains, and EDTA alone rarely removes this cell type.

3.5.1.2. Proteolytic Enzymes

Enzymes are normally used for routine harvesting of a wide variety of cells from Cytodex microcarriers. In general trypsin, VMF Trypsin (Worthington), Pronase® (Sigma), or Dispase® (Boehringer) are used with all microcarriers and collagenase is used in combination with Cytodex 3. Trypsin is the most commonly used general protease, although Pronase has advantages for harvesting cells from primary cultures, and Dispase can be used for cells that are sensitive to trypsin.

Trypsin

1. Stirring is stopped and the microcarriers allowed to settle.
2. The medium is drained from the culture, and the microcarriers are washed for 5 min in Ca^{2+}, Mg^{2+}-free PBS containing 0.02% (w/v) EDTA, pH 7.6. The amount of EDTA-PBS solution should be 50–100 mL/g Cytodex.

3. The EDTA-PBS is removed and replaced by trypsin-EDTA.
4. Incubate at 37°C with occasional agitation.
5. After 15 min, the action of the trypsin is stopped by addition of culture medium containing 10% (v/v) serum (20–30 mL medium/g Cytodex). An alternative method for inactivating the trypsin is to add soybean trypsin inhibitor (0.5 mg/mL). Chicken serum does not contain trypsin inhibitors (*see* **Note 16**).

The procedures for harvesting cells with Dispase or Pronase are similar to those used for trypsin. The activity of Dispase is not inhibited by serum, and thus harvesting must be accompanied by thorough washing of the cells.

Collagenase: Standard cell culture harvesting procedures using trypsin and chelating agents alter cell viability and remove large amounts of surface-associated molecules from the cells ***(23)***. If subsequent studies require intact cell membranes, or if rapid harvesting with maximum yields without impairment of cell viability is required, an alternative method of harvesting must be used.

Collagenase digests the culture surface rather than the surface of the cell. Thus, cells harvested with collagenase are generally more viable and have greater membrane integrity than those harvested with trypsin. This method can only be applied for collagen coated beads like Cytodex 3.

1. Stirring is stopped and the microcarriers allowed to settle.
2. The medium is drained from the culture and the microcarriers washed for 5 min in two changes of Ca^{2+}, Mg^{2+}-free PBS containing 0.02% (w/v) EDTA, pH 7.6 (50 mL PBS/g Cytodex 3). Standard PBS can be used instead for this step if chelating agents are to be avoided.
3. The PBS is removed and replaced by collagenase solution. Approximately 30–50 mL of this solution should be used per gram of Cytodex 3.
4. The microcarriers are then mixed well in the collagenase solution and incubated with occasional agitation at 37°C.
5. After approx 15 min, the collagenase solution is diluted with fresh culture medium (50 mL medium/g Cytodex 3), and any cells remaining on the microcarriers are dislodged by aspiration with a pipet or by gentle agitation.
6. The detached cells can be separated from the microcarriers as described in **Note 17**.

3.5.1.3. Hypotonic Treatment

Incubation in hypotonic solution can be used for harvesting cells that do not have strong adhesion properties, e.g., some established and transformed cell lines. The osmotic shock associated with the hypotonic solution causes the cells to adopt rounded morphology, and they can then be shaken from the microcarriers.

1. Cytodex microcarriers with cells attached are washed twice in hypotonic saline (8 g NaCl, 0.4 g KCI, 1 g glucose in 1 L distilled water).
2. Incubate in fresh hypotonic saline (50 mL/g Cytodex) at 37°C for 15 min with gentle agitation.

Lai et al. ***(70)*** used hypotonic treatment to harvest Chinese hamster ovary CHO cells from Cytodex 1. Cell recoveries are usually less with this method than when enzymes are used. However, an advantage of using hypotonic saline is that harvesting does not involve exogenous protein.

3.5.1.4. Cold Treatment

Incubation at low temperatures causes many types of cells to detach from culture surfaces. When Cytodex microcarriers with cells attached are incubated in culture medium without serum at 4°C for 8 h, a significant proportion of cells detach. The sudden fall in temperature associated with a change from warm to cold culture medium often leads to a more rounded cell morphology, and the cells can then be gently shaken from the microcarriers. However, the use of temperature shifts for harvesting cells is generally associated with low viability, and the method is best reserved for established cell lines when other methods are not desirable.

3.5.1.5. Sonication

Sonication alone cannot be used to harvest intact cells from Cytodex microcarriers. However, in combination with the above methods, low-intensity sonication can increase cell yields. Sonication can be used to rupture cells and leave membrane fragments attached to intact microcarriers ***(24)***.

3.5.2. Harvesting from Macroporous Microcarriers

All reagents and procedures described in the sections above can also be used for Cytopore and Cytoline. Incubation times and protease concentration must be optimized for these microcarriers because the cells grow inside the pores in multilayers. Quantitative removal of cells from a completely dense culture will therefore not be possible; cell loss will be 10–30% of the total cells. If the culture is in the exponential growth phase and not confluent, you will get most of the cells out of the pores. The ideal situation for inoculating macroporous microcarriers is to use cells growing in suspension as well.

3.5.3. Separating Detached Cells from Microcarriers

3.5.3.1. Differential Sedimentation

Recovering cells by differential sedimentation takes advantage of the fact that cells and microcarriers sediment at different rates. For routine harvesting and subculturing when maximum recovery is not required, differential sedimentation is the simplest way of obtaining a preparation of cells essentially free from microcarriers.

1. After completing harvesting, culture medium is added (50–100 mL/g Cytodex) and the microcarriers allowed to settle.

2. After approx 5 min, the culture vessel is tilted to an angle of 45° and the cells collected into the supernatant.
3. Better recovery can be achieved if the microcarriers are washed once more with medium and the supernatant collected. Pooled supernatants can be used directly to inoculate the next culture (*see* **Note 18**).

Using these techniques, it is possible to recover more than 80% of cells in the harvest suspension. A short period of centrifugation (200 *gav*, 2 min) can be used to hasten sedimentation. If greater recoveries are required, use filtration.

3.5.3.2. Filtration

Filtration can be used when it is important to obtain very high recoveries of harvested cells without contamination from microcarriers. Any sterilizable filter with a mesh of approximately 100 µm that is nontoxic for animal cells is suitable (e.g., nylon or stainless steel filters). Sintered glass filters may also be used, but full recovery of cells may not be possible with such a filter.

3.5.3.3. Density Gradient Centrifugation

Provided a difference in density exists between the cells and the microcarriers, density gradient centrifugation can be used to obtain a preparation of cells free from microcarriers. Manousos et al. ***(20)*** used discontinuous density gradient centrifugation in Ficoll-Paque™ (density 1.077 g/mL) to achieve efficient separation with no contamination of the cells by microcarriers. Ficoll-Paque, supplied sterile, ready for use by GE Healthcare can be used for this application.

3.5.3.4. Fluidized Bed Separation

A fluidized bed system can be used to separate cells or virus from microcarriers or microcarrier with cells. The best reactor design would be a conical shape. The fluid stream comes from the bottom and harvest occurs from the top. A top filter with mesh size around 100 µm for Cytodex and Cytopore or 1 mm for Cytoline reduces the risk of losing microcarriers from the system.

3.5.3.5. Vibromixer

A vibromixer consists of a disk with conical perforations, perpendicularly attached to a shaft moving up and down with a controlled frequency and amplitude. Increasing the amplitude and the frequency of the translatory movement increases the turbulence in the reactor (the shaft is sealed with a membrane or bellows). Vibromixers do not require a dynamic sealing, are a closed system, and are therefore considered as very safe regarding containment. Mixing is efficient and gentle (low shear force). Vibromixers have found application in the production of tetanus toxin and animal cell culture.

At larger scale, the bioreactor contents can be emptied into a special harvesting reactor developed by Van Wezel (available from B. Braun). This vessel is divided into two compartments by a stainless steel 60- to120-μm mesh filter. The upper compartment contains a Vibromixer, a reciprocating plate with 0.1- to 0.3-mm holes moving at a frequency of 50 Hz. The microcarriers are collected on top of the mesh. Washing is done by adding washing buffer and draining it through the mesh. Prewarmed trypsin is added so that it just covers the microcarriers and is left for some minutes (depending on the cell line). The Vibromixer is then used for a short while to help detach cells. After detaching, the cells separate from the used carriers by draining through the mesh. Additional washing will improve the yield. The filters work most efficiently with hard carriers (plastic, glass) that do not readily block the mesh. It is also possible to only detach the cells and to transfer the whole mixture of used beads and cells to the new reactor.

3.5.4. Subculturing Techniques

Cells harvested from one microcarrier culture can be used directly to inoculate the next culture containing fresh microcarriers. Transferring a few microcarriers in the inoculum from the previous culture has no effect on subsequent culture development.

3.5.4.1. Colonization

For one subculture cycle, it is possible when scaling up to harvest cells from the microcarriers with trypsin, inactive the trypsin with medium containing serum, and then to add fresh microcarriers. In this way, the culture contains old and new microcarriers.

An alternative method for scaling up is to simply add fresh microcarriers when the culture approaches confluence. Manaus et al. ***(20)*** demonstrated that adding a further 1 mg of microcarriers per milliliter of culture prolonged the life of RD cell cultures and improved production of oncornavirus. The success of this method of scaling up depends on the ability of cells to move from the confluent microcarriers to inoculate the fresh ones.

3.5.4.2. Bead-to-Bead Transfer

Cultures of cells that do not attach very well to carriers or that detach easily during mitosis can be scaled up just by diluting the microcarrier culture with fresh microcarriers and more media ***(25,26)***. A 1:1000 dilution has been achieved with CHO cells.

If transfer by adding microcarriers does not work, Landauer ***(27)*** described a method for enhancing bead-to-bead transfer by releasing a certain portion of cells for recolonization.

3.5.5. Suspension Cells for Inoculation

Cells that grow in suspension can first be scaled up by increasing the volume of the suspension cultures. During final production, cells can be immobilized at high densities in carriers and the product continuously harvested via perfusion.

3.5.6. Reusing Microcarriers

Using microcarriers for more than one culture/harvest cycle is not recommended. The reuse of surfaces for cell culture requires alternate washing in strongly acidic and basic solutions. These steps are required to remove the debris remaining after the harvesting steps. Such procedures for Cytodex 1 are not recommended since extreme pH may alter both the microcarrier matrix and the degree of substitution. Used microcarriers can be washed in sterile PBS directly after harvesting and used for a further culture step, but attachment of cells is poor and yields are less than 70% of those obtained with fresh microcarriers. Reuse of microcarriers for a third culture step has not been feasible with all cell types tested. For some cell strains, reusing microcarriers is impossible. Cytodex 3 microcarriers cannot be reused when the cells have been harvested by enzymatic methods. For large-scale production, it is more costly to make a cleaning validation procedure for the reuse of microcarriers than to use new ones.

3.5.7. Waste

Disposal of microcarriers is very much a local issue, and costs vary according to country or region. For example, it can be advantageous to dissolve the microcarriers rather than incinerate them.

3.5.7.1. Dissolving Microcarriers

It is possible to dissolve Cytodex because it is made of dextran. Two alternatives are:

1. For smaller scale: dextranase from Sigma (500 U/mg). Use 10 mg dextranase for 50 mL microcarrier. Incubate at 37°C and pH 6.0 for 20 min. For large scale: dextranase 50 L from NovoNordisk (Novozymes) with 50 KDU (dextranase units). Concentration is 5 mL in 1 L. In our experiments, the concentration of Cytodex was 53 g/L dry weight, incubation time 2 h. Temperature and pH (pH optimum is 4.5–6.0) are optimized, but not the other concentrations, stirring, and incubation time.
2. It is possible to hydrolyze Cytodex with strong acids, but not inside stainless steel reactors. Drain the reactor contents first. Neutralize dissolved Cytodex with sodium hydroxide and dispose of the whole amount of liquid.

Microcarrier Cell Culture Technology

Theoretically, you can dissolve Cytopore with cellulase, but this microcarrier is so strongly crosslinked that this is hard to do in practice. The concentration of cellulose would be 50 U/mL cellulose solution. Add 50 mL for 20 mL of microcarrier.

Cytoline is made of high-density polyethylene and cannot be dissolved.

3.5.7.2. Incinerating Microcarriers

Let the microcarriers sediment, pump them to a collection vessel, and send to a waste treatment plant for incineration.

3.6. Example: Vaccine Production on Microcarriers in Disposable Wave Bioreactor

Procedure for the cultivation of Vero cells on Cytodex 1 microcarriers in a 2-L wave bag followed by a poliovirus culture (**Fig. 1**):

1. Prepare 3 g sterile Cytodex 1 microcarriers in 950 mL Serum Containing Medium 199 (SC-M199) (concentration in bag: 3 g/L).
2. Bring the 950-mL Cytodex suspension in the 2-L wave bag to stabilize the medium.
3. Set the rocker at 17 rpm and angle with a temperature of 37°C and an air/5% CO_2 flow of 0.200 L/min.
4. Trypsinize the Vero cell monolayer cultures and resuspend the cells in SC-M199 + 0.1% Pluronic F-68.
5. Inoculate the 2-L bag with 200×10^6 cells in 50 mL of SC-M199 + 0.1% Pluronic F-68 (cell concentration in bag: 0.2×10^6 cells/mL).
6. After inoculation rock the Vero cells and Cytodex 1 suspension for 5 min at 17 rpm and angle 7 to obtain a homogeneous suspension.
7. Take a sample for cell count. During sampling always set the rocker at 17 rpm and angle 7.
8. After 5 min set the rocker at 6 rpm and angle 3 for 30 min.
9. After 30 min set the rocker for 2 min at 12 rpm and angle 6.
10. Repeat **steps 7** and **8** during the first 3 h of incubation.
11. After 3 h of incubation check whether most of the Vero cells are attached to the microcarriers and set the rocking speed at 10 rpm and angle 5.
12. Set the rocking speed 1 d after inoculation at 17 rpm and angle 7.
13. Refresh 500 mL of medium at day 3.
14. At day 4 let the microcarriers settle by stopping the rocker and wash the microcarriers twice with Virus medium (Medium 199 without serum).
15. Fill the wave bag with virus medium up to a working volume of 1 L.
16. Set the rocking speed at 17 rpm and angle 7 at a temperature of 36°C.
17. Inoculate the suspension with poliovirus with a multiplicity of infection (number of virus particles per cell) appropriate for the virus of choice.
18. Harvest the culture when the cythopathic effect is complete and the cells are detached from the microcarriers.

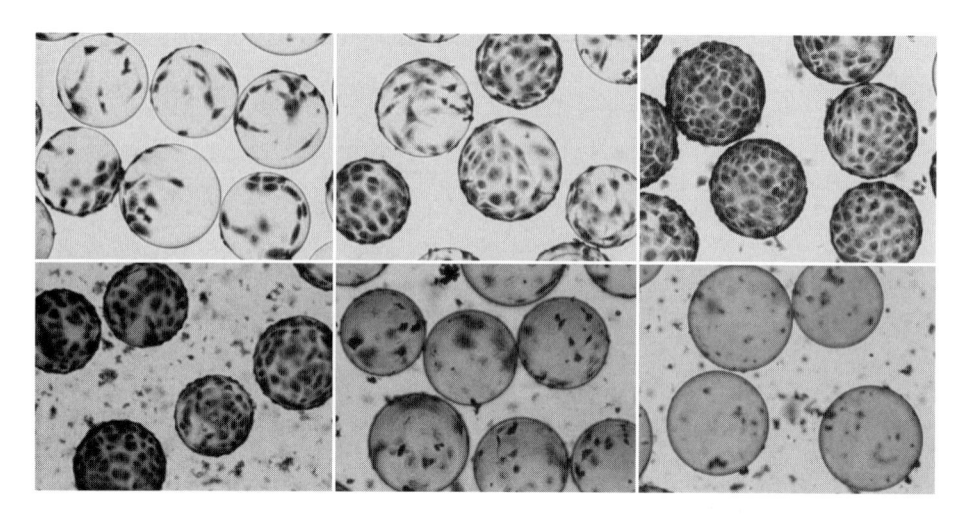

Fig. 1. Cultivation and infection of Vero cells on Cytodex 1 in wave bags. Upper three pictures are, from left to right, day 1, day 3, day 5 of Vero cell cultivation. Lower three pictures are, from left to right, day 1, day 3, day 4 postinfection with poliovirus. (With kind permission of NVI, The Netherlands.)

3.7. Troubleshooting

3.7.1. Stirred Microcarrier Cultures

Some problems may arise when working with stirred microcarrier cultures for the first time. The following list summarizes typical areas of difficulty and the most likely solutions. These points also form a useful checklist when culturing new types of cells.

1. Medium turns acidic when microcarriers added.
   - Check that the microcarriers have been properly prepared and hydrated.
2. Medium turns alkaline when microcarriers added.
   - Gas the culture vessel and equilibrate with 95% air, 5% CO_2.
3. Microcarriers lost on surface of culture vessel.
   - Check that the vessel has been properly siliconized.
4. Poor attachment of cells and slow initial growth.
   - Ensure that the culture vessel is nontoxic and well washed after siliconization.
   - Dilute culture in PBS remaining after sterilization and rinse microcarriers in growth medium.
   - Modify initial culture conditions, increase length of static attachment period, reduce initial culture volume, or increase the size of the inoculum.
   - Check the condition of the inoculum and ensure it has been harvested at the optimum time with an optimized procedure.

- Eliminate vibration transmitted from the stirring unit.
- Change to a more enriched medium for the initial culture phase.
- Check the quality of the serum supplement.
- If serum-free medium is used, increasing attachment protein concentration may be necessary (fibronectin, vitronectin, laminin).
- Check for contamination by mycoplasma.

5. Microcarriers with no cells attached.
   - Modify initial culture conditions, increase length of static attachment period, reduce initial culture volume.
   - Improve circulation of the microcarriers to keep beads in suspension during stirring.
   - Check the condition of the inoculum, especially if it is a single cell suspension.
   - Check that the inoculation density is correct (number of cells/bead).
6. Aggregation of cells and microcarriers.
   - Modify initial culture conditions, reduce the time that the culture remains static.
   - Increase stirring speed during growth phase, improve circulation of microcarriers.
   - Reduce the concentration of serum supplement as the culture approaches confluence.
   - Reduce the concentration of Ca^{2+} and Mg^{2+} in the medium.
   - Prevent collagen production by adding proline analogs to the culture medium.
7. Rounded morphology of cells and poor flattening during growth phase.
   - Replenish the medium.
   - Check the pH and osmolality of the culture medium.
   - Reduce the concentration of antibiotics if low concentrations of serum are used.
   - Check for contamination by mycoplasma.
8. Rounding of cells when culture medium is changed.
   - Check temperature, pH, and osmolality of replenishment medium.
   - Reduce the serum concentration.
9. Cessation of growth during culture cycle.
   - Replenish the medium or change to a different medium.
   - Check that pH is optimal for growth.
   - Regas the culture vessel or improve supply of gas.
   - Reduce stirring speed.
   - Check for contamination by mycoplasma.
10. Difficulties in controlling pH.
    - Check that the buffer system is appropriate.
    - Improve the supply of gas to the culture vessel, lower the concentration of CO_2 in the headspace, or increase the supply of oxygen.
    - Improve the supply of glutamine, supplement the medium with biotin, or use an alternative carbon source, e.g., galactose.
11. Difficulties in maintaining confluent monolayers.
    - Check that pH and osmolality are optimal.

- Reduce the concentration of serum supplement.
- Improve the schedule for medium replenishment.
- Reduce the concentration of antibiotics.
- When culturing cell lines that produce proteases in a serum-free medium, it may be necessary to add protease inhibitors to prevent the cells from detaching (CHO cells have been shown to secrete proteases!).

12. Broken microcarriers.
    - Ensure that dry microcarriers are handled carefully.
    - Check the design of the culture vessel/impeller and ensure that the bearing is not immersed in the culture.
13. Difficulty in harvesting cells from microcarriers.
    - Ensure that the carriers have been washed extensively together with mixing.
    - Check that approximately the same amount of protease (U/cell) is used as when harvesting from flasks.
    - Check that the trypsin has not been thawed for too long (loss of activity).
    - Check that sufficient shear force is used in addition to trypsinization.
14. Microcarriers float in foam because of sparging.
    - Reduce the serum/protein concentration as much as possible.
    - Add pluronic F68 to decrease foaming.
    - Add polymers to increase viscosity.
    - Aerate via silicone tubing (Diesel, bubble-free aeration), via spin filter (New Brunswick, Celligen), via external loop (vessel, hollow fiber).

3.7.2. Fluidized Bed

1. Culture medium too acidic.
    - Expel CO_2 by using a sparger that creates large bubbles.
    - Add sodium hydroxide to titrate the pH, observe the osmolarity.
    - Try to increase buffer capacity.
    - Optimize the culture medium and oxygen support to avoid production of lactic acid.
2. Bridging of microcarriers.
    - Use higher circulation rates. This shortens contact time between the microcarriers. Bed expansion should be between 150 and 200%.
3. No attachment.
    - Initial cell density too low.
    - pH level and dO_2 concentration were incorrect during inoculation.
    - Contact time between cells and the microcarriers was too short. More time is needed in packed bed mode.
    - Microcarriers were not washed properly.
    - Microcarriers were not equilibrated.
    - Cell inoculum was in stationary phase and not in the exponential growth phase.

4. Oxygen supply too low.
   - Use microsparging technique with a sparger that creates small bubbles (pore size around 0.5 μm).
   - Use pure oxygen for gassing. Cells inside the pores of the macroporous microcarriers are protected against the toxicity of oxygen.
   - Increase the circulation rate.

4. Notes

1. It is a big advantage if the carriers are supplied dry. The exact amount needed can be weighed and then prepared *in situ* by autoclaving. If a mistake is made during preparation, the carriers can be resterilized. Presterilized microcarriers always have to be handled under sterile conditions and can normally not be resterilized by autoclaving.
2. When hydrating Cytodex 3, initial surface tension may occasionally prevent wetting and sedimentation of the microcarriers. Should this occur, add Tween 80 to the PBS used for the first hydration rinse (two to three drops, Tween 80/100 mL PBS). Alternatively autoclave followed by rinsing and finally autoclave to sterility. Cytodex 3 does not swell to the same extent as Cytodex 1.
3. At large scale, it is advantageous to prepare the carriers directly inside the reactor because of the problem of heat transfer into vessels standing in autoclaves. The correct amount of carrier is added to an already cleaned and sterilized reactor. A volume of distilled water large enough for the microcarriers to swell properly (hydrated microporous microcarriers) and be covered after settling is then added to the fermenter. (Note that for soft microcarriers, the swelling factor in water is much larger than in PBS!) The reactor is then closed and the vessel resterilized while the carriers are stirred.
4. Cytopore is very stable, and even three autoclavings will not affect cell culture performance. The carriers can be sanitized with 70% ethanol. They also withstand 1% NaOH solution.
5. Carriers for packed or fluidized beds are also normally prepared inside the reactor. At small scale, the entire reactor can be sterilized by autoclaving. At larger scale, carriers are sterilized *in situ*.
6. Make every attempt to ensure that the culture pH is within the limits optimal for cell attachment (usually pH 7.0–7.4). Alkaline pH decreases cell attachment. Gas mixtures used during the initial stage of culture should be allowed to exchange with the culture medium before inoculation. These factors are particularly important at large culture volumes (more than 500 mL) when it takes longer for equilibration. Small culture volumes (500 mL or less) can be equilibrated by incubating the culture vessel containing medium at 37°C and in an atmosphere of 95%:5% CO_2.
7. It is a general cell culture phenomenon that the survival and growth of cells depends to a large extent on the inoculation density and conditioning effects. These conditioning effects are dependent on the density of the culture; a low density leads to relatively poor growth. With respect to anchorage-dependent cells, one of the most critical parameters at inoculation is the number of cells per cm^2 of culture

surface area. Since the efficiency of cell attachment to Cytodex is similar to that observed in Petri dishes, the microcarrier cultures should be inoculated with approximately the same number of cells per cm^2 as when starting other types of monolayer cultures. The number of cells per cm^2 used to inoculate the culture will depend on the plating efficiency of the cells (**Subheading 3.4.5.**). When inoculating a culture, it is generally necessary to use more primary or normal diploid cells than established cells.

8. Consider from the beginning how to create the large quantities of cells needed to inoculate a production scale fluidized bed. Imagine you need 1 kg of product and your anchorage-dependent cell line produces 100 mg/L carrier and day. A fluidized bed with 100 L of carriers has to be operated over a period of 100 d; 2×10^{11} cells would be needed to inoculate these carriers (2×10^6 per mL of carrier). This would require 2000 roller bottles of 850 cm^2.

   The problem could be overcome by:
   - Preparing the inoculum as aggregates in a stirred tank.
   - Scaling up on the same carriers via trypsinization.
   - Scaling up on different carriers (smooth carriers to macroporous).
   - Carrier-to-carrier transfer.
   - For suspension cells, the easiest way is always to create the inoculum in stirred tanks before transfer to macroporous carriers. The attachment phase after inoculation is finished after 2 h on macroporous carriers in fluidized bed applications. Fluidization can thus be started very soon after introducing the cells into the reactor.

9. The specific details of a modified initial culture procedure depend on the cell being cultured. A preliminary experiment with a stationary microcarrier culture in a bacteriological Petri dish will help estimate the time required for attachment to the microcarriers and will also reveal any tendency for cells and microcarriers to aggregate under static culture conditions.
10. For cells with low plating efficiency (less than 10%) or for cultures started with suboptimal numbers of cells, the culture volume can be maintained at 50% of the final volume for the first 3 d, after which fresh medium is added to the final volume. In this way, the population of cells can be cultured at greater densities during the period the culture is most susceptible to dilution.
11. The rate of stirring influences greatly the growth and final yield of cells, an effect related to the integrated shear factor. Slower stirring speeds reduce shearing forces on cells attached to the microcarriers, but if the rate is too slow, growth is reduced. This effect is mainly a result of inadequate gas diffusion, sedimentation, and aggregation. If stirring is too fast, less strongly attached cells (mainly mitotic cells) dislodge from the microcarriers and there is no net increase in cell number (such a phenomenon can be used to advantage for harvesting mitotic cells). Excessive stirring speed causes a general loss of cells from the microcarriers and thereby poor cell yields. When cultures are to be maintained at high densities, a slight increase in stirring speed will improve gas exchange and can improve the

supply of oxygen to the cells. However, if the culture is still dividing, an increase in stirring speed must be limited and should not result in detachment of mitotic cells. Some cells (mainly transformed cells and some fibroblast strains, e.g., chicken fibroblasts) tend to form aggregates during the later stages of culture. In these cases, a slight increase in stirring speed will reduce this effect.

12. With all types of stirring equipment, it is important to avoid sedimentation of the microcarriers, particularly during later stages of the culture. The accumulation of microcarriers that often occurs under the stirring axis in magnetic spinner vessels can be reduced by positioning the stirrer as close to the base as possible, while still leaving clearance for circulation of microcarriers. Such sedimentation does not occur with cultures stirred by a bulb-shaped rod, especially if contained in culture vessels with convex bases.
13. Persistent difficulties with controlling pH at high cell densities can be overcome by modifying the carbon source in the medium, by increasing the oxygen tension slightly, or by daily addition of glutamine in the presence of reduced concentrations of glucose. If excessive aggregation of the culture occurs and cannot be controlled by adjusting stirring speed, the calcium and magnesium concentrations in the medium can be reduced. A simple method for reducing these ions is to use mixtures of media that include suspension culture versions (e.g., Spinner MEM). A 50:50 mixture of this medium and the usual culture medium normally overcomes difficulties with aggregation without affecting cell growth.
14. Better preservation of morphology will be achieved when aldehyde fixatives are used. Use either 10% (v/v) formalehyde in PBS or 2–5% (v/v) glutaraldehyde in PBS and fix the material overnight at 4°C. Glutaraldehyde fixation results in the best preservation of morphology, and the fixed material can be used for electron microscopy studies.
15. The main task is to break the cell–surface and cell–cell interactions (the cells have to be harvested prior to confluence) and to round up the cells. As cell binding is dependent on divalent ions, the carriers have to be washed with citric acid or EDTA (0.2% w/v) containing buffers (PBS) to help detach the cells. If the media contains serum, α_1-antitrypsin (a trypsin-blocking protein abundant in serum) also has to be washed out. Microporous carriers require more extensive washing compared with solid carriers—normally two washings (optimize depending on the cell/microcarrier/medium) with a volume of washing solution equal to the sedimented microcarrier volume. As trypsin has a pH optimum close to pH 8.0, it may be necessary to increase the buffer capacity of the PBS used for harvesting to maintain a pH of 7.4 throughout the procedure. Note that the substrate concentration for trypsin increases drastically for sedimented beads compared with ordinary flask cultures as the cell density becomes very high. Try to maintain the same ratio of units of trypsin to number of cells. Normally, this means increasing trypsin concentration ten-fold (i.e., 0.25% w/v, but this may depend on the supplier) compared with harvesting flasks. To speed up harvesting, prewarm both the washing solution and the trypsin solution to 37°C. As trypsin activity is dependent on Ca^{2+} ions, it is advantagous to separate the EDTA washes and the final trypsinization. Screen

different suppliers for a suitable trypsin for a particular cell line. Do not use crystalline trypsin. Normally, it is better to use a trypsin contaminated with other proteases. This is usually more effective. Some shear force may have to be applied to quickly detach the cells. In spinners, stirring speed can be increased during this step or flasks shaken. At larger scale, the bioreactor can be emptied into a special harvesting reactor developed by Van Wezel (available from B. Braun). The trypsin has to be inactivated before the cells can be used as inoculate. Add either serum to the harvest or aprotinin, soybean trypsin inhibitor, if serum-free cultures are required. If serum is used, add it at the same time as the cells.

16. The success of harvesting with trypsin depends on complete removal of medium and serum from the culture and the microcarriers before the trypsin is added (serum contains trypsin inhibitors). pH is critical when harvesting with trypsin, and care must be taken to ensure that harvesting is done between pH 7.4 and 8.0. It is important to expose cells to trypsin for as short a period as possible. For sensitive types of cells, trypsinization at 4°C may be preferable, but the advantages of such a procedure must be weighed against the increased time for detachment and the occasional tendency for aggregation. With some types of cell, attachment to Cytodex is very strong, e.g., FS-4 human fibroblasts, and an additional wash with EDTA-PBS is required before the EDTA trypsin is added. Detachment of cells from the microcarriers can be enhanced by continuous stirring in the enzyme solution at a speed slightly greater than that used for normal culture.
17. Collagenase requires Ca^{2+} and Mg^{2+}, and therefore chelating agents should not be used during harvesting. When using procedures containing collagenase, inactivation of collagenase is usually not required. The dilution factor plus cysteine in the medium are sufficient to reduce the enzyme activity. If collagenase must be removed completely, washing cells by centrifugation is the most convenient method.
18. Alternatively, the products of harvesting can be transferred into a narrow container with a high head, e.g., test tube or measuring cylinder. After 5 min, the microcarriers settle to the bottom of the container and the cells can be collected in the supernatant.

References

1. van Wezel, A. L. (1967) Growth of cell strains and primary cells on microcarriers in homogeneous culture. *Nature* **216**, 64–65.
2. Nilsson, K., Buzsaky, F., and Mosbach, K. (1986) Growth of anchorage dependent cells on macroporous microcarrier. *Biol. Technol.* **4**, 989–990.
3. Young, M. W. and Dean, R. C. (1987) Optimization of mammalian-cell bioreactors. *Biotechnology* **5**, 835–837.
4. Swamy, J. (2003) Genzyme. The development of long term, microcarrier-based perfusion process for the production of recombinant proteins. IBC Conference, Cell Culture and Upstream Processing, San Diego.
5. Müller, D., Unterluggauer, F., Kreismayr, G., et al. (1999) Continuous perfused Fluidized bed technology, Increased productivities and product concentration. *Proceedings of the 16th ESACT Meeting: Animal Cell Technology: Products from Cells, Cells as Products* (Bernard, Griffiths, Noe, Wurm, eds.), pp. 319–321.

6. Reiter, M. (2004) Vaccine production with a contiuous perfusion system. Cell culture and upstream processing conference, Sept 13–14, Berlin, Germany.
7. Kasche, V., Probst, K., and Maass, J. (1980) The DNA and RNA content of crude and crystalline trypsin used to trypsinize animal cell cultures: kinetics of trypsinization. *Eur. J. Cell Biol.* **22**, 388.
8. Dehm, P. and Prockop, D. I. (1971) Synthesis and extrusion of collagen by freshly isolated cells from chick embryo tendon. *Biochim. Biophys. Acta* **250**, 358–369.
9. Griffiths, B., Thornton, B., and McEntee, I. (1980) Production of herpes viruses in microcarrier cultures of human diploid and primary chick fibroblast cells. *Eur. J. Cell Biol.* **22**, 606.
10. Meignier, B. (1979) Cell culture on beads used for the industrial production of foot-and-mouth disease virus. *Dev. Biol. Standard* **42**, 141–145.
11. Meignier, B., Mougeot, H., and Favre, H. (1980) Foot and mouth disease virus production on microcarrier-grown cells. *Dev. Biol. Standard* **46**, 249–256.
12. BlümL, G., Reiter, M., Zach, N., et al. (1998) Fluidized bed technology: Influence of fluidization velocity on nutrient consumption and product expression. *Proceedings of the 15th ESACT Meeting: New Developments and New Applications in Animal Cell Technology)*, pp. 385–387.
13. van Wezel, A. L. (1985) Monolayer growth systems: homogeneous unit processes, in *Animal Cell Biotechnology* (Spier, R. E. and Griffiths, J. B., eds.), Academic Press Inc., London. pp 266–281.
14. Mosmann, T. (1983) Rapid colorimetric assay for cellular growth and survival: application to proliferation and cytotoxicity assays. *J. Immunol. Methods* **65**, 55.
15. Capiaumont, J., Legrand, C., Gelot, M. A., Straczek, J., Belleville, F., and Nabet, P. (1993) Search for cell proliferation markers suitable for cell count in continuous immobilized cell bioreactors. *J. Biotechnol.* **31**, 147–160.
16. Sanford, K. K., Earle, W. R., Evans, V. J., et al. (1951) The measurement of proliferation in tissue cultures by enumeration of cell nuclei. *J. Nat. Cancer Inst.* **11**, 773–795.
17. van Wezel, A. L., Kruse, P. F., and Patterson, M. K., eds. (1973) Microcarrier cultures of animal cells, in *Tissue Culture: Methods and Applications*, Academic Press, New York, pp. 372–377.
18. Patterson, M. K. (1979) Measurement of growth and viability of cells in culture. *Methods Enzymol.* **58**, 141–152.
19. Horng, C.-B. and McLimans, W. (1975) Primary suspension culture of calf anterior pituiary cells on a microcarrier surface. *Biotechnol. Bioeng.* **17**, 713–732.
20. Manousos, M., Ahmed, M., Torchio, C., et al. (1980) Feasibility studies of oncornavirus production in microcarrier cultures. *In Vitro* **15**, 507–515.
21. Pawlowski, K. R., Loyd, R., and Przybylski, R. (1979) Skeletal muscle development in culture on beaded microcarriers, Cytodex 1. *J. Cell Biol.* **83**, 115 A.
22. Prokop, A. and Rosenberg, M. Z. (1989) Bioreactor for mammalian cell culture. *Adv. Biochem. Eng.* **39**, 29.
23. Anghileri, L. J. and Dermietzel, P. (1976) Cell coat in tumour cells—effects of trypsin and EDTA; a biochemical and morphological study. *Oncology* **33**, 11–23.

24. Lai, C., Hopwood, L. E., and Swartz, H. M. (1980) ESR studies on membrane fluidity of Chinese hamster ovary cells groown on microcarriers and in suspension. *Exp. Cell Res.* **130**, 437–442.
25. Kamiya, K., Yanagida, K., and Shirokaze, J. (1995) Subculture method for large scale cell culture using macroporous microcarrier. In: Beuvery, Griffiths, Zeiglemaker (eds.) *Developments toward the 21st Century.* Kluwer Academic Publisher, Dordrecht, pp. 759–763.
26. Dürrschmid, M., Landauer, K., Simic, G., BlümL, G., and Doblhoff-Dier, O. (2003) Scaleable inoculation strategies for microcarrier-based animal cell bioprocesses. *Biotechnol. Bioeng.* **83**, 681–686.
27. Landauer, K., Dürrschmid, M., Klug, H., Wiederkum, S., BlümL, G. and Doblhoff-Dier, O. (2002) Detachment factors for enhanced carrier to carrier transfer of CHO cell lines on macroporous microcarriers. *Cytotechnology* **39**, 37–45.

6

Cell Encapsulation

Holger Hübner

Summary

Encapsulation of cell cultures is a universally applicable tool in cell culture technology. Basically it can be used to protect cells against hazardous environmental conditions as well as the environment against disadvantageous effects triggered by the immobilized cultures. The three basic encapsulation systems, described here, are the bead, the coated bead, and the membrane-coated hollow sphere. Another two universally applicable methods to quantify the number of living cells immobilized in those encapsulation systems could be used.

Key Words: Immobilization; beads; coated beads; hollow microspheres.

1. Introduction

Several situations may make the immobilization of cell cultures necessary. The setup of a microenvironment for the cells, as favorable as possible, is probably one of the most important factors. The forced cell–cell contact within the capsules, which also leads to increasing concentrations of autocrine growth factors, is part of it. These growth factors can also be fed during immobilization to be available for the cells in a sufficient dose during their first, often critical growth phase. Ideally, the cells change directly to the exponential growth phase or stay in it, although being stressed by the immobilization procedure. This is an important aspect, as preferably cells of the exponential phase are to be used for experiments.

Using immobilized cells in vitro provides a further advantage, such as simplified handling of cells. The medium exchange can easily be performed in ways so that no cells are lost. Also, the cells are safely protected against shear stress in the immobilisates; hence any cultivating system with sufficient mixing performance can be chosen. Even direct aeration in combination with higher aeration rates, high agitation rates, or high stirrer speeds, usually only applicable for bacterial fermentations, can be used without impairment of

From: *Methods in Biotechnology, Vol. 24: Animal Cell Biotechnology: Methods and Protocols, 2nd Ed.* Edited by: R. Pörtner © Humana Press Inc., Totowa, NJ

cells. This protective action ceases as soon as the cells emigrate off the immobilisates or if the immobilisates are destroyed. It is therefore important to know the physicochemical characteristics of the immobilization materials and their interaction with media components and products of the cell metabolism. The determination of the mechanical stability is a parameter that can be determined fairly easily. Such a parameter gives information about any changes of the immobilisate during cultivation. In this context, using Ca alginate beads often leads to problems when the cells release organic acids such as lactate into the medium or when phosphate ions as buffering agents are existent in the medium and compete for binding with calcium ions. This difficulty led to the development of coated beads where additional polyelectrolyte layers, which are also stable in presence of other ions, are applied to the beads. The subsequent dissolving of the bulb core could be done without any problem, and the production of pure polyelectrolyte immobilisates without a solid bead core was a consequent improvement.

The preceding methods produce immobilisates meant to stay mechanically intact as long as possible, ideally forever. They are not applicable for releasing cells slowly, e.g., to colonize destroyed cutaneous areas with intact cells. Here a substance has to be used which is decomposed by the cells within a certain time. A typical example for the production of those immobilisates for the burn medicine is based on the use of fibrinogen, also existing naturally in the human body.

In case the immobilized cells are not autologous but have to be used in vivo anyway, it must first be ensured that the cells are protected against nonspecific immune defense of the recipient and then that no direct immune response of the recipient is mobilized. This can be caused by the immobilization material as well as by emigrating immunogenic cell products or cell fragments.

Because any immobilization means stress for the cells and the determination of vital cells in the immobilisates cannot be carried out by simple sampling and counting, two further indirect determination methods are important: the 1-methyltetrazole-5-thiol (MTT) test and the resazurin test. Both methods take advantage of the conversion of colored substrates through dehydrogenases of cells. The sensibility can be increased by addition of diaphorase. During the MTT test a water-soluble substrate changes into an insoluble product, a formazan salt, which crystallizes within the cells. Hereby, cell membranes are destroyed by growing crystals and metabolization stops. The amount of gained product thereby is proportional to the number of vital cells in the gross sample. This test is only suitable if a sufficient amount of biological material is available and an aliquot can be taken, as all cells in the aliquot sample die during this experiment. If this is not the case, the cell number can also be determined by resazurin. The difference will be that the gained product, the resorufin, will also be water-soluble and will not injure the cells directly. A certain amount of additional work has to be carried out then to investigate the conversion of resazurin online. Nevertheless, the application of resazurin is not unproblematic *(1,2)*, contrary to the success described in literature. Resorufin reacts to resazurin spontaneously, whereas the reaction velocity is dependent on the pH value. During offline measuring, this leads to higher deviations in CO_2-buffered media as the pH value changes as a result of leaking CO_2 during the test. Further on, the cells will be injured during a longer exposition with resazurin (Jurkat about 72 h, Sf21 about 4 h), as the resazurin conversion acts as an energy-consuming secondary metabolism *(3)*. Finally, some cells, e.g., insect cells, are able to convert the produced resorufin into dehydrofurin, which adversely affects cells at lower concentrations.

Common polymers and appropriate encapsulation devices as well as their application have already been described *(4)*. Here, only the practical problems will be analyzed: any material used for cell cultivation has to be sterile. A common practice is autoclaving at 121°C for 15–30 min, when the chain length of the polymers is decreasing. Because this depends primarily on the time of heat exposition, it is important to keep this as constant as possible between different experiments. This also includes the heating and cooling phase of the autoclave.

In case cationic polymers are used, e.g., for polyelectrolyte capsules (**Subheading 3.3.**) or for additional polyelectrolyte membranes (**Subheading 3.2.**), it has to be assured that they do not adversely affect the cells. Polycations in general, even poly-L-lysine in solution, are cytopathogenic by nature, as they form irreversible complexes with molecules of the cytoplasm membrane, proteins, and DNA and RNA under physiological conditions *(5,6)*. Their interaction with proteins can be used for the protection of the cells, however. Adding pure serum, pure albumin, or other proteins to the immobilized cell culture provides an effective protection from polycations.

The salt concentration in polyelectrolyte solutions plays another role. Polyvalent oppositely charged ions (e.g., Mg^{2+} in sodium cellulose sulfate solution) should be avoided as they can lead to unintentional crosslinking. This always leads to an increase in viscosity, which is problematic for the most common droplet-generation methods.

Charge density, chain length, and branching of the polyelectrolyte play a decisive role in their adsorption structure. The salt concentration and pH value of the solution are also relevant (*see* **Fig. 1**). Generally flat adsorption will occur at high-charge density and/or medium molar mass of the polyelectrolytes (**Fig. 1A**). Low charge density and/or high molar mass as well as branched structures lead to partial adsorption (**Fig. 1B**). The last example is especially favored if several polyelectrolyte layers are to be applied for increasing the capsule stability. In extreme cases colloidal particles are combined to form aggregates

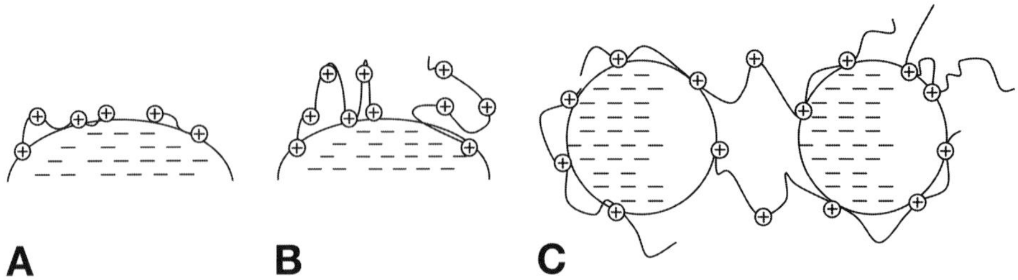

Fig. 1. Different forms of adsorption to charged surfaces.

by polyelectrolyte bond bridge growth (**Fig. 1C**) or as a result of a partial adsorption and charge masking (mosaic charging). In the majority of cases one can watch the forming of fluffy voluminous flocs that separate again during the subsequent cultivation.

Immobilization has been successful when the cells within the capsules can grow in tissue density before their growth seems to be stopped by space limitations. Cell densities of 10^8 cells per milliliter are standard. These values can only be reached, however, when the nutrient supply of the cells is adequate and the diameter of the capsules is not too large. Sufficient supply of dissolved oxygen, generally the compound with lowest solubility in the culture medium compared to other substrates, is normally the only critical factor. If an incomplete oxygen supply occurs, those cells in the center of the capsules suffer most. A chronic incomplete oxygen supply leads to a start of apoptosis for some cell types, followed by a release of apoptosis signals and finally necrobiosis of all cells in the affected capsule (**Fig. 2**). This incomplete supply can only be avoided by increasing the partial oxygen pressure in the culture system (e.g., aeration with pure oxygen). It cannot be prevented totally, however, as there is also a maximum oxygen tolerance for cells and therefore now the outer cells are affected. This means that it is important to find the ideal partial pressure range for the cells, while the maximum tolerable partial pressure is not to be exceeded and the critical partial pressure in the center of the cells must be maintained. This can only be guaranteed by thorough sizing of the capsules. Cells with a high consumption rate (Sf21 in $d = 1.2$ mm capsules) can only be cultivated in relatively small capsules. Cells with a lower consumption rate can be used for larger capsules (Jurkat in $d = 3$ mm capsules). In case the oxygen consumption rate of a cell culture is not clear, the conversion rate of MTT or resazurin can act as an indicator. In this context not only must the diameter be considered, but also the proportion between the surfaces and volumes of the immobilisates, if they are not spherical but lentoid shaped.

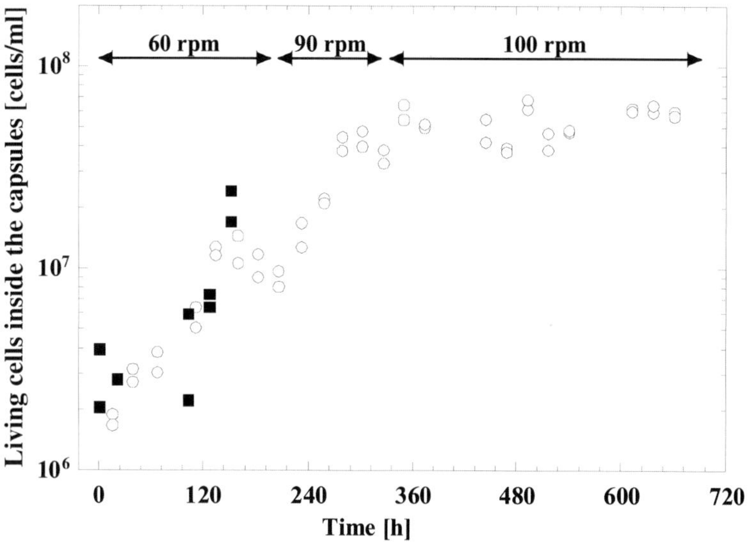

Fig. 2. Growth of immobilized baby hamster kidney (BHK) cells. An exponentially growing culture of BHK cells has been split and immobilized according to **Subheading 3.3.** (open circles) and **Subheading 3.4.** (black squares). Thereafter, the immobilized cells were transferred into shaker flasks and the living cell density was monitored by MTT test.

2. Materials

2.1. Solid Capsules

1. Alginate solution: dissolve 1.5% (w/v) sodium alginate (low viscosity) in 0.9% (w/v) NaCl solution or buffer (*see* **Note 1**). Make sure that the alginate dissolves completely before autoclaving. The sterile solution can be stored at room temperature for several weeks. Alginate will undergo partial hydrolysis during autoclaving, thereby transforming the low viscosity alginate to an unpredictable lower viscosity.
2. Precipitation bath: 100 mM $CaCl_2$ in deionized water or buffer (*see* **Note 2**). The volume of the precipitation bath should be at least five times the volume of the alginate solution.

2.2. Membrane-Coated Solid Capsules

1. Polycation solution: 0.05% (v/v) polyethyleneimine (PEI, MW 15,000–30,000, Sigma) in 0.9% (w/v) NaCl solution or buffer (*see* **Note 3**).
2. Polyanion solution: 0.03% (w/v) sodium alginate (low viscosity) in 0.9% (w/v) NaCl solution or buffer (*see* **Note 3**).
3. Sterile filter or autoclave both solutions; thereafter they can be stored at 8°C for a few days.

2.3. Membrane-Coated Hollow Spheres

1. NaCS solution: 3.5% (w/v) cellulose sulfate (NaCS, Euroferm) in 0.9% (w/v) NaCl solution or buffer. Check the pH and adjust (HCl or NaOH) to a neutral pH prior to autoclaving, if necessary (*see* **Note 4**).
2. pDADMAC solution: 2.2% (v/v) polydiallyldimethyl ammonium chloride (pDADMAC, low molecular weight, Sigma) in 0.9% (w/v) NaCl solution or buffer.
3. Autoclave both solutions, and use them soon after cooling to room temperature.

2.4. Solid Capsules from Fibrinogen

1. Fibrinogen solution: dissolve 100 mg/mL human plasma fibrinogen (Sigma), 8 mg/mL human fibronectin in 0.9% (w/v) NaCl solution. Add 45 plasma equivalent units (~4860 Loewy units) human Factor XIII (Enzyme Research Laboratories), and allow dissolving at 37°C for at least 45 min. Prepare this solution fresh from sterile products.
2. Thrombin solution: prepare 5 mg/mL $CaCl_2$ in 0.9% (w/v) NaCl solution and sterilize by autoclaving. Add 40 IU/mL human thrombin (Sigma) and allow dissolving at 37°C for at least 15 min. Prepare this solution freshly from sterile products (*see* **Note 5**).
3. Polymerization bath: add one volume Hostinert 216 (Hoechst AG, Germany) and one volume Miglyol 812 (*see* **Note 6**). These two liquids will not mix but will form two phases and can be stored at room temperature after autoclaving.

2.5. Quantification of Living (Immobilized) Cells

2.5.1. MTT Test

1. MTT solution: dissolve 5 mg/mL MTT (Sigma) in cell culture medium or buffer. Produce a larger batch of frozen aliquots. MTT solutions should be stored frozen and in the dark. It can be exposed to light for a short time and repeatedly thawed and frozen again, but this is not recommended.
2. SDS solution: 20% (w/v) sodium dodecylsulfate (SDS) in water. Store at room temperature.
3. Lysis buffer: 405 mL 2-propanol, 20 mL 1 *M* HCl, 75 mL SDS solution. Store at room temperature in a well-sealed flask in order to avoid evaporation of the 2-propanol.
4. Spectrometer is able to determine extinctions at 570 nm.

2.5.2. Resazurin Test

1. Resazurin solution: dissolve 40 mg/L resazurin (Sigma) in cell culture medium or buffer (*see* **Note 7**). Produce a larger batch of frozen aliquots. Resazurin solution should be stored frozen and in the dark. It can be exposed to light for a short time and repeatedly thawed and frozen again, but this is not recommended.

2.6. Determination of Mechanical Stability of the Capsules

1. Chatillon penetrometer equipped with a 3-mm-diameter plunger (J. Chatillon and Sons, New York).

2.7. Release of Cells From Membrane-Coated Hollow Spheres

1. Cellulase solution: prepare a 0.5% (w/v) endocellulase, 10% (w/v) Pluronic F-68 solution in the appropriate cell culture medium (depending on cell line). The optimum activity of most cellulases is around 37°C at pH values between 4 and 7; hence conditioned medium might be used.
2. Sterile-filter this solution if necessary.

3. Methods

The growth rate of immobilized cell cultures depends on maintaining suitable conditions for the cells rather than on the selection of the immobilization system. **Figure 2** shows a comparison of growth kinetics of baby hamster kidney cells in membrane-coated hollow spheres and solid fibrin capsules.

Initially both cultures did grow at maximum growth rate. The number of living cells in the hollow spheres decreased considerably between 120 and 216 h because of insufficient oxygen supply. This was countered by increasing the agitation rate to 90 rpm, thus improving oxygen supply for the cells. The second increase of the agitation rate did not show any effect because peak cell density, determined by shortage of space within the immobilisates, was reached already.

Fibrin capsules disintegrated rapidly as a result of the release of proteases from dying cells when the culture conditions became unsuitable, as outlined above. Although the disintegration of fibrin capsules was intended for a later stage of the process, at this moment it was not.

3.1. Solid Capsules

1. Prepare the alginate solution and the precipitation bath at a volume ratio of 1:10.
2. Spin down the cells of the cell culture and remove the medium.
3. Resuspend the cells in the alginate solution.
4. Use droplet generator device or a simple syringe to produce droplets and let them fall into the well-stirred precipitation bath.
5. Let the beads harden in the precipitation bath for at least 3 min.
6. Remove the precipitation bath and transfer the beads to the culture vessel of your choice containing an appropriate cell culture medium.
7. Incubate at cell culture conditions (*see* **Note 8**).

3.2. Membrane-Coated Solid Capsule

1. Prepare a batch of beads according to **Subheading 3.1., steps 1–5**.
2. Prepare a polyanion and a polycation solution of the volume of the precipitation bath for each additional layer, respectively.
3. Remove the precipitation bath.
4. Wash the beads carefully with buffer.
5. Remove the buffer and incubate the beads in the polycation 1 solution for 5 min.
6. Wash the beads carefully with buffer.

7. Remove the buffer and incubate the beads in the polyanion solution 1 for 5 min.
8. Repeat the steps 3–6 to add further double layers on your bead. Each additional double layer will increase the mechanical resistance, diffusion resistance, and molecular weight cut-off of the coated beads.
9. Wash the beads with buffer and transfer them to the culture vessel of your choice containing an appropriate cell culture medium.
10. Incubate at cell culture conditions (*see* **Note 8**).

3.3. Membrane-Coated Hollow Spheres

1. Prepare the NaCS solution and the pDADMAC solution at volume ratio of 1:4.
2. Spin down the cells of the cell culture and remove the medium.
3. Resuspend the cells in the NaCS solution (*see* **Note 9**).
4. Use droplet generator device or a simple syringe to produce droplets (*see* **Note 10**) and let them fall into the well-stirred precipitation bath containing the polycation pDADMAC.
5. Let the beads harden in the precipitation bath for at least 10 min, preferably 15 min, but not longer than 30 min (*see* **Note 11**).
6. Remove the precipitation bath and wash the hollow spheres very carefully with buffer at least 2 times in order to completely remove any traces of the polyanion. Keep in mind that polyanions are toxic for the cells.
7. Transfer the beads to the culture vessel of your choice containing an appropriate cell culture medium.
8. Incubate at cell culture conditions (*see* **Note 8**).
9. Change the medium after 4–8 h. Again, this is necessary to remove the last traces of the pDADMAC solution.

3.4. Solid Capsules from Fibrinogen

1. Connect two syringes in such way that their outlet flow merges in one short capillary, such as the DUPLOJECT system (Baxter).
2. Prepare the fibrin solution, the thrombin solution, and the polymerization bath at volume ratio of 1:1:100.
3. Heat up all solutions and the polymerization bath to 37°C.
4. Spin down the cells of the cell culture and remove the medium.
5. Resuspend the cells in the fibrin solution.
6. Produce drops that consist of equal volumes of fibrin solution and thrombin solution and let them fall into the well-stirred polymerization bath.
7. Let the beads harden in the polymerization bath for at least 30 min.

3.5. Quantification of Living (Immobilized) Cells

3.5.1. MTT Test

1. Add four volumes of MTT solution to one volume of the cell culture sample (*see* **Note 12**).
2. Incubate the MTT containing sample at cultivation conditions for 4 h. Ensure that the sample is well aerated and gently agitated throughout this period (*see* **Note 13**).

3. Add 20 volumes of lysis buffer. From now on the samples have to be sealed in order to avoid evaporation of 2-propanol (*see* **Note 14**).
4. Supersonicate the sample until all the formazan crystals are dissolved and the sample is homogeneous.
5. Centrifuge the sample to remove debris from the samples. The pellet should be plain white and contain no dark spots. If dark spots can be observed, these are very likely not yet dissolved formazan crystals. In this case repeat **step 4** of this protocol.
6. Determine the extinction (*see* **Note 15**) of the formazan in the supernatant using a spectrometer (wavelength 570 nm). For the necessary calibration curve, correlate the extinction vs sample cell density (*see* **Note 16**).

3.5.2. Resazurin Test

1. Add one volume of resazurin solution to 10 volumes of the cell culture sample.
2. Incubate under cell culture conditions.
3. Monitor the changes of the extinction in the sample using a fluorescence spectrometer (excitation wavelength 530–560 nm; emission wavelength 590 nm) (*see* **Note 16**).

3.6. Determination of Mechanical Stability of the Capsules

1. Take a sample of at least 10 capsules.
2. Place the capsules under the plunger of the penetrometer one by one and start the automated capsule compression process (*see* **Note 17** and **18**).

3.7. Release of Cells from Membrane-Coated Hollow Spheres

1. Remove old medium.
2. Add approx 10 volumes of the cellulase solution to one volume of capsules.
3. Incubate for 30 min under culture conditions. The optimum activity of most cellulases is around 37°C and pH 4–7, hence conditioned medium might used to prepare. Gentle agitation increases mass transport and hence the oxygen transport to the cells and the conversion rate of the cellulose.
4. Pipet the solution and capsules gently up and down a few times for further mechanical disintegration. Use a pipet tip with a diameter of about 1.0–1.5 times the capsule size.
5. Remove the large membrane fragments by sieving using an appropriate mesh (e.g., SpectraMesh 53 µm for capsules with a diameter of 500 µm).

4. Notes

1. The term “buffer” refers to any aqueous salt solution that can be used to wash your cell cultures with. However, make sure that it contains no polyvalent (metal) cations such as Mg^{2+}, Ca^{2+}, etc.
2. The optimum concentration of the alginate solution and the corresponding precipitation bath may vary with the alginate source. In most cases the alginate concentration should be around 1.0–1.5% (w/v) and 0.1–.5 *M* $CaCl_2$ for the precipitation bath. In general, use of the higher concentrated solution leads to beads with a higher mechanical resistance with a distinct superficial crust. This may lead to diffusion limitations of

nutrients. Furthermore, those capsules are generally less homogeneous on a supra-macromolecular level: many heterogeneous structures such as shafts and cavities will be formed inside the bead. Another strategy to increase the bead strength is to replace the $CaCl_2$ with $BaCl_2$ in the precipitation bath.

3. For the coating of Ca alginate beads, nearly any combination of polycations and polyanions can be used. Successfully applied other charged polymers are poly-L-lysine (PLL), poly-*N*-vinylamine (PNVA), poly-acrylic acid (PAA), and the polymers mentioned in **Subheading 3.3**.
4. All polymers undergo degradation if exposed to heat such as during autoclaving. However, NaCS is very prone to increased hydrolytic cleavage of the β-glycosidic bond between its carbohydrate subunits of the cellulose backbone. Even if the cell culture to be immobilized favors a pH other than 7, such as in insect cells, it is strongly recommended to adjust the pH of any NaCS solution to pH 7. Keep in mind that polycations have H^+ exchange properties like any ion exchange materials; hence the adjustment of the pH has to be checked again after 30 min.
5. The actual concentrations may vary from lot to lot. Generally, the homogeneity of the microstructure of the fibrin capsules depends mainly on the mixing of fibrinogen and thrombin solution. This can rarely be accomplished during the dropping process but happens inside the well-stirred polymerization bath. Hence, it is important that the fibrin polymerization reaction not proceed too quickly. This can be controlled by the thrombin concentration used. The mechanical strength of the capsules increases with rising initial fibrinogen concentrations. However, proteases released from dead cells may weaken the fibrin matrix. This effect can be slowed down by adding protease inhibitors to the fibrinogen solution and to the following culture medium.
6. Miglyol is a saturated oil swimming on top of Hostinert, a perfluorcarbon compound. Neither takes part in the reaction or mixes with the other or the aqueous drops that will eventually form the fibrin immobilisates. The purpose of the polymerization bath components is to form an inert environment where the clotting reaction can proceed until a homogeneous fibrin matrix will form. Thus, Hostinert can be replaced by any other liquid that is immiscible with water and oil, such as Galden and Foblin (Solvay Solexis), FC 40 (3M), or that used as an inert oxygen supporter in medicine. Likewise, Miglyol can be replaced by any other commercially available oil. However, it is highly recommended not to use oils containing unsaturated fatty acids, since those may form toxic peroxides.
7. As mentioned in the introduction briefly, resazurin is converted by the cells to resafurin, which in turn converts back to resazurin. In contrast to the pseudo-first-order behavior of the MTT conversion that leads to linear calibration curves, the resazurin test leads to the typical results of equilibrium reactions. The apparent conversion rate of this reaction slightly slows down as a result of increased resorufin concentrations and the in turn increasing speed of the backreaction. Unfortunately, this backreaction is pH dependent, too. At acidic pH values below pH 6, the backreaction is very slow and hardly affects the experiment. However, at a neutral or slightly basic pH value common for human and animal cell cultures, the speed of the backreaction increases

considerably. As long as the pH of the medium stays constant, one can deal with that problem. But if CO_2-buffered media are used, the evaporation of the dissolved CO_2 during measurement shifts the pH toward unwanted values and the rate of the back-reaction becomes unpredictable.

8. The term "culture conditions" refers to the temperature, humidity, and CO_2 content of the air under which a certain cell line grows best in the lab. The term "conditioned medium" refers to cell culture medium that has been used to grow cells in, but is not depleted from nutrients.
9. If the cells suffer from toxic effects caused by the polycation after the immobilization, this effect can be reduced by adding proteins to the cells. In this case suspend the cells in a small amount of serum or albumin solution before mixing into the NaCS solution.
10. Keep in mind that using a simple syringe for the droplet generation will inevitably lead to incorporation of cells into the membrane of the spheres. Some of these cells will protrude the membrane, leaving empty spaces and holes in the membrane after cell death. This may be desired because these pores enable the transport of even very large molecules, plasmids, or viruses across the membrane.
11. Other polycation–polyanion pairs can be used to produce these types of capsules. Because of the toxicity of polyanions, the inner polyion must be a polyanion, preferably a highly charged polyanion with a MW in the range of 10^5–10^6 and a pH-dependent charge. As the precipitation bath polymer, a polycation in the MW range of 10^3–10^6 is suited best. In most cases these polycations posses either a permanently charged quaternary ammonium group or a tertiary amine group.
12. Both the MTT- and resazurin-based methods to quantify the number of living cells in beads, coated beads, or hollow spheres make it necessary to produce a calibration curve with suspended cells. A calibration experiment with a new cell line in the range of $5e^5$ cells/mL–$2e^6$ cells/mL should produce a linear calibration curve. To apply the calibration results to the results obtained with the immobilized cells, one must determine the volume of the capsule as well as any remaining cultivation medium that dilutes the sample.
13. Especially when this kind of test is applied to immobilized cells, pay special attention not to limit the oxygen and nutrient supply of the cells throughout the incubation time. The accuracy of these tests depends greatly on optimal culture conditions of the cells. Frequently Eppendorf tubes are used incubation flasks. However, in Eppendorf tubes the cells or immobilisates sediment to the bottom, where they are no longer supplied with sufficient oxygen. This leads to conversion rates that are too low for the number of cells. To avoid this error, it is recommended to use open and gently agitated glass tubes for the MTT test instead. At this stage of the experiment, the sample does not need to be sterile. The few contaminating microorganisms of the environmental air that may fall into the glass tube hardly contribute to the MTT conversion, since the incubation time is too short for any considerable growth and the conversion of the MTT will kill the microorganisms fairly quickly. Another frequent error is the use of nutrient-depleted cell culture medium. To avoid this error, it is recommended to add about 25% fresh medium to the sample before adding the MTT solution.

In principle, the above is valid for the resazurin test, too. The only exceptions are experiments with slowly converting cells. Here one should maintain a sterile environment throughout the experiment. Contaminating microorganisms would be able to contribute a considerable amount to the resazurin conversion of the cells because of both the generally longer experimental time and the nontoxic nature of the involved compounds.

14. Most polymers used to form immobilisates, especially the here-described NaCS, precipitate if brought into contact with the 2-propanol of lysis buffer. The polymer precipitate forms a shell around the formazan crystals, dramatically increasing the time necessary to completely dissolve them. Waiting overnight is not recommended, because MTT converts slowly to formazan in aqueous solutions. Especially in the case of the NaCS spheres, it is recommended to treat them with warm SDS solution for about 15 min and homogenize them mechanically thereafter. The homogenate may be treated like a standard cell culture for the remaining steps of the protocol.
15. In the case of the extinction exceeding the linear range of the spectrometer and thus of the calibration curve, it is suggested to dilute the sample with a mixture of cell culture medium (one volume) and lysis buffer (four volumes). This mixture contains the same 2-propanol:water ratio as the sample and avoids slight deviations as a result of volume contraction effects if pure medium, water, or lysis buffer is used as a thinner.
16. For each cell line an individual calibration curve is necessary. However, for some cell types the conversion rate depends on the metabolic state of the cells caused by the previous culture conditions. For instance, hybridomas do change the number of mitochondria per cell. If hybridomas are short on oxygen or nutrients for a time, the number of mitochondria decreases.
17. The apparent stability of a solid capsule or apparent bursting force of a hollow sphere determined by any penetrometer device is a function of the plunger's speed: the higher the plunger's speed, the higher the apparent values. Unfortunately, there exists no valid formula to compare the apparent forces obtained from different plunger speeds.
18. Since penetrometers are quite expensive, one may choose a less accurate method for determining the bursting force of hollow spheres. Place one sphere on a balance and start compressing it with punching tool. Watch the increasing weight on the balance. The weight (m) shown at the moment when the sphere bursts can be used to calculate the bursting force (F) according to $F = m * g$.

Acknowledgments

The author would like to thank Anette Amtmann for technical assistance and Katja Beyerlein for her help with this script.

References

1. Batchelor, R. H. and Zhou, M. (2004) Use of cellular glucose-6-phosphate dehydrogenase for cell quantitation: applications in cytotoxicity and apoptosis assays. *Anal. Biochem.* **329**, 35–42.

2. Zhang, H.-X., Du, G.-H., and Zhang J.-T. (2004) Assay of mitochondrial functions by resazurin in vitro. *Acta Pharmacol. Sin.* **25(3)**, 385–389.
3. Maeda, H., Matsu-Ura, S., Yuji Yamauchi, Y., and Ohomori, H. (2001) Resazurin as an electron acceptor in glucose oxidase-catalyzed oxidation of glucose. *Chem. Pharm. Bull.* **49(5)**, 622–625.
4. Huebner, H. and Buchholz, R. (1999) Microencapsulation, in *Encyclopedia of Bioprocess Technology: Fermentation, Biocatalysis and Bioseparation.* (Flickinger, M. C. and Drew, S. W., eds.), John Wiley & Sons Inc., Hoboken, NJ, pp. 1785–1798.
5. Kabanov, V. A. (1994) Basic properties of soluble interpolyelectrolyte complexes applied to bioengineering and cell transformations, in *Macromolecular Complexes in Chemistry and Biology* (Dubin, P., Bock, J., Davis, R., Schulz, D. N., and Thies, C., eds.), Springer-Verlag, Berlin, pp. 151–174.
6. Kabanov, V. A., Yaroslavov, A. A., and Sukhishvili, S. A. (1996) Interaction of polyions with cell-mimetic species: physico-chemical and biomedical aspects. *J. Controlled Release* **39**, 173–189.

7

Tools for High-Throughput Medium and Process Optimization

Martin Jordan and Nigel Jenkins

Summary

There is an increasing need for high-throughput scale-down models that are representative of commercial bioprocesses, particularly in the field of animal cell biotechnology. In this chapter we describe two protocols for small-scale disposable plastic vessels: 125-mL Erlenmeyer shake flasks and 50-mL centrifuge tubes. We also describe two common applications for these scale-down platforms: (1) satellite cultures derived from conventional bioreactor runs for optimizing the feed strategy for fed-batch cultures and (2) testing multifactorial experimental designs for designing the components of a base medium.

Key Words: Cell culture; Chinese hamster ovary; high-throughput; fed-batch; media; factorial.

1. Introduction

The use of animal cells as the leading host system for the production of recombinant proteins requires the use of high-throughput technologies in order to quickly establish production processes. Process development timelines can only be met with intensive screening for high-producing clones followed by medium screening and process optimization experiments, typically carried out in small-scale culture systems. The classical spinner flask culture vessel is still widely used today *(1)*. However, spinner flasks and small-scale bioreactors need relatively large culture volumes (0.5–3 L) and are not ideal for a truly high-throughput platform. With the availability of (1) new shaking incubators, providing humidity and a CO_2-controlled environment, and of (2) sterile disposable culture vessels **(Figs. 1–3)**, shake technology (used for decades in microbiology research) has rapidly become a standard approach in process development for mammalian cells *(2,3)*.

From: *Methods in Biotechnology, Vol. 24: Animal Cell Biotechnology: Methods and Protocols, 2nd Ed.* Edited by: R. Pörtner © Humana Press Inc., Totowa, NJ

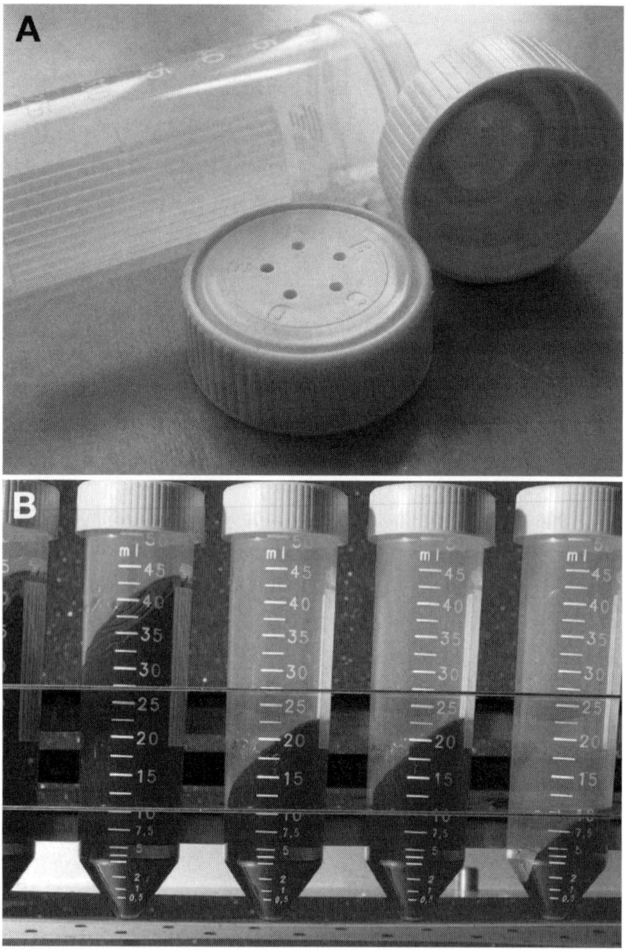

Fig. 1. Disposable 50-mL culture tubes. **(A)** Optimized filter caps with defined permeability for gases and humidity (similar caps are available for shake flasks). **(B)** Tubes with different volumes of culture, shaken at 200 rpm.

Two disposable culture vessels will be described in this chapter: polycarbonate Erlenmeyer flasks and polypropylene tubes. While the 125-mL Erlenmeyer flasks are well established in many companies, the 50-mL tubes are not well known by most users. Both culture vessels have many features in common and show similar performance characteristics. However, the 50-mL tube is the most flexible culture vessel available today **(Fig. 2)** because the working volume can be varied within the range 1–40 mL. A working volume of 5 mL is most convenient since it facilitates many experiments with only 100 mL of cell culture inoculation, and there is still enough culture volume available for sampling and

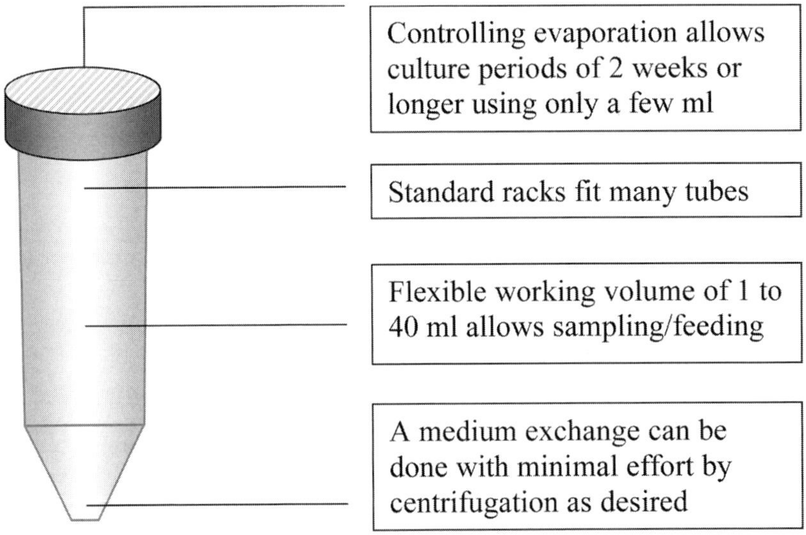
Controlling evaporation allows culture periods of 2 weeks or longer using only a few ml
Standard racks fit many tubes
Flexible working volume of 1 to 40 ml allows sampling/feeding
A medium exchange can be done with minimal effort by centrifugation as desired

Fig. 2. Advantages of the 50-mL disposable tube as a cell culture system.

analysis *(4)*. Typical shake flask sizes are 125 mL (with 30- to 40-mL working volume) or 250 mL (with 70- to100-mL working volume); both are available presterilized with vented caps for gas exchange.

2. Materials

2.1. Shake Flask Experiments

1. A shaking incubator with CO_2, temperature, humidity control, and with clamps for 125-mL shake flasks, e.g., from Kuhner A.G., Switzerland **(Fig. 3)**. Set the orbit radius (throw) at 2.5 cm (*see* **Note 1**), rotation at 130 rpm, temperature at 37°C, atmosphere of 5% CO_2 in air, and the relative humidity at 80% (*see* **Note 1**).
2. A centrifuge with a rotor arm and buckets for 50-mL centrifuge tubes.
3. Corning sterile polycarbonate shake flasks (125 mL) with vented caps to allow gas exchange and prevent contamination.
4. Basal medium (e.g., Dulbecco's modified Eagle's medium [DMEM]/F12).
5. A cell-counting device (e.g., Neubauer hemacytomer and microscope or an automated device such as the Beckman Vi-Cell).
6. For large experiments, software for experimental design and analysis, e.g., Design Expert from Stat-Ease *(5)*.

2.2. Shaken Tubes

1. A shaking incubator with CO_2, temperature, and humidity control with holders for 50-mL centrifuge tubes (e.g., from Kuhner A.G., Switzerland). Set the orbit radius

Fig. 3. 125-mL shake flasks in the Kuhner shaking incubator with temperature, humidity, and CO_2 control.

(throw) at 5 cm **(Note 1)**, rotation at 200 rpm, temperature at 37°C, atmosphere of 5% CO_2 in air, and relative humidity at 80%.

2. A centrifuge with a rotor arm and buckets for 50-mL centrifuge tubes.
3. Techno Plastic Products (TPP) 50-mL sterile polypropylene tubes with vented filter caps **(Fig. 1)**.
4. A cell-counting device (e.g., Neubauer hemacytomer and microscope or an automated device such as the Beckman Vi-Cell).

3. Methods

3.1. Multifactorial Experiments in Shake Flasks

Choosing the right factors to be tested in each experiment is recognized as the most difficult part of the experiment. The actual number of factors tested depends on the objectives of the experiment and the knowledge that is already available for a given cell line. A large number of factors (>10) is typically used to screen different media or clones, whereas for optimization purposes (e.g., dose-response curves and optimization of factors) typically <10 factors are evaluated. A factor can be one single chemical compound (e.g., glucose, glutamine) or a group of compounds (e.g., essential amino acids, vitamins) or physical parameters (e.g., pH, temperature). Many different factors have been described in the literature that have positive effects on productivity and/or cell growth ***(6)***, and a comprehensive literature search is recommended before embarking on this type of study.

1. Design the experimental treatments and flask plan. Although this can be done manually, it is easier to use specialist software for experimental design such as Design Expert from Stat-Ease (*see* **Note 2** and **ref.** ***5***). The experimental plan in **Table 1** shows an example of a factorial experiment testing five factors and their interactions, which uses 16 shake flasks.
2. Prepare concentrated solutions of each factor according to the flask plan and add the base medium to achieve a constant volume of 31.5 mL per shake flask. Incubate at room temperature overnight to check for sterility.
3. Take cells from the seed-train culture for inoculation at working day 0 (*see* **Note 3**). Centrifuge cells at 200*g* for 10 min at room temperature and resuspend cells in the base medium to make a 10-fold cell concentrate; e.g., for a cell seeding density of 0.2 million viable cells per mL make a concentrated solution of 2 million viable cells per mL. Inoculate each flask with 3.5 mL of this 10-fold cell concentrate to achieve a final volume of 35 mL per shake flask.
4. Incubate for the duration of the experiment in the shaking incubator (set at 130 rpm, 80% humidity, 5% CO_2, and 37°C). For experiments where only cell growth is measured 7 d is usually sufficient, but a longer period (e.g., 10–12 d) may be required to measure protein production or cell viability at the end of a culture.
5. Samples are taken under sterile conditions for cell counts and other measurements every 2–3 d for the duration of the experiment. For each sample (typically 1–2 mL of culture) perform cell counts, either manually using trypan blue staining or using an automated device such as the Beckman Vi-Cell.
6. If protein productivity data are important throughout the culture, cells are separated from the culture supernatant by centrifugation (200*g*, 10 min at room temperature) and either enzyme-linked immunosorbent assay or an alternative protein quantitation assay (e.g., Biacore) is performed on the supernatants. Metabolite data may also be measured at this stage, if required (*see* **Note 4**). We have found that up to 10 mL of samples can be taken from each flask without adversely affecting the culture.
7. Alternatively, in simple experiments protein and metabolite data are only measured on the final day of the experiment.
8. At the end of the experiment the quantitative data (e.g., cell counts, protein levels, and metabolite levels) are fed back into the Design Expert or equivalent software. Note that that each response requires a separate data column and is analyzed independently. An example of one such response is shown in **Fig. 4**, and a normal

Table 1
Experimental Plan and Viable Cell Results for a Multifactorial Experiment in Shake Flasks, Testing the Interactions Between Five Media Components (Factors A-E)

Flask	Factor A	Factor B	Factor C	Factor D	Factor E	Viable cells (million per mL)
1	1	1	1	1	1	3.15
2	0	1	0	0	0	2.04
3	0	1	1	0	1	3.41
4	0	0	0	1	0	3.14
5	1	1	0	1	0	1.30
6	0	0	1	0	0	2.31
7	1	0	0	1	1	3.63
8	1	0	0	0	0	1.57
9	1	1	1	0	0	1.09
10	0	0	0	0	1	3.92
11	1	1	0	0	1	2.32
12	0	1	0	1	1	4.08
13	0	0	1	1	1	3.97
14	0	1	1	1	0	2.40
15	1	0	1	1	0	2.09
16	1	0	1	0	1	2.50

0 = factor is absent; 1 = factor is present. Viable cell counts were obtained using the Beckman Vi-Cell on working day 9 of the experiment.

plot of the five factors analyzed in a multifactorial experiment on CHO cells using the flask plan is shown in **Table 1** (*see* **Note 5**).

9. It is good practice to perform a repeat part of experiment in triplicate shake flasks to ensure reproducibility. Typically, we repeat the best two to three flasks from the original experiment in triplicate and also include triplicate flasks of the best combination predicted by the Design Expert model. A negative control (i.e., base medium only) is also included in triplicate flasks.

3.2. Optimization of a Fed-Batch Process in Shaken Tubes

A useful application of the 50-mL shaken tube platform is the optimization of a fed-batch process (*see* **Fig. 5**). The batch process is started in a single bioreactor, from which the cells are taken and distributed into 50-mL tubes where candidate feeds are tested in parallel. Feeds are usually added toward the end of the exponential growth phase. However, it is recommended to test different time points in order to define the optimal feeding strategy.

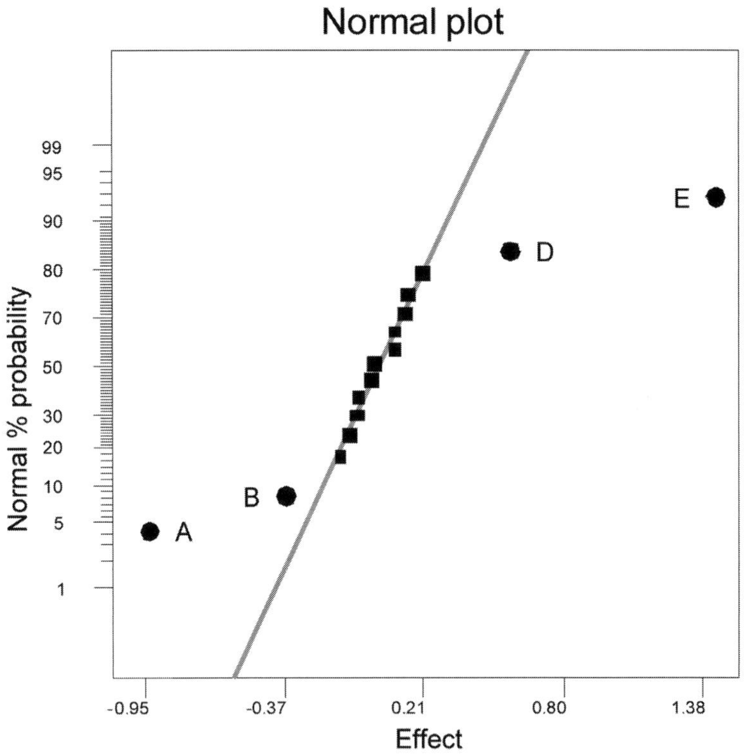

Fig. 4. Normal plot of the results shown in **Table 1**. Significant factors are shown in circles (**A** and **B** are negative, **D** and **E** are positive). Nonsignificant factors **(C)** and their interactions are shown in squares.

1. Define the 20 feeds and prepare the different feed stock solutions. Fed volumes should be no more than 10% of the total volume (e.g., 0.5 mL in a final volume of 5.5 mL). Include a negative control, to which no feed is added.
2. The feed stock solutions are loaded into sterile 50-mL centrifuge tubes that are clearly labeled. Three series of 20 tubes are prepared in order to test the effects of adding the feeds at three different time points (*see* **Note 6**).
3. The first set of feeds is tested after a few days in the bioreactor with cells that are still growing. One hundred milliliters of cell culture is taken from the bioreactor and distributed in 5-mL aliquots into the 20 tubes (*see* **Note 7**).
4. Tubes are shaken at 200 rpm in the Kuhner incubator (80% humidity, 5% CO_2, 37°C) until the end of the experiment (*see* **Note 8**).
5. A second set of 20 tubes is launched when the cells are at the transition between the exponential growth and stationary phase of culture, and the third set of tubes is launched 24 h later.

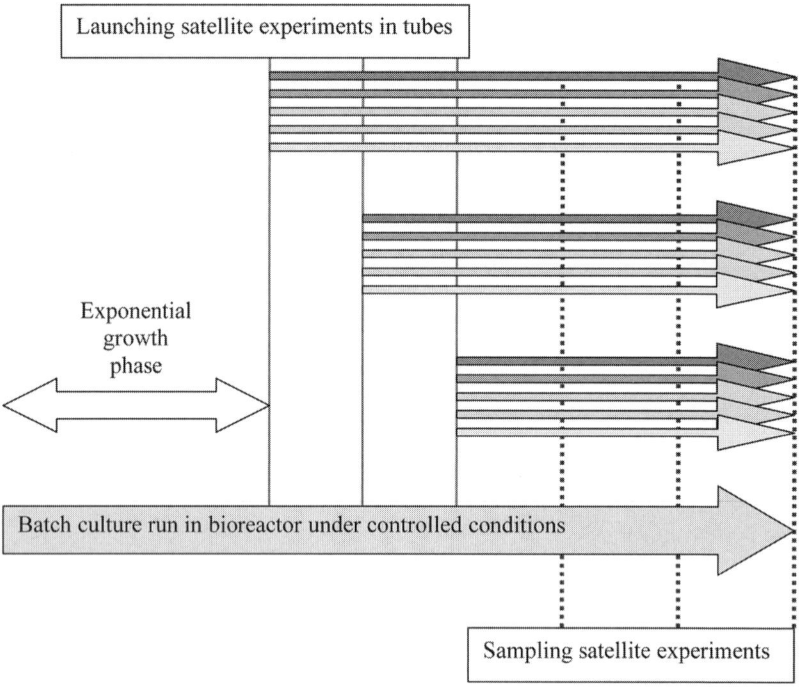

Fig. 5. Process optimization by satellite experimentation in shaken tubes, as described in **Subheading 3.2**.

6. Two to three days before the end of the experiment a sample is taken and analyzed for viable cell number, product concentration, and other parameters of interest, such as metabolites (*see* **Note 4**). The sample volume can be up to 2 mL without inducing any adverse effects in the remaining culture.
7. At the end of the experiment the whole tube is harvested and used for cell counts and other types of analysis (*see* **Note 9**).

3.3. Conclusions

In addition to the shake flask and shaken tube platforms described here for high-throughput screening, other types of vessels can be used. At the micro scale, animal cells have been successfully grown in conventional plastic plates, e.g., in 96- or 24-well plates *(7)*. Alternatively, cells can be grown to high densities in shaken deep-well microtiter plates, although precautions need to be taken to avoid excessive evaporation *(8)*. Working at small volumes (less than 1 mL per well) also limits the amount of sample that can be taken for analysis, so identical plates are often set up in parallel in order to overcome this constraint. Process development can also be performed in small-scale bioreactors of working volume 1-5 L *(9)*, although such a platform cannot be run in a high-throughput mode.

4. Notes

1. Various orbit radii (throws) were tested using the 50-mL tubes, and we observed that a throw of 5 cm was optimal for keeping the cells perfectly in suspension. Similarly, a throw of 2.5 cm proved optimal for cells in 125-mL shake flasks.
2. The statistical techniques of multifactorial analysis and response surface design are explained elsewhere ***(10,11)***. There are a number of software packages available to assist the researcher in experimental design and analysis, such as Design Expert from Stat-Ease *(5)* or MODDE from Umetrics ***(12)***. We highly recommend these techniques for large experiments (when more than three variables are tested simultaneously).
3. If the medium used for growing the cells for inoculation is a richer formulation than the ones being evaluated, one should be aware of potential carryover effects, where cells are still responding to factors in the previous medium. Such carryover effects may mask a lack of essential components in the test medium for two to three cell passages (about 5–7 d). Therefore, for testing lean media we recommend growing the cells in the test medium for 5–7 d prior to the start of the experiment.
4. In addition to cell counts and protein productivity measurements, useful information can be gained about the cell's metabolism by testing the supernatants on a metabolite analyzer, e.g., a Nova Bioanalyzer, which measures glucose, lactate, glutamine, and ammonia.
5. In this analysis (**Fig. 4**) factors that do not lie on the median line (circles) have a significant effect on the viable cell concentrations measured on day 9 of the experiment. Factors to the right of this line have positive effects, and factors to the left of the line have negative effects. Note that not all the responses will be the same, e.g., factors that are positive for cell growth may not be positive for protein production.
6. Depending on the stability of the feed solutions, this can be done at the beginning of the experiment. The solutions are then stored at 4°C until they are needed. Alternatively, unstable feed solutions are distributed into the tubes immediately before the cells are added.
7. Since the tubes already contain the feed solutions, the distribution of the cells should be fast. It is important that this does not take too much time, otherwise the cells will start settling, and it is critical that all the tubes get the same number of cells.
8. Five percent CO_2 in air will guarantee a physiological pH if the bicarbonate concentration of the medium is within the range 12–18 m*M*. In small-scale bioreactors (with a few liters working volume), bicarbonate can be completely utilized. This occurs when CO_2 is stripped out of the medium via aeration and the pH controller adds HCl or cells produce equivalent amounts of lactate. In bicarbonate-deficient media the pH drops below 7 in the presence of 5% CO_2. In such cases (if the bicarbonate can be measured) it can be easily added back or the CO_2 level of the vessel can be lowered. Alternatively, 20–30 m*M* hydroxyethyl piperazine ethane sulfonate

(HEPES) can be included from the beginning of the experiment to maintain pH. Although the pH might still change in the presence of HEPES, the medium is much less sensitive to changes in bicarbonate or CO_2.

9. In general, both shake flask and shaken tube data correlate well with bioreactor data. Nevertheless, these platforms should be regarded as high-throughput screening tools that facilitate the testing of many different media and feed components in parallel. For industrial purposes, all results should be confirmed in a bioreactor that is representative of the final process.

References

1. Backstrom, M., Graham, R., Essers, R., et al. (2004) Bioprocess development for the production of a recombinant MUC1 fusion protein expressed by CHO-K1 cells in protein-free medium. *J. Biotechnol.* **110**, 51–62.
2. Muller, N., Girard, P., Hacker, D. L., Jordan, M., and Wurm, F. M. (2005) Orbital shaker technology for the cultivation of mammalian cells in suspension. *Biotechnol. Bioeng.* **89**, 400–406.
3. Girard, P., Jordan, M., Tsao, M., and Wurm, F. M. (2001) Small-scale bioreactor system for process development and optimization. *Biochem. Eng. J.* **7**, 117–119.
4. De Jesus, M. J., Girard, P., Bourgeois, M., and Wurm, F. M. (2004) Tubespin satellites: a fast track approach for process development with animal cells using shaking technology. *Biochem. Eng. J.* **17**, 217–223.
5. http://www.statease.com/
6. Hunt, L., Batard, P., Jordan. M., and Wurm, F. M. (2002) Fluorescent proteins in animal cells for process development: optimization of sodium butyrate treatment as an example. *Biotechnol. Bioeng.* **77**, 528–537.
7. Balcarcel, R. R. and Clark, L. M. (2003) Metabolic screening of mammalian cell cultures using well-plates. *Biotechnol. Prog.* **19**, 98–108.
8. Strobel, R., Bowden, D., Bracy, M., et al. (2001) High-throughput cultivation of animal cells using shaken microplate techniques, in *Animal Cell Technology: From Target to Market* (Lindner-Olsson, E. ed.), Kluwer Academic Press, Dordrecht, pp. 307–311.
9. Chu, L., Blumentals, I., and Maheshwari, G. (2005) Production of recombinant therapeutic proteins by mammalian cells in suspension culture, in *Therapeutic Proteins: Methods and Protocols,* Humana Press, Totowa, NJ, pp. 107–122.
10. Anderson, M. J. and Whitcomb, P. J. (2000) *DOE Simplified: Practical Tools for Effective Experimentation,* Productivity Press, University Park, IL, pp 236.
11. Anderson, M. J. and Whitcomb, P. J. (2005) RSM Simplified. *Optimizing Processes Using Response Surface Design Methods for Design of Experiments.* Productivity Press, University Park, IL.
12. http://www.umetrics.com/

III

Cell Characterization and Analysis

8

Cell Counting and Viability Measurements

Michael Butler and Maureen Spearman

Summary

The accurate determination of cell growth is pivotal to monitoring a bioprocess. Direct methods to determine the cells in a bioprocess include microscopic counting, electronic particle counting, biomass monitoring, and image analysis. These methods work most simply when a fixed volume sample can be taken from a suspension culture. Manual microscopic counting is laborious but affords the advantage of allowing cell viability to be determined if a suitable dye is included. Electronic particle counting is a rapid method for replicate samples, but some data distortion may occur if the sample has significant cell debris or cell aggregates. The use of a biomass probe detects cells by the dielectric properties and can be used as a continuous monitor of the progress of a culture. Image analysis based on the use of digital camera images acquired through a microscope has advanced rapidly with the availability of several commercially available software packages.

Indirect methods of cell determination involve the chemical analysis of a culture component or a measure of metabolic activity. These methods are most useful when it is difficult to obtain intact cell samples. However, the relationship between these parameters and the cell number may not be linear through the phases of a cell culture. The determination of nucleic acid (DNA) or total protein can be used as an estimate of biomass, while the depletion of glucose from the media can be used as an estimate of cellular activity. The state of cellular viability may be measured by the release of an enzyme such as lactate dehydrogenase (LDH) or more directly from the intracellular adenylate energy charge from cell lysates. Alternatively, radioactive techniques may be used to for an accurate determination of cellular protein synthesis.

Key Words: Hemocytometer; crystal violet; microcarrier; biomass monitor; Coulter counter; Coomassie blue; DAPI (4′,6-diamidino-2-phenylindole); Hoechst reagent; glucose oxidase; hexokinase; tetrazolium dye; MTT (3-(4,5-dimethylthiazol-2-yl)-2,5-diphenyltetrazolium bromide); lactate dehydrogenase; trypan blue.

1. Introduction

The growth of mammalian cells in culture can be monitored by a number of parameters related to the increase of cellular biomass over time. The simplest

From: *Methods in Biotechnology, Vol. 24: Animal Cell Biotechnology: Methods and Protocols, 2nd Ed.* Edited by: R. Pörtner © Humana Press Inc., Totowa, NJ

method is by cell counting at regular intervals. In routine cultures this would be performed once a day, which corresponds to the approximate doubling time of mammalian cells during the exponential growth phase. This would establish an overall growth profile of a culture. More frequent counts would be required to follow more subtle changes that may, for example, be associated with the cell growth cycle.

The two direct cell-counting methods that are used routinely are based on visual examination through a microscope or electronically by a particle counter. Both methods depend upon obtaining a sample of an even distribution of cells in suspension. Therefore, it is extremely important to ensure that the culture is well mixed by stirring or shaking before taking a sample.

Indirect methods of estimating cell growth rely on the measurement of an intracellular cell component such as DNA or protein or, alternatively, an extracellular change such as nutrient depletion or an enzyme activity released by the cells. Indirect methods of growth estimation depend upon a relationship between the measured parameter and cell concentration. However, it is important to realize that these relationships are rarely linear over the course of a culture. It is well documented that the total protein content and specific enzyme activity levels measured on a per cell basis vary substantially over the course of a culture as a result of changes in the growth rate and composition of the culture medium.

In some situations as may occur, for example, in immobilized cell bioreactor systems, an indirect measurement of cell growth may be the only option available. This can be used to monitor the progress of a culture. However, care must be taken if such data are used in comparative analysis between cultures, as differences may be a reflection of changes in metabolic or functional activity rather than of cell concentration.

Viability is a measure of the metabolic state of a cell population which is indicative of the potential of the cells for growth. One of the simplest assay types is dye exclusion, which is an indication of the ability of the cell membrane to exclude a dye. This may be included in the protocol for microscopic cell counting. More sophisticated measures involve the ability of cells to perform DNA or protein synthesis. A further metabolic assay measures the intracellular adenylate nucleotide concentrations. This allows the determination of the energy charge, which is an index of the metabolic state of the cells. These viability assays are described in **Subheading 1.2**.

1.1. Direct Methods of Cell Counting

1.1.1. Cell Counting by Hemocytometer

The improved Neubauer hemocytometer consists of a thick glass plate that fits onto the adjustable stage of a microscope. A grooved calibrated grid is

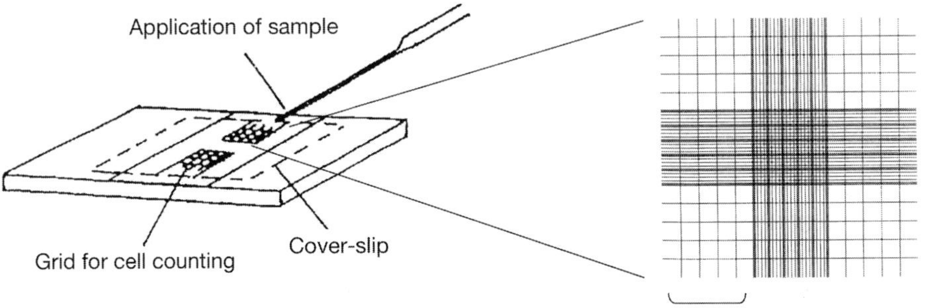

Fig. 1. The hemocytometer. A sample of cells is loaded between the plate and the coverslip. The cells are then counted over five large grids, each of which corresponds to a sample volume of 0.1 µL.

observed through the microscope on the hemocytometer surface (**Fig. 1**). A cell suspension is put onto the grid by touching the end of a capillary tube (can be a pipet tip or Pasteur pipet) containing the cell suspension at the edge of a coverslip placed on the upper surface of the hemocytometer. The cells are then counted in a standard volume (usually 5 × 0.1 µL) as defined by the area of the grid. A hand-held tally counter helps in counting.

Trypan blue is often added to the cell suspension before counting *(1)*. The dye penetrates the membrane of nonviable cells, which are stained blue and can therefore be distinguished from viable cells.

1.1.2. Nuclei Counting by Hemocytometer

A modification of the hemocytometer method involves counting nuclei. Incubation of cell samples in a mixture of citric acid and crystal violet causes cells to lyse and the released nuclei to stain purple *(2)*. Nuclei counting is well suited to the determination of anchorage-dependent cells, for example, when attached to microcarriers.

1.1.3. Nuclei Counting in Macroporous Microcarriers

Macroporous microcarriers such as Cytopore (Amersham) entrap the cells, making whole cell removal difficult. Under these conditions nuclei counting is possible *(3)*. The nuclei are stained with crystal violet, which is added as a hypertonic reagent containing detergent. The reagent causes lysis of cells, and the nuclei are stained during continuous, gentle agitation. To facilitate removal of nuclei from the beads, hydraulic pressure may be applied by forcing the treated microcarriers through a syringe needle. This should be performed in multiple steps until a minimal number of nuclei remain within the microcarriers as observed through a light microscope. The "empty" microcarriers sediment to the bottom of the sample tube. The stained nuclei may be counted in the liquid suspension that remains.

1.1.4. Particle Counter

The Coulter counter (Beckman Coulter Inc., Fullerton, CA) has been used for several decades for rapid cell counting. The principle of this electronic cell counter is that a predetermined volume (usually 0.5 mL) of a cell suspension diluted in buffered saline is forced through a small hole (diameter 70 μm) in a tube by suction. The cells interrupt the current flow between two electrodes: one inside and one outside the glass tube. This produces a series of pulses recorded as a signal on the counter. Particles smaller than cells can be eliminated from the count by setting a lower threshold of detection. The largest particle size is determined by the size of the hole in the tube.

The CASY particle counter (Scharfe System, Reutlingen, Germany) is an alternative high-resolution counter that combines the resistance measurements of the traditional Coulter counter with pulse area signal processing. Here each cell passes through a capillary in which a change of electrical resistance is recorded. The electrical signal is scanned, and the data from each particle are processed into a series of values that can be interpreted by pulse area analysis. This allows a particle size range to be recorded for each sample that can be interpreted as viable cell, dead cells, cell debris, and cell aggregates.

1.1.5. On-Line Cell Concentration Determination by a Biomass Monitor

The Biomass monitor (Aber Instruments, Aberystwyth, Wales) consists of a sterilizable probe that can be inserted through a head-plate port of a bioreactor. The probe incorporates four platinum electrode pins held in an inert polymeric matrix. During operation the probe applies a low current radio-frequency field within 20–25 nm of the electrodes. Within this field cells with intact plasma membranes can accumulate an electrical charge and act as tiny capacitors. The radio-frequency impedance that can be measured from this probe can be converted to measurements of capacitance (pF/cm) or relative permittivity (dimensionless) that have been shown to correlate well with the viable cell concentration in the culture.

The value of this technique is that the dielectric properties monitored are dependent upon cells with an intact plasma membrane. Therefore, the measurements of changes in capacitance can be related directly to the viable cell concentration of the culture. This allows growth profiles to be determined in animal or insect cell cultures where the viability can vary and even if there is a high degree of cell clumping ***(4)***. Furthermore, measurements can be made in microcarrier cultures when off-line cell concentration determinations may be difficult ***(5)***.

A detailed study of this type of capacitance measurement shows that the cell size and cell metabolism can also influence the values obtained. For hybridomas a good correlation has been shown between the specific capacitance and the specific amount of nucleotide triphosphates in the cell ***(6)***. Thus, during a batch culture the cell-specific capacitance changes by as much as 45%, with the highest value occurring at the maximum growth rate.

1.1.6. On-Line Fluorescence Probe

The presence of reduced NAD (NADH) in the cell enables fluorescence measurements to be made following excitation with a UV source of light. This property allows cells to be monitored in culture by use of an *in situ* autoclavable fiberoptic-based probe. This provides information about the metabolic activity as well as changes in cell concentration or biomass. Custom Sensors & Technology (CST, Fenton, MO) produces a photometric transmitter and *in situ* probe for such measurements. The monitor delivers excitation energy from a pulsed UV lamp and monitors the amount of fluorescence from the cell culture. The reading is based on the amount of fluorescence from the viable cell suspension compared to a reference signal. In this device the ratio of fluorescence intensity to excitation intensity is scaled over a 4- to 20-mA output. This probe and photometric transmitter allows continuous monitoring of fluorescence during cell culture and enables a profile of cell growth to be determined.

1.1.7. Image Analysis

Several systems are available that are capable of analyzing particles or shapes in a preset field of view through a microscope. This is based on image capture through a charge-coupled device (CCD) camera and transfer of data to suitable software on a computer. Several systems are available commercially. These include the following:

1. Sorcerer Automatic Image Analysis System (Perceptive Instruments, Steeple Bumpstead, Suffolk, UK).
2. Metamorph Imaging System (Molecular Devices Corp., Sunnyvale, CA).
3. Northern Eclipse Image Analysis (Empix Imaging, Cary, NC).
4. Axovision (Carl Zeiss Microimaging Inc., Thornwood, NY).

The standard equipment for image acquisition is a color or monochrome CCD camera with variable zoom and fixed-focus lenses adapted to a standard light microscope. The image is transmitted to a computer screen and displayed through the specific software-containing processing tools, which allows the operator to choose various parameters for analysis including the particle size and shape for analysis. The images are recognized by virtue of their contrast with the background. Information from such a system includes number, size, and shape of any particles (or cells) in a predefined field of view. The measurements are rapid and can be suitable for replacing manual counting from the field of view of a hemocytometer. The commands through the software allow specific images to be selected or deselected from analysis based upon a number of size or shape parameters. Therefore, such a system can be suitable for the analysis of specific cell types in a mixed population.

1.2. Indirect Methods of Cell Determination

There are a number of colorimetric methods based on the measurement of cell components. These are relatively simple methods and suitable for multiple samples. However, the contents of cells can vary dramatically during culture. For example, the protein and enzyme content per cell will be high during exponential growth but lower in the lag or stationary phases.

1.2.1. Protein Determination

Total cell protein can be used as a measure of biomass (total cellular material). The protein content of a mammalian cell is typically 100–500 pg/cell. These measurements are also useful in the determination of specific enzyme activities, which are commonly expressed as the maximum measured reaction velocity of an enzyme per total cell protein.

The most common colorimetric assays are the Lowry and Bradford methods. Of these, the Bradford assay is favored because of speed, sensitivity, and negligible interference from other cell components *(7)*. By this method lysed cells are added to the reagent, Coomassie blue. A blue color that develops within 10 min can be measured by a colorimeter or spectrophotometer and compared with standard proteins.

1.2.2. DNA Determination

A commonly used protocol involves treatment of the solubilized cells with fluorescent reagents that bind to DNA. Fluorescence detection offers high sensitivity with reagents such as Hoechst 33258 *(8)* or 4′,6-diamidino-2-phenylindole (DAPI; ***9***) from Sigma Chemical Co.

1.2.3. Glucose Determination

Cell growth can be monitored by changes in the concentration of key components of the culture medium. The rate of change in the glucose content of the medium may be suitable for such an assay as an indirect measure of cell concentration. Alternatives include measurement of lactic acid production or oxygen consumption.

Correlations have been shown between cell concentration and rates of consumption or production of these components. This relationship may be constant for a particular cell line under a given set of conditions. However, if the cell line or any of the culture conditions are altered, the relationship between substrate consumption or product formation and cell number will change.

1.2.3.1. Glucose Oxidase Assay

Glucose can be determined by a colorimetric assay utilizing the two enzymes, glucose oxidase and peroxidase ***(10)***:

$$\text{D-glucose} + H_2O + O_2 \leftrightarrow \text{D-gluconic acid} + H_2O_2 \quad (1)$$

$$H_2O_2 + \text{reduced o-dianisidine} \leftrightarrow 2H_2O + \text{oxidized o-dianisidine (brown)} \quad (2)$$

$$\text{Oxidized o-dianisidine (brown)} + H_2SO_4 \leftrightarrow \text{oxidized o-dianisidine (pink)} \quad (3)$$

Equation 1 is catalyzed by glucose oxidase (GOD) and Eq. 2 by peroxidase (POD). The dye, o-dianisidine hydrochloride, is reduced by hydrogen peroxide to a product that has a pink color in the presence of sulfuric acid (Eq. 3) and is measured colorimetrically. The glucose oxidase kit from Sigma Chemical Co. contains glucose oxidase/peroxidase reagent and o-dianisidine reagent.

1.2.3.2. Hexokinase Assay

Glucose can also be measured enzymatically in the following two reactions catalyzed by hexokinase (HK) and glucose 6-phosphate dehydrogenase (G6PDH; ***11***):

$$\text{D-glucose} + \text{ATP} \leftrightarrow \text{glucose-6-P} + \text{ADP} \quad (1)$$

$$\text{Glucose-6-P} + \text{NAD} \leftrightarrow \text{6-phosphogluconate} + \text{NADH} + H^+ \quad (2)$$

HK converts glucose into glucose 6-phosphate in the presence of ATP (Eq. 1). The glucose-6-P is immediately converted into 6-phosphogluconate by G6PDH (Eq. 2). The associated formation of NADH is monitored by the change in absorbance at 340 nm, and this is proportional to the concentration of glucose originally present. The HK kit from Sigma Chemical Co. contains a HK/G6PDH reagent. The kit includes a glucose standard solution (1 mg/mL).

1.2.3.3. The Glucose Analyzer

A modification of the glucose oxidase assay system is used in an analyzer such as the YSI model 27 Industrial Analyzer (Yellow Spring Instrument Co., Inc., Yellow Springs, OH). The instrument is provided with various membranes containing immobilized enzymes appropriate for measuring a particular analyte such as glucose or lactic acid. The sample is injected into a membrane, which converts the glucose to hydrogen peroxide, which can be determined by a sensor system based on a Clark electrode. The latter consists of a platinum electrode, which measures the hydrogen peroxide amperometrically.

$$H_2O_2 \leftrightarrow 2H^+ + O_2 + 2e^- \quad (1)$$

$$AgCl + e^- \leftrightarrow Ag + Cl^- \quad (2)$$

Current flow in the platinum anode is linearly proportional to the local concentration of hydrogen peroxide. This electrode is maintained at an electrical potential of 0.7 V with respect to a silver/silver chloride reference electrode, the potential of which is determined by Eq. 2. The signal current, which is proportional to the quantity of injected glucose, is converted to a voltage by the instrument circuitry.

1.3. Viability Measurements

Viability is a measure of the proportion of live, metabolically active cells in a culture, as indicated by the ability of cells to divide or to perform normal metabolism. The viability is measured by an indicator of the metabolic state of the cells (such as energy charge) or by a functional assay based on the capacity of cells to perform a specific metabolic function.

A viability index may be determined from simple assays such as dye exclusion, where cells are designated as either viable or nonviable. The index is usually expressed as a percentage of viable cells in a population:

$$\text{Viability index} = (\text{number of viable cells/total number of cells}) \times 100.$$

1.3.1. Dye Exclusion

Cell counting by hemocytometer as described earlier in this chapter can be adapted to measure viability. The most common is the dye exclusion method in which loss of viability is recognized by membrane damage resulting in the penetration of the dye trypan blue. Other dyes that can be used include erythrosin B, nigrosin, and fluorescein diacetate.

1.3.2. Tetrazolium Assay

The tetrazolium assay is a measure of cellular oxidative metabolism. The tetrazolium dye MTT (3-(4,5-dimethylthiazol-2-yl)-2,5-diphenyltetrazolium bromide) is cleaved to a colored product by the activity of NAD(P)H-dependent dehydrogenase enzymes, and this indicates the level of energy metabolism in cells ***(12)***. The color development (yellow to blue) is proportional to the number of metabolically active cells. The assay response may vary considerably between cell types. Cells of biotechnological importance such as Chinese hamster ovary (CHO) cells and hybridomas can be monitored with the MTT assay.

1.3.3. Colony-Forming Assay

The most precise of all the methods of viability measurement is the colony-forming assay. Here the ability of cells to grow is measured directly

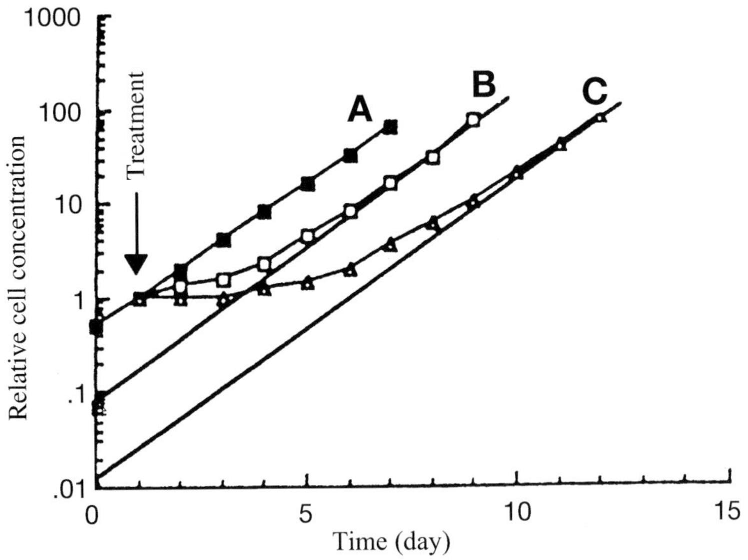

Fig. 2. Cell viability by extrapolation from a growth curve. Curve A is of an untreated control. Cultures B and C have different treatments at day 1, such as the addition of a toxic compound. The relative surviving fraction of the treated cultures can be used as a measure of the effect of the treatment on cell viability. B = 0.1/0.8 = 0.125; C = 0.01/0.8 = 0.0125. (From **ref. *13*.**)

(13). A known number of cells at low density is allowed to attach and grow on the surface of a Petri dish. If the cell density is kept low, each viable cell will divide and give rise to a colony or cluster of cells. From this the plating efficiency is determined (number of colonies scored per 100 cells plated × 100). Although the colony-forming assay is time-consuming, it has been widely used in cytotoxicity studies.

A less precise method of determining the viability by the cellular reproductive potential is from the lag phase of a growth curve. **Figure 2** shows that by extrapolation from the linear portion of a growth curve to time zero, the derived cell number can be compared with the original cell count. This method can be easily adapted to determine how a particular treatment (such as addition of a toxic compound) affects cell viability.

1.3.4. LDH Determination

A decrease in the viability of cells is usually associated with a damaged cell membrane, which causes the release of large molecules such as enzymes from the cell into the medium. Thus, the loss of cell viability may be followed by an increase in enzyme activity in the culture medium ***(14)***. LDH is the enzyme most commonly measured in this technique. The enzyme activity can be measured easily by a simple spectrophotometric assay involving the oxidation of NADH in the presence of pyruvate. The reaction is monitored by a decrease in UV absorbance at 340 nm.

$$\text{Pyruvate} + \text{NADH} + \text{H}^+ \overset{\text{LDH}}{\leftrightarrow} \text{lactate} + \text{NAD}^+$$

NADH absorbs at $\lambda = 340$ nm.

1.3.5. Intracellular Energy Charge

The energy charge is an index based on the measurement of the intracellular levels of the nucleotides AMP, ADP, and ATP. The energy charge is ([ATP] + 0.5*[ADP])/([ATP] + [ADP] + [AMP]) and is based on the interconversion of the three adenylate nucleotides in the cell: AMP ↔ ADP ↔ ATP. This index varies between the theoretical limits of 0 and 1. For normal cells, values of 0.7–0.9 would be expected, but a gradual decrease in the value gives an early indication of loss of viability by a cell population.

These nucleotide concentrations can be measured by chromatography (such as high-performance liquid chromatography) or by luminescence using the luciferin-luciferase enzyme system ***(15,16)***. The luminescence assay is dependent upon the emission of light resulting from the enzymic oxidation of luciferin, a reaction requiring ATP:

$$\text{ATP} + \text{LH}_2 + \text{O}_2 \leftrightarrow \text{AMP} + \text{PPi} + \text{CO}_2 + \text{L} + \text{light}$$

ADP and AMP is also be measured by the luciferase assay after conversion to ATP by coupled enzymic reactions—(1) pyruvate kinase:

$$\text{ADP} + \text{PEP} \leftrightarrow \text{ATP} + \text{pyruvate}$$

and (2) myokinase:

$$\text{AMP} + \text{CTP} \leftrightarrow \text{ADP} + \text{CDP}$$

1.3.6. Rate of Protein Synthesis

The rate of protein synthesis of intact cells can be measured by incubation in standard culture medium to which is added a radioactively labeled amino acids. Any radioactive amino acid is suitable, but those most commonly used are ^{3}H-leucine and ^{35}S-methionine.

2. Materials

2.1. Cell Counting by Hemocytometer

1. Phosphate-buffered saline (PBS): 0.1 *M* NaCl, 8.5 m*M* KCl, 0.13 *M* Na_2HPO_4, 1.7 m*M* KH_2PO_4, pH 7.4.

2. Trypan blue reagent: 0.2% w/v trypan blue in PBS. This reagent should be filtered through paper to remove any precipitates or particles in the solution.

2.2. Nuclei Counting by Hemocytometer

Crystal violet reagent: 0.1% w/v crystal violet in 0.1 *M* citric acid.

2.3. Nuclei Counting in Macroporous Microcarriers

Crystal violet reagent: 0.2% w/v crystal violet and 2% w/v Triton/X-100 in 0.2 *M* citric acid.

2.4. Particle Counter

Saline solution: 0.7% NaCl, 1.05% citric acid, 0.1% mercuric chloride in distilled water.

2.5. Protein Determination

Bradford's reagent: dissolve 100 mg of Coomassie brilliant blue G (Sigma Chemical Co.) in 95% ethanol (50 mL) and 85% phosphoric acid (100 mL). After the dye dissolves make the solution up to 1 L with distilled water. Alternatively, a dye (Coomassie) reagent liquid concentrate can be purchased from Biorad.

2.6. DNA Determination/Hoechst Method

1. Buffer: 0.05 *M* $NaPO_4$, 2.0 *M* NaCl, 2 m*M* EDTA, pH 7.4.
2. Hoechst reagent: 0.1 μg/mL Hoechst 33258 in buffer.
3. Standard DNA solution: 8 mg/mL of calf thymus DNA (Sigma Chemical Co.) in distilled water.

2.7. DNA Determination/DAPI Method

1. Buffer: 5 m*M* hydroxyethyl piperazine ethane sulfonate, 10 m*M* NaCl, pH 7.
2. DAPI reagent: a stock solution (×100) contains 300 mg DAPI (4′,6-diamidino-2-phenylindole) in buffer.
3. Standard DNA solution: 8 mg/mL of calf thymus DNA (Sigma Chemical Co.) in distilled water.

2.8. Glucose Determination/Glucose Oxidase Assay

1. Glucose oxidase/peroxidase reagent: dissolve the contents of a reagent capsule from Sigma in 39.2 mL of distilled water. Each capsule contains 500 units of glucose oxidase and 100 units of peroxidase.
2. O-dianisidine reagent: dissolve the contents of a vial of o-dianisidine (Sigma Chemical Co.) in 1 mL of dissolved water. Each vial contains 5 mg of o-dianisidine dihydrochloride.
3. Assay reagent: mix 0.8 mL of o-dianisidine reagent with 39.2 mL of glucose oxidase/peroxidase reagent.

4. Glucose standard solution: 1 mg/mL of D-glucose.
5. Sulfuric acid, 12 *M*.

2.9. Glucose Determination/Hexokinase Assay

1. Glucose (HK) assay reagent. Dissolve the contents of a reagent vial (Sigma) into 20 mL of distilled water. The dissolved reagent contains 1.5 m*M* NAD, 1.0 m*M* ATP, 1 unit/mL HK, and 1 unit/mL G6PDH.
2. Glucose standard solution. 1 mg/mL of D-glucose.

2.10. The Glucose Analyzer

1. YSI model 27 Industrial Analyzer (Yellow Spring Instrument Co., Inc., Yellow Springs, OH).
2. Glucose standard solution: 2–5 g/L of D-glucose.

2.11. LDH Determination

1. Tris HCl (0.2 *M*), pH 7.3.
2. NADH (6.6 m*M*).
3. Sodium pyruvate (30 m*M*).
4. Standard LDH enzyme (Sigma Chemical Co.).
5. Spectrophotometer with a UV wavelength of 340 nm.

2.12. Viability/Tetrazolium Assay

1. MTT reagent: 5 mg/mL of the tetrazolium dye MTT (3-(4,5-dimethylthiazol-2-yl)-2,5-diphenyltetrazolium bromide; Sigma Chemical Co.) in PBS pH 7.4.
2. SDS reagent: 10% w/v sodium dodecyl sulfate (SDS), 45% w/v *N,N*-dimethyl formamide in water adjusted to pH 4.5 with glacial acetic acid.

2.13. Intracellular Energy Charge

1. ATP monitoring reagent/ATP-MR (LKB/BioOrbit, Turku, Finland) contains a lyophilized mixture of firefly luciferase, D-luciferin, bovine serum albumin, magnesium acetate, and inorganic pyrophosphate. Reconstitute each vial with 4 mL buffer plus 1 mL potassium acetate (1 *M*).
2. ATP standards (LKB): ATP (0.1 μ*M*) and magnesium sulfate (2 μ*M*).
3. Buffer: 0.1 *M* Tris-acetate, pH 7.75.
4. PK-PEP reagent: 55 μL tricyclohexylammonium salt of phosphoenolpyruvate (0.2 *M*) + 50 μL pyruvate kinase (500 U/mg) in Tris buffer.
5. MK-CTP reagent: 95 μL myokinase (2500 U/mg) + 10 μL CTP (110 m*M*) in Tris buffer.

2.14. Rate of Protein Synthesis

1. ^{3}H-leucine or ^{35}S-methionine (Amersham) at 200–400 μCi/mL.
2. Trichloroacetic acid (TCA) 5%.
3. PBS.

4. NCS™ tissue solubilizer (Amersham).
5. Radioactivity scintillation counter.

3. Methods

3.1. Cell Counting by Hemocytometer

1. Add an equal volume of trypan blue reagent to a cell suspension and leave for 2 min at room temperature.
2. Introduce a sample into the hemocytometer chamber by a Pasteur pipet.
3. Count cells on each of five grid blocks defined by triple lines in the hemocytometer chamber (*see* **Note 1**).
4. Determine the cell concentration (cells/mL) in the original sample = $(2 \times \text{total count}/5) \times 10^4$.

(The calculation is based upon the volume of each grid block = 0.1 μL.) The percentage of cells that are not stained with trypan blue is a measure of the viability.

3.2. Nuclei Counting by Hemocytometer

1. Allow microcarriers from a culture sample (1 mL) to settle to the bottom of a centrifuge tube.
2. Remove clear supernatant by aspiration. The supernatant can be checked microscopically for any nonattached cells.
3. Add 1 mL of crystal violet reagent.
4. Incubate at 37°C for at least 1 h.
5. Introduce a sample into the hemocytometer chamber and count the purple stained nuclei as for whole cells (*see* **Note 2**).

3.3. Nuclei Counting in Macroporous Microcarriers

1. Remove 0.5 mL of a microcarrier suspension from the culture while stirring.
2. Add 0.5 mL of crystal violet reagent.
3. Incubate at 37°C for up to 2–3 h with gentle agitation to lyse the cells and stain the exposed nuclei.
4. Remove the nuclei from macroporous microcarriers by aspirating the whole solution through a 25g (1.5") needle with a 1- or 3-mL syringe. Repeat this between 20 and 25 times until all the stained nuclei are cleared form the microcarriers (*see* **Note 3**).
5. **Step 4** is continued until all nuclei are removed from the microcarriers. This is determined by taking a sample (0.1 mL) of the suspension of reagent-treated microcarriers. The reagent is removed by decanting, and the microcarriers are washed two or three times with 1 mL of PBS, allowing the microcarriers to settle between washes. A sample of the microcarrier suspension is visually inspected under the microscope on a slide to ensure that there are no remaining nuclei (*see* **Note 4**).
6. Allow the beads to settle. Then introduce a portion of the remaining suspension of nuclei into the hemocytometer chamber and count the purple stained nuclei as for whole cells.

3.4. Particle Counter

1. Add 0.5 mL of a cell suspension (10^5–10^6 cells/mL) to 19.5 mL of the saline solution.
2. Introduce the suspension into a Coulter counter (Beckman-Coulter Inc., Bedfordshire, UK or Hialeah, FL).
3. From standard settings of the counter 0.5 mL of the suspension is counted. Multiply this count by 40 to give the original cell concentration (*see* **Note 5**).

3.5. Protein Determination

1. Homogenize or sonicate a cell suspension (10^6 cells/mL).
2. Add 5 mL Bradford's reagent to 100 μL of the lysed cell sample (0–0.5 mg/mL protein).
3. Incubate for 10 min at room temperature.
4. Measure the absorbance at 595 nm.
5. Determine the sample concentrations from a standard curve, which is established from standard solutions of bovine serum albumin at 0–0.5 mg/mL protein.

3.6. DNA Determination/Hoechst Method

1. Homogenize or sonicate to lyse a cell suspension (10^5 cells/mL) in buffer.
2. Dilute lysate or standard DNA solution 1 in 10 in Hoechst reagent.
3. Measure fluorescence with an excitation λ of 356 nm and emission λ of 492 nm.
4. Determine DNA concentration by reference to standard DNA.

3.7. DNA Determination/DAPI Method

1. Homogenize or sonicate to lyse a cell suspension (10^5 cells/mL).
2. Dilute 150 μL lysed cell suspension with 850 μL buffer.
3. Prepare a DAPI solution (×10) by diluting 100 μL of DAPI stock solution with 900 μL of buffer and mix well. Prepare a DAPI working solution by adding 0.5 mL of DAPI (×10) to 4.5 mL of buffer.
4. Add 50 μL of DAPI working solution to each cell suspension or standard DNA (up to 0.8 μg) in a tube which is kept dark by a foil cover.
5. Vortex the tubes and let stand for 30 min.
6. Measure fluorescence with an excitation λ of 372 nm and emission λ of 454 nm.
7. Determine DNA concentration by reference to the standard DNA (*see* **Note 6**).

3.8. Glucose Determination/Glucose Oxidase Assay

1. Start the reaction by adding 2 mL of assay reagent to glucose standard or culture media supernatant (0.01–0.1 mL). Make the assay volume up to 3 mL with distilled water.
2. Allow the reaction to proceed for 30 min at 37°C.
3. Stop the reaction by adding 2 mL of 12 *M* H_2SO_4.
4. Measure the absorbance at 540 nm.
5. Determine the glucose concentration of the media samples against a standard values obtained with the glucose solution.

3.9. Glucose Determination/Hexokinase Assay

1. Mix 10–200 μL of standard glucose solution or sample of culture media with 1 mL of assay reagent. Make the total assay volume up to 2 mL with distilled water.
2. Incubate at room temperature for 15 min.
3. Measure the absorbance at 340 nm.
4. Determine the glucose concentration of the media samples against a standard values obtained with the glucose solution (*see* **Note 7**).

3.10. The Glucose Analyzer

1. Fit the appropriate membrane into the analyzer for glucose analysis.
2. Calibrate the instrument with standard glucose solutions (2–5 g/L).
3. Inject 25 μL of a cell-free sample of culture supernatant into the instrument and compare with standard readings (*see* **Notes 8** and **9**).

3.11. Viability/Tetrazolium Assay

1. Remove the media from adherent cells in a multiwell plate and add 0.1 mL MTT reagent. Alternatively, add 0.1 mL MTT reagent to a 1-mL cell suspension in PBS.
2. Incubate for 2 h at 37°C.
3. Add 600 μL of SDS reagent and mix to solubilize the formazan crystals.
4. Measure the absorbance at 570 nm (*see* **Notes 10–12**).

3.12. LDH Determination

1. Mix 2.8 mL Tris HCl (0.2 *M*), pH 7.3, 0.1 mL NADH (6.6 m*M*), and 0.1 mL sodium pyruvate (30 m*M*) in a cuvette.
2. Preincubate for 5 min at the desired reaction temperature (25 or 37°C).
3. Start reaction by adding 50 μL of sample or standard LDH enzyme (Sigma Chemical Co.).
4. Record enzyme activity as an absorbance decrease at 340 nm (*see* **Note 13**).

3.13. Intracellular Energy Charge

1. Extract soluble nucleotides by addition of 0.1 mL perchloric acid (20% v/v) to 1 mL of a cell culture sample (10^6 cells/mL).
2. Place on ice for 15 min and centrifuge for 5 min at 10,000*g*.
3. Remove supernatant and neutralize with 5 *M* KOH.
4. For ATP determination: mix 860 μL buffer, 100 μL ATP-MR, and 10 μL sample.
5. For ADP determination: add a further 10 μL PK-PEP.
6. For AMP determination: add a further 10 μL MK-CTP.
7. For standardization: add a further 10 μL ATP standard.
8. Measure the light emission in a luminometer (e.g., LKB 1250) after 1 min of each stage of addition (*see* **Note 14**).

3.14. Rate of Protein Synthesis

1. Add ^{3}H-leucine or ^{35}S-methionine (Amersham) at a final specific activity of 20–40 μCi/mL to cell suspension at 5–10 × 10^6 cells/mL.

2. Remove 5–10 × 10^5 cells at each time point up to 4–6 h.
3. Isolate cell pellet by centrifugation in a microcentrifuge tube and wash in PBS.
4. Precipitate protein by addition of 500 μL TCA (5%) containing unlabeled amino acids.
5. Wash the protein precipitate three times in the TCA solution.
6. Add 30 μL of NCS™ tissue solubilizer (Amersham) to the pellet and leave for 60 min.
7. Cut tip of tube and place in scintillation fluid for radioactive counting (*see* **Note 15**).

4. Notes

1. The hemocytometer counting method is the most commonly used assay for cell viability.
   The method is simple and effective but can be laborious for multiple samples. At least 100 cells should be counted for statistical validity of the final value.
2. Care must be taken in interpreting nuclei counts as cells can become binucleated, particularly when growth is arrested. As a result the nuclei concentration may be higher than the cell concentration ***(17)***.
3. The number of aspirations required to remove the nuclei from the microcarriers is dependent on the cell density within each bead. Cultures of CHO cells with 1 × 10^6 nuclei/mL (or 300 nuclei/bead) required only 20 aspirations for 99% removal of nuclei from the beads. At a later stage of the same culture with a higher cell density of 1.75 × 10^6 nuclei/mL (or 515 nuclei/bead), 25 aspirations were required to remove 98% of the nuclei. Therefore, higher cell densities may require more stringent conditions for removal ***(3)***.
4. The incubation time and number of needle aspirations needed to completely remove all nuclei may vary between cell types. The method described was used with CHO cells, but other cell types may require more or less stringent conditions ***(3)***.
5. The major advantage of the Coulter counter method is the speed of analysis, and it is therefore suitable for counting a large number of samples. The method is based upon the number of particles contained in suspension, and, consequently, the proportion of viable cells in the sample cannot be determined. It must be ensured that cell aggregates are not present in the sample; otherwise the cell count will be underestimated. The Coulter counter can also be used to determine the size distribution of a cell population by careful control of the threshold settings of the instrument.
6. The DNA content of diploid cells is usually constant, although variations can occur as a result of the distribution of cells through the cell cycle. Cells in the G_1 phase have the normal diploid content of DNA, which is typically 6 pg per cell. DNA measurement is probably one of the best indicators of cell concentration in solid tissue ***(18)***.
7. The sensitivity of the HK assay for glucose can be increased by measuring the rate of increase of absorbance at 340 nm. This can be achieved with a recording spectrophotometer or using the kinetic mode of a multiwell plate reader.

8. The glucose analyzer is particularly suitable for the analysis of glucose in multiple samples of culture medium.
9. By the selection of the appropriate membrane in this instrument, various analytes can be determined such as glucose, sucrose, starch, lactose, galactose, glycerin, lactate, or ethanol.
10. The tetrazolium method is particularly convenient for the rapid assay of replicate cell cultures in multiwell plates. Plate readers are capable of measuring the absorbance of each well of a standard 96-well plate.
11. It is important to ensure that the colored formazan salt formed from MTT is completely dissolved in the SDS reagent.
12. Alternative tetrazolium salts can be used in this assay such as XTT and WST-1, which are available from Boehringer-Mannheim. These form soluble colored products.
13. The LDH assy is well suited for the determination of multiple samples, particularly if a multiwell plate reader is available. Care must be taken when interpreting the results by this method because the LDH content per cell can change considerably during the course of batch culture. The loss of cell viability can be expressed as the activity of LDH in the medium as a proportion of total LDH in the culture.
14. The measurement of energy charge is more time-consuming than the routine counting procedures discussed earlier but can allow a means of monitoring the decline in the energy metabolism of a cell culture that occurs during the loss of viability.
15. The cells should be incubated in the medium long enough to measure radioactivity in the extracted cell pellet. Normally 4–6 h is sufficient but this may be longer ***(19)***. The rate of DNA synthesis of a cell population can be determined in a assay similar to that described for protein synthesis but using a radioactively labeled nucleotide precursor such as tritiated thymidine (^{3}H-TdR) or deoxycytidine (^{3}H-CdR; Amersham). The exposure period may be short (30–60 min) for DNA synthesis rate determinations, and a specific activity of 1 μCi/mL of culture is sufficient. Higher specific activities may be required if using culture media contain the corresponding nonradioactive components such as methionine or thymidine.

References

1. Patterson, M. K. (1979) Measurement of growth and viability of cell in culture. *Methods Enzymol.* **58**, 141–152.
2. Sanford, K. K., Earle, W. R., Evans, V. J., Waltz, H. K., and Shannon, J. E. (1951) The measurement of proliferation in tissue cultures by enumeration of cell nuclei. *J. Natl. Cancer Inst.* **11**, 773–795.
3. Spearman, M., Rodriguez, J., Huzel, N., and Butler, M. (2005) Production and glycosylation of recombinant beta-interferon in suspension and cytopore microcarrier cultures of CHO cells. *Biotechnol. Prog.* **21(1)**, 31–39.
4. Zeiser, A., Bedard, C., Voyer, R., Jardin, B., Tom, R., and Kamen, R. R. (1999) On-line monitoring of the progress of infection in Sf-9 insect cell cultures using relative permittivity measurements. *Biotechnol. Bioeng.* **63(1)**, 122–126.

5. Guan, Y. and Kemp, R. B. (1997) The viable cell monitor: a dielectric spectroscope for growth and metabolic studies of animal cells on macroporous beads. *ESACT Proc.* **15**, 321–328.
6. Noll, T. and Biselli, M. (1998) Dielectric spectroscopy in the cultivation of suspended and immobilized hybridoma cells. *J. Biotechnol.* **63**, 187–198.
7. Bradford, M. (1976) A rapid and sensitive method for the quantitation of microgram quantities of protein using the principle of protein-dye binding. *Anal. Biochem.* **72**, 248–254.
8. Labarca, C. and Paigen, K. (1980) A simple, rapid and sensitive DNA assay procedure. *Anal. Biochem.* **102**, 344–352.
9. Brunk, C. F., Jones, K. C., and James, T. W. (1979) Assay for nanogram quantities of DNA in cellular homogenates. *Anal. Biochem.* **92**, 497–500.
10. Bergmeyer, H. U. and Bernt, E. (1974) *Methods of Enzymatic Analysis*, 2nd ed., Vol. 3 (Bergmeyer, H. U., ed.), VCH Weinheim, pp. 1205–1212.
11. Kunst, A., Draeger, B., and Ziegenhorn, J. (1984) *Methods of Enzymatic Analysis*, 3rd ed., Vol. 8 (Bergmeyer, H. U., ed.), VCH Weinheim, pp. 163–172.
12. Mosmann, T. (1983) Rapid colorimetric assay for cellular growth and survival: application to proliferation and cytotoxicity assays. *J. Immunol. Methods* **65**, 55–63.
13. Cook, J. A. and Mitchell, J. B. (1989) Viability measurements in mammalian cell systems. *Anal. Biochem.* **179**, 1–7.
14. Wagner, A., Marc, A., and Engasser, J. M. (1992) The use of lactate dehydrogenase (LDH) release kinetics for the evaluation of death and growth of mammalian cells in perfusion reactors. *Biotechnol. Bioeng.* **39**, 320–326.
15. Holm-Hansen, O. and Karl, D. M. (1978) Biomass and adenylate energy charge determination in microbial cell extracts and environmental samples. *Methods Enzymol.* **57**, 73–85.
16. Lundin, A., Hasenson, M., Persson, J., and Pousette, A. (1986) Estimation of biomass in growing cells lines by adenosine triphosphate assay. *Methods Enzymol.* **133**, 27–42.
17. Berry, J. M., Huebner, E., and Butler, M. (1996) The crystal violet nuclei staining technique leads to anomalous results in monitoring mammalian cell cultures. *Cytotechnology* **21**, 73–80.
18. Kurtz, J. W. and Wells, W. W. (1979) Automated fluorometric analysis of DNA, protein and enzyme activities: application of methods in cell culture. *Anal. Biochem.* **94**, 166–175.
19. Dickson, A. J. (1991) Protein expression and processing, in *Mammalian Cell Biotechnology: A Practical Approach* (Butler, M., ed.), Oxford University Press, New York, pp. 85–108.

9

Monitoring of Growth, Physiology, and Productivity of Animal Cells by Flow Cytometry

Silvia Carroll, Mariam Naeiri, and Mohamed Al-Rubeai

Summary

The development of flow cytometric techniques has greatly advanced our ability to identify and characterize the morphological and biochemical heterogeneity of cell populations, permitting a rapid and sensitive cell analysis. In this chapter, various flow cytometric applications in cell culture are described for cell lines grown on a laboratory scale, but the principles can also be employed for the monitoring of cell lines up to the large industrial scale.

Key Words: Flow cytometry; monitoring; growth; productivity; cell cycle; stability.

1. Introduction

Flow cytometry is a powerful and effective tool for analyzing the structural and functional characteristics of individual cells or particles in suspension as well as analyzing different populations of cells within a culture. It has been particularly useful for tracking cellular productivity with changes in culture environment and culture age. With the aid of fluorescently labeled probes, it is possible to collect quantitative data on specific cells and their distribution in the cell population. Researchers have attempted to correlate surface-associated antibody, total cell-associated antibody, RNA levels, total protein levels, growth rate, and amount of secreted antibody using the flow cytometer ***(1–9)***. Studies correlating antibody production with cell cycle stage have also been reported ***(10–15)***. These studies have contributed to a better understanding of factors determining protein production and how to manipulate cultures of growing cells to overproduce biopharmaceuticals.

Along with product yield, one of the most fundamental parameters is metabolic changes within the cell culture. Changes in the internal pH of the cells may result in changed production rates, while the cell cycle analysis can provide a

From: *Methods in Biotechnology, Vol. 24: Animal Cell Biotechnology: Methods and Protocols, 2nd Ed.* Edited by: R. Pörtner © Humana Press Inc., Totowa, NJ

reliable index for the prediction of potential changes in cell number with time during the cultivation period *(16)*. Furthermore, it has been found that because of instability as generation number of the culture increases, productivity decreases, hence the metabolic state of the culture is a key factor in determining productivity *(17)*. Cultures with high mitochondrial activity have been found to produce higher antibody titers *(17)*. Likewise, cultures with higher antibody productivity were found to have a higher biomass *(18)*.

Furthermore, flow cytometry provides the opportunity to monitor the amount of intracellular antibody a culture is capable of producing. It has been well documented that in certain cell lines the levels of intracellular antibody correlate well with extracellular antibody *(17,18)*. In hybridoma cultures, it is also possible to select for high producer cells by flow sorting based on the amount of surface antibody. It is also possible to capture the secreted product of the cell using a biotin- or avidin-matrix-based secretion assay, which allows for the selection of high producing cells using flow cytometry (for review, *see* **ref. *19***).

2. Materials

2.1. Cell Cycle Analysis

1. 70% Ethanol (ice cold).
2. DNase solution: dissolve 100 mg DNase (Sigma, Poole, UK) in 200 mL SMT (2.43 g Tris, 85.6 g sucrose, and 1.01 g $MgCl_2$ in 1 L distilled water, pH 6.5). Store at –20°C in 1-mL aliquots.
3. Phosphate-buffered saline (PBS), pH 7.2.
4. RNase solution: dissolve 5 mg RNase (Sigma) in 50 mL distilled water. Store at –20°C in 500-µL aliquots. Incubate in a boiling water bath for 10 min before use to eliminate contaminated DNase. Otherwise, use chromatographically purified RNase.
5. Propidium iodide solution (PI; Sigma): 1 mg/mL stock solution dissolved using PBS. Store at 4°C and in the dark.

2.2. Monoclonal Antibody Productivity by Measuring Intracellular Antibody

1. 4% Paraformaldehyde stock solution (Sigma): dilute in PBS to make a final concentration of 1%. Store at 4°C.
2. 70% Ethanol (ice-cold).
3. Fluorescein isiothiocyanate (FITC) conjugated goat anti-human (anti-mouse) (H+L) IgG (Sigma): 1:10 dilution in PBS.

2.3. Selection of High Producers of Monoclonal Antibody by Affinity Capture Surface Display

1. PBS adjusted to pH 8.0 with 1 *M* NaOH.
2. NHS-LC biotin (Pierce): 0.05 mg/mL in PBS pH 8.0 filter-sterilized.
3. Secretion medium: normal medium with 10% w/v gelatin (Sigma).

Table 1
Components Needed for High-K^+-Calibration Buffers

Buffer A	Buffer B
135 m*M* KH_2 PO_4	135 m*M* K_2H PO_4
20 m*M* NaCl	20 m*M* NaCl
1 m*M* MgCl	1 mM MgCl
1 m*M* $CaCl_2$	1 m*M* $CaCl_2$
10 m*M* glucose	10 m*M* glucose

Table 2
Combinations of Buffers[a] A and B to Give a Range of pHi Values

pH	Volume of A (mL)	Volume of B (mL)
6.4	18.40	6.60
6.8	12.75	12.25
7.0	9.75	15.25
7.4	9.75	20.25
8.0	1.30	23.70

[a]The volume of each buffer is made up to 50 mL with distilled water and nigericin is added to 2 µg/mL to allow the buffers to permeate the cell membranes and equilibrate with the cells.

4. Biotinylated rabbit antihuman IgG, Fc specific (Pierce).
5. Neutravidin (Pierce): 1 mg/mL diluted using PBS.
6. FITC conjugated goat anti-human κ light chain antibody (Sigma).

2.4. Intracellular pH

1. 1 m*M* 5 (and 6)-Carboxy SNARF-1, acetoxymethyl ester, acetate (Molecular Probes, Inc.). 1 m*M* in dimethyl sulfoxide (DMSO; Sigma) to give a final concentration of 10 µ*M*.
2. High-K^+ calibration buffers in the pH range of 6.5–8.0. **Tables 1** and **2** illustrate the high-K^+ calibration buffers and the combinations of these buffers used to give the desired pH values, respectively.
3. Hydroxyethyl piperazine ethane sulfonate (HEPES)-buffered medium (Sigma).
4. Nigericin (Sigma): dissolved in ethanol, used at final concentration of 2 µg/mL.
5. Buffers: **Tables 1** and **2** show components and combinations of buffers for various pHi values.

2.5. Mitochondrial Mass

For mitochondrial mass analysis, dilute the 1 m*M* MitoTracker Green (MTG; Molecular Probes) stock solution to the final working concentration in growth medium, e.g., Dulbecco's modified Eagle medium (DMEM), with or without serum to match the medium that the cells were grown in. At higher concentrations, these probes tend to stain other cellular structures.

3. Methods

3.1. Cell Cycle Analysis

In continuously proliferating cell culture, cells progress through the cell cycle and cell division, which eventually leads to an increase in cell number and biomass. Since cell division begins and ends at S (DNA synthetic) phase, it is possible to use the measurement of cells in S phase as an indicator of change in the specific growth rate of the cell culture. This can be achieved by staining fixed cells with a variety of fluorochromes, and analyzing the stained cells with a flow cytometer. Several stains are available, such as PI, ethidium bromide, and mythramycin, to name a few. For the purpose of this chapter we will concentrate on using PI, which is widely available and excites at 488 nm. It intercalates into DNA to produce fluorescence with a maximum excitation of 620 nm. A typical pattern of cell cycle distribution during batch cultivation of animal cells shows a substantial decrease in the percent of cells in the S and G_2 fractions during the stationary and decline phases of the culture ***(20)*** **(Fig. 1).**

The accuracy of the cell cycle is dependent on the coefficient of variation (CV), which is measured across each peak ***(21)***. Doublets aggregates and debris also need to be excluded from the analysis by proper gating, and the G_1 peak must have a CV of 3 or lower ***(22)***. DNA distribution in the selected gate can be mathematically analyzed to yield the percentages of cells in G_1, S, and G_2/M compartments of the cell cycle by using computer programs such as the Multicycle (Phoenix Flow Systems, San Diego, CA), which is based on a polynomial S-phase algorithm with an iterative, nonlinear least-squares fit.

3.1.1. Analysis of DNA Distributions of Suspension Cells in Batch Culture

1. Remove cell samples at frequent intervals from the batch culture (approx 1×10^6 cells/mL are required). Centrifuge at 100*g* for 5 min and wash in 5 mL PBS.
2. Centrifuge and resuspend the cells in 1 mL stock DNase solution, which removes DNA from necrotic cells, leaving DNA intact in live cells, and therefore eliminates nonviable cells from analysis. Incubate at 37°C for 15 min and then place on ice for 5–10 min.
3. Centrifuge and resuspend on ice-cold 70% ethanol with vigorous shaking. At this stage the cells may be kept at –20° prior to analysis.
4. Centrifuge and wash in 5 mL PBS.
5. Centrifuge and resuspend in 4.5 mL PBS. Add 500 µL stock RNase to make a final concentration of 10 µg/mL. RNase is used to eliminate double-stranded RNA,

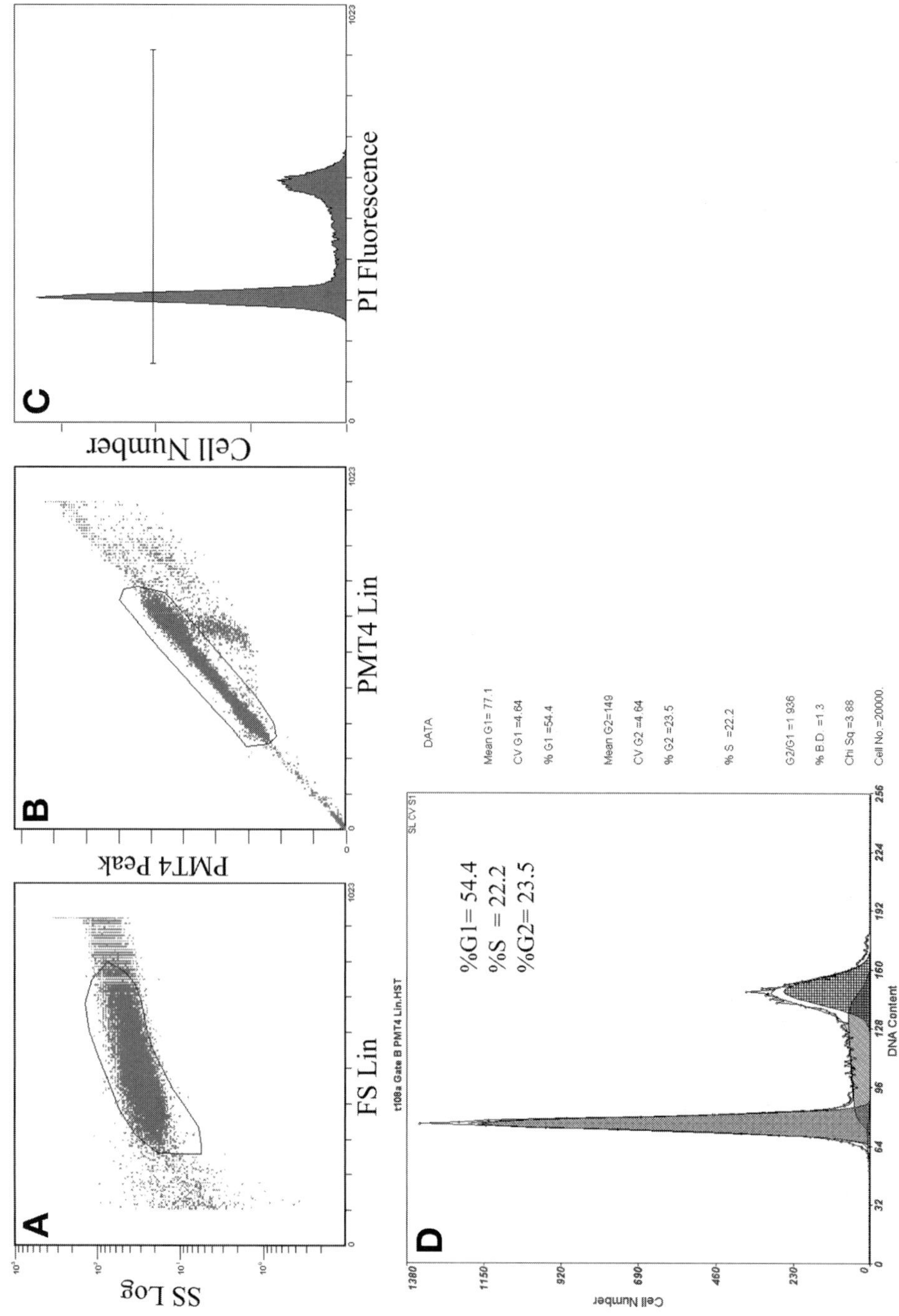

A
SS Log
FS Lin
B
PMT4 Peak
PMT4 Lin
C
Cell Number
PI Fluorescence
D
t108a Gate B PMT4 Lin.HST
SL CV S1
%G1= 54.4
%S = 22.2
%G2= 23.5
Cell Number
DNA Content
DATA
Mean G1= 77.1
CV G1 =4.64
% G1 =54.4
Mean G2=149
CV G2 =4.64
% G2 =23.5
% S =22.2
G2/G1 =1.936
% B.D. =1.3
Chi Sq =3.88
Cell No.=20000.

Fig. 1.

which can be stained by intercalating dyes such as PI and cause distortion of cell cycle distribution. Incubate at 37°C for 20 min.
6. Centrifuge and wash in 5 mL PBS.
7. Centrifuge and resuspend in 4.75 mL PBS. Add 250 µL stock PI solution and incubate for 15 min at room temperature.
8. Centrifuge and resuspend in 1 mL PBS.
9. Analyze the cells using flow cytometry. Argon ion laser with excitation at 488 nm is used to excite the dye, and a 620-nm bandpass interference filter is used to collect the emission of PI fluorescence.

3.2. Monoclonal Antibody Productivity by Measuring Intracellular Antibody

The precise mechanisms of cellular control of antibody synthesis, secretion, and the relationship between growth and antibody production remain largely unclear. The nature of intracellular antibody distribution in the population and whether antibody synthesis and secretion reflect the secretion rates within a cell culture is knowledge that may make it possible to optimize the cell culture process and significantly increase production of antibody. Several authors have used flow cytometry to monitor the production of antibody in recombinant cell lines *(23,24)*, because it is a fast and effective tool that allows the distinction between high producers and low producers.

Intracellular antibody content has also been used to monitor cell line stability in recombinant cell lines. Several authors have reported a bimodal distribution when a mixed population of producers and nonproducers was analyzed for intracellular antibody content *(23,24)*. The bimodal distribution represents cells that may have still harboring the recombinant gene and are secreting antibody and those that may have lost the recombinant gene and therefore are not secreting antibody (**Fig. 2**).

3.2.1. Monitoring IgG Productivity

1. Remove cell samples at intervals from the culture, centrifuge at 100*g* for 5 min, and resuspend in 1% paraformaldehyde.

Fig. 1. Gating analysis of Chinese hamster ovary (CHO) cells from batch culture to enhance DNA distributions. CHO cells were stained with propidium iodide and analyzed by means of dual-parameter analysis of peal height vs. integral fluorescence (**B**). Selective gating of cells to exclude debris and aggregates produces a typical DNA distribution (**C**). **A** is side scatter vs forward scatter of ungated population. The percentages of cells in G1, S, and G2 phases were mathematically analyzed by the Multicycle program from Phoenix Flow Systems (**D**). (Courtesy of Priya Santhalingam, Animal Cell Culture Group, University of Birmingham.)

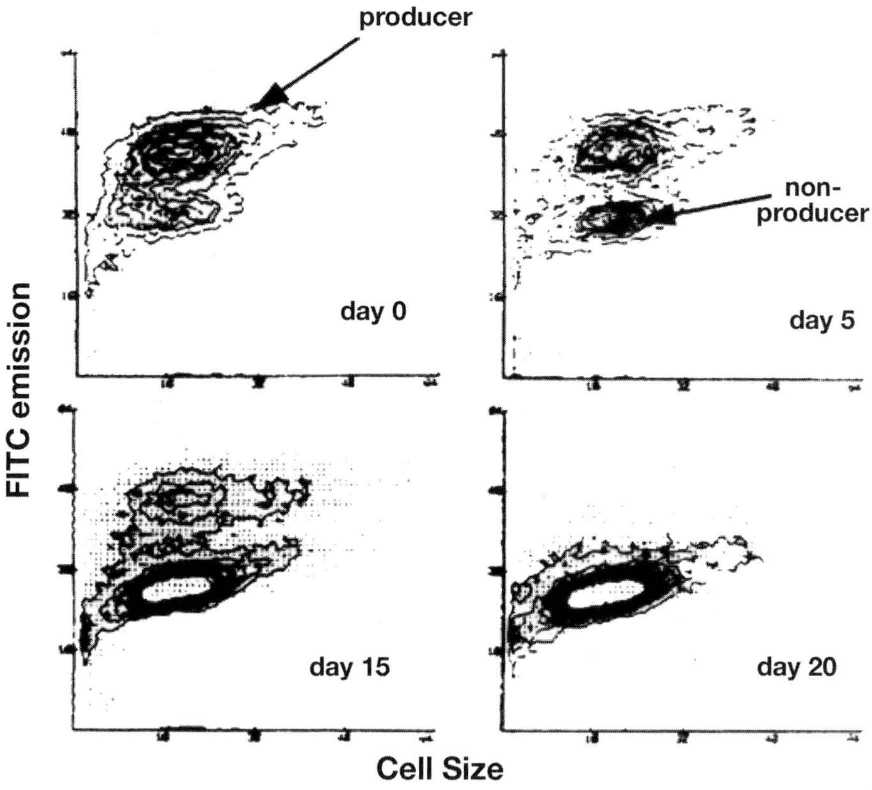

Fig. 2. Monitoring of instability of antibody production in hybridoma culture. The dot plots represent intracellular IgG vs cell size measured for cell samples removed at days 0, 5, 15, and 20. FITC emission indicates the log fluorescence emission of FITC-conjugated goat antimouse IgG. (From **ref. *20***.)

2. Wash in 5 mL PBS, and resuspend in 5 mL ice-cold 70% ethanol with rigorous agitation. The cells can be stored at –20°C until analysis for up to 3 wk.
3. Wash in 5 mL PBS and reuspend in 90 µL PBS with 10 µL FITC conjugated antibody and incubated in the dark for 1 h at room temperature.
4. Wash with 0.5% Tween-20 in PBS. Resuspend in 1 mL PBS.
5. Analyze samples by collecting the fluorescence readings on the logarithmic scale using forward scatter and side scatter to gate intact single cells.

3.3. Selection of High Producers of Monoclonal Antibody by Affinity Capture Surface Display

Traditionally, the selection of high-producing cell lines is time- and labor-consuming, and only a few clones with the ability to produce large amounts of recombinant protein are selected for expansion. Since recombinant protein production stability is an important issue, several rounds of screening for high producers have to be performed. Moreover, many of the high-producing cells are frequently overgrown by the faster growing low- or nonproducing cells. Thus, it is important to select for stable high producers accurately, efficiently, and at an early stage of the cell line development process.

The only commonly used application of this technology has been the selection of hybridoma cells for high IgG productivity because of the correlation found in this cell line between secretion rate and surface-associated protein ***(8,25)***. For other cell lines several methods based on the affinity of biotin and avidin have been described in which it is possible to capture the secreted protein on the cell surface or in the vicinity of the secreting cell ***(26–29)***. Biotin/avidin matrix systems have been described for the enrichment of populations of cells with relatively high producers and the removal of relatively low-producing cells.

3.3.1. Affinity Surface Capture Display

This procedure is cell line independent, and the process needs to be optimized for each cell line used. For the purpose of this chapter we will describe affinity capture surface display (ACSD) used on an antibody producer mouse myeloma cell line, NS0 6A1 **(Fig. 3)**.

1. A minimum of 1×10^7 cells are used for the assay. The cells are washed in PBS, followed by washing with PBS adjusted to pH 8 with 1 *M* NaOH, which removes any primary amines present on the cell surface.
2. The cells are then biotinylated by incubation in 1 mL of filter-sterilized NHS-LC Biotin (0.05 mg/mL in PBS pH 8; Pierce) at room temperature for 20 min.
3. Following incubation the cells are washed twice in PBS (pH 7) and resuspended in 1 mL of complete medium containing gelatin (10% w/v; Sigma), 26 μL biotinylated rabbit antihuman IgG, Fc specific (Pierce) and 5 μL neutravidin (1 mg/mL; Pierce). The cells are incubated in a viscous medium to prevent cross-talk, i.e., antibody secretion needs to remain localized to the cell surface.
4. The cells are allowed to secrete product for 20 min at 37°C, after which the gelatin was removed by washing twice in PBS.
5. The bound product is detected by incubating the cells in ice for 15 min in the dark with 1 mL of complete media together with 20 μL FITC conjugated goat anti-human κ light chain antibody (Sigma).
6. After washing twice in PBS, the cells are suspended in complete medium.
7. Two controls are used in the assay: the NS0 WT (nonsecretors) stained with ACSD labeling, and NS0 6A1 stained with ACSD labeling without the addition of the biotinylated antibody (negative control).
8. The cells can be gated based on the fluorescence that they emit and sorted using a cell sorter.

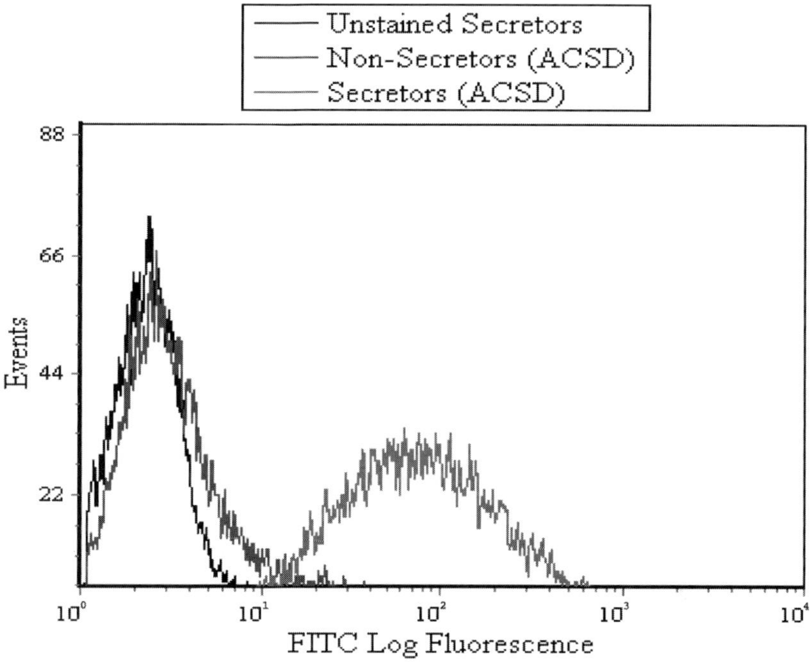

Fig. 3. NS0 secretor and NS0 wild-type (nonsecretor) cells stained by the affinity capture surface display (ACSD) method. On the basis of fluorescence intensity, high secretors are selected.

3.3.1.1. Flow Cytometry Analysis

1. Flow cytometric analysis of the ACSD cells can be performed using Coulter EPICS Elite Analyzer (Coulter Electronics, Luton, UK) equipped with an argon laser emitting 15 mW at 488 nm.
2. Cells are passed through the focus of the beam and cause fluorescent emissions. Forward scattered light (10°) is collected by the forward angle light scatter (FLS) photometer, and the 90° emissions are routed to photomultiplier tubes (PMT).
3. The forward scatter can be used as a measurement for cell size, while the side scatter can be used to determine the granularity of the cells.
4. The laser light is blocked by a 488-nm dichroic long-pass filter. FITC emissions (550 nm) are collected at PMT2 (using a 550-nm dichroic filter), and PI emissions (550–700 nm) are collected at PMT4 (635-nm dichroic filter).

3.3.1.2. Modified Method

The ACSD protocol can be modified by replacing capture antibody with Protein A as described by the Lonza Group ***(30)***: Following biotinylation of the cells **(step 2)**, the cells are washed twice with PBS pH 7, then resuspended in 1 mL PBS pH 7 with 128 μL of a 1 mg/mL solution of neutravidin (Pierce) and incubated at room temperature for 15 min. The cells are washed in PBS pH 7 and resuspended in 1 mL PBS with nonbiotinylated Protein A (Sigma) to a final concentration of 1 μg/mL and incubated at room temperature for 5 min. 52 μL of 1 mg/mL biotinylated protein A is added, and the cells are incubated for a further 15 min at room temperature. The cells were allowed to secrete in the gelatinous medium described in **step 3** with 8 μg/mL of nonbiotinylated protein A (Sigma) in order to capture antibody that diffuses away from the producer cells and thereby prevent it from binding to nonproducer cells.

3.4. Intracellular pH

Cytoplasmic pH can be used to monitor the physiological state of cell cultures. In hybridoma cultures, the pHi increases after inoculation, peaks in midexponential phase, and drops during the later phases *(31)*. Furthermore, Al-Rubeai and Walsh *(32)* found that the pHi values increase during the phases of the cell cycle noted for increased cell activity; that is, the S and G2 phases. Clearly, the pHi has a critical role in the regulation of both DNA and protein synthesis and, thereby, progression through the cell cycle. It is also clear that the external environment and specifically external pH have a greater influence on pHi.

To make direct measurement of pHi, cells are stained with pH-sensitive fluoroprobes such as SNARF-1 and analyzed by flow cytometry **(Fig. 4)**. The main advantage of SNARF as a pHi indicator is the useful ratioing properties that it exhibits. A calibration curve is needed to link the fluorescence ratio to pHi. This is accomplished by allowing cells stained with SNARF to equilibrate in high-concentration potassium buffers of different pH values containing the H^+ ionophore nigericin.

1. Centrifuge 6×10^6 cells in a sterile 25 mL vial at 450*g* for 5 min at 4°C.
2. Remove supernatants and resuspend in 2 mL HEPES media.
3. Add 20 μL from the stock solution of 1m*M* SNARF-1 to the cell suspension to give a final concentration of 10 μ*M*.
4. Incubate at 37°C for 30 min with frequent mixing to ensure homogeneity of the sample.
5. While incubating cells add 2 μg/mL Nigericin to the tubes containing the calibration buffers in the pH range of 6.5–8.0.
6. Centrifuge cell suspension at 450*g* for 5 min at 4°C.
7. Discard supernatant. A pink pellet is observed.
8. Resuspend pellet in 3 mL of fresh HEPES-buffered medium.
9. Split up into six flow cytometry tubes and spin down.
10. Carefully remove the supernatant.
11. Resuspend cells in the range of buffers to create the calibration curve or in fresh medium to make up the sample to be tested.
12. Allow cells to equilibrate in the buffers for 5 min before analyzing them in the flow cytometer.

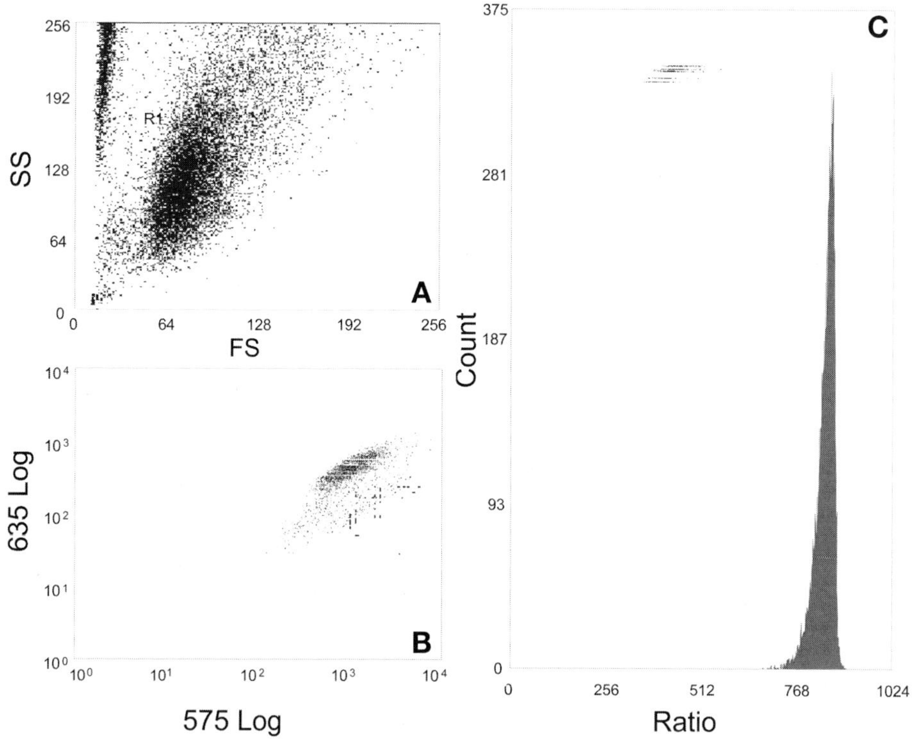

Fig. 4. pHi measurement by flow cytometry of Chinese hamster ovary cells growing in batch culture. The live cell population is gated from the side scatter vs forward scatter dot plot **(A)** for subsequent analysis of fluorescence at 575 and 635 nm **(B)** and ratio of fluorescence **(C)**.

3.4.1. Flow Cytometry Analysis

1. Flow cytometers used for pHi measurement are usually equipped with an argon ion laser (488 nm).
2. Collect fluorescence emission at 635 and 575 nm. Use forward angle scatter vs side scatter to select the live cells. Fluorescent emission gives rise to a number of useful measurements, as shown in **Fig. 4**.

3.5. Mitochondrial Mass

For mitochondrial mass analysis, MTG can be used. Cells stained with MTG exhibit bright green fluorescence with 516 nm emission. MTG is a mitochondrion-selective stain that contains a mildly thiol-reactive chloromethyl moiety that appears to be responsible for keeping the dye associated with the mitochondria, even when there is a loss of membrane potential. When the stain enters an actively respiring cell, it is oxidized and sequestered in the mitochondria,

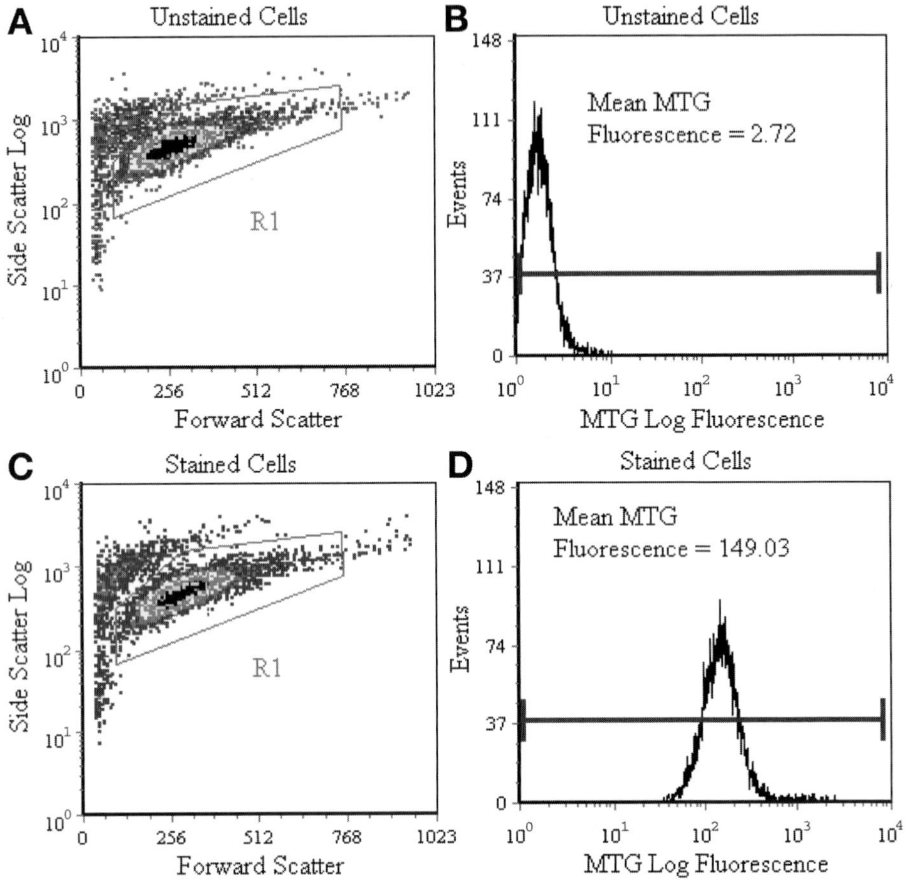

Fig. 5. NS0 WT and NS0 6A1 cells stained using MitoTracker green (MTG). Forward scatter against side scatter is used to determine the viable cells shown in region 1 of the unstained **(A)** and stained **(C)** cells. The histograms show the distributions and mean MTG fluorescence of the viable unstained **(B)** and stained **(D)** cells.

where it reacts with thiols on proteins and peptides to form an aldehyde-fixable conjugate. This can only be used with live cells at concentrations between 20 and 200 n*M* **(Fig. 5)**.

1. A minimum of 1×10^6 cells is required for this experiment. The cells are washed in prewarmed complete medium and resuspended in different concentrations of MTG diluted in fresh complete medium.
2. The concentration of MTG varies in different cell lines, and it is important to find the optimum concentration. For the purpose of this chapter we will concentrate on the NS0 6A1 cell line which uses a concentration of 150 n*M*.

3. The cells are incubated in the dark at 37°C for 45 min and analyzed using the flow cytometer.
4. The control used is cells without the MTG stain, and the measurement of mitochondrial mass is the mean fluorescence of the MTG emission, which is calculated by the flow cytometer.

4. Notes

1. PI is a suspected carcinogen. Gloves should be worn when using all DNA-binding fluorochromes. It is possible to observe a bell-shaped region overlapping the G1 peak, which extends to merge with debris region, towards the left-hand side of the histogram. The degree of DNA leakage can be manipulated by the extent of cell washing after fixation so the overlap with cells that undergo apoptosis is minimal and separation between these two populations is adequate.
2. Cell biotinylation in the ACSD protocol is dependent on the amount of amines present on the cell surface of the cells, the viability of the cells, and their growth phase. It is therefore important to carry out certain biotinylation studies to determine the optimum biotin concentration for the cell line. This can be achieved by washing the cells in PBS followed by another wash in PBS pH 8.0. The cells are then biotinylated by incubation in different concentrations of biotin at room temperature for 20 min. Following incubation the cells are washed twice in PBS and resuspended in 1 mL PBS with 10 µL streptavidin FITC (Sigma) for 10 min in the dark. Following this the cells are washed twice in PBS and resuspended in 2 mL PBS with 10 µg PI to distinguish between viable and nonviable cells. The control used for this experiment is unbiotinylated cells stained with streptavidin FITC.
3. It is advisable to also create a saturation curve for ACSD labeling to determine the sensitivity of the assay on the cell line used. This can be done by following **Subheading 3.3.1., step 4** using different secretion times.
4. Flow cytometry results should first be analyzed by plotting forward scatter against side scatter, which will show the viable cells, which can be gated and a histogram plotted from this gate. The histogram should include the stain (1) in the case of cell cycle analysis—PI against events, and (2) in the case of intracellular antibody concentration or ACSD—FITC against count. Flow cytometry analysis software or the flow cytometer itself should calculate the amount of mean fluorescence and coefficient of variation for the gated population.
5. Nigericin does exhibit a degree of cytotoxicity in addition to the possible SNARF leakage from cells; therefore, the flow cytometric analysis of pHi should be performed within 30 min after the addition of nigericin.

References

1. Leibson, P. J., Loken, M. R., Runem, S., and Schreiber, H. (1999) Clonal evolution of myeloma cells leads to quantitative changes in immunoglobulin secretion and surface antigen expression. *Proc. Natl. Acad. Sci. USA* **76(6)**, 2937–2941.

2. Meilhoc, E., Witturp, K. D., and Bailey, J. E. (1989) Application of flow cytometric measurement of surface IgG in kinetic analysis of monoclonal antibody and secretion by murine hybridoma cells. *J. Immunol. Methods* **121**, 167–174.
3. Dalili, M. and Ollis, D. F. (1990) A flow cytometric analysis of hybridoma growth and monoclonal antibody production. *Biotechnol. Bioeng.* **36(1)**, 64–73.
4. Marder, P., Maciak, R. S., Fouts, R. L., Baker, R. S., and Starling, J. J. (1990) Selective cloning of hybridoma cells for enhanced immunoglobulin production using flow cytometric cell sorting and automated laser nephelometry. *Cytometry* **11(4)**, 498–505.
5. Sen, S., Hu, W. S., and Srienc, F. (1990) Flow cytometric study of hybridoma cell culture: correlation between cell surface fluorescence and IgG production rate. *Enzyme Microb. Technol.* **12**, 571–576.
6. Al-Rubeai, M., Emery, A. N., and Chalder, S. (1991) Flow cytometric study of cultured mammalian cells. *J. Biotechnol.* **19**, 67–82.
7. Leno, M., Merten, O.-W., Vuiller, F., and Hache, J. (1991) IgG production in hybridoma batch culture: kinetics of IgG mRNA, cytoplasmic-, secreted- and membrane-bound antibody levels. *J. Biotechnol.* **20**, 301–312.
8. McKinney, K. L., Dilwith, R., and Belfort, G. (1991) Manipulation of heterogeneous hybridoma cultures for overproduction of monoclonal-antibodies. *Biotechnol. Prog.* **7(5)**, 445–454.
9. Park, S. H. and Ryu, D. Y. (1994) Cell cycle kinetics and monoclonal antibody productivity in hybridoma cells during perfusion culture. *Biotechnol. Bioeng.* **44**, 361–367.
10. Altshuler, G. L., Dilwith, R., Sowek, J., and Belfort, G. (1986) Hybridoma analysis at cellular level. *Biotechnol. Bioeng. Symp.* **17**, 725–736.
11. Al-Rubeai, M. and Emery, A. N. (1990) Mechanisms and kinetics of monoclonal-antibody synthesis and secretion in synchronous and asynchronous hybridoma cell culture. *J. Biotechnol.* **16(1–2)**, 67–86.
12. Ramirez, O. T. and Mutharason, R. (1990) Cell cycle and growth phase dependent variations in size distributions, antibody productivity, and oxygen demand in hybridoma cultures. *Biotechnol. Bioeng.* **36**, 839–848.
13. Kromenaker, S. J. and Srienc, F. (1991) Cell cycle-dependent protein accumulation by producer and non-producer murine hybridoma cell lines: A population analysis. *Biotechnol. Bioeng.* **38**, 665–677.
14. Hayter, P. M., Kirkby, N. F., and Spier, R. E. (1992) Relationship between hybridoma growth and monoclonal antibody production. *Enzyme Microbiol. Technol.* **14**, 454–461.
15. Linardos, T. I., Kalogerakis, N., and Behie, L. A. (1992) Cell cycle model for growth rate and death rate in continuous suspension hybridoma cultures. *Biotechnol. Bioeng.* **40**, 359–368.
16. Leelavatcharamas, V., Emery, A. N., and Al-Rubeai, M. (1996) Monitoring the proliferative capacity of cultured animal cells by cell cycle analysis, in *Flow Cytometry Application in Cell Culture* (Al-Rubeai M. and Emery A. N., eds.), Marcel Dekker Inc., New York, pp.1–15.
17. Kearns, B., Lindsay, D., Manahan, M., McDowall, J., and Rendeiro, D. (2003) NS0 batch cell culture process characterization: a case study. *Biol. Proc. J.* **2(1)**, 52–57.

18. Carroll, S. (2004) The selection and monitoring of high producing recombinant cell lines using flow cytometry, PhD thesis, The University of Birmingham.
19. Carroll, S. and Al-Rubeai, M. (2005) ACSD labeling and magnetic cell separation: A rapid method of separating antibody secreting cells from non-secreting cells. *J. Immunol. Methods* **(1–2)**, 171–178.
20. Al-Rubeai, M. (1999) Monitoring animal cell growth and productivity by flow cytometry, in *Methods in Biotechnology,* Vol. 8: *Animal Cell Biotechnology* (Jenkins, N., ed.), Humana Press Inc., Totowa, NJ, pp. 145–153.
21. Ormerod, M. G. and Imrie, P. R. (1990) Flow cytometry, in *Methods in Molecular Biology*, Vol. 5. (Pollard, J. W. and Walker, J. M., eds.), Humana Press Inc., Totowa, NJ, pp. 543–558.
22. Grogan, W. M. and Collins, J. M. (1990) *Guide to Flow Cytometry Methods*, Marcel Dekker, New York.
23. Merritt, S. E. and Palsson, B. O. (1993) Loss of antibody productivity is highly reproducible in multiple hybridoma subclones. *Biotechnol. Bioeng.* **42**, 247–250.
24. Borth, N., Strutzenberger, K., Kunert, R., Steinfellner, W., and Katinger, H. (1999) Analysis of changes during subclone development and ageing of human antibody-producing heterohybridoma cells by Northern blot and flow cytometry. *J. Biotechnol.* **67**, 57–66.
25. Chuck, A. S. and Palsson, B. O. (1992) Population balance between producing and non-producing hybridoma clones is very sensitive to serum level, state of innoculum, and medium composition. *Biotechnol. Bioeng.* **39**, 354–360.
26. Manz, R., Assenmacher, M., Pfluger, E., Miltenyi S., and Radbruch, A. (1995) Analysis and sorting of live cells according to secreted molecules, relocated to a cell-surface affinity matrix. *Proc. Natl. Acad. Sci. USA* **92**, 1921–1925.
27. Holmes, P. and Al-Rubeai, M. (1999) Improved cell line development by a high throughput affinity capture surface display technique to select for high secretors. *J. Immunol. Methods* **230**, 141–147.
28. Borth, N., Zeyda, M., and Katinger, H. (2001) Efficient selection of high-producing subclones during gene amplification of recombinant Chinese hamster ovary cells by flow cytometry and cell sorting. *Biotechnol. Bioeng.* **71(4)**, 266–273.
29. Carroll, S. and Al-Rubeai, M. (2004) The selection of high-producing cell lines using flow cytometry and cell sorting. *Expert Opin. Biol. Ther.* **4(11)**,1821–1829.
30. Lonza Group AG. (2004) Method for selecting antibody expressing cells. EP1415158B1.
31. Al-Rubeai, M., Kloppinger, M., Fertig, G., Emery, A. N., and Miltenburger, H. G. (1992) Monitoring of biosynthetic and metabolic activity in animal cell culture using flow cytometric methods, in *Animal Cell Technology* (Spier R., Griffiths, J., and MacDonald, C., eds), Butterworth-Heinemann, pp. 301–307.
32. Welsh, J. P. and Al-Rubeai, M. (1996) The relationship between intracellular pH and cell cycle in cultured animal using SNARF-1 indicator, in *Flow Cytometry Applications in Cell Culture* (Al-Rubeai, M. and Emery, A. N., eds.), Marcel Dekker, New York, pp. 163–175.

10

Nuclear Magnetic Resonance Methods for Monitoring Cell Growth and Metabolism in Intensive Bioreactors

André A. Neves and Kevin M. Brindle

Summary

Magnetic resonance imaging (MRI) and spectroscopy (MRS) are powerful noninvasive techniques that can be used to monitor the behavior of cells in intensive bioreactor systems. We describe here a number of nuclear magnetic resonance (NMR)-based techniques that have been used successfully to investigate cell growth and distribution, cellular energetics and the porosity to medium flow and linear flow velocity profiles around and across cell layers in perfusion bioreactors. These are important parameters to determine when designing bioreactor systems and operation protocols that optimize cell productivity.

Key Words: Nuclear magnetic resonance; magnetic resonance imaging; magnetic resonance spetroscopy; cell growth; cell metabolism; bioreactors.

1. Introduction

Nuclear magnetic resonance (NMR) was first described in 1946 by Felix Bloch and Edward Purcell, who received the Nobel Prize in Physics in 1952 for their discovery. NMR was originally a spectroscopic technique used mainly by organic chemists for elucidating the structure of relatively small molecules. The advent of Fourier transform NMR and the development of superconducting magnets with higher field strengths opened the technique to a range of biological and clinical applications. This included the development of NMR methods in the early 1980s for solving the structures of proteins in solution, for which Richard Ernst and Kurt Wuthrich were awarded the Nobel Prize in Chemistry in 1991 and 2002, respectively, and the development of methods in the early 1970s for magnetic resonance imaging (MRI), for which Paul Lauterbur and Sir Peter Mansfield shared the Nobel Prize for Physiology and Medicine in 2003. The 1970s also saw the first demonstrations that NMR spectroscopy can be used to obtain chemical information noninvasively from intact biological systems

From: *Methods in Biotechnology, Vol. 2: Animal Cell Biotechnology: Methods and Protocols, 2nd Ed.* Edited by: R. Pörtner © Humana Press Inc., Totowa, NJ

(1–4) (the word nuclear is frequently dropped and the technique called magnetic resonance spectroscopy [MRS]). The technique has since become a proven and established technique for monitoring metabolism in a range of biological systems, including cells, perfused organs, bioartificial tissues, as well as intact animals and humans *(2,5–7)*. A major analytical advantage of NMR is its ability to provide extensive information on a wide range of biologically important low molecular weight species simultaneously. This has been exploited recently in the field of metabolomics *(8,9)*, where the aim is to define metabolite profiles that characterize and can be used to identify particular disease states. Although many NMR-detectable nuclei exist, studies of cell metabolism have, for sensitivity reasons, been restricted mainly to ^{1}H, ^{31}P, ^{13}C, and ^{15}N nuclei *(6)*. The basic principles of MRS and magnetic resonance imaging (MRI) are beyond the scope of this chapter and have been described extensively elsewhere (*see*, for example, **refs.** *7* and ***10***). The ability of the technique to obtain chemical information noninvasively and in a spatially resolved manner from intact biological systems offers considerable scope for the development of novel applications in the pharmaceutical and biotechnology industries. This chapter will focus on various NMR-based approaches for studying the growth and metabolism of mammalian cells in culture flasks and in intensive bioreactor systems. The application of NMR to intensive bioreactor systems has been reviewed *(11)*.

1.1. Intensive Bioreactor Systems

There has been considerable interest in the biotechnology industry in the development of systems for maintaining high-density mammalian cell cultures, mainly because the majority of biopharmaceuticals currently on the market are produced using mammalian expression systems *(12)*. In addition to the intensive culture of mammalian cells as systems for producing monoclonal antibodies, vaccines, and other proteins with diagnostic or therapeutic applications, intensive bioreactors have also been used recently as vehicles for the production of human cell lines for cell- and tissue-based therapies *(13)*.

The examination of the growth and metabolism of intact cells by NMR requires the use of cell perfusion systems that are compatible with the geometry of the magnet/spectrometer to be used and are also able to sustain a viable cell population at high density under well-defined and homogeneous conditions. Various cell perfusion methods have been developed for NMR applications. The type of metabolic data obtainable can be found in a number of publications *(11,14)*. The hollow-fiber bioreactor (HFBR), in particular, is an important intensive cell culture system, which has been adapted by a number of groups for use with an NMR spectrometer. This has allowed noninvasive NMR studies of cellular metabolism at various stages of cell growth *(15,16)*.

Thus, intensive bioreactors are a valuable tool for studying cell metabolism by NMR; conversely, NMR is a valuable tool for investigating the performance

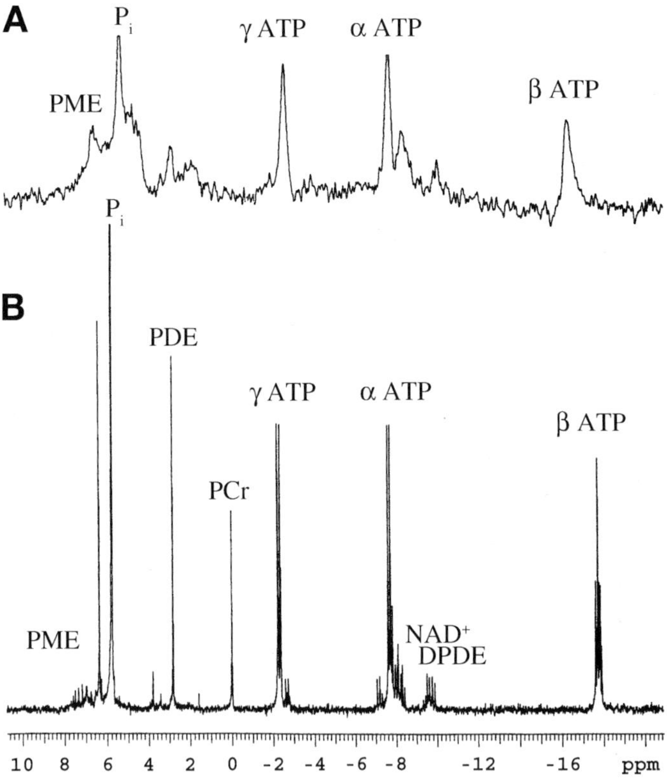

Fig. 1. ^{31}P NMR spectrum of Chinese hamster ovary (CHO-K1) cells growing in a hollow-fiber bioreactor **(A)** and of a perchloric acid extract **(B)** of the same cells. Note the improvement in spectral resolution and signal-to-noise ratio in the extract spectrum. Signal assignments are PME, phosphomonoesters; P_i, inorganic phosphate; PDE, phosphodiesters; PCr, phosphocreatine; ATP, adenosine triphosphate; NAD^+, nicotinamide adenine dinucleotide; DPDE, diphosphodiesters.

of these systems. NMR spectroscopy and imaging **(Figs. 1** and **2)** can provide important information that assists in the design, operation, and optimization of these systems. This is likely to be of particular importance in the area of tissue engineering, where the growth and metabolic activity of the cells and the physical properties of the scaffold material are closely related to their function as bioartificial tissues ***(17,18)***.

1.2. Assessing Cellular Energetics Using ^{31}P MRS

^{31}P MRS is a particularly useful technique for studying cellular metabolism, as several phosphorylated metabolites are present in sufficiently high concentration to yield detectable NMR signals within a reasonable time frame. The energy status of a cell can be assessed by monitoring the levels of adenosine

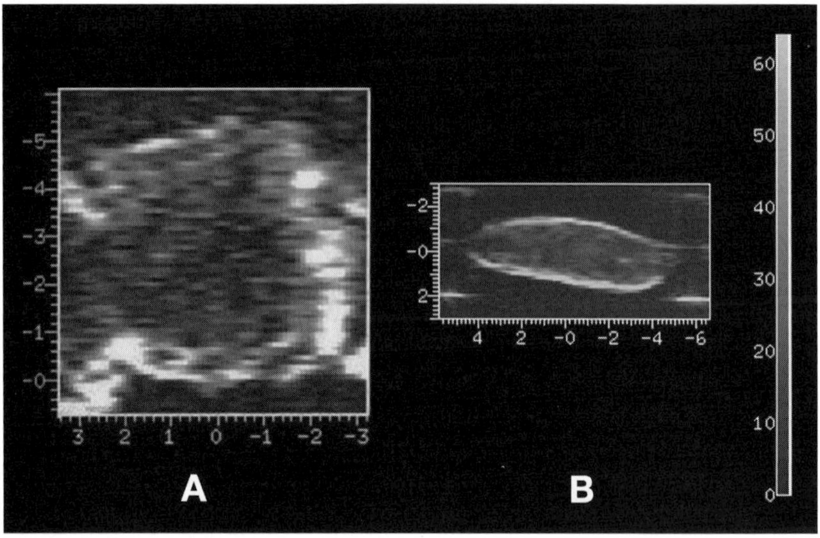

Fig. 2. Diffusion-weighted ^{1}H MR image of Chinese hamster ovary (CHO-K1) cells growing in a single carrier **(A)** in the perfusion bioreactor (*see* **Subheading 2.2.2.**) and from meniscal cartilage cells growing in a tissue-engineered construct **(B)** cultivated in a perfusion bioreactor (*see* **Subheading 2.2.3.**). The two images were acquired 17 and 14 d, respectively, after seeding of the bioreactors. The image plane was saggital to the carrier and construct and parallel to the direction of medium flow. In these diffusion-weighted images, signal intensity, expressed in arbitrary units (right-hand-side scale), is proportional to cell density, with the highest cell densities being found at the edges of the carrier **(A)** and construct **(B)**. The scales of the x and y axes are in millimeters. (Panel **A** from **ref. *26*.**)

triphosphate (ATP), phosphocreatine (PCr), and inorganic phosphate (Pi). The chemical shift of Pi can be used to measure the intracellular pH ***(1,19,20)***. Information on a number of other phosphorylated metabolites that are important in intermediary metabolism (e.g., glycolytic intermediates), phospholipid metabolism (phosphocholine, glycerophosphocholine, phosphoethanolamine, and glycerophosphoethanolamine), and protein glycosylation (uridine diphosphate sugars) can also be obtained from ^{31}P NMR spectra ***(5,6)***.

1.3. Quantifying and Visualizing Cell Growth in Bioreactors Using MRS and MRI

The content and distribution of cells in a NMR-compatible bioreactor can be assessed throughout the cultivation period using diffusion-weighted MRS and MRI, respectively. The introduction of a pair of pulsed magnetic field gradients into a spin-echo experiment sensitizes the NMR signal to diffusion.

By measuring the decrease in signal intensity with increasing gradient amplitudes, the diffusion coefficient of a molecule can be estimated ***(21)***. This technique has been used to measure the diffusion coefficient of water and a variety of small molecule metabolites in tissues ***(22,23)***. In a cellular system, the diffusion of intracellular species is restricted by the cell membrane, and this can reduce the observed or apparent diffusion coefficient (ADC). Quantitation of the fractions of fast- and slow-diffusing water in a bioreactor can be used to provide a measure of the total cell density ***(24)***. Similarly, the incorporation of diffusion-weighting gradients into an imaging sequence and the use of relatively large gradient amplitudes, which suppress signal from fast-diffusing water, can be used to map cell distribution in bioreactors ***(15,25–27)***. By using a variety of gradient amplitudes, these imaging experiments can also be used to produce maps of the ADC of the molecule or ion of interest, normally water, for reasons of sensitivity.

Other NMR-based techniques can also be used to estimate cell density and viability in bioreactors. ^{1}H MRS measurements of choline concentration were found to be an accurate and nondestructive way of monitoring viable cell numbers in tissue-engineered constructs in vitro ***(28)***. Mancuso and coworkers developed a technique for determining cell concentration in a HFBR using ^{23}Na MRS measurements of extracellular sodium concentration. The values obtained were in good agreement with concentrations calculated from ^{31}P NMR measurements of nucleoside triphosphate concentration and measurements of oxygen consumption rate ***(29)***. ^{31}P MRS has also been used to measure cell viability and the concentrations of phosphorus-containing metabolites in neocartilage ***(30)*** and in hepatocytes ***(31)*** growing in HFBRs and to study the effects of short-term hypoxia on transformed cell-based bioartificial pancreas constructs ***(32)***.

1.4. Nutrient Metabolism and Metabolite Fluxes From Isotope Labeling Experiments

This NMR-based strategy is analogous to the use of labeled compounds in radiotracer-based studies. Cells are incubated with ^{13}C- or ^{15}N-labeled nutrients, and the redistribution of the label among cellular metabolites is monitored using ^{13}C or ^{15}N MRS experiments, respectively. In contrast to radiotracer studies, in which separation of the labeled metabolites is normally required, all those labeled metabolites that are present at sufficiently high concentration can be detected simultaneously in the NMR spectrum. Furthermore, NMR can reveal not only which molecules are labeled, but also the positions in the molecules that are labeled. It is therefore possible to focus on the metabolic fate of a labeled precursor and to measure specific metabolic fluxes within cells ***(4,33–35)***. For example, the fluxes in the glycolytic pathway and tricarboxylic acid cycle have been studied using ^{13}C-labeled substrates ***(36,37)***. The sensitivity of label detection can be improved by observing the label indirectly *via* the ^1H nucleus in $^1H/^{13}C$ or $^1H/^{15}N$-NMR experiments. In such cases, quantitative information on fractional labeling is also obtained as both the labeled and unlabeled species can be detected in the ^1H spectrum. This type of strategy has been applied to monitor the ^{15}N labeling of cellular metabolites in cultures of mammalian cells ***(38)***.

2. Materials

2.1. Cell Culture

1. Medium for adherent or suspension culture cells (e.g., DMEM or RPMI, Invitrogen, Carlsbad, CA), typically supplemented with 2 mM L-glutamine, 100 U/mL penicillin, 100 mg/mL streptomycin, and 10% fetal bovine serum.
2. Phosphate-buffered saline (PBS, Invitrogen) and trypsin (0.05%)–ethylenediamine tetraacetic acid (EDTA, Invitrogen) (0.02%) solution for routine subculturing of adherent cell lines.
3. T175 tissue culture flasks and 140 × 15 mm tissue culture dishes (Nunclon, Nunc, Rochester, NY) for adherent cultures; siliconized spinner flasks, and magnetic stirrer (Techne, Burlington, NJ) for suspension cultures.

2.1.1. Perchloric Acid Extraction of Cells

1. 6% Perchloric acid (PCA; Fisons, Beverly, MA).
2. 2 *M* K_2CO_3.
3. Ion exchange resin; Chelex™ 100 resin (200–400 mesh sodium form; Bio-Rad, Hercules, CA).

2.1.2. ^{31}P NMR Analysis of Cell Extracts

1. NMR tubes (typically 10-mm diameter; Wilmad, Buena, NJ).
2. Extract buffer: 50 m*M* triethanolamine, pH 8.4, containing 15 m*M* EDTA.
3. Deuterium oxide (D_2O; Sigma-Aldrich, Milwaukee, WI) (*see* **Note 1**).
4. NMR chemical shift and quantitation standard: 30 m*M* methylene diphosphonic acid (MDP) in 50 m*M* triethanolamine buffer, pH 8.4, with 1 m*M* EDTA, contained in a sealed coaxial capillary tube. These tubes can be obtained from Wilmad (*see* **Note 2**).
5. High-field, high-resolution NMR spectrometer (*see* **Note 3**).

2.2. Culture of Intact Cell Systems in Perfusion Bioreactors for NMR

1. Culture medium (e.g., Dulbecco's modified Eagle's medium, or DMEM), typically supplemented with 2 mM L-glutamine, 100 U/mL penicillin, 100 mg/mL streptomycin, and 10% fetal bovine serum.
2. T175 tissue culture flasks (Nunclon), spinner flasks, and magnetic stirrer (Techne).
3. Microcarrier beads, e.g., Cytodex-1™ (GE Healthcare, Waukesha, WI).

2.2.1. Hollow-Fiber Bioreactor

1. HFBR (NMR-compatible, e.g., CellFlo™ Plus, Spectrum Laboratories, Rancho Dominguez, CA).

2. Stirred-tank fermenter (e.g., Applicon, Clinton, NJ), used as a conditioning vessel, with a geometric volume of 2 L and a working volume of 1.25 L.
3. NMR spectrometer (*see* **Subheading 2.1.2.**) (*see* **Note 4**).

2.2.2. Fixed-Bed Bioreactor With Cell Carriers

1. Same stirred-tank fermenter as described in **Subheading 2.2.1**.
2. A fixed-bed bioreactor. We constructed one in-house, which consisted of a screw-capped polysulfone tube (RS Components/Allied Electronics, Fort Worth, TX), 20 mm internal diameter ***(26)***.
3. Polyethyleneimine-coated cellulose sponge carriers (Cellsnow™ FN-S05T, Kirin Brewing Company Ltd., Tokyo, Japan).

2.2.3. Fixed-Bed Bioreactor With Tissue-Engineering Scaffolds

1. A stirred-tank fermenter, as described in **Subheading 2.2.1.**, and bioreactor, as described in **Subheading 2.2.2**.
2. We have used a fixed bed consisting of three scaffolds in the form of disk-shaped meshes 12 mm in diameter and 4 mm thick. The meshes were produced from polyethylene terephthalate (PET; Dow Chemicals, Piscataway, NJ), which had been extruded in 13-μm fibers to produce scaffolds with a void volume of 97% and a density of 45 mg/cm^3. The lower section of the bioreactor was fitted with ultra-high-density polyethylene spacers (RS Components/Allied Electronics), which allowed for separation of the scaffolds and for flow of medium both through and around the scaffolds ***(17)***.

3. Methods

3.1. Cell Culture and Preparation of Cell Extracts for NMR Analysis

The intrinsically low sensitivity of NMR requires cells to be grown to very high densities (of the order of 10^8 cells/mL) if spectra are to be acquired from intact cells. For this reason many studies are performed on protein-free cell extracts. These have the advantage of being magnetically more homogeneous and therefore yield better resolved spectra (compare **Figs. 1A** and **B**). The extracts can be concentrated prior to NMR analysis, and because there is, in principle, no limit on the NMR measurement time, the signal-to-noise ratio obtainable in the spectra is, therefore, much better. The most commonly used aqueous extraction procedure employs perchloric acid (PCA), although trichloroacetic acid and chloroform/methanol can also be used. A ^{31}P MRS spectrum of a typical PCA extract is shown in **Fig. 1A** (*see* **Note 5**).

3.1.1. Perchloric Acid Extraction of Cells

1. For anchorage-dependent cells, remove cell culture medium from the dishes and scrape monolayer (typically 0.5–1.0 × 10^9 cells, protein content approx 90 mg) into 30 mL of 6% PCA. Leave at 4°C for 20 min. For suspension cultures, centrifuge cell

suspension (typically 0.5–1.0 × 10^9 cells in total) at 500*g* for 5 min, discard the supernatant culture medium, and resuspend cell pellet in 30 mL of 6% PCA for 20 min at 4°C.

2. Centrifuge extract at 1500*g* for 5 min and neutralize supernatant with 2 *M* K_2CO_3.
3. Centrifuge to remove the potassium perchlorate precipitate and remove any paramagnetic ions present by treating the supernatant with 2 g of Chelex™ 100 resin. Leave at 4°C for 30 min.
4. Centrifuge to remove the Chelex resin from the solution. Place the supernatant on ice.
5. Wash the Chelex resin with an equal volume of water and leave at 4°C for a further 30 min. Centrifuge and pool the supernatant with the first supernatant.
6. Snap-freeze the final sample in liquid nitrogen.
7. Lyophilize the sample and store at –20°C prior to NMR analysis.

3.1.2. ^{31}P MRS Analysis of Cell Extracts

1. Dissolve the cell extract in 5 mL of extract buffer and remove undissolved material by centrifugation.
2. Transfer 3.0 mL of the supernatant to a 10-mm-diameter NMR tube containing 0.6 mL D_2O. MDP contained in a coaxial capillary tube can be used as a chemical shift and quantitation standard (*see* **Note 2**).
3. Acquire the NMR spectrum, maintaining the sample at a fixed temperature, typically 30°C. A 5-s interpulse delay and a 60° flip angle pulse are normally sufficient to ensure full relaxation of the metabolite resonances (*see* **Note 6**).

3.2. Intact Cell Systems

3.2.1. Cell Culture for Perfusion Bioreactor Systems

1. The MR-compatible bioreactors described here have been operated with adherent cell lines: CHO-K1 cells ***(15,24,26,27)*** and meniscal fibrochondrocytes ***(17)***. The different cell lines are propagated in T-flasks (175 cm^2 surface area) and split when they reach 85–95% confluence. They are then washed with PBS and treated with 5 mL of a solution containing 0.05% trypsin and 1 m*M* EDTA in PBS for 5 min at 37°C. The resulting cell suspension is then mixed with an equal volume of medium and centrifuged at 240*g* for 5 min. The cell pellets are resuspended and seeded at a density of 4 × 10^4 cells/cm^2.
2. In the case of the HFBR ***(15)***, CHO-K1 cells were further propagated in spinner flasks containing Cytodex™ beads (1 g/L) until a cell density of approx 5 × 10^8 cells/mL was reached. Cells were inoculated in these flasks at 2 × 10^5 cells/mL, and after propagation the cell-loaded beads were placed in the extracapillary space of the HFBR.
3. In the case of the perfusion bioreactor operated with cell carriers ***(26)***, the seeding was performed on-line. Approximately 150 randomly oriented carriers (5 × 5 × 5 mm), that constitute the 28-mL fixed-bed, were soaked with PBS, prior to autoclaving. Seeding of the carriers in the bioreactor was performed by recirculating a suspension of 8 × 10^7 cells in 300 mL of culture medium, through the fixed-bed, providing a final seeding density of 2.6 × 10^5 cells/mL of bed volume (the seeding efficiency was >95% ***[24]***).

4. The tissue-engineering perfusion bioreactor *(17)* was loaded off-line with three scaffolds that had been previously seeded with cells. The scaffolds were seeded with sheep fibrochondrocytes in well-mixed 250-mL spinner flasks (Fisher Scientific, Pittsburgh, PA). Each flask, containing eight scaffolds, was inoculated with 1×10^8 cells, corresponding to 1.25×10^7 cells per scaffold. Over a period of 3 d cells attach to the surface of the scaffolds with no significant cell loss and an adhesion yield >95%. The scaffolds can then be transferred aseptically to the bioreactor.

3.2.2. Bioreactor Perfusion Systems and NMR Analysis

1. Medium is pumped from the batch-fed stirred tank fermenter containing 1.25 L of medium through the perfusion bioreactor. The flow rate required depends on the bioreactor. For the HFBR a flow rate of 50 mL/min was used. The fixed-bed bioreactor with cell carriers, or tissue-engineering scaffolds, was operated at flow rates between 5 and 45 mL/min or 30–60 mL/min, respectively.
2. The medium in the fermenter was gassed by passing a mixture of O_2 (21%), N_2 (79%), and CO_2 (the required amount to keep the pH of the medium at the desired pH) through silicone rubber tubing wound on a former in the mixing vessel (*see* **Note 7**).
3. The pO_2 and pH of the medium in the fermenter were measured using a polaro graphic oxygen electrode and a glass pH electrode, respectively.
4. The dissolved oxygen tension, pH, and temperature of the perfusate are regulated.
5. Cells can be maintained in this type of system for several weeks and can achieve densities of $>10^8$/mL.
6. Imaging techniques are as follows:
   a. Spin-echo images can be used to monitor fiber distribution in the HFBR *(15)* or the position of the cell carriers *(24)* or tissue-engineering scaffolds *(17)* in the other perfusion bioreactors.
   b. Diffusion-weighted MRS and MRI can be used to estimate cell density *(24)* and to map cell distribution in the three systems *(15,17,26)* (*see* **Fig. 2**).
   c. Flow imaging can be used to map flow rates in both the fibers and the extra-capillary space of the HFBR *(39)* or the linear flow velocities in and around the tissue-engineering constructs *(17)*.
   d. Chemical shift imaging may also be used to map the distribution of cellular metabolites *(40)*.
   e. Contrast-agent enhanced MRI can be used to monitor the diffusivity of a small molecular weight contrast agent inside the cell carriers *(26)* or inside the tissue-engineering constructs *(17)*, providing information on tissue density and permeability, which affect directly cell density and viability.
   f. ^{19}F MRI of a perfluorocarbon probe molecule can be used to map the dissolved oxygen concentration in these systems *(27)*.
   g. ^{31}P MRS can be used to evaluate cell viability and energetics *(17,24)*.

4. Notes

1. The deuterium lock stabilizes the spectrometer against variation in its static magnetic field or operating frequency by maintaining the ratio of the two constant. This is

achieved by monitoring the resonance frequency of the deuterium signal and varying the static magnetic field so that the lock frequency remains constant.
2. The inclusion of a standard allows the conversion of signal intensities into concentrations and also provides a reference for chemical shift. The chemical shifts in the ^{31}P spectra shown in this chapter are expressed relative to the resonance of MDP, which was contained in a coaxial capillary tube. Phosphocreatine is also frequently used as an internal chemical shift standard.
3. Relatively high magnetic fields (typically >7 T) are required in order to obtain sensitivity and spectral resolution (i.e., machines that operate at proton frequencies >300 MHz).
4. The instrument should be high field in order to obtain the best sensitivity. A wide-bore magnet is preferable as this can accommodate relatively large HFBRs and other perfusion bioreactors (up to 45 mm outer diameter in a standard vertical wide-bore [89-mm] magnet). A wide-bore magnet can also be equipped with a magnetic field gradient set for micro-imaging and localized MRS experiments.
5. Although cell extraction is frequently adopted, information about molecules affected by the extraction procedure is necessarily lost. In the case of ^{31}P MRS, the level of inorganic phosphate in a cell extract can be significantly higher than that measured in the corresponding intact cells because of hydrolysis of organic phosphates during the extraction procedure.
6. The inherent insensitivity of NMR imposes limitations on the concentrations of compounds that can be detected. This is particularly relevant for nuclei for which NMR is rather insensitive (e.g., ^{31}P). Repeated data acquisitions are therefore required to obtain adequate signal-to-noise ratios in the resulting spectra. However, to determine concentrations from resonance intensities, either the relaxation times of the resonances must be known or the acquisition conditions must be such as to allow the complete relaxation of the resonances between successive acquisitions, i.e., the delays between successive acquisitions must be relatively long (normally in the range of seconds). This frequently leads to lengthy data-acquisition times for high-resolution ^{31}P MRS.
7. The medium cannot be sparged directly with the gases as this would cause extensive foaming because of the presence of serum.

Acknowledgments

The work performed in the K.M.B. laboratory, which is described here, was supported by the Biotechnology and Biological Sciences Council, UK, and the European Community Framework IV Programme (Biotechnology-950207). A.A.N. thanks F.C.T. (Portugal) for his PhD scholarship (PRAXIS XXI/BD 19519/99). The authors acknowledge Smith & Nephew Plc, UK, for sponsorship of the tissue-engineering research.

References

1. Gadian, D. G. and Radda, G. K. (1981) NMR studies of tissue metabolism. *Annu. Rev. Biochem.* **50**, 69–83.

2. Radda, G. K. (1986) The use of NMR spectroscopy for the understanding of disease. *Science* **233(4764)**, 640–645.
3. Radda, G. K. (1992) Control, bioenergetics, and adaptation in health and disease: non-invasive biochemistry from nuclear magnetic resonance. *FASEB J.* **6(12)**, 3032–3038.
4. Shulman, R. G., Brown, T. R., Ugurbil, K., Ogawa, S., Cohen, S. M., and den Hollander, J. A. (1979) Cellular applications of ^{31}P and ^{13}C nuclear magnetic resonance. *Science* **205(4402)**, 160–166.
5. Avison, M. J., Hetherington, H. P. and Shulman, R. G. (1986) Applications of NMR to studies of tissue metabolism. *Annu. Rev. Biophys. Chem.* **15**, 377–402.
6. Cohen, J. S., Jaroszewski, J. W., Kaplan, O., Ruiz-Cabello, J., and Collier, S. W. (1995) A history of biological applications of NMR spectroscopy. *Prog. NMR Spectrosc.* **28(1)**, 53–85.
7. Gadian, D. S. (1995) *NMR and Its Applications to Living Systems*, 2nd ed., Oxford Science Publications, Oxford.
8. Griffin, J. L. and Shockcor, J. P. (2004) Metabolic profiles of cancer cells. *Nat. Rev. Cancer* **4(7)**, 551–561.
9. Lindon, J. C., Holmes, E., and Nicholson, J. K. (2001) Pattern recognition methods and applications in biomedical magnetic resonance. *Prog. Nucl. Mag. Res. Sp.* **39(1)**, 1–40.
10. Sanders, J. K. M. (1987) *Modern NMR Spectroscopy: A Guide for Chemist*, Oxford University Press, Oxford.
11. Brindle, K. M. (1998) Investigating the performance of intensive mammalian cell bioreactor systems using magnetic resonance imaging and spectroscopy. *Biotechnol. Genet. Eng. Rev.* **15**, 499–520.
12. Molowa, D. T. and Mazanet, R. (2003) The state of biopharmaceutical manufacturing. *Biotechnol. Annu. Rev.* **9**, 285–302.
13. Martin, I., Wendt, D., and Heberer, M. (2004) The role of bioreactors in tissue engineering. *Trends Biotechnol.* **22(2)**, 80–86.
14. McGovern, K. A. (1994) *Bioreactors* in *NMR in Physiology and Biomedicine* (Gillies, R.J. ed.), Academic Press, London.
15. Callies, R., Jackson, M. E., and Brindle, K. M. (1994) Measurements of the growth and distribution of mammalian cells in a hollow-fiber bioreactor using nuclear magnetic resonance imaging. *Biotechnology (NY)* **12(1)**, 75–78.
16. Galons, J. P., Job, C., and Gillies, R. J. (1995) Increase of GPC levels in cultured mammalian cells during acidosis. A ^{31}P MR spectroscopy study using a continuous bioreactor system. *Magn. Reson. Med.* **33(3)**, 422–426.
17. Neves, A. A., Medcalf, N., and Brindle, K. (2003) Functional assessment of tissue-engineered meniscal cartilage by magnetic resonance imaging and spectroscopy. *Tissue Eng.* **9(1)**, 51–62.
18. Potter, K., Butler, J. J., Horton, W. E., and Spencer, R. G. (2000) Response of engineered cartilage tissue to biochemical agents as studied by proton magnetic resonance microscopy. *Arthritis Rheum.* **43(7)**, 1580–1590.
19. Arnold, D. L., Matthews, P. M., and Radda, G. K. (1984) Metabolic recovery after exercise and the assessment of mitochondrial function in vivo in human skeletal muscle by means of ^{31}P NMR. *Magn. Reson. Med.* **1(3)**, 307–315.

20. Kintner, D. B., Anderson, M. K., Fitzpatrick, Jr., J. H., Sailor, K. A., and Gilboe, D. D. (2000) ^{31}P-MRS-based determination of brain intracellular and interstitial pH: its application to in vivo H+ compartmentation and cellular regulation during hypoxic/ischemic conditions. *Neurochem. Res.* **25(9–10)**, 1385–1396.
21. Stejskal, E. O. (1965) Spin diffusion measurements: spin-echoes in the presence of a time-dependent field gradient. *J. Chem. Phys.* **42**, 288–292.
22. Hakumaki, J. M., Poptani, H., Puumalainen, A. M., et al. (1998) Quantitative ^{1}H nuclear magnetic resonance diffusion spectroscopy of BT4C rat glioma during thymidine kinase-mediated gene therapy in vivo: identification of apoptotic response. *Cancer Res.* **58(17)**, 3791–3799.
23. Van Zijl, P. C., Moonen, C. T., Faustino, P., Pekar, J., Kaplan, O., and Cohen, J. S. (1991) Complete separation of intracellular and extracellular information in NMR spectra of perfused cells by diffusion-weighted spectroscopy. *Proc. Natl. Acad. Sci. USA* **88(8)**, 3228–3232.
24. Thelwall, P. E. and Brindle, K. M. (1999) Analysis of CHO-K1 growth in a fixed bed bioreactor using magnetic resonance spectroscopy and imaging. *Cytotechnology* **3**, 121–132.
25. Neves, A. A., Medcalf, N., and Brindle, K. M. (2002) Tissue engineering of meniscal cartilage using perfusion culture. *Ann. NY Acad. Sci.* **961**, 352–355.
26. Thelwall, P. E., Neves, A. A., and Brindle, K. M. (2001) Measurement of bioreactor perfusion using dynamic contrast agent-enhanced magnetic resonance imaging. *Biotechnol. Bioeng.* **75(6)**, 682–690.
27. Williams, S. N., Callies, R. M., and Brindle, K. M. (1997) Mapping of oxygen tension and cell distribution in a hollow fiber bioreactor using magnetic resonance imaging. *Biotechnol. Bioeng.* **56**, 56–61.
28. Stabler, C. L., Long, R. C., Sambanis, A., and Constantinidis, I. (2005) Noninvasive measurement of viable cell number in tissue-engineered constructs in vitro, using ^{1}H nuclear magnetic resonance spectroscopy. *Tissue Eng.* **11(3–4)**, 404–414.
29. Mancuso, A., Fernandez, E. J., Blanch, H. W., and Clark, D. S. (1990) A nuclear magnetic resonance technique for determining hybridoma cell concentration in hollow fiber bioreactors. *Biotechnology (NY)* **8(12)**, 1282–1285.
30. Petersen, E. F., Fishbein, K. W., McFarland, E. W., and Spencer, R. G. (2000) ^{31}P NMR spectroscopy of developing cartilage produced from chick chondrocytes in a hollow-fiber bioreactor. *Magn. Reson. Med.* **44(3)**, 367–372.
31. Macdonald, J. M., Grillo, M., Schmidlin, O., Tajiri, D. T., and James, T. L. (1998) NMR spectroscopy and MRI investigation of a potential bioartificial liver. *NMR Biomed.* **11(2)**, 55–66.
32. Papas, K. K., Long, Jr., R. C., Constantinidis, I., and Sambanis, A. (2000) Effects of short-term hypoxia on a transformed cell-based bioartificial pancreatic construct. *Cell Transplant.* **9(3)**, 415–422.
33. Callies, R. and Brindle, K. M. (1996) Nuclear magnetic resonance studies of cell metabolism in vivo, in Bittar, E. E. and Bittar, N., eds., *Principles of Medical Biology*; *Cell Chemistry and Physiology*: Part III, Vol. 4. Elsevier Science, New York, NY, pp. 241–269.

34. Jeffrey, F. M., Rajagopal, A., Malloy, C. R. and Sherry, A. D. (1991) ^{13}C-NMR: a simple yet comprehensive method for analysis of intermediary metabolism. *Trends. Biochem. Sci.* **16(1)**, 5–10.
35. Sherry, A. D. and Malloy, C. R. (1996) Isotopic methods for probing organization of cellular metabolism. *Cell Biochem. Funct.* **14(4)**, 259–268.
36. Blank, L. M., Kuepfer, L., and Sauer, U. (2005) Large-scale ^{13}C-flux analysis reveals mechanistic principles of metabolic network robustness to null mutations in yeast. *Genome Biol.* **6(6)**, R49.
37. Mancuso, A., Sharfstein, S. T., Tucker, S. N., Clark, D. S., and Blanch, H. W. (1994) Examination of primary metabolic pathways in a murine hybridoma with carbon-13 nuclear magnetic resonance spectroscopy. *Biotechnol. Bioeng.* **44**, 563–585.
38. Street, J. C., Delort, A. M., Braddock, P. S. H., and Brindle, K. (1993) A ^{1}H/^{15}N NMR study of nitrogen metabolism in cultured mammalian cells. *Biochem. J.* **291**, 485–492.
39. Hammer, B. E., Heath, C. A., Mirer, S. D., and Belfort, G. (1990) Quantitative flow measurements in bioreactors by nuclear magnetic resonance imaging. *Biotechnology (NY)* **8(4)**, 327–330.
40. Constantinidis, I. and Sambanis, A. (1995) Towards the development of artificial endocrine tissues: ^{31}P NMR spectroscopic studies of immunoisolated, insulin-secreting AtT-20 cells. *Biotechnol. Bioeng.* **47**, 431–443.

11

Methods for Off-Line Analysis of Nutrients and Products in Mammalian Cell Culture

Heino Büntemeyer

Summary

Mammalian cell culture has obtained high relevance for the production of recombinant therapeutic and diagnostic proteins from established cell lines. Furthermore, great effort has also been made to grow primary and undifferentiated cells in vitro. In each case it is of great importance for the establishment of a certain culture to know the demands of the cells as exactly as possible. Apart from global parameters like temperature, pH, oxygen tension, and osmolarity, the availability of substrates and the concentration of metabolic waste products play a major role in the performance of the cell culture. In order to monitor the concentrations of these substances and to evaluate the state of the culture, some different methods have to be applied. There is a broad spectrum of well-defined procedures available to estimate the concentration of each of the important substances. In general, it is very much recommended to use standardized working instructions and automatic systems to obtain trustworthy analysis results. Some such reliable methods are listed in this chapter. These procedures are all established in our laboratory and employed routinely during daily work. They should be easily transferred and installed in another lab.

Key Words: Glucose; amino acids; pyruvate; lactate; ammonia; biosensor; ammonia electrode; chromatographic separation; high-performance liquid chromatography; assay kits.

1. Introduction

The main substrates of mammalian cells are glucose *(1)*, amino acids *(2)*, and in some cases pyruvate *(3,4)*. From these substrates the cells obtain enough energy, carbon, nitrogen, and sulfur to propagate and synthesize protein. Furthermore, the cells need many other substances, which are either present in very low concentrations (e.g., vitamins, precursors) or serve as cell-stabilizing agents (e.g., salts, proteins). Most of these compounds have to be supplied to the cells because they are not able to synthesize them. This leads to medium formulations that often contain more than 40 ingredients. On the other hand, some products

From: *Methods in Biotechnology, Vol. 24: Animal Cell Biotechnology: Methods and Protocols, 2nd Ed.* Edited by: R. Pörtner © Humana Press Inc., Totowa, NJ

are synthesized by the cells during primary metabolism. These substances are particularly lactate *(5)*, alanine, and ammonia *(6)*, which are released into the culture fluid. It was shown that these metabolites can influence cell behavior severely. Reduction of growth *(7,8)*, cell death, and apoptosis or low and modified protein productivity *(9,10)* were often observed.

For the development of optimized media and well-engineered processes, it is very desirable to be able to monitor the concentrations of substrates and metabolic products. In addition, it can be of vital importance to control the concentrations of these substances to obtain a defined and stable process. The following descriptions present some available and reliable methods for the analysis of the main substrates and waste products of mammalian cells. Although these procedures are designated for the examination of extracted samples, many of them can be adapted for online analysis. Other authors have already shown the functionality of flow injection systems and high-performance liquid chromatography (HPLC) couplings to bioreactors.

2. Materials

2.1. Removal of Cells

Cell centrifuge, up to 200*g*, capable of handling 10- to 50-mL disposable centrifugation tubes.

2.2. Photometric Glucose Assay (Hexokinase Assay)

1. Cell-free sample, 50–100 µL.
2. Glucose assay kit (e.g., GAHK20, Sigma-Aldrich; 10 716 251 035, R-Biopharm, Roche Diagnostics).

2.3. Enzyme Electrode-Based Glucose Biosensor System

1. Cell-free sample, approx 50 µL.
2. Biosensor YSI 2700S (Yellow Springs Instruments, Yellow Springs, OH) or BioProfile basic (Nova Biomedical, Waltham, MA).

2.4. Precolumn Derivatization of Amino Acids With O-Phthaldialdehyde (OPA)

1. Microliter centrifuge, up to 16,000*g*, capable of handling 0.5- and 1-mL cups.
2. Protein precipitation reagent: 3% perchloric acid, 300 µ*M* internal standard (e.g., 5-aminopentanoic acid).
3. 0.6 *M* Potassium borate buffer, pH 10.4.
4. Derivatization reagent: 25 mg OPA, 0.5 mL methanol, 50 µL mercaptopropionic acid, 4.5 mL 0.4 *M* potassium borate buffer, pH 10.4.
5. Elution buffer A: 89% 85 m*M* sodium acetate, pH 7.5, 10% methanol, 1% tetrahydrofuran; elution buffer B: 80% methanol, 20% 85 m*M* sodium acetate, pH 5.2.
6. HPLC column: ODS material (C_{18}), 5 µ, 150 mm × 4 mm (e.g., Kromasil).

7. Binary HPLC system, autosampler with derivatization functionality, spectrofluorimeter, column oven (*see* **Note 4**).

2.5. Precolumn Derivatization of Amino Acids With Phenylisothiocyanate (PITC)

1. Microliter centrifuge, up to 16,000*g*, capable of handling 0.5- and 1-mL cups.
2. Vacuum concentrator.
3. Internal standard solution, 300 μ*M* norleucine.
4. Derivatization mixture 1: methanol:water:triethylamine 2:1:1 (v/v).
5. Derivatization mixture 2: methanol:water:triethylamine:PITC 7:1:1:1 (v/v).
6. Elution buffer A: 98% 85 m*M* sodium acetate, pH 5.2, 2 % acetonitrile; elution buffer B: 70% acetonitrile and 30% 85 m*M* sodium acetate, pH 5.2.
7. HPLC column: ODS material (C_{18}), 5 μ, 250 mm × 4 mm (e.g., Kromasil).
8. Binary HPLC system, autosampler, UV detector (*see* **Note 4**).

2.6. Precolumn Derivatization of Pyruvate With ortho-Phenylenediamine (OPD)

1. Microliter centrifuge, up to 16,000*g*, capable of handling 0.5- and 1-mL cups.
2. Protein precipitation reagent: 3% perchloric acid, 1 m*M* internal standard (e.g., 2-oxoglutarate).
3. 0.4 *M* Potassium borate buffer, pH 9.5.
4. Derivatization reagent: 100 mg OPD, 100 mL 2 *M* HCl, 250 μL mercaptopropionic acid.
5. Elution buffer A: 85% 50 m*M* sodium acetate, pH 6.5, 14% methanol, 1% tetrahydrofuran; elution buffer B: 80% methanol, 20% 50 m*M* sodium acetate, pH 6.0.
6. HPLC column: ODS material (C_{18}), 5 μ, 150 mm × 4 mm (e.g., Kromasil).
7. Binary HPLC system, autosampler, spectrofluorimeter (*see* **Note 4**).

2.7. Photometric Lactate Assay

1. Cell-free sample, 100 μL.
2. Lactate assay kit (e.g., 10 139 084 035, R-Biopharm, Roche Diagnostics).

2.8. Enzyme Electrode-Based Lactate Biosensor System

1. Cell-free sample, approx 50 μL.
2. Biosensor YSI 2700S (Yellow Springs Instruments, Yellow Springs, OH) or BioProfile basic (Nova Biomedical, Waltham, MA).

2.9. Photometric Ammonia Assay (Glutamate Dehydrogenase Assay)

1. Cell-free sample, 50–100 μL.
2. Ammonia assay kit (e.g., AA0100, Sigma-Aldrich).

2.10. Ammonia Electrode

1. Cell-free sample, 20 mL.
2. Ammonia electrode (Mettler-Toledo, Giessen, Germany, www.mt.com).

3. Measuring amplifier for electrode.
4. 10 *M* Sodium hydroxide.

2.11. Derivatization and Fluorescence Detection of Ammonia

1. Cell-free sample, 20 µL.
2. Half microquartz cuvette.
3. Derivatization reagent: dissolve 1.85 mmol OPA and 4 mmol mercaptoacetic acid (thioglycolic acid) in 2 mL methanol and add 100 mL 0.4 *M* sodium borate buffer pH 10.4.
4. Spectrofluorimeter.

2.12. Precolumn Derivatization of Pyroglutamic Acid With p-Bromophenacylbromide

1. Vacuum concentrator.
2. Heating block.
3. Acetonitrile, potassium bicarbonate.
4. Derivatization reagent: 50 m*M* *p*-bromophenacylbromide, 5 m*M* 18-crowne-6.
5. Elution buffer A: 20% acetonitrile in water; elution buffer B: 90% acetonitrile in water.
6. HPLC column: ODS material (C_{18}), 5 µ, 150 mm × 4 mm (e.g., Kromasil).
7. Binary HPLC system, autosampler, UV detector (*see* **Note 4**).

3. Methods

3.1. Removal of Cells and Sample Storage

It is strongly advised to remove all cells and cell debris before further analysis of substrates and products, even if it is possible to perform some methods without prior cell removal. This is especially recommended when samples are to be stored frozen. Samples should be aliquotted into portions of 0.5–1 mL, depending on further analysis, and kept frozen at least at –20°C. For long-term storage a temperature below –70°C should be considered, especially when the protein concentration or quality has to be determined. In any case, avoid thawing and freezing the same sample several times—a new aliquot should be used instead (*see* **Note 1**).

3.2. Glucose

Glucose is the main substrate for mammalian cells. It is utilized by the cells in the highest amount of all available substrates. A limitation in glucose normally leads to a cessation in cell growth, cell death, loss of productivity, or other severe impact on cellular behavior. Although the applied glucose is not metabolized completely to carbon dioxide, the cells recruit most of its energy from it. For special process strategies it may be essential to control glucose concentration in the medium. In any case, it is strongly recommended to consider monitoring the glucose concentration during the whole process. Many established methods are available. Most convenient for this purpose are photometric enzyme-based complete assay kits or fully automated biosensor systems.

3.2.1. Photometric Glucose Assay (Hexokinase Assay)

In the presence of ATP, D-glucose is converted by the enzyme hexokinase to D-glucose-6-phosphate. This product is further converted by the second enzyme glucose-6-phoshate-dehydrogenase in the presence of $NADP^+$ to D-gluconate-6-phosphate. During this last reaction NADPH is formed. The concentration of NADPH can be easily monitored at a wavelength of 340 nm. The concentrations of produced NADPH and assayed glucose are proportional. This assay kit is designed to be used in a cuvette UV spectrophotometer and needs no further special laboratory equipment (glucose assay kits: GAHK20, Sigma-Aldrich, 10 716 251 035, R-Biopharm, Roche Diagnostics) (*see* **Note 5**).

3.2.2. Enzyme Electrode-Based Glucose Biosensor Systems

The enzyme electrode biosensor consists of an amperometric hydrogen peroxide-sensitive electrode combined with immobilized glucose oxidase trapped between synthetic membranes ***(11)***. The electrode itself consists of a platinum anode and a Ag/AgCl cathode. The platinum anode is polarized at +700 mV relative to the cathode. At the electrodes, a current results from decomposition of hydrogen peroxide to water and oxygen. The membrane covering the electrodes consists of three layers. The outer layer, made of polycarbonate, is in direct contact with the sample. The next layer is the immobilized glucose oxidase enzyme followed by a cellulose acetate membrane, which readily passes H_2O_2 but excludes chemical compounds with a molecular weight above approx. 200 Da. This last cellulose acetate film also protects the electrode from fouling caused by proteins, detergents, lipids, and other substances. After applying a sample that contains glucose, a dynamic equilibrium is achieved very rapidly when the rate of peroxide production and peroxide diffusion to the electrode is equivalent. This is indicated by a steady-state current response from the electrode. The current is proportional to the hydrogen peroxide conversion and the glucose oxidation. Fully automated systems are provided by several vendors like Yellow Springs Instruments (YSI 2700 S, Yellow Springs, OH) or Nova Biomedical (BioProfile basic/100, Waltham, MA). Calibration and cleaning routines are self-controlled. The machines only need to be fed with the cell-free sample (*see* **Note 2**).

3.3. Amino Acid Analysis

The amino acids are the major group of medium compounds from which cells generate energy and carbon, except glucose. They can be divided into a group of essential amino acids and another group of nonessential amino acids.

This principal classification arises from their relevance for human beings. For other mammalian cells this might not be completely valid. For rodent-derived cells, for instance, it is known that there is another graduation. Therefore, almost all amino acids are present in the common basal medium formulations, some in high and some in low concentrations. Glutamine is the amino acid with the highest medium concentration. Although it is not classified as an essential amino acid, it is indispensable for all cell lines that do not express the glutamine synthetase gene.

The amino acids are a very heterogeneous group of chemicals ***(12)***. Aliphatic, aromatic, polar, nonpolar, acidic, basic, and cyclic amino acids are present. Common for all acids are the carboxy and the primary (except proline and hydroxyproline) amino groups at the α-carbon. Except for the three aromatic amino acids phenylalanine, tyrosine, and tryptophan, amino acids cannot be detected with adequate sensitiity by photometric means. Therefore, a quantitative analysis method must include a derivatization procedure to introduce a detectable group into the amino acid molecule. This can be done with several reagents depending on the chromatographic procedure chosen for separation technique.

A multitude of reagents have been developed for precolumn and postcolumn derivatization. Using the precolumn method the amino acids are chemically modified first to introduce a chromophore or a fluorophore ***(13)***. The commonly used reagents are OPA ***(14)***, 9-fluorenyl-methoxycarbonyl chloride (FMOC), and PITC ***(15–18)***. The amino acid derivatives can be subsequently separated on a reversed-phase column and monitored with a photometer or a fluorimeter (*see* **Note 4**). In the postcolumn method the amino acids are first separated on an ion exchange column and then modified just before detection by ninhydrin or OPA. Standard amino acid analyzer systems usually employ the postcolumn method. Whereas resolution and substance identification are very good in these systems, the analysis period for one sample is very long. For the determination of the free amino acids in cell culture medium the precolumn methods are more appropriate because of their shorter analysis time and higher sensitivity. It should be mentioned that some authors are successfully using ion exchange chromatography with direct amperometric detection or gas chromatography after derivatization with a volatile reagent. Prior to each successful routine analysis of samples, an elution profile has to be optimized in respect to separation quality for a chosen method, column, and HPLC system. Some applicable procedures are described here.

3.3.1. Precolumn Derivatization With OPA for Primary Amino Acids

1. Take 50 µL of cell-free sample, add 50 µL of protein precipitation reagent, and vortex 1 min vigorously.

2. Centrifuge 5 min at 16,000*g*.
3. Take 50 µL of mixture, add 100 µL of potassium borate buffer, and vortex 1 min vigorously.
4. Centrifuge 5 min at 16,000*g*.
5. Fill 20 µL in appropriate autosampler vial and place vial in sampler.
6. Set column temperature to 30°C.
7. Set exication wavelength of fluorimeter to 340 nm and emission wavelength to 450 nm, respectively.
8. Program an appropriate elution gradient for chosen column from 0% to 100 B% in approx 30 min.
9. Program autosampler to conduct following sequence for each sample prior to injection.
   a. Add 70 µL of derivatization reagent to sample.
   b. Mix briefly.
   c. Await 1 min reaction time (total).
   d. Inject 20 µL onto HPLC column.
10. Start elution immediately.

Figure 1 shows a chromatogram obtained for a standard mixture. All amino acids are baseline separated and can be quantified by peak area integration with high accuracy.

3.3.2. Precolumn Derivatization With PITC

1. Centrifuge 100 µL of cell-free sample 5 min at 16,000*g*.
2. Take 50 µL of centrifuged supernatant, and add 50 µL of internal standard solution.
3. Lyophilize in a vacuum concentrator.
4. Resuspend in 20 µL of derivatization mixture 1, vortex, and lyophilize again.
5. Resuspend in 20 µL of derivatization mixture 2 and vortex.
6. Await 20-min reaction time at room temperature (20–25°C).
7. Lyophilize again to remove excess PITC.
8. Resuspend in 100 µL of elution buffer A.
9. Set wavelength of UV detector to 254–269 nm.
10. Program an appropriate elution gradient for chosen column from 0% to 100 B% in approx 45 min.
11. Inject 20–40 µL onto HPLC column.
12. Start elution immediately.

Figure 2 shows a chromatogram obtained for a standard mixture. All amino acids are baseline separated and can be quantified by peak area integration.

3.4. Pyruvate

Pyruvate is an intermediate in the glycolysis, but it is also present in many basal medium formulations. Although the cells synthesize pyruvate from glucose, it was also observed that this substance was incorporated from the medium by

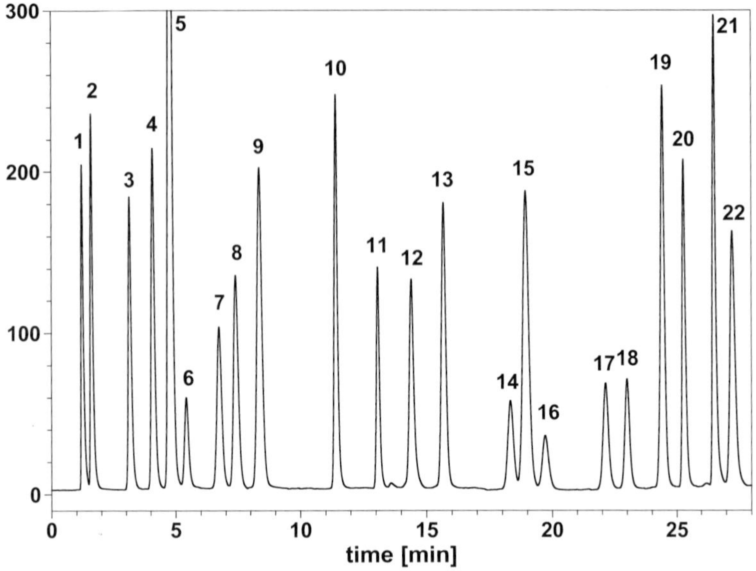

Fig. 1. Standard chromatogram of amino acids by processing the *o*-phthaldialdehyde (OPA) method. List of amino acids: 1, aspartic acid; 2, glutamic acid; 3, asparagine; 4, serine; 5, glutamine; 6, histidine; 7, glycine; 8, threonine; 9, arginine; 10, alanine; 11, tyrosine; 12, ethanolamine; 13, δ-amino-n-valeric acid (internal standard); 14, methionine; 15, valine; 16, tryptophan; 17, phenylalanine; 18, by-product; 19, isoleucine; 20, leucine; 21, lysine; 22, double derivative of lysine.

several cell lines. Therefore, it can be of special interest to monitor the pyruvate concentration in the medium. Pyruvate is a α- keto carbonic acid, which is reactive with vicinal primary amino groups. Such a constellation is found in chemicals like OPD ***(19)*** or 1,2-diamino-4,5-methylenedioxybenzene (DMB) ***(20)***. Whereas DMB gives a higher fluorescence yield and a better sensitivity, the OPD reagent is less expensive. The diamines react under acid conditions to form fluorescent derivatives, which can be separated on reversed-phase columns. With this derivatization method all other α- keto carbonic acids like α-ketoglutarate, ascorbic acid, *N*-acetylneuraminic acid, or *N*-glycolylneuraminic acid can be determined as well. **Figure 3** shows a medium containing pyruvate and ascorbic acid (vitamin C).

3.4.1. Precolumn Derivatization With OPD

1. Take 50 µL of cell-free sample, add 50 µL of protein precipitation reagent, vortex, and centrifuge 5 min at 16,000*g*.
2. Take 50 µL of mixture, add 500 µL of OPD reagent, and vortex.
3. Incubate mixture for 2 h at 95°C in a thermal block.
4. Take 100 µL of mixture, add 100 µL of borate buffer, and vortex.

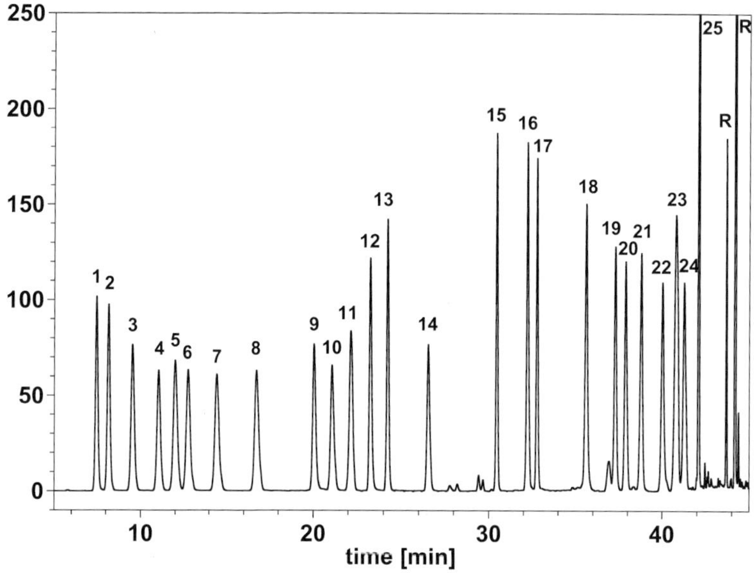

Fig. 2. Separation of a standard mixture of amino acids using the phenylisothiocyanate (PITC) method. Amino acids were identified as follows: 1, aspartic acid; 2, hydroxyproline; 3, glutamic acid; 4, serine; 5, asparagine; 6, glycine; 7, glutamine; 8, taurine; 9, histidine; 10, threonine; 11, alanine; 12, arginine; 13, proline; 14, ethanolamine; 15, tyrosine; 16, valine; 17, methionine; 18, cytine; 19, isoleucine; 20, leucine; 21, norleucine (internal standard); 22, phenylalanine; 23, ornithine; 24, tryptophan; 25, lysine; R, reagent.

5. Set exication wavelength of fluorimeter to 340 nm and emission wavelength to 412 nm, respectively.
6. Program an appropriate elution gradient for chosen column from 0% to 100 B% in approx 20 min.
7. Inject 40–80 μL onto HPLC column.
8. Start elution immediately.

3.5. Lactate (Lactic Acid)

Lactate is one of the predominant metabolic waste products of mammalian cells. In many cell types glucose is not metabolized completely to acetyl coenzyme A because of a very low activity of some enzymes. This leads to an accumulation of pyruvate. To reduce this holdup, pyruvate is converted to lactate by the enzyme lactate dehydrogenase and released into the medium. A maximum of 2 mol of lactate can be produced from 1 mol of glucose even under aerobic conditions. Additionally, lactate may also be synthesized by other pathways, especially amino acid metabolism. High lactate concentrations in the medium can affect cell viability and product formation. Therefore, it may

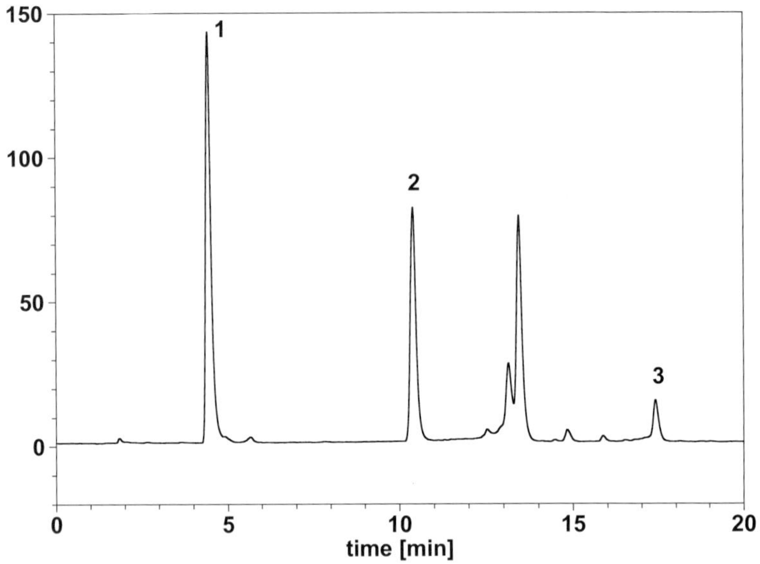

Fig. 3. Separation of α-keto acids in cell culture medium (knockout Dulbecco's modified Eagle's medium, Invitrogen) using the *o*-phenylenediamine (OPD) derivatization method. Keto acids were identified as follows: 1, α-ketoglutarate; 2, pyruvate; 3, ascorbic acid.

be necessary to monitor the lactate concentration of the culture to be able to maintain low concentrations by use of special means or process procedures. Two methods for the determination of lactate are common.

3.5.1. Photometric Lactate Assay

Lactate is converted in the presence of NAD^+ by the enzyme L-lactate dehydrogenase to pyruvate. During this reaction NADH is formed, which can be monitored at a wavelength of 340 nm. The reaction is reversible, and the equilibrium is more on the lactate side than on the pyruvate side. Therefore, pyruvate has to be removed. This can be achieved by adding the second enzyme glutamate-pyruvate-transaminase, which converts pyruvate in the presence of glutamate to alanine and 2-oxoglutarate. The assay is desired to be used in a cuvette UV spectrophotometer and needs no further laboratory equipment. Several commercial kits are available from different vendors. They can be used as described in the instruction guidelines provided with the chemicals (*see* **Note 5**).

3.5.2. Enzyme Electrode-Based Biosensor Systems

The principle of these systems has been described above ***(11)***. In case of lactate determination, the immobilized enzyme is lactate oxidase, which oxidizes lactate and generates hydrogen peroxide (*see* **Note 2**).

3.6. Ammonia

In animal cell culture other than in mammals, ammonia is the main end product of nitrogen metabolism. Beside ammonia, alanine is also produced in a remarkable amount, while only little urea is secreted by the cells. Especially when high glutamine and other amino acid concentrations must be used to supply the cells, a high ammonia concentration can be detected. High ammonia concentrations can drastically affect cell growth, productivity, and product integrity. Several analytical methods have been used for the quantification of ammonia, including enzyme assay systems, gas-sensitive electrodes (e.g., Mettler-Toledo), ion chromatography (e.g., Metrohm, Dionex), and derivatization techniques ***(21,22)***. Each of these methods has advantages and disadvantages (*see* **Note 3**).

3.6.1. Enzymatic Ammonia Assay

2-Oxoglutarate and ammonium are converted in the presence of NADH by the enzyme glutamate dehydrogenase to glutamate, water, and NAD^+. The decreasing concentration of NADH can be easily monitored at a wavelength of 340 nm. The concentrations of reacted NADH and assayed ammonium are inversely proportional. The necessity of the coenzyme NADH and the consumption of enzyme make this method a cost-intensive analysis procedure. This assay test kit is desired to be used in a cuvette UV spectrophotometer and needs no further laboratory equipment (Assay kit AA0100, Sigma-Aldrich) (*see* **Note 5**).

3.6.2. Gas-Sensitive Ammonia Electrode

An ammonia electrode works by the principle that under alkaline conditions ammonia degasses from aqueous solutions, passes through a hydrophobic membrane (PTFE), and dissolves in an acidic solution, where the resulting pH shift can be measured and quantitatively evaluated. In most cases, when no volatile amines are present this is a specific detection method for ammonia. The main disadvantages are the size of the electrode and the fragile membrane adapter. Depending on the ammonia concentration in the sample and a predilution possibility, a sample volume of up to 10–20 mL may be necessary for analysis.

1. Take 20 mL of cell-free sample or dilute 2 mL of cell-free sample with water to a volume of 20 mL.
2. Place the ammonia electrode into the solution.
3. Wait until a stable value is obtained.
4. Add 0.5 mL of 10 *M* NaOH.
5. Monitor increasing voltage signal and register the highest value.
6. Calculate ammonium concentration from calibration curve.
7. Prepare calibration curve in the same way by analyzing different known concentrations of ammonium chloride.

3.6.3. Derivatization and Fluorescence Detection

OPA reacts with primary amines, including ammonia, in the presence of mercaptoacetic acid (thioglycolic acid) at basic conditions to unstable, fluorescent isoindole derivatives. The reaction product is analogous to the products formed with 3-mercaptopropionic acid and amino acids, but its fluorescent property is quite different. Whereas the fluorescence intensity at excitation and emission wavelengths of 330 and 450 nm (amino acids), respectively, is very poor, the fluorescence yield can be highly increased by changing to wavelengths of 415 and 485 nm, respectively. On the other hand, the fluorescence intensity of the amino acid derivatives is negligible at those wavelengths. Therefore, this method has good selectivity for ammonium quantification.

1. Set excitation wavelength of spectrofluorimeter to 415 nm and emission wavelength to 485 nm.
2. Fill 1.3 mL derivatization reagent into a half micro quartz cuvette.
3. Place cuvette into fluorimeter.
4. Perform an autozero at fluorimeter.
5. Add 20 µL of cell-free sample to reagent, mix briefly.
6. Read fluorescence signal until highest emision value is reached (after approx 70 s).
7. Calculate concentration from calibration curve.
8. Prepare calibration curve in the same way by analyzing different known concentrations of ammonium chloride.

3.7. Pyroglutamic Acid

Pyroglutamic acid (pyroglutamate, 5-oxo-2-pyrollidone carboxylic acid) and ammonia are formed from glutamine in a spontaneous thermal intramolecular degradation reaction. Depending on medium composition and cultivation conditions, glutamine has a half-life of about 6–8 d. Therefore, when high glutamine concentrations are used and maintained under such conditions, a considerable amount of pyroglutamate is produced. It was shown that pyroglutamic acid can influence cell growth at high concentrations ***(23)***. Pyroglutamate can be analyzed by using the same derivatization procedure, which was originally established for the HPLC analysis of fatty acids, but sample preparation and elution conditions distinguish clearly. The formed pyroglutamate derivative is much more hydrophilic and can only be separated on a very hydrophobic reversed phase.

1. Adjust the cell-free sample to pH 3 with sulfuric acid.
2. Take 50 µL and lyophilize.
3. Dissolve residue in 180 µL acetonitrile.
4. Add 20 µL of derivatization reagent and 1 mg KHCO3.
5. Heat for 90 min at 85°C.

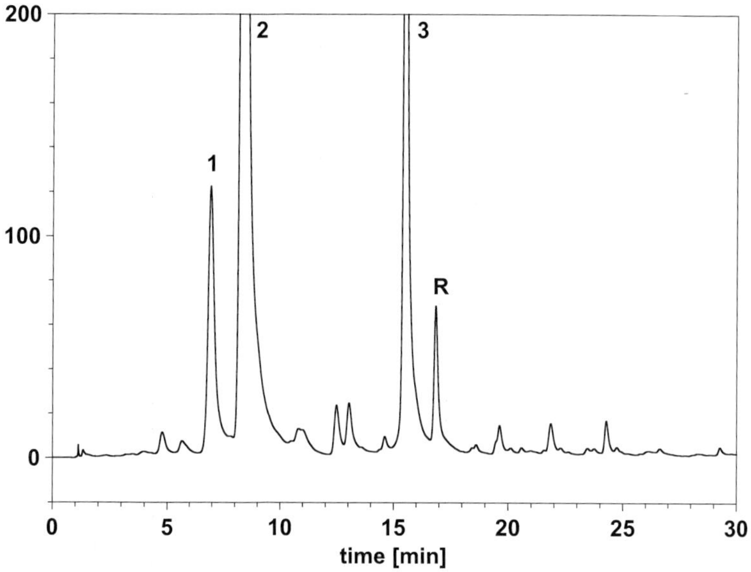

Fig. 4. Chromatogram obtained by following the pyroglutamate derivatization procedure. Derivatization and separation conditions apply as described in the text. Identified substances are: 1, pyroglutamate; 2, lactate; 3, unresolved amino acids (mainly glutamine); R, reagent.

6. Program a linear gradient from 0% B to 100% B in 30 min.
7. Set flow to 1.3 mL/min.
8. Set UV detector to a wavelength of 254 nm.
9. Cool down reaction mixture in ice water.
10. Inject 15–30 µL.
11. Start elution.

Besides pyroglutamate, lactate can also be analyzed quantitatively with this method. **Figure 4** shows an example chromatogram.

3.8. Conclusions

Several methods were presented here for the main substrates and waste products of mammalian cells. All methods were successfully adapted to the requirements of mammalian cell culture. Concentration ranges were investigated especially carefully. Furthermore, the need for additional equipment was kept as low as possible. Only one binary HPLC system in addition to the standard laboratory equipment, such as photometer, centrifuge, heater, and vacuum concentrator, is indispensable to perform all analysis. Of course, some devices like the biosensor systems, a spectrofluorimeter, and the ammonia electrode are very useful and convenient. With these methods any mammalian cell culture lab will be able to monitor the needs of the cells in small- and large-scale culture.

4. Notes

1. It is very beneficial to establish a good sample management system. This includes a consistent sample labeling format. All samples should be stored under the same conditions. Repeated thawing and freezing of the same sample should be avoided. Use parallel aliquots instead. Always keep a very last sample in case of an unforeseen repetition of an analysis.
2. The mentioned biosensor systems need daily attendance, although they work automatically. Calibrating solutions should not be used beyond their expiration dates. The immobilized enzymes on the membranes lose their sensitivity with time. It is recommended to perform response tests at regular intervals to keep the system in good order.
3. Analyses of the ammonia concentration in the sample should be performed as soon as possible after sampling and before freezing. The samples change their pH to an alkaline pH because of outgassing carbon dioxide. This shifts the ammonium equilibrium to a higher portion of ammonia, which degasses as well. Therefore, the ammonia concentration will be underestimated later.

   The use of the ammonia electrode is quite simple and needs no special precautions. But in between sample analyses and during storage, the gas-sensitive membrane at the bottom of the electrode should be kept wet at all times. Once the membrane has dried out, it is fragile and must be handled very carefully. The electrode should be stored in a regeneration solution, and the electrolyte inside should be changed at regular intervals.
4. This chromatographic analysis can be performed with one complete HPLC system. The system should be capable of conveying two elution buffers (binary system). For at least one method it is indispensable to use an autosampler with derivatization functionality. Both detectors, the spectrophotometer and the spectrofluorimeter, can be attached in serial and used as necessary. Although all presented separations are performed on ODS material, it is strongly recommended to use a different column for each method. Furthermore, one should consider protecting the main column by a guard. This will improve the stability of the separation and the durability of the main column.
5. Commercial photometric assay kits can be used as described by the vendor. In any case, it is recommended to check the detection limits and to consider special sample treatment before analysis. Special attention should be paid to interfering substances that may be present in the sample.

5. Websites

Mettler-Toledo, Giessen, Germany, www.mt.com
Nova Biomedical, Waltham, MA, www.novabiomedical.com
R-Biopharm, Darmstadt, Germany, www.r-biopharm.com

Sigma-Aldrich, St. Louis, MO, www.sigmaaldrich.com
Yellow Springs Instruments, Yellow Springs, OH, www.ysi.com

References

1. Fitzpatrick, L., Jenkins, H. A., and Butler, M. (1993) Glucose and glutamine metabolism of a murine B-lymphocyte hybridoma grown in batch culture. *Appl. Biochem. Biotechol.* **43**, 93–116.
2. Büntemeyer, H., Lütkemeyer, D., and Lehmann, J. (1991) Optimization of serum-free fermentation processes for antibody production. *Cytotechnology* **5**, 57–67.
3. Conaghan, J., Handyside, A. H., Winston, R. M., and Leese, H. J. (1993) Effects of pyruvate and glucose on the development of human preimplantation embryos *in vitro*. *J. Reprod. Fertil.* **99**, 87.
4. Genzel, Y., Ritter, J. B., König, S., Alt, R., and Reichl, U. (2005) Substitution of glutamine by pyruvate to reduce ammonia formation and growth inhibition of mammalian cells. *Biotechnol. Prog.* **21**, 58–69 .
5. Miller, W. M., Wilke, C. R., and Blanch, H. W. (1988) Transient response of hybridoma cells to lactate and ammonia pulse and step changes in continuous culture. *Bioproc. Eng.* **3**, 113–122.
6. Hassel, T., Gleave, S., and Butler, M. (1991) Growth inhibition in animal cell culture: the effect of lactate and ammonia. *Appl. Biochem. Biotechnol.* **30**, 29–41.
7. Ryll, T., Valley ,U., and Wagner, R. (1994) Biochemistry of growth inhibition by ammonium ions in mammalian cells. *Biotechnol. Bioeng.* **44**, 184–193.
8. Newland, M., Kamal, M. N., Greenfield, P. F., and Nielsen, L. K. (1994) Ammonia inhibition of hybridomas propagated in batch, fed-batch, and continuous culture. *Biotechnol. Bioeng.* **43**, 434–438.
9. Andersen, D. C. and Goochee, C. F. (1992) Cell culture effects on the O-linked glycosylation of granulocyte colony-stimulating factor produced by CHO cells. *Abstr. Pap. Am. Chem. Soc.* **203**, 113.
10. Borys, M. C., Linzer, D. I. H., and Papoutsakis, E. T. (1994) Ammonia affects the glycosylation patterns of recombinant mouse placental lactogen-I by Chinese hamster ovary cells in a pH-dependent manner. *Biotech. Bioeng.* **43**, 505–514.
11. Operations manual, Biochemical Analyser YSI 2700S, Yellow Springs Instrument Inc., Yellow Springs, OH.
12. Smith, A. J. (1997) Amino acid analysis. *Methods Enzymol.* **289**, 419–426.
13. Roth, M. (1971) Fluorescence reaction for amino acids. *Anal. Chem.* **43**, 880–882.
14. Cooper, J. D. H., Lewis, M. T., and Turnell, D. C. (1984) Precolumn o-phthalaldehyde derivatization of amino acids and their separation using reversed-phase high-performance liquid chromatography. I. Detection of the imino acids hydroxyproline and proline. *J. Chromatogr.* **285**, 484–489.
15. Bidlingmeyer, B. A., Cohen, S. A., and Tarvin, T. L. (1984) Rapid analysis of amino acids using precolumn derivatization. *J. Chromatogr.* **336**, 93–104.
16. Cohen, S. A. and Strydom, D. J. (1988) Amino acid analysis utilizing phenyliso-thiocyanate derivatives. *Anal. Biochem.* **174**, 1–16.

17. Reid, S., Randerson, D. H., and Greenfield, P. F. (1987) Amino acid determination in mammalian cell culture supernatants. *Austral. J. Biotech.* **2**, 69–72.
18. Koop, D. R., Morgan, E. T., Tarr, G. E., and Coon, M. J. (1982) Purification and characterization of a unique isozyme of cytochrome P-450 from liver microsomes of ethanol-treated rabbits. *J. Biol. Chem.* **257**, 8472–8480.
19. Hayashi, T., Tsuchiya, H., and Naruse, H. (1983) HPLC-determination of α-keto acids in plasma with fluorometric detection. *J. Chromatogr.* **273**, 245–252.
20. Hara, S., Takemori, Y., Yamaguchi, M., and Nakamura, M. (1985) Determination of α-keto acids in serum and urine by high performance liquid chromatography with fluorescence detection. *J. Chromatogr*. **344**, 33–39.
21. Campmajo, C., Cairo, J. J., Sanfeliu, A., Martinez, E., Alegret, S., and Godia, F. (1994) Determination of ammonium and L-glutamine in hybridoma cell cultures by sequential flow injection analysis. *Cytotechnology* **14**, 177–182.
22. Wakisaka, S., Tachiki ,T., Sung, H.-C., Kumagai, H., Tochikura, T., and Matsui, S. (1987) A rapid assay method for ammonia using glutamine synthetase from glutamate-producing bacteria. *Anal. Biochem.* **163**, 117–122.
23. Büntemeyer, H., Iding, K., and Lehmann, J. (1998) The influence of pyroglutamic acid on the metabolism of animal cells. Cell Culture Engineering VI, Engineering Foundation, San Diego, CA.

12

Application of Stoichiometric and Kinetic Analyses to Characterize Cell Growth and Product Formation

Derek Adams, Rashmi Korke, and Wei-Shou Hu

Summary

Stoichiometric and kinetic analyses are critical tools for developing efficient processes for the production of animal cell culture products such as therapeutic proteins or vaccines. This chapter describes the common analytical methods for quantifying animal cell growth and metabolism. It also presents sample calculations to illustrate the utility and limitations of various analytical methods. The application of calculated stoichiometric and kinetic parameters in mathematical models is also discussed. Hence, this chapter provides the basic tools for obtaining quantitative description of current cell culture state as well as applying the information for process optimization.

Key Words: kinetics; stoichiometry; specific rates; integral of viable cell density; Monod.

1. Introduction

The growth state and productivity of animal cells in culture can be monitored and characterized by applying the appropriate stoichiometric and kinetic analytical methods. These methods use knowledge of the stoichiometric requirements of cells at various growth states as well as familiarity with mathematical modeling principles to describe kinetics of cell growth, metabolic activity, and product formation. Stoichiometric and kinetic analyses are critical tools for developing efficient processes for the production of high-value cell culture-based products such as therapeutic proteins or vaccines. These analyses are necessary for cell culture scientists to determine the culture environment for optimum cell growth and nutrient utilization without performing numerous cycles of empirical experimentation. This chapter focuses on the applications of these methods for typical situations encountered while analyzing animal cell cultures.

From: *Methods in Biotechnology, Vol. 24: Animal Cell Biotechnology: Methods and Protocols, 2nd Ed.* Edited by: R. Pörtner © Humana Press Inc., Totowa, NJ

Although the principles are similar, the stoichiometric and kinetic analyses of cell culture processes are somewhat different than for microbial cultures. The typical animal cell volume is a few pL (~1000 times larger than bacteria), and cell sizes exhibit a distribution within the culture that changes with the growth stage *(1)*. The size of cells varies with cell line, medium, growth stage, pH, osmolality, etc. For anchorage-dependent cells, the size is also dependent on cell density. The size distribution can be obtained along with cell number using instruments that automatically image, count, and size viable and nonviable cells (such as the Cedex from Innovatis or the ViCell from Beckman-Coulter). In general, the kinetic parameters to characterize cell properties are calculated based on cell number, rather than on dry cell weight. Expressing the kinetic parameters this way does not take into account the differences in cell size among different cell lines or the same cell line under different culture conditions. Nevertheless, all calculations in this chapter are based on cell number.

This chapter describes the common analytical methods for quantifying cell growth and metabolism. It also presents sample calculations to illustrate the utility and limitations of various analytical methods. Finally, the application of calculated stoichiometric and kinetic parameters in mathematical models is discussed. Hence, this chapter provides the basic tools for obtaining quantitative description of current cell culture state as well as applying the information for process optimization.

2. Methods

2.1. Quantitative Description of Cell Growth and Product Formation

Typically a cell culture process is characterized by its cell growth curve, nutrient consumption curves, and product concentration profile. Three classes of quantities contribute to a complete description of the process: concentrations (cell, nutrient, metabolite, or product), kinetic parameters (specific rates and stoichiometric ratios of growth, nutrient consumption, and product formation), and the stage of the culture (lag, exponential, stationary, or declining phase). This section includes how the concentrations and calculated kinetic parameters interact and influence each other at different stages of the culture. A typical batch mode cell growth curve is shown in **Fig. 1**. The different stages of the culture are annotated on the figure.

2.1.1. Stages of Cell Growth

1. Lag phase: Newly inoculated cells may experience hour-long or even 1- to 2-d long periods of very slow or no growth. This phase can be caused by inoculation of the culture with cells that were in stationary or decline phase or if the cells are being inoculated into medium with significant changes in factors such as pH, temperature, osmolality, or dissolved oxygen concentration. However, in most cases this lag phase is not experienced by the culture.

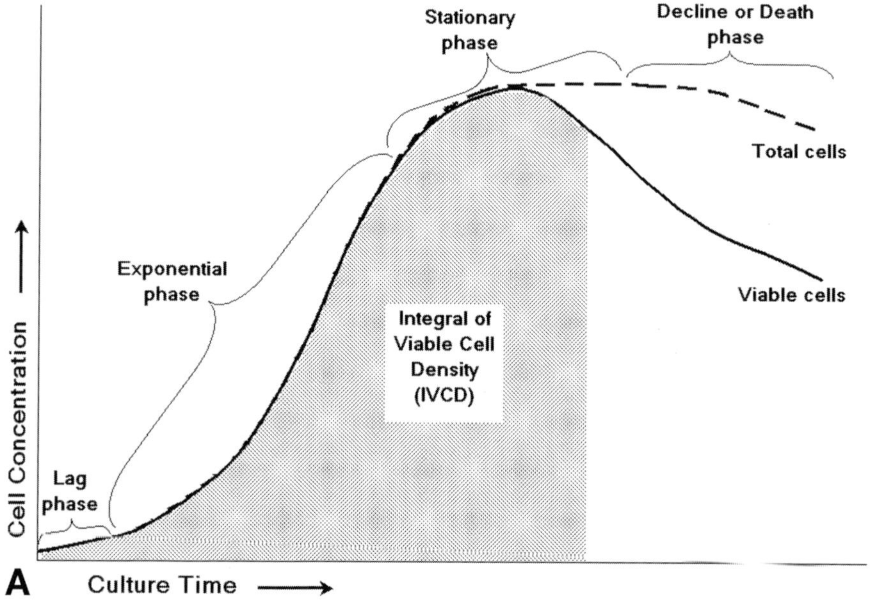

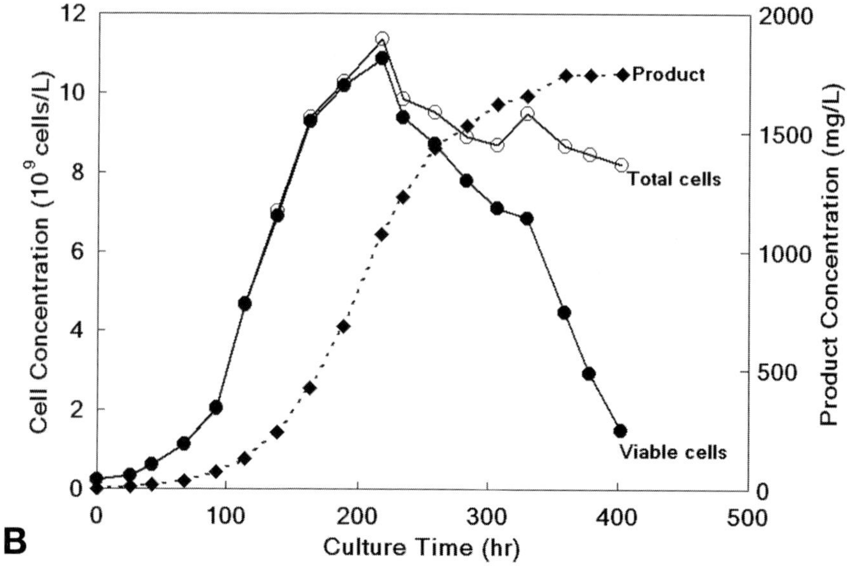

Fig. 1. Typical cell growth and productivity curves. **(A)** Traces for both viable cells (solid line) and total cells (dashed line) as functions of culture time. Each phase of the culture is shown with braces along the length of the curves. **(B)** Graph of experimental data obtained from a fed-batch culture of Chinese hamster ovary (CHO) cells producing a monoclonal antibody (previously unpublished data presented for illustration of typical growth and production rates). Each culture stage illustrated in **A** is visible in the cell growth curves in **B**.

2. Exponential phase: This phase is the classic, first-order growth stage for cells. This phase is marked by the greatest and steadiest specific cell growth rate (defined below). It is during this phase that cell-doubling times are typically measured and reported.
3. Stationary phase: After the period of rapid cell growth during the exponential phase, cultures exhibit reductions in cell growth rate. This may be induced by exhaustion (or reduction to rate-limiting levels) of key nutrients in the culture and/or an increase in the concentration of metabolites. The latter part of this stage is marked by the flat cell growth curve, indicating a balance between cell growth and death rates.
4. Decline or death phase: The cell death rate is higher than the growth rate during the death phase and the viable and total cell concentration declines because of exhaustion of key nutrients or accumulation of inhibitory levels of metabolites or other toxic species.

Figure 1 also shows the integral of viable cell concentration (represented by the shaded area underneath the viable cell curve) over the culture time to that time point, called integral of viable cell density (IVCD), or cumulative cell time. In comparing two cultures of equal specific productivity, the one with the greater IVCD will produce more product. There are many other ways to describe the state of cultures, including physiological state (e.g., a quiescent state vs actively growing state), cell morphology (e.g., well-spread-out state on a flat surface vs a cuboidal shape on a high-curvature surface), clumping, or aggregation (e.g., single cell suspension or spheroid state). These methods are beyond the scope of this chapter. Changes in culture state may be indicated by changes in specific production or consumption rates. Specific rates describe how active cells are and the various specific rates in a culture change over time.

2.1.2. Volumetric and Specific Kinetic Parameters

Concentrations or amount per unit volume are typically denoted as:

x = total (or viable) cell concentration (number of cells/L)

s = substrate (e.g., glucose) concentration (g substrate/L)

p = product concentration (g product/L)

Normally the concentration is expressed per unit volume of culture, which is essentially the same as basing it on per unit volume of liquid because the volume occupied by cells is very small. However, in some cases (such as high-density solid microcarrier culture), the volume occupied by the solid beads is significant and the liquid volume and culture volume (i.e., liquid volume plus bead volume) are not the same. In all instances, the "volume" used must be well-defined and consistent because the concentration data are used for calculation of kinetic parameters.

Kinetic parameters characterize rates of change for cell growth, nutrient consumption, or product formation. Animal cell growth is most often characterized by the specific growth rate, μ, and the cell doubling time, t_d. Cell growth or the increase in cell concentration is autocatalytic, meaning that the rate is dependent on the cell concentration at a given time and is as shown below (assuming cell size does not matter):

$$G = \frac{dx}{dt} = \mu x \tag{1}$$

where G is the volumetric cell growth rate (number of cells/L/h) and $\mu (= \frac{1}{x}\frac{dx}{dt})$ is the specific cell growth rate (number of cells/cell/h or h^{-1}) and x is viable cell concentration.

In the case that nonviable cells are present in significant number, as often is the case in cell culture, a specific death rate, α, can be incorporated into Eq. 1:

$$\frac{dx}{dt} = \mu x - \alpha x \tag{2}$$

and

$$\frac{dx_d}{dt} = \alpha x \tag{3}$$

where x_d is the dead cell concentration. The cell doubling time is derived from integrating Eq. 1 and rearranging:

$$\ln \frac{x_2}{x_1} = \mu (t_2 - t_1) \tag{4}$$

$$t_d = \frac{\ln 2}{\mu} = \frac{0.693}{\mu} \tag{5}$$

The doubling time ranges from 15 to 60 h for animal cells. A culture started with an inoculum cell concentration of x_0 at $t = 0$ will have a cell concentration of $x_0 e^{\mu t}$ at t, when μ is constant during the exponential growth phase. For optimum growth, a minimum inoculum concentration is required. Low inoculum concentrations often cause poor initial cell growth because of the lack of sufficient conditioning factors in the culture medium. In general, 10^5 cells/mL (or 10^4 cells/cm^2 for anchorage-dependent cells) can be used as the minimum inoculum concentration. For microcarrier cultures, one should start with an average of six to eight cells per bead to ensure that most beads are occupied by at least one cell. Cells will not migrate from bead to bead, and thus empty beads will remain so for the culture duration.

Specific nutrient consumption and product formation are characterized similarly to cell growth:

$$q_s = -\frac{1}{x}\frac{ds}{dt} \tag{6}$$

$$q_p = \frac{1}{x}\frac{dp}{dt} \tag{7}$$

where s and p represent substrate (nutrient) and product concentration and q_s and q_p are the respective specific rates. As with the volumetric cell growth, the corresponding volumetric rates of nutrient consumption and product formation ($Q_s = \frac{ds}{dt}$ and $Q_p = \frac{dp}{dt}$) denoted by capital letters, while the specific rates are typically denoted by lower case letters. Concentrations are denoted by upper or lower case letters. Just be consistent.

2.1.2.1. Fed-Batch Mode Considerations

The volumetric rate and specific rate described above are intrinsic properties of the cell in a culture. For batch culture, the calculation of those parameters is straightforward. However, when deriving the values of those volumetric or specific rates in a fed-batch culture, volume adjustment will have to be performed as the volume is not a constant value. Equation 1, used for describing the volumetric cell growth rate in batch mode, needs to be modified to account for the changing volume in fed-batch cultures as follows:

$$G = \frac{d}{dt}(xV) = \mu(xV) \tag{8}$$

where V is the culture volume, which changes as medium is added over time. The product xV, the total cell number in the reactor, is used rather than just the concentration.

2.1.3. Stoichiometric Analysis

Culture state can also be characterized by yield on amount of substrate used and the stoichiometric relationships among different nutrients and products (metabolites). The yield coefficient on cells, i.e., cells produced per unit mass of substrate consumed is expressed as

$$Y_{x/s} = \frac{\Delta x}{\Delta s} = \frac{dx}{ds} \tag{9}$$

and mass of protein product produced per mass of nutrients consumed (e.g., total amino acids) gives a yield coefficient for product,

$$Y_{p/s} = \frac{\Delta p}{\Delta s} = \frac{dp}{ds} \tag{10}$$

These yields can be used to measure process efficiency. Stoichiometric ratios are also used to characterize process efficiency cell physiological state. Important stoichiometric ratios for cell culture include the mass of metabolites (e.g., lactic acid or ammonia) produced per mass of substrate (e.g., glucose, glutamine, or oxygen) consumed. Changes in stoichiometric ratios can signal changes in cell physiological state during a culture run or between culture runs. Stoichiometric ratios calculated over the course of a culture are important metabolic indicators because combinations of stoichiometric ratios can reflect the metabolic state of cells. They help determine if the culture is in a favorable metabolic regime. If not, the analyses and mathematical models described in this chapter can help make decisions to drive the culture to a favorable regime.

Medium design is an integral part of developing an optimal process for animal cell culture, and stoichiometric analyses play a key role. Key nutrients must be maintained at an optimal range of concentration and at an acceptable ratio to other nutrients that will support their consumption at a desired rate to sustain growth and production. The stoichiometric analyses help determine the demand, i.e., specific uptake rate, and how it varies as a function of time, stage of culture, and concentration as well as interaction between the nutrients. Hence, these analyses can help determine optimal concentration of medium components.

2.1.4. Calculation of Parameter Values From Experimental Data

This section describes the methods for calculating kinetic parameters from experimental data. The calculations in this section are easily performed with a spreadsheet program. The different ways to calculate specific growth and production rates are described. In general, upon obtaining experimental data, it is always good practice to:

1. Plot all measurements (raw data) to ensure that data are within normal ranges.
2. Calculate specific rates and stoichiometric ratios while the culture is still ongoing. Although the rates calculated may be "noisy," they provide valuable information for in-process diagnosis for any potential abnormality.
3. Perform regression analyses on the full set of data to obtain a better estimate of key parameters upon completion of the culture and the chemical analysis of various medium components (postprocess analysis).

It is also good practice to plot data in both normal and semi-log scales, especially when performing regression analyses for exponential curve fits. This practice allows one to see the same data from two different perspectives and will minimize the tendency to infer patterns in the plotted data where there are none.

2.1.4.1. Use of Differential vs Integral Values

There are several methods for calculating specific rates for either in-process or postprocess analyses. Calculations of specific rates can be performed in a differential way, i.e., take the differential at a given time (e.g., $\mu = \frac{1}{x}\frac{dx}{dt}$), which is generally calculated over a short time interval (e.g., $\mu = \frac{1}{x}\frac{\Delta x}{\Delta t}$). This method is accurate over the time interval and easy to perform in-process, but it is also sensitive to experimental errors and varies with culture stage. Calculations are also performed by regression of smoothed cell growth, nutrient and product concentration curves to obtain the equations of cell, nutrient and product concentrations as functions over time. The slopes ($\frac{dx}{dt}, \frac{ds}{dt}$, or $\frac{dp}{dt}$) are then calculated to obtain the specific rates. The difference between the short time interval approach and regression for calculating specific growth rates is illustrated in **Table 1** and **Fig. 2**. The regression method complements the more rapid and variable alternatives illustrated in **Table 1**. An exponential curve fit for specific growth rate for all data (solid line) is a reasonable fit except that the plot indicates an approx 1-d lag phase in addition to a transition to stationary phase in the final two data points. The specific rate can be estimated with all the data, but it is clear that the specific rate is changing from lag to exponential to stationary phase. Thus, a better fit is to drop the first data point and the last two data points from the regression to yield a more accurate estimation of the specific growth rate in exponential phase. The calculated specific rates are the fitted constants in the exponents in units of time^{-1} (h^{-1} in this case).

Specific rates can also be calculated over a long interval (e.g., over an exponential growth stage). Calculating q_p by

$$\bar{q}_p = \frac{1}{x_{ave}}\frac{\Delta p}{\Delta t} = \frac{p_2 - p_1}{t_2 - t_1}\left(\frac{2}{x_2 + x_1}\right) \tag{11}$$

will reduce the effects of single point variability; however, as the time interval becomes large, this method becomes less accurate. The use and limitations of this method are also illustrated in **Table 1**.

Another paramter, integral cell numbers (i.e., IVCD), is used for characterization of volumetric product production rate,

$$\frac{dP}{dt} = q_p x \tag{12}$$

rearranging and integrating,

Table 1
Illustration of Specific Rate Calculations[a]

Time, t (h)	Viable cell density, x (10^9 cells/L)	Product concentration, p (mg/L)	IVCD[b] (10^9cells*hr/L)	Specific growth rate, μ (h^{-1}) Short time interval[c]	Specific production rate, q_p (pg/cell/d) Short time interval[d]	Long interval[e]	by IVCD[f]
0	0.2	0	0	–	–	–	–
26	0.3	10	7	0.012	34	34	34
43	0.6	18	15	0.036	23	24	28
68	1.1	36	37	0.024	19	18	23
92	2.0	71	75	0.023	22	16	23
115	4.6	130	151	0.035	19	11	21
139	6.9	241	289	0.016	19	12	20
163	9.3	423	489	0.012	22	13	21

[a]The table is populated with the data from the first 8 data points of **Fig. 1B**. The first three columns (time, viable cell density, and product concentration) are the measured values, and the remaining columns are calculated values from the measured values (with the formulae used highlighted in footnotes). One method for calculating specific cell growth rate is shown in the table and an alternative method is illustrated in **Fig. 2A**. Three different methods for calculating specific productivity are shown along with a fourth method illustrated in **Fig. 2B**.

[b]Calculated by $(x_t+x_{t-1})/2*\Delta t+\text{IVCD}_{t-1.}$

[c]Calculated by $\Delta x/\Delta t*2/(x_1+x_{t-1})$ at each time interval.

[d]Calculated by $\Delta p/\Delta t*2/(x_t+x_{t-1})$ at each time interval.

[e]Calculated by $\Delta p/\Delta t*2/(x_t+x_0)$ over the cumulative time interval, t-t_0.

[f]Calculated by $\Delta p/\Delta\text{IVCD}$ over the cumulative time interval, t-t_0.

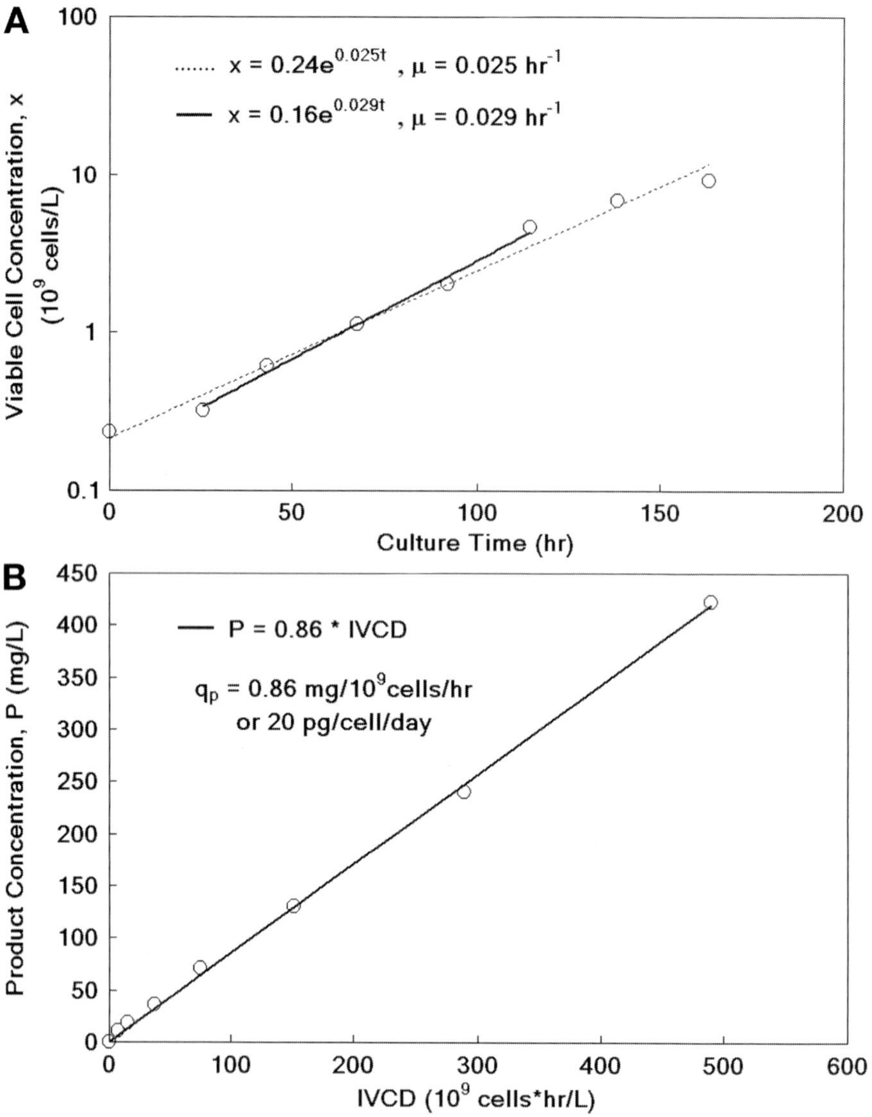

Fig. 2. Regression analysis for specific rates. The regression trendlines for both specific growth rate, μ (**A**), and specific productivity, q_p (**B**), are shown based on the data from **Table 1**.

$$\int_0^{P_f} dP = \int_{t_0}^{t_f} q_p x dt \tag{13}$$

Assuming that q_p is constant, the equation becomes:

$$P = \int_0^{P_f} dP = q_p \int_{t_0}^{t_f} x dt \tag{14}$$

where

$$IVCD = \int_{t_0}^{t_f} x dt \tag{15}$$

Under the constant q_p assumption, IVCD is proportional to the product titer at each time. In **Fig. 2B**, the production concentration is accurately represented as a linear function of IVCD, indicating a constant q_p over the time interval. The q_p is the slope of the line and is typically reported in units of pg/cell/d. Often one may see two production modes with different slopes, especially if the culture conditions are altered specifically to increase productivity later in the culture (e.g., shifting the temperature lower to slow growth, maintain viability, and enhance productivity).

In many situations, measurement of product concentration takes time, whereas integral cell numbers can be obtained quickly. Although *qp* is not always constant and integral cell number is not necessarily proportional to product concentration, it is often used as a first approximation estimate of product concentration. This method of estimating product concentration is not ideal for analyzing data from extended duration cultures because the specific productivity is certain to change over a long culture period.

2.1.5. Important Considerations

Some important factors need to be considered when applying the above analyses. The highest specific rates could occur in early exponential growth or in late exponential growth (before cell concentration reaches its peak) for different cell lines. Specific rates and yield coefficients may vary widely among different cell lines. They are also affected by the medium composition and metabolic states of cells. In a typical culture, the fraction of carbon atoms going into metabolites (CO_2, lactate, and excreted nonessential amino acids) is much higher than that incorporated into biomass and product. Under different metabolic conditions that fraction may change, and the stoichiometric ratios for different nutrients may also change. Thus, as culture conditions change, the extent of change among different specific rates may be rather different. Nevertheless, specific rates and stoichiometric ratios characterize the cell culture at various stages of the culture. They can be used to facilitate our understanding of the behavior of cells in culture. In addition, they can also be used for predictive purposes, as discussed in the next section.

2.2. Kinetic Description of Cell Growth and Product Formation Via Mathematical Models

2.2.1. Utility of Mathematical Models

Mathematical models are useful in many ways, such as for summarizing a large number of experimental data, for probing concepts and testing hypothesis, and, most importantly, for predicting and optimizing processes. The various motivations for making mathematical models of biological systems have been captured in a review article *(2)* as follows: to organize disparate information into a coherent whole, to think and calculate logically what components and interactions are important in a complex system, to discover new strategies to control traditional processes, to make important corrections in the conventional thinking and to understand the essential, qualitative features of a system. This section highlights the salient features of what is desirable in a mathematical model of cell growth and productivity.

The mathematical models that have been explored for describing animal cell growth range from simple Monod-type unstructured models to more complicated structured or segregated models. These models, being empirical in nature rather than mechanistic, are limited to conditions under which the kinetic parameters are obtained. No single such model can describe animal cell growth, metabolism and kinetics, and product formation under all possible conditions. This situation could be a result of the complex nature of animal cell metabolism as compared to simpler microbial systems. Animal cells exhibit more regulatory control because they have been derived from tissues and are not free living like microbes. There is interaction of multiple metabolic pathways, which makes animal cell metabolism difficult to understand. In spite of these limitations, these models can help organize data from experiments and apply them to future experiments.

2.2.2. Building Blocks for Mathematical Description of a Cell Culture Process

In a way, a mathematical model is a mathematical description of a physical model or hypothesis. For example, the Monod model is basically a mathematical version of the statement (hypothesis) that "cell growth increases with increasing limiting nutrient concentration until the increase slows down at high nutrient concentration and eventually reaches a maximum value asymptotically." Several elements are required to construct a mathematical model:

1. Hypothesis or constraints that will help design the framework of the model:
   - Growth model to account for dependence of growth rate on a "controlling factor" such as glucose concentration.
   - Nutrient utilization relationship for relating the dependence of specific consumption rate to the "controlling variable" (growth rate, nutrient concentration, etc.).

- Product formation equation for describing the dependence of specific production rate on the "controlling variable."

2. Experimental data—the model will have some parameters (such as half saturation constants) where experimental data are required to determine their value.
3. Material balance equations for "state variables" (the concentrations of cell, nutrients, product, inhibitors, etc.).

2.2.3. Monod Model and Its Variants

A majority of the models found in literature for the growth of non-anchorage-dependent cells are unstructured and based on Monod's model ***(3–5)***. The Monod relationship between specific cell growth rates and substrate (nutrient) concentration is:

$$\mu = \frac{\mu_{\max} S}{K_S + S} \quad (16)$$

where $\mu_{\max}$ is the maximum specific growth rate, S is the concentration of the controlling substrate such as glucose, and K_S is the concentration of the controlling substrate where the specific rate is half the maximum rate. If $S >> K_S$ then $\mu \rightarrow \mu_{\max}$. Note that in the Monod model description S is the controlling substrate concentration. Thus, the other substrates are abundantly supplied and their concentrations do not affect the cell growth rate. The specific growth rate is also a function of pH, temperature, nutritional status (i.e., the presence of serum), and waste products; it may also depend upon cell density for anchorage-dependent cells, which are subject to contact inhibition. The specific growth rate can also be described with dual substrate limitation:

$$\mu = \mu_{MAX} \frac{S_1}{K_1 + S_1} \frac{S_2}{K_2 + S_2} \quad (17)$$

or with multiple inhibitors as controlling factors as well:

$$\mu = \mu_{MAX} \frac{S_1}{K_1 + S_1} \frac{S_2}{K_2 + S_2} \frac{K_{iA}}{K_{iA} + A} \frac{K_{iB}}{K_{iB} + B} \quad (18)$$

where S_1 and S_2 are growth-limiting substrates, while A and B are inhibitory metabolites. Most animal cell growth models consider glucose and glutamine as the limiting substrates and lactate and ammonia as inhibitors.

2.2.3.1. Concept of Limiting Substrate

The term limiting substrate has been used in two different contexts. A stoichiometric limiting nutrient refers to the nutrient that is depleted first in a batch culture whose depletion causes the cessation of growth, and a rate-limiting nutrient is the nutrient whose concentration restricts the growth rate of cells. Most of the modeling work on cell growth has been based on results obtained in batch culture, largely relying on the limiting subtrate defined by stoichiometry. This approach, although not ideal, is generally accepted because the slow growth rate and the high expense of performing continuous culture experiments make a more thorough and precise study prohibitively costly.

The latter definition of rate-limiting substrate is best demonstrated in a continuous culture at steady state: a step increase in the rate-limiting substrate concentration in the feed results in an increase in the steady-state cell concentration. When a unique rate-limiting substrate is identified in a culture, the specific growth rate of the cells is governed by the concentration of this substrate, and the state of the system is identifiable by the operating parameters (dilution rate and feed concentrations of the rate-limiting substrate). Upon a step increase of the limiting nutrient concentration in the feed while the culture is at steady state, the nutrient concentration increases in the culture because the supply rate is greater than the consumption rate, resulting in an increased growth rate and gradual increase in cell concentration. This increase in cell concentration ceases eventually, and the specific growth rate returns to its steady-state value.

Although cell culture medium consists of a large number of nutrients, it is possible to impose a substrate to be the sole rate-limiting nutrient in a culture as shown previously by Frame and Hu ***(5)***. This was largely achieved by deliberately decreasing the concentration of a nutrient essential for cell growth in the feed. However, in some simple continuous culture studies, the rate-limiting substrate is not apparent. Pulse additions of many essential nutrients, individually or combined, and step changes of one or more of their concentrations in the feed all resulted in perturbations in the steady-state cell concentration ***(6–8)***. Such behavior is not commensurate with rate limitation by a single substrate. This type of model states that the effect of decreasing the concentration of one substrate can be compensated by increasing the concentration of the other. It also implies that the possibility exists that more than one combination of the concentrations of two substrates give the same specific growth rate.

2.2.4. Equation for Product Formation

The kinetics of product formation in microbial systems are frequently categorized according to the relationship to growth rate such as:

$$\frac{dp}{dt} = q_P x = \alpha \mu x + \beta x \tag{19}$$

Such mathematical representations of product formation are useful for microbial fermentations of amino acids and antibiotics. However, their applicability to animal systems is limited. The production of a majority of recombinant protein products are only weakly affected by specific growth rate. In general, animal cell culture products fall into the following categories:

- Product formation results in cell destruction.
- Product formation is initiated by induction or infection, and, after a relatively short production stage, cell destruction occurs.
- Product formation is initiated by infection, but it does not result in immediate cell destruction.
- Product formation is primarily during the growth stage.
- Production formation is "constitutive," and an extended production period can be achieved.

The product formation equation can be set up differently depending on the categories described above correlating it to growth instead of using more detailed models discussed earlier in this section.

2.2.5. Kinetic Parameters

There are various variants of Monod model based descriptions of mammalian cell growth.

A simple Monod model with glucose as a limiting nutrient:

$$\mu = \mu_{MAX} \frac{C_{GLC}}{K_{GLC} + C_{GLC}} \tag{20}$$

where $\mu_{MAX} = 0.0251\ \text{h}^{-1}$ and $K_{GLC} = 0.059$ m*M* ***(9)***.

A model with lactate inhibition:

$$\mu = \mu_{MAX} \frac{C_{GLC}}{K_{GLC} + C_{GLC}} \frac{K_{LAC}}{K_{LAC} + C_{LAC}} \tag{21}$$

where $\mu_{MAX} = 0.033\ \text{h}^{-1}$, $K_{GLC} = 0.05$ m*M*, and $K_{LAC} = 14$ m*M* ***(10)***.

A model describing glucose and glutamine dependent growth and lactate and ammonium inhibition:

$$\mu = \mu_{MAX} \frac{C_{GLC}}{K_{GLC} + C_{GLC}} \frac{C_{GLN}}{K_{GLN} + C_{GLN}} \frac{K_{LAC}}{K_{LAC} + C_{LAC}} \frac{K_{NH_3}}{K_{NH_3} + C_{NH_3}} \tag{22}$$

where $\mu_{MAX} = 0.0625\ \text{h}^{-1}$, $K_{GLC} = 0.15$ m*M*, $K_{GLN} = 0.15$ m*M*, $K_{LAC} = 140$ m*M*, and $K_{NH_3} = 20$ m*M* ***(4)***.

It is important to use relevant experimental data to calculate these parameters. In the absence of actual experimental data, parameters can be calculated on the basis of data available in the literature. The scope of the model is limited and dictated by the experimental conditions used to calculate the parameters. Continuous cultures with appropriate environmental parameters like pH and appropriate metabolic state can provide more reliable experimental data as compared to short batch cultures in uncontrolled environment.

3. Final Remarks

The characterization of an animal cell culture process is aided by the calculation of relevant specific rates (cell growth, nutrient consumption, and product formation). These calculations can be performed in several ways where a method can be chosen on the basis of the need for accuracy versus speed. Yield coefficients and stoichiometric ratios, together with the specific rates, give a detailed quantitative assessment of culture performance. Mathematical models that use appropriately calculated kinetic parameters can be used to further analyze cell culture processes to allow inference of unmeasured parameters (e.g., internal metabolic rates) or, in some cases, predict the set of conditions necessary to drive the process to the desired production state. In all cases, stoichiometric and kinetic analyses, with proper consideration of the limitations of each method, are critical tools for the cell culture scientist.

References

1. Sen, S., Srienc, F., and Hu, W. S. (1989) Distinct volume distribution of viable and non-viable hybridoma cells: a flow cytometric study. *Cytotechnology* **2**, 85–94.
2. Bailey, J. E. (1998) Mathematical modeling and analysis in biochemical engineering: past accomplishments and future opportunities. *Biotechnol. Progr.* **14**, 8–20.
3. Glacken, M. W., Adema, E., and Sinskey, A. J. (1988) Mathematical descriptions of hybridoma culture kinetics: I. Initial metabolic rates. *Biotechnol. Bioeng.* **32**, 491–506.
4. Miller, W. M., Blanch, H. W., and Wilke, C. R. (1988) A kinetic analysis of hybridoma growth and metabolism in batch and continuous suspension culture: effect of nutrient concentration, dilution rate, and pH. *Biotechnol. Bioeng.* **32**, 947–965.
5. Frame, K. K. and Hu, W.-S. (1991) Kinetic study of hybridoma cell growth in continuous culture. I. A model for non-producing cells. *Biotechnol. Bioeng.* **37**, 55–64.
6. Miller, W. M., Wilke, C. R., and Blanch, H. W. (1988) Transient responses of hybridoma cells to lactate and ammonia pulse and step changes in continuous culture. *Bioproc. Eng.* **3**, 113–122.
7. Miller, W. M., Wilke, C. R., and Blanch, H. W. (1988) Transient responses of hybridoma metabolism to changes in the oxygen supply rate in continuous culture. *Bioproc. Eng.* **3**, 103–111.
8. Miller, W. M., Wilke, C. R., and Blanch, H. W. (1989) The transient responses of hybridoma cells to nutrient additions in continuous culture: II. Glutamine pulse and step changes. *Biotechnol. Bioeng.* **33**, 487–499.
9. Hayter, P. M., Curling, E. M. A., Gould, M. L., Baines, A. J., Jenkins, N., Salmon, I., Strange, P. G., and Bull, A. T. (1993) The effect of the dilution rate on CHO cell physiology and recombinant interferon-gamma production in glucose-limited chemostat culture. *Biotechnol. Bioeng.* **42**, 1077–1085.
10. Kurokawa, H., Park, Y. S., Iijima, S., and Kobayashi, T. (1994) Growth characteristics in fed-batch culture of hybridoma cells with control of glucose and glutamine concentrations. *Biotechnol. Bioeng.* **44**, 95–103.

13

Measurement of Apoptosis in Cell Culture

Adiba Ishaque and Mohamed Al-Rubeai

Summary

Apoptosis is a genetically regulated process by which cells can be eliminated in vivo in response to a wide range of physiological and toxicological signals. Cells in vitro may be induced to die by apoptosis, e.g., by depletion of nutrients or survival factors from the culture media. Described here are several biochemical and cytometric techniques of varying complexity that have been developed to detect cellular and subcellular changes occurring during apoptosis. For all procedures, cells can be obtained from either monolayer or suspension culture.

Key Words: Apoptosis; flow cytometry; annexin V; TUNEL; caspase.

1. Introduction

Cell death can be divided into two distinct categories: apoptosis and necrosis. Apoptosis is a genetically regulated process by which cells are eliminated in response to a wide range of physiological and toxicological signals *(1)*. It occurs during embryogenesis, immune system homeostasis, and tumor regression and in response to a number of cytotoxic stimuli. Necrosis is a passive form of cell death in which the cells die by plasma membrane injury. In vitro necrosis results in the release of cellular constituents and harmful proteolytic enzymes, whereas in vivo an inflammatory reaction is triggered. Apoptotic cells, on the other hand, are neatly disposed of by professional phagocytes to help maintain overall tissue architecture. In vitro environment cells with apoptotic morphological characteristics ultimately undergo membrane injury, making them distinguishable from primary necrotic cells. The end result of apoptotic cell death in industrial cell culture processes is cellular fragmentation.

Monitoring commercial cell culture processes for the induction of apoptotic cell death or simply for studying apoptotic mechanisms has placed flow

From: *Methods in Biotechnology, Vol. 24: Animal Cell Biotechnology: Methods and Protocols, 2nd Ed.* Edited by: R. Pörtner © Humana Press Inc., Totowa, NJ

cytometry (FC) at the forefront of all detection methods ***(2–8)***. FC is the most rapid, sensitive, and accurate analytical means of studying individual cells in a heterogeneous cell culture ***(9)***. By FC it is possible to determine apoptosis at the level of individual cells to highlight subpopulations with specific apoptotic characteristics that can often pass undetected by other methods. For instance, apoptotic cells in culture maintain their ability to exclude membrane-impermeable dyes like Trypan blue before undergoing secondary necrosis and hence becoming Trypan blue positive. Assessment of cell viability by these methods therefore overestimates the viability of cultures. Other conventional biochemical and microscopic analyses of apoptosis are too labor intensive, time consuming, and subjective. Unlike assessment of apoptosis by microscopy, the choice of cells analyzed by FC is objective and unbiased by their visual perception ***(10,11)***. Biochemical analyses are also less useful as they yield average population values. Moreover, FC analysis permits a physical sorting of cells for further analysis, making it possible to directly study apoptosis in rare cell populations ***(11)***.

This chapter begins with two conventional biochemical and morphological assays that are established methodologies for studying apoptosis in mammalian cell cultures. Their respective advantages and limitations are also highlighted. We then focus on several FC assays. Most of the fluorescent assays of apoptosis focus on two features of apoptosis, which distinguish this process from necrosis: specific DNA fragmentation and intact membrane integrity. By FC a variety of cell-associated or cell organelle properties of apoptosis for which there is a fluorescent probe, such as nuclear condensation, nuclear fragmentation, loss of mitochondrial membrane potential, and externalization of phosphatidylserine ***(12–16)***, can be determined. There are a number of FC techniques that measure these apoptotic parameters, and some of the most commonly used are discussed below.

2. Materials

1. 70% Ice-cold ethanol.
2. Phosphate-buffered saline (PBS), without Ca^{2+} and Mg^{2+}, pH 7.2, and (1X) Hanks buffered saline solution (HBBS).
3. EDTA solution (10 m*M*).
4. RNase solution: dissolve 5 mg RNase (Sigma, Poole, UK) in 50 mL distilled water. Store at –20°C in 0.5-mL aliquots. Incubate in a boiling water bath for 10 min before use. Otherwise, use chromatographically purified RNase.
5. Extraction buffer: 10 m*M* EDTA, 50 m*M* Tris-HCl, pH 7.8, 0.5.% sodium lauryl sarkosinate, 0.5 mg/mL Proteinase K.
6. Loading buffer: 0.25% bromophenol blue and 30% glycerol.
7. Low-melting agarose (Invitrogen).
8. Tris-borate–ethylene diamine tetraacetic acid (EDTA) (Sigma).

9. Propidium iodide (PI; Sigma): 10 mL of 1 mg/mL in PBS.
10. RNase A: prepared in PBS without Mg^{2+} and Ca^{2+} to give a final concentration of 250 µg/mL.
11. Ethidium bromide.
12. Cytofix/Cytoperm™ solution (BD Pharmingen) or 4% paraformaldehyde stock solution (Sigma). Dilute in PBS to make a final concentration of 1%.
13. 10% Perm/Wash™ buffer (BD Pharmingen). Dilute in distilled H_2O to make a final concentration of 1%.
14. Flow cytometer (e.g., FACS calibur, Becton Dickinson).
15. Fluorescence microscope.
16. Centrifuge.
17. Mix of acridine orange (AO) (10 µg/mL) and PI (10 µg/mL) in PBS.
18. Annexin V–fluorescein isothiocyanate (FITC) (BD Pharmingen).
19. Binding buffer: 10 m*M* hydroxyethyl piperazine ethane sulfonate/NaOH, pH 7.4, 140 m*M* NaCl, 2.5 m*M* $CaCl_2$.
20. TUNEL reaction mixture: 0.1*M* Na-cacodylate, pH 7.0, 0.1 m*M* dithiothreitol, 0.05 mg/L bovine serum albumin (BSA), 2.5 m*M* $CaCl_2$, 0.4 m*M* bio-16-d uridine triphosphate (UTP), and 0.1 U/mL terminal deoxynucleotidyl transferase (TdT) enzyme (Boehringer-Mannheim).
21. TUNEL staining buffer: 2.5 mg/mL fluoresceinated avidin, 4X concentrated saline-sodium citrate buffer; 0.1% Triton X-100 and 5% low-fat dry milk (Boehringer-Mannheim).
22. Caspase-3–phycoerythrin (PE) (BD Pharmingen).
23. Rhodamine 123 (Sigma): make a stock solution (1 mg/mL) in distilled H_2O.

3. Methods

3.1. DNA Gel Electrophoresis

During apoptosis DNA is cleaved at the internucleosomal sections to produce low molecular DNA fragments of 180–200 pb ***(17)***. These smaller fragments can be detected by a characteristic "laddering" pattern on agarose gels after electrophoresis (**Fig. 1**). One of the drawbacks of this technique is that high levels of necrosis may mask lower levels of apoptosis.

1. Take cells (5×10^6) and centrifuge at 180*g* for 5 min.
2. Store pellet at –20°C.
3. Thaw pellet and extract DNA by resuspending pellet in 20 µL extraction buffer and incubate for 1 h at 50°C.
4. Add 10 µL of DNase free RNase (0.5 µg/mL) and incubate for 1 h at 50°C.
5. Add 10 µL of loading buffer.
6. Load samples immediately into dry wells of an agarose gel (2%) + 0.5 µg/mL ethidium bromide.
7. Use Tris-borate-EDTA as running buffer.
8. Run gel for 3 h at 50 V.
9. Detect DNA fragments by eithidium bromide stain under UV light.

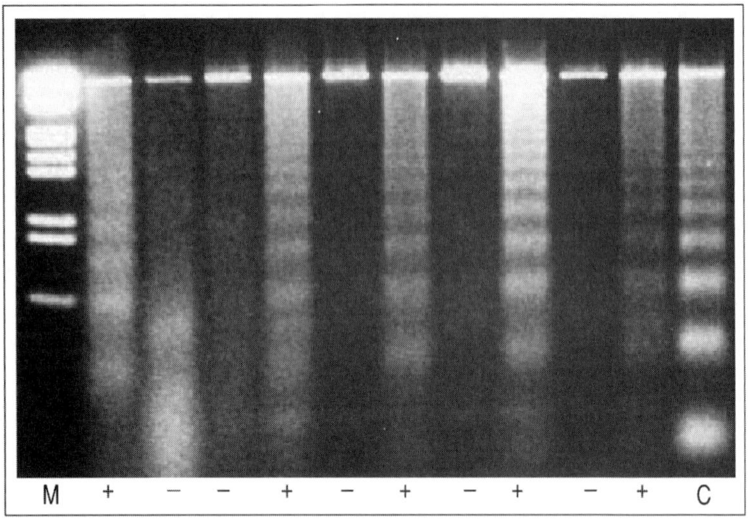

Fig. 1. DNA ladder of apoptotic cells induced by camptothecin. M, size marker; –, controls cells without camptothecin; +, cells treated with campothecin; C, positive control. (From **ref.** ***8***.)

3.2. Fluorescence Microscopy Analysis

Microscopy analysis provides a quantitative assessment of apoptosis. A combination of fluorescent probes can be used that are either impermeable or permeable to the plasma membrane such as AO and PI.

1. Take a small aliquot (~50 μL) of live cells (≤1 × 10^6cells).
2. Add 10 μL of AO/PI solution (10 μg/mL).
3. Incubate dye and cell suspension mix in the dark and analyze within 3 min.

AO is membrane permeable and intercalates with DNA to produce a green fluorescent emission **(Fig. 2)**. Once PI enters cells with damaged membranes, it emits a red fluorescence following its interaction with nucleic acids. At the same time early membrane-intact apoptotic cells, excluding PI, will be distinguished from viable and necrotic cells by their spherical-shaped, condensed chromatin, highlighted by AO staining.

3.3. Sub-G1 Peak

Apoptotic cells can be quantified by a reduction in DNA stainability with a variety of high-affinity DNA binding fluorochromes such as PI, AO, or Hoeschst dyes, which intercalate into DNA ***(12,18)***.

1. Fix 1 × 10^6 cells taken from a cell suspension in ice-cold ethanol (70%). Cells can be stored in fixative for up to 2 wk.

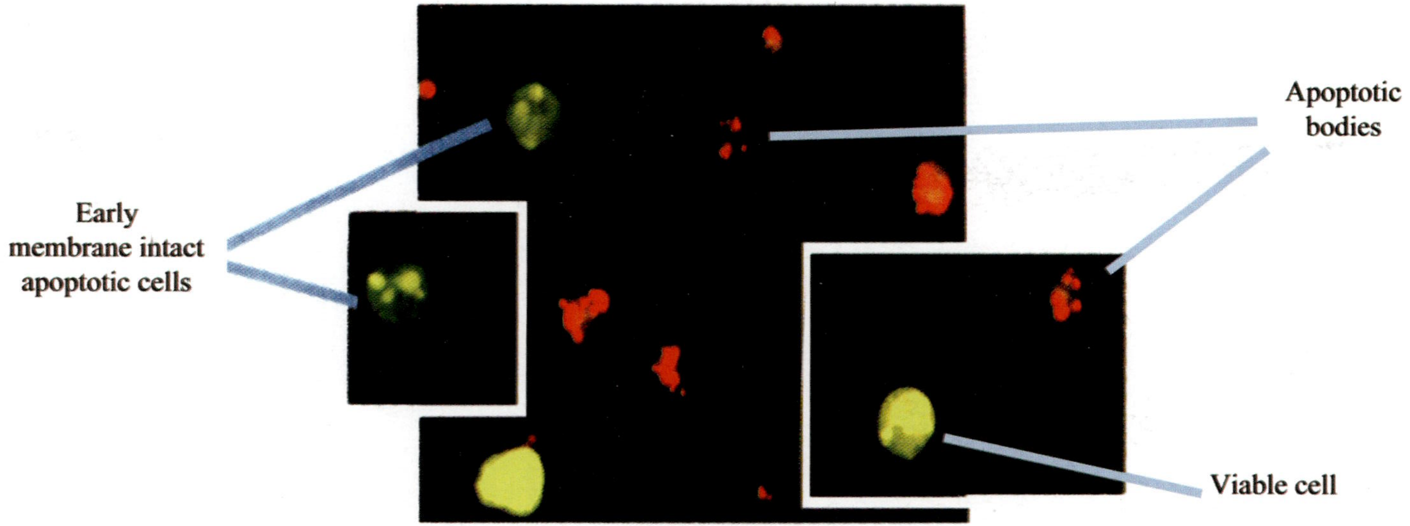

Fig. 2. Fluorescence microcopy analysis of live hybridoma cells stained with a combination of acridine orange and propidium iodide. (From **ref. *2***.)

2. Centrifuge fixed cells at high speed (180*g*) for 10 min to adequately pellet cells suspended in ethanol.
3. Decant supernatant. Resuspend pellet in ice-cold PBS. Centrifuge cell suspension in at 180*g* for 10 min.
4. Resuspend final pellet in 1 mL of RNase A.
5. Incubate mixture at 37°C for 30 min.
6. At the end of incubation add DNA binding dye such as PI (1 mg/mL) to give a final concentration of 50 μg/mL.
7. Incubate mixture at room temperature for 10 min in the dark.
8. Analyze cells by flow cytometry with 488 nm excitation and measurement of PI fluorescence at 620–640 nm.

Low molecular weight apoptotic DNA fragments arising from proteolytic digestion is extracted during cell rinsing and washing following fixation in ethanol 70% ***(12)***. The level of apoptotic DNA staining is therefore below that of G0/G1, resulting in the appearance of a distinct region on the DNA histogram, referred to as a sub-G1 peak, shown in **Fig. 3**. The size of the peak increases with the number of apoptotic cells. Often, the increasing sub-G1 population merges into the cell debris, which suggests that apoptotic cells disintegrate into apoptotic bodies and small fragments (**Fig. 3**). The degree of DNA degradation varies depending on the stage of apoptosis, cell type, and the nature of the apoptosis-inducing agent. The extractability of DNA during the staining procedure also varies.

Measurement of DNA provides the information about the cell cycle position of the nonapoptotic cells. Another advantage of this approach is the simplicity and applicability to any DNA fluorochrome. However, there are three potential drawbacks to these techniques. First, necrotic cells can also exhibit a reduction in DNA content and therefore a sub-G0/G1 peak. Second, the technique is most effective under conditions in which large numbers of cells enter apoptosis in a relatively short period, and therefore is not very sensitive in defining a time-dependent induction of apoptosis. Third, the sub-G1 region represents the number of nuclear fragments and provides no information about the number of apoptotic cells. Moreover, it has been shown that after G2 cell cycle arrest induced by the anticancer agent oracin, flow cytometric analysis of the DNA distribution revealed the presence of a "sub-G2" region, which represented the apoptotic cells ***(19)***, indicating that cells in any phase of the cell cycle may undergo apoptosis with the consequent reduction in PI stainability.

3.4. Annexin V–FITC

Apoptotic cells lose their membrane phospholipid asymmetry, resulting in the exposure of phosphatidylserine (PS) (normally present on the inner surface of an intact membrane) to the outer layer of the cell surface ***(20)***. Externalization of PS

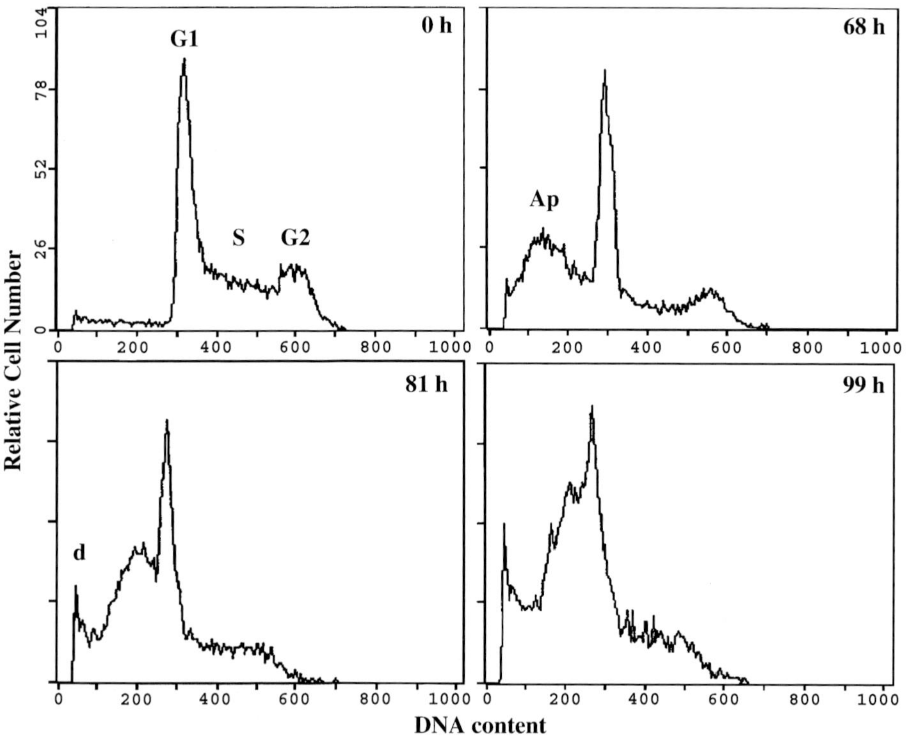

Fig. 3. DNA distribution of hybridoma cells during batch culture. Apoptotic (Ap) cells are induced as a result of stress from the depletion of nutrients to produce a level of staining below the G1 peak (sub-G1 region). As conditions deteriorate further (81 h), the sub-G1 combines with the cellular debris. (From **ref. *2***.)

can be probed by recombinant annexin V conjugated to the green fluorescent dye FITC, which binds with high affinity to phosphatidylserine in a Ca^{2+}-dependent manner ***(9)***. Because necrotic cells lose their membrane integrity, they will also stain positive with annexin V–FITC. Dual staining with PI enables these to distinguish the necrotic cells.

1. Take live cells (1×10^5 cells) from suspension culture and centrifuge at 180*g* for 5 min.
2. Decant supernatant and resuspend pellet in 100 mL of binding buffer.
3. Add 5 µL of annexin V–FITC (1–3 mg/mL) and 10 µl of PI (50 µg/mL).
4. Gently vortex the cells and incubate cell suspension for 15 min at room temperature in the dark.
5. At end of incubation add 400 µL of binding buffer. Analyze by flow cytometry as soon as possible (within 1 h) using 515–545 nm (FITC detection) and 620–640 nm (PI emission).

Expression of PS on the cell surface plays an important role in the recognition and removal of apoptotic cells by macrophages *(20)*. In vitro these phagocytic cells are not present, and therefore the cells accumulate and subsequent cell lysis exposes the internal phospholipids *(16,21)*. **Figure 4** shows that dual staining with PI and annexin V–FITC enables identification of viable (annexin V–FITC$^+$/PI$^-$), early membrane intact (annexin V–FITC$^+$/PI$^-$) and necrotic membrane-disrupted (annexin V–FITC$^+$/PI$^+$) cells. Secondary necrotic are apoptotic cells that have lost membrane integrity and therefore cannot be distinguished from primary necrotic cells and are included in the annexin V–FITC$^+$/PI$^+$ necrotic cell fraction *(21)*. Cells that are annexin V–FITC$^-$ and PI$^-$ are alive and not undergoing measurable apoptosis.

To define the quadrants set in **Fig. 4**, the following controls are necessary: (1) unstained cells, (2) cells stained with annexin V–FITC alone, and (3) cells stained with PI alone. Additionally, it is useful to confirm the annexin V–FITC population by inducing cells using an established cell line and chemical stimulus. Also, the lysed cell population and the necrotic cell fraction may be defined by using 0.1% saponin-treated cells.

Annexin V–FITC staining is a simple and reliable method of monitoring a transition in cell viability *(16,21)*. Importantly, it identifies cells at an earlier stage of apoptosis than assays based on DNA fragmentation because PS externalization is executed prior to nuclear digestion *(21)*. There is also a very rapid generation of results from the time of sampling to within approx 30 min. However, not all cell types exhibit PS exposure during cell culture. Even if there is induction, it may only be confined to a small fraction of the apoptotic cells, thereby underestimating the level of apoptotic cell induction. Nevertheless, the annexin V assay is a useful indicator of cell death by apoptosis.

3.5. TUNEL

DNA fragmentation can also be detected by labeling 3′OH termini of DNA fragments with biotin-conjugated nucleotides. This reaction is catalyzed by exogenous deoxynucleotidyl terminal transferase (TdT) *(22)* or by DNA polymerase I (nick translation) *(23)*. Because of these reactions, the general method is referred to as TUNEL (TdT-mediated dUTP Nick End Labeling). The net result is identical, in that fluorescein, digoxigen, or biotin conjugated nucleotides are incorporated at the DNA breaks. TdT-catalyzed reactions are more rapid than the nick translation assays. Fluoresceinated dUTP is detected directly, whereas biotin and digoxigenin-dUTP are detected using labeled streptavidin and labeled anti-digoxigenin antibodies, respectively.

1. Take 5×10^5 cells and centrifuge at 180*g* for 5 min.
2. Resuspend pellet in 1 mL PBS and centrifuge again.
3. Resuspend pellet in 1 mL of Cytofix/Cytoperm™ solution for 20 min at 4°C.

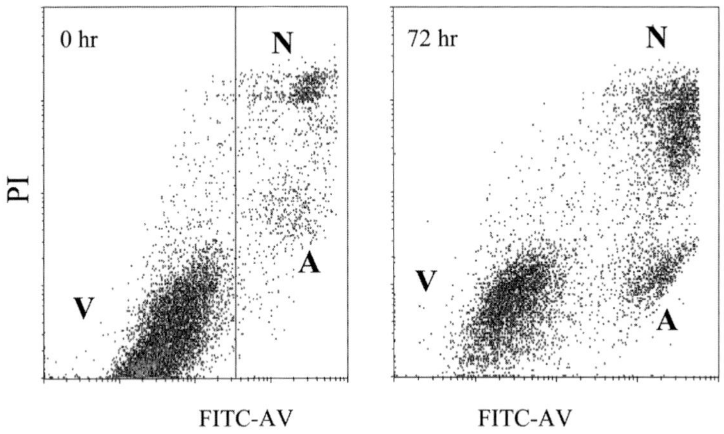

Fig. 4. Analysis of phosphatidylserine (PS) exposure and plasma membrane integrity during hybridoma batch culture. Cells were double labeled with fluorescein isothiocyanate–annexin V (FITC-AV) and propidium iodide (PI). The viable, early membrane intact apoptotic, and necrotic cell fractions were analyzed with flow cytometry. By 72 h a significant fraction of cells with PS exposure, i.e., FITC-AV (+ve)/PI (–ve), were induced. V, viable cells; A, apoptotic cells; N, secondary necrotic cells.

4. Centrifuge fixed cells at 1200 rpm for 5 min and resuspend pellet in 50 μL of reaction mixture.
5. Incubate at 37°C for 30 min.
6. Centrifuge cells at 1200 rpm for 5 min and resuspend pellet in 100 μL of staining buffer and incubate for 30 min at room temperature in the dark.
7. Centrifuge stained cells at 1200 rpm for 5 min and resuspend pellet in 500 μL PBS prior to FC analysis to measure green fluorescence (~515–545 nm).

Combining the green fluorescence of FITC-dUTP or biotin-dUTP, for example, with PI simultaneously stains DNA, and its fragments. This allows a correlation of apoptosis with the phase of the cell cycle or DNA ploidy ***(23)***.

The TUNEL assay was initially regarded as a very specific assay for apoptosis, especially in the occurrence of a high proportion of DNA strand breaks. Although TUNEL is a quantitative assay applicable to all cell types, there are a number of difficulties in the interpretation of the TUNEL technique. First, the emergence of DNA strand breaks may not be unique to apoptosis. DNA strand breaks were also detected in primary necrotic cells ***(17)*** and in cells that were devoid of internucleosomal cleavage, as confirmed by the absence of low-molecular DNA fragmentation following DNA gel electrophoresis ***(24)***. The method itself can be technically quite difficult, and many reaction steps are involved. These may account for a loss in the level of apoptotic cells. Primarily, the cells have to be fixed by appropriate concentrations of crosslinking fixatives such as formaldehyde *(22)*. Failure of the DNA fragments to crosslink to various intracellular proteins will ultimately result in their leakage from the cells, possibly during the washing procedures. Other detrimental factors may be the assay components themselves, e.g., a possible loss in TdT activity, and degradation of the nucleotides. A number of appropriate negative and positive controls should accompany this assay.

3.6. Rhodamine-123 Uptake and PI Exclusion

An early-event apoptosis is the loss of mitochondrial transmembrane potential (Ψm) *(15)*. One method for determining this reduction uses a charged cationic green fluorochrome, Rhodamine 123 (Rh123) *(14)*. Uptake of Rh123 is specific to functionally active mitochondria. Live cells, therefore, with intact plasma membrane and charged mitochondria concentrate this dye and exhibit strong green fluorescence. In contrast, dead cells with uncharged mitochondria fail to stain significantly with this dye. Simultaneous staining with PI distinguishes dead cells.

1. Take live cells (1×10^6 cells) from suspension culture and centrifuge at 180*g* for 5 min.
2. Resuspend pellet in 1 mL of tissue culture medium or HBBS.
3. Add 5 µL of Rh123 (1 mg/mL).
4. Incubate in the dark for 15 min at 37°C.
5. Add 20 µL of PI (50 µg/mL) 5 min prior to analysis.
6. Analyze by FC. Rh123 fluoresces green (~530 nm) and PI fluoresces red (~620 nm).

Viable and early membrane intact apoptotic cells with intact mitochondria take up Rh123 (**Fig. 5**, Day 2). Cells with lysed membranes are positive for PI staining and negative for Rh123. As apoptosis is induced, there is a loss in the Ψm, and the viable cell fraction decreases and more cells become necrotic, as highlighted in **Fig. 5** (Day 3).

3.7. Detection of Caspase-3 Activation

Caspases are a family of cysteine proteases that cleave a variety of substrates responsible for classic apoptotic morphological phenotype *(25)*. Caspase-3 is the most terminal caspase in the apoptotic cascade and is activated by self-proteolysis and/or cleavage by another caspase in response to the apoptotic stimuli. Active caspase-3 can be determined by a monoclonal rabbit antibody raised against the active fragment of caspase-3 conjugated to either fluorescent PE or FITC *(26)*.

1. Take 1×10^6 cells and centrifuge at 180*g* for 5 min.
2. Decant supernatant and resuspend pellet in PBS.
3. Repeat centrifugation.
4. Fix cells in 500 µL of Cytofixf/CytoPerm™ fixative (BD Pharmingen) for 20 min on ice.

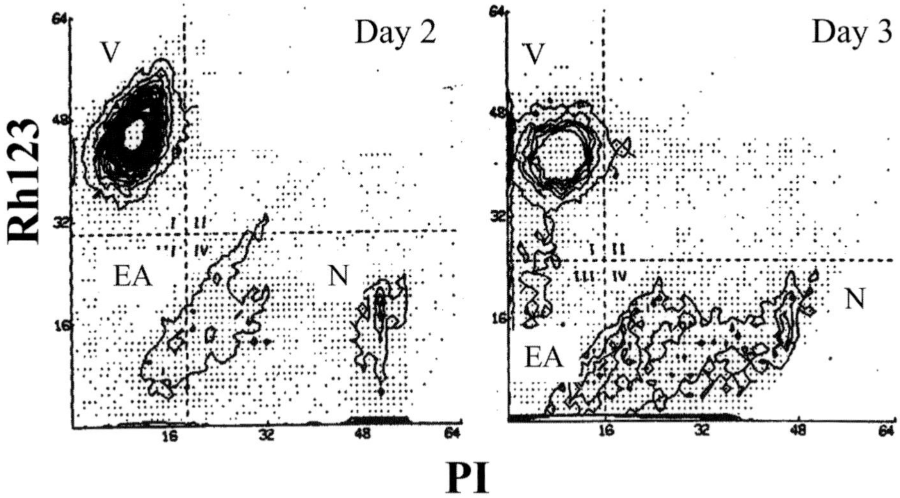

Fig. 5. Measurement of mitochondrial integrity by Rh123 and propidium iodide (PI) staining of hybridoma cells during batch culture. Rh123 staining of viable (V) cells is reduced and necrotic (N) cells are increased from day 2 to day 3. Early apoptotic (EA) cell fraction shows low R123 and PI fluorescence.

5. Can store cells at 4°C at this point for later analysis by resuspending pellet in 500 μL staining solution (PBS composed of 2% fetal bovine serum and 0.09% NaAzide). Otherwise proceed to **step 6** for analysis.
6. At end of incubation, centrifuge fixed cells at 1000 rpm for 5 min.
7. Make 1X Perm/Wash buffer from the 10X buffer and keep 1X Perm/Wash buffer on ice.
8. Resuspend pellet in Perm/Wash buffer and centrifuge at 4°C.
9. Resuspend pellet in 500 μL of Perm/Wash buffer ready for incubation with probe.
10. Add 20 μL of anti-caspase-3 PE antibody or anti-caspase-3 FITC.
11. Incubate for 30 min at 4°C in dark.
12. Analyze by flow cytometry to measure emission at 670 nm (PE) or 515–540 nm (FITC).

Figure 6 shows that a subpopulation of cells is induced that has increased caspase-3 activation, as shown by a shift in the histogram of cell count vs active caspase-3 PE fluorescence to the right of the main viable population. The monoclonal antibody specifically recognizes the active form of caspase-3 in human and mouse cells.

3.8. Concluding Remarks

FC detection of apoptosis provides a rapid, quantitative, and objective assessment of apoptosis. Ideally there should also be confirmation of apoptosis by the inspection of cells under the fluorescence microscope. Morphological

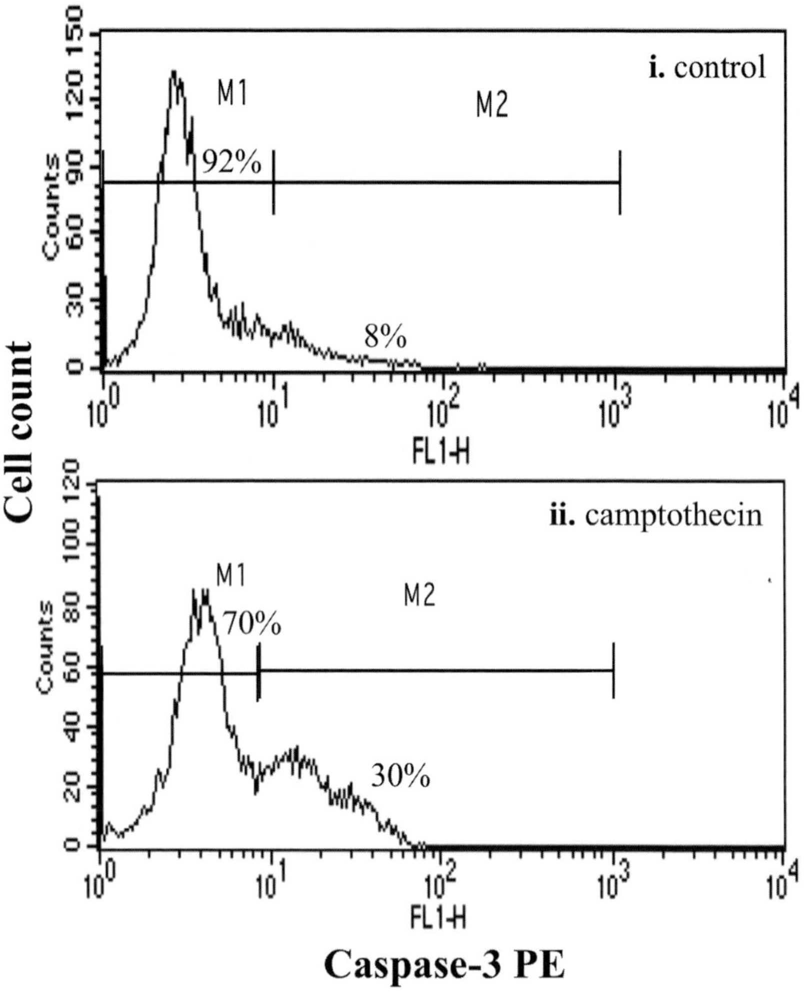

Fig. 6. Level of caspase-3 activation as determined by anti-caspase-3 phycoerythrin (PE) fluorescence. Untreated (control) and camptothecin (4 μ*M*)-treated cells were fixed, permeabilized, and stained for caspase-3 activation. A subpopulation of cells (30%) was induced by camptothecin treatment in comparison to only 8% in the untreated control.

changes during apoptosis are unique, and they should be the deciding factor when ambiguity arises regarding the mechanism of cell death. Detection of apoptosis is essential for monitoring of cell culture processes to obtain reliable measurements of viable and dead cell number.

FC, more than any other technique, is used for the detection and assaying of apoptosis. A more imperative business necessity is the utility of a high-throughput cell screening platform. The GuavaEasyCyte™ Personnel Cell Analysis (PCA) 96-well format is close to meeting these needs. It has capabilities of automation, miniaturization, and multiplexing of apoptosis assays that will inevitably streamline drug discovery and animal cell technology optimization strategies. Its unique microcapillary flow system minimizes reagent consumption, achieves a reasonable concentration of limited reagents, and uses small assay volumes. The machine itself is less expensive than the much larger FC, also minimizing space requirements currently required for a larger traditional FC.

4. Notes

1. Agarose gel electrophoresis can only be qualitative and cannot be representative of the frequency of apoptotic cell death. A high number of cells is required to reveal this apoptotic event.
2. PI is a suspected carcinogen. Gloves should be worn when using all DNA-binding fluorochromes. Although fluorescence microscopy is a reliable quantitative, morphological assessment of apoptosis, it is time consuming, labor intensive, and subjective to operator use.
3. It is possible to observe a bell-shaped region overlapping the G1 peak, which extends to merge with the debris region (late necrosis), toward the left-hand side of the histogram. The degree of DNA leakage can be manipulated by the extent of cell washing after fixation so the overlap with cells that undergo apoptosis is minimal and separation between these two populations is adequate.
4. In addition to the green fluorescent (FITC) probe, a red PE conjugate of annexin V is commercially available. In this case the DNA-binding dye 7-aminoactinomycin D (7-AAD) should be used in combination with annexin V–PE. 7-AAD fluorescence is detected in the far-red range of the spectrum (650 nm long-pass filter) and separated from PE fluorescence.
5. One of the main assay limitations of Rh123 uptake is its inability to show a distinct separation between the viable and early membrane intact apoptotic population. Because the Rh123 dye has affinity for the cytoplasmic membrane potential, it is important to establish the lowest possible appropriate concentrations of Rh123 and incubation times to minimize this effect.
6. For all procedures, cells can be obtained from either monolayer or suspension culture. In monolayer culture, the culture medium should be added back to the cells after trypsin treatment, as it may contain detached late apoptotic or necrotic cells.

References

1. Kerr, J. F. R., Wyllie, A. H., and Currie, A. R. (1972) Apoptosis: a basic biological phenomenon with wide ranging implications in tissue kinetics. *Br. J. Cancer* **26**, 239–257.
2. Al-Rubeai, M. (1998) Apoptosis and cell culture technology, in *Advances in Biochemical Engineering/Biotechnology* (Scheper, T., ed.), Springer-Verlag. New York, pp. 225–249.

3. Ishaque, A. and Al-Rubeai, M. (1997) Detection and analysis of apoptosis by flow cytometry, in *Protocols in Cell and Tissue Culture* (Griffiths, J. B., Doyle, A., and Newell, D. G., eds.), John Wiley, New York.
4. Al-Rubeai, M. and Singh, R. (1998) Apoptosis in cell culture. *Curr. Opin. Biotechnol.* **9**, 152–156.
5. Singh, R. and Al-Rubeai, M. (1998) Analysis of apoptosis by fluorescence microscopy, in *Protocols in Cell and Tissue Culture, Techniques for Cell Characterization and Analysis* (Griffiths, J. B., Doyle, A., and Newell, D. G., eds.), John Wiley, New York.
6. Ishaque, A. and Al-Rubeai, M. (1999) Role of Ca, Mg and K ions in determining apoptosis and extent of suppression afforded by bcl-2 during hybridoma cell culture. *Apoptosis* **4/5**, 335–355.
7. Ishaqe, A. and Al-Rubeai, M. (2002) Role of vitamins in determining apoptosis and extent of suppression by bcl-2 during hybridoma cell culture. *Apoptosis* **7**, 231–239.
8. Ishaque, A. and Al-Rubeai, M. (2004) Monitoring of apoptosis, in *Cell Engineering IV*. Kluwer Academic Publishers, Dordrecht.
9. Al-Rubeai M. and Emery A.N. (1993) Flow cytometry in animal cell culture. *Bio/technology* **11**, 573–579.
10. Dive, C., Gregory, C. D., Phipps, D. J., Evans, D. L., Milner, A. E., and Wyllie, A. H. (1992) Analysis and discrimination of necrosis and apoptosis (programmed cell death) by multiplarameter flow cytometry. *Biochim. Biophys. Acta* **1133**, 275–285.
11. Darzynkiewicz, Z., Robinson, J. P., and Crissman, H. A. (1994) Assays of cell viability: discrimination of cells dying by apoptosis. *Methods Cell Biol.* **41**, 15–38.
12. Nicoletti, I., Migiliorati, G., Pagliacci, M. C., Grignani, F., and Riccardi, C. (1991) A rapid and simple method for measuring thymocyte apoptosis by propidium iodide staining and flow cytometry. *J. Immunol. Methods* **139**, 271–280.
13. Darzynkiewicz, Z., Bruno, S., Del Bino, G., et al. (1992). Features of apoptotic cells measured by flow cytometry. *Cytometry* **13**, 795–808.
14. Darzynkiewicz, Z., Juan, G., Li, X., Gorczyca, W., Murakami, T., and Traganos, F. (1997) Cell necrobiology: analysis of apoptosis and accidental cell death (necrosis). *Cytometry* **27**, 1–20.
15. Ormerod, M. G., Sun, X. M., Snowden, R. T., Davies, R., Fearnhead, H., and Cohen, G. M. (1993) Increased membrane permeability of apoptotic thymocytes: a flow cytometric study. *Cytometry* **14**, 595–602.
16. Koopman, G., Reutlingsperger, C. P. M., and Kuijten, G. A. M. (1994) Annexin V for flow cytometric detection of phosphatidylserine expression on B cells undergoing apoptosis. *Blood* **84**, 1415.
17. Collins, M. K., Marvel, J., Malde, P., and Lopez-Rivas, A. (1992) Interleukin 3 protects murine bone marrow cells from apoptosis induced by DNA damaging agents. *J. Exp. Med.* **176**, 1043–1051.
18. Elstein, K. H. and Zucker, R. M. (1994) Comparison of cellular and nuclear flow cytometric techniques for discriminating apoptotic sub-populations. *Exp. Cell Res.* **211**, 322–331.
19. Klucar, J. and Al-Rubeai, M. (1997) G2 cell cycle arrest and apoptosis are induced in Burkitt's lymphoma cells by the anticancer agent oracin. *FEBS Lett.* **400**, 127–130.

20. Fadok, V. A., Voelker, D. R., Campbell, P. A., Cohen, J. J., Bratton, D. L., and Henson, P. M. (1992) Exposure of phosphatidylserine on the surface of apoptotic lymphocytes triggers specific recognition and removal by macrophages. *J. Immunol.* **148**, 2207.
21. Ishaque, A. and Al-Rubeai, M. (1998) Use of intracellular pH and annexin V flow cytometric assays to monitor apoptosis and its suppression by *bcl-2* over-expression in hybridoma cell culture *J. Immunol. Methods* **221**, 43–57.
22. Gorczyca, W., Bruno, S., and Darzynkiewicz, Z. (1992) DNA strand breaks occuring during apoptosis: Their early in situ detection by the terminal deoxynucleotidyl transferase and nick translation assays and prevention by serine protease inhibitors. *Int. J. Oncol.* **1**, 639–648.
23. Gold, R., Schmeid, M., and Rothe, G. (1993) Detection of DNA fragmentation in apoptosis-application of in situ nick translation to cell culture systems and tissue-sections. *J. Histochem. Cytochem.* **41**, 1023–1030.
24. Green, D. R. and Reed, J. C. (1998) Mitochondria and apoptosis. *Science* **281**, 1309–1313.
25. Thornberry, N. A. and Lazebnik. (1998) Caspases: enemies within. *Science* **281**, 1312–1316.
26. Ishaque, A., Sorokin, A., and Dunn, M. J. (2001) Cyclooxygenase-2 inhibits TNF-alpha mediated apoptosis in renal glomerular mesangial cells. *J. Biol. Chem.* **278(12)**, 10,629–10,640.

14

Metabolic Flux Estimation in Mammalian Cell Cultures

Chetan T. Goudar, Richard Biener, James M. Piret, and Konstantin B. Konstantinov

Summary

Metabolic flux analysis with its ability to quantify cellular metabolism is an attractive tool for accelerating cell line selection, medium optimization, and other bioprocess-development activities. In the stoichiometric flux estimation approach, unknown fluxes are determined using intracellular metabolite mass balance expressions and measured extracellular rates. The simplicity of the stoichiometric approach extends its application to most cell culture systems, and the steps involved in metabolic flux estimation by the stoichiometric method are presented in detail in this chapter. Specifically, overdetermined systems are analyzed because the extra measurements can be used to check for gross measurement errors and system consistency. Cell-specific rates comprise the input data for flux estimation, and the logistic modeling approach is described for robust specific rate estimation in batch and fed-batch systems. A simplified network of mammalian cell metabolism is used to illustrate the flux estimation procedure, and the steps leading up to the consistency index determination are presented. If gross measurement errors are detected, a technique for determining the source of gross measurement error is also described. A computer program that performs most of the calculations described in this chapter is presented and references to flux estimation software are provided. The procedure presented in this chapter should enable rapid metabolic flux estimation in any mammalian cell bioreaction network by the stoichiometric approach.

Key Words: Cell culture; computer software; metabolite balancing; metabolic flux analysis; overdetermined systems.

1. Introduction

Flux is defined as the rate with which material is processed through a bioreaction pathway *(1)*. While a reaction flux does not contain information on the activity of enzymes in a particular reaction, it does contain information on the functional extent of that reaction. For this reason it has been argued that metabolic fluxes constitute a fundamental determinant of cell physiology and metabolic flux estimation

From: *Methods in Biotechnology, Vol. 24: Animal Cell Biotechnology: Methods and Protocols, 2nd Ed.* Edited by: R. Pörtner © Humana Press Inc., Totowa, NJ

is the preferred method for characterizing the physiological state of a cell *(2,3)*. Despite the ability of mammalian cells to effectively fold and glycosylate proteins, their low productivities call for improvement, and metabolic flux analysis with its quantitative insights into cellular metabolism is a promising approach. During process development, flux data can contribute to the selection of cell lines, optimal medium formulations, and bioreactor operating conditions. Flux analysis can also be applied to an established manufacturing process to quantify and archive cellular metabolism. Comparison of data from multiple runs can provide valuable information, such as the reproducibility of cell performance and long-term cell bank stability.

Metabolite balancing *(4–6)* and isotope tracer studies *(7–11)* are the two methods used for flux estimation. For the metabolite-balancing approach, intracellular fluxes are estimated from measured extracellular rates and intracellular metabolite mass balance expressions. Experimental and computational techniques for the metabolite-balancing approach are relatively simple, and mammalian cell applications include baby hamster kidney (BHK) *(12,13)*, Chinese hamster ovary (CHO) *(14,15)*, hybridoma *(4,5,10,16–19)*, and human *(20)* cells. For the isotope tracer approach, labeled substrate (usually ^{13}C glucose) is added to the cultivation medium and the label distribution resulting from cellular metabolism is detected by nuclear magnetic resonance (NMR) or gas chromatography-mass spectrometry (GC-MS) analysis. Label distribution is a function of the intracellular metabolic fluxes, and several methods have been proposed to determine fluxes from NMR and GC-MS data *(8,9,21–24)*. Isotope tracer experiments are expensive and require substantial experimental and analytical effort.

A simplified mammalian cell network is used in this chapter to illustrate the steps involved in metabolic flux analysis by the metabolite-balancing approach. Procedures for error analysis and data consistency testing that are applicable to overdetermined systems are also presented.

2. Materials

1. The cell line of interest and a chemically defined, serum-free medium such as EX-CELL™ (JRH Biosciences Inc., Lanexa, KS; *see* **Note 1**).
2. A batch, fed-batch, chemostat, or a perfusion cultivation system.
3. A computer program that calculates metabolic fluxes.

3. Methods

3.1. Establishing the Metabolic Network

A simplified representation of metabolism is formulated with groups of serial reactions lumped into single reactions and only including pathways carrying significant flux. These typically include reactions in glycolysis, the tricarboxylic acid (TCA) cycle, amino acid biosynthesis and catabolism, along with biomass and product formation (*see* **Note 2**). Several groups ***(4,5,13,14,17,25)*** have established mammalian cell metabolic networks that typically include 20–40 reactions and provide a good starting point for most studies. Additional refinements can be made by consulting biochemistry texts or online databases ***(26)***. However, computational constraints must be considered when selecting reactions to include in the biochemical network (*see* **Note 3**).

3.2. Experimental and Analytical Techniques

Chemostat and perfusion cultivations are best suited for metabolic flux analysis since they can be operated at steady state. Samples from the exponential phase of batch and fed-batch cultivations can provide pseudo-steady-state conditions (*see* **Note 4**). The following measurements are necessary to generate the input data necessary for metabolic flux analysis:

1. Viable cell concentration using a hemocytometer or an automated system such as the CEDEX (Innovatis, Bielefeld, Germany) for growth-rate determination.
2. Biomass composition to determine amino acid requirements (*see* **Note 5**).
3. Glucose and lactate concentrations in the culture using YSI 2700 analyzer (Yellow Springs Instruments, Yellow Springs, OH) or equivalent for determining cell-specific uptake and production rates.
4. Oxygen and carbon dioxide concentrations in the outlet gas streams using a mass spectrometer (MGA-1200, Applied Instrument Technologies, Pomona, CA) or equivalent to determine oxygen uptake rate (OUR) and carbon dioxide evolution rate (CER; *see* **Note 6**).
5. Amino acid concentrations using the AminoQuant method (Agilent Technologies, Palo Alto, CA) or equivalent to determine biosynthetic and catabolic rates.

3.3. Estimation of Cell-Specific Rates

Cell-specific rates must be accurately determined from the above experimental measurements of prime variables since they constitute the input data for flux estimation. While batch and fed-batch specific rates can be obtained in multiple ways ***(27–31)***, logistic equation modeling is recommended since it logically constrains the fit ***(32)***. This approach includes the generalized logistic equation:

$$X = \frac{A}{\exp(Bt) + C\exp(-Dt)} \tag{1}$$

the monotonically decreasing logistic decline equation:

$$N = \frac{A}{\exp(Bt) + C} \tag{2}$$

and the more widely used logistic growth equation:

$$P = \frac{A}{1 + C\exp(-Dt)} \tag{3}$$

where *X*, *N*, and *P* are the cell density, nutrient, and product (metabolite) concentrations, *t* is time, and *A, B, C*, and *D* model parameters that in limiting cases correspond to conventional biological variables *(32)*. Parameters *A* and *C* are related to the initial and final values of *X*, *N*, and *P* in Eqs. 1–3, while *B* and *D* correspond to the maximum death (k_{dmax}) and growth rate (μ_{max}), respectively, in Eq. 1. Parameters *B* and *D* in Eqs. 2 and 3 are indicative of the rate of nutrient decay and product (metabolite) formation, respectively *(32)*.

The first step in logistic equation modeling involves nonlinear parameter estimation in Eqs. 1–3. Once parameters *A–D* are known, derivatives of the variables in Eqs. 1–3 can be readily estimated since these are analytically derived. Finally, specific rates are estimated from the derivatives and cell density data. For instance **(Fig. 1)**, the apparent growth rate, μ, is estimated as $\mu = \frac{1}{X_v}\frac{dX_v}{dt}$ with $\frac{dX_v}{dt}$ determined by fitting the time course of cell density (X_v) data to Eq. 1 *(32)*. A similar approach for specific rate determination can be applied to the other measured variables **(Fig. 1)**.

Perfusion specific rates are obtained from the prime variable mass balance expressions across the bioreactor and the cell-retention device **(Table 1)**. Accurate prime variable measurement and specific rate estimation are critical for reliable flux determination since their errors will propagate into the calculated fluxes.

3.4. Estimation of Metabolic Fluxes

Mammalian cell biochemical networks are either underdetermined, determined, or overdetermined depending on the number of experimental measurements and the unknown intracellular fluxes. The procedure for flux estimation in overdetermined systems (extra measurements available) is presented below since these systems allow error and consistency checking to obtain more robust estimates (*see* **Note 7**). **Figure 2** shows a simplified bioreaction network that was originally proposed by Balcarcel and Clark *(33)* for flux analysis from well plate cultivations where a limited number of measurements were available. The corresponding reactions are shown in **Table 2**. Glycolytic reactions have been lumped into a single reaction (glucose → pyruvate; flux v_{c1}), as have those for the TCA cycle (pyruvate → CO_2; flux v_{c3}). Conversion of pyruvate to lactate is an important reaction in most mammalian cell cultures and has been included in the network (flux v_{c2}) along with the oxidative phosphorylation reactions (NADH and $FADH_2$ for v_{c4} and v_{c5}). Consumption rates of glucose (v_{m1}) and oxygen (v_{m4}) along with production rates of lactate (v_{m2}) and CO_2

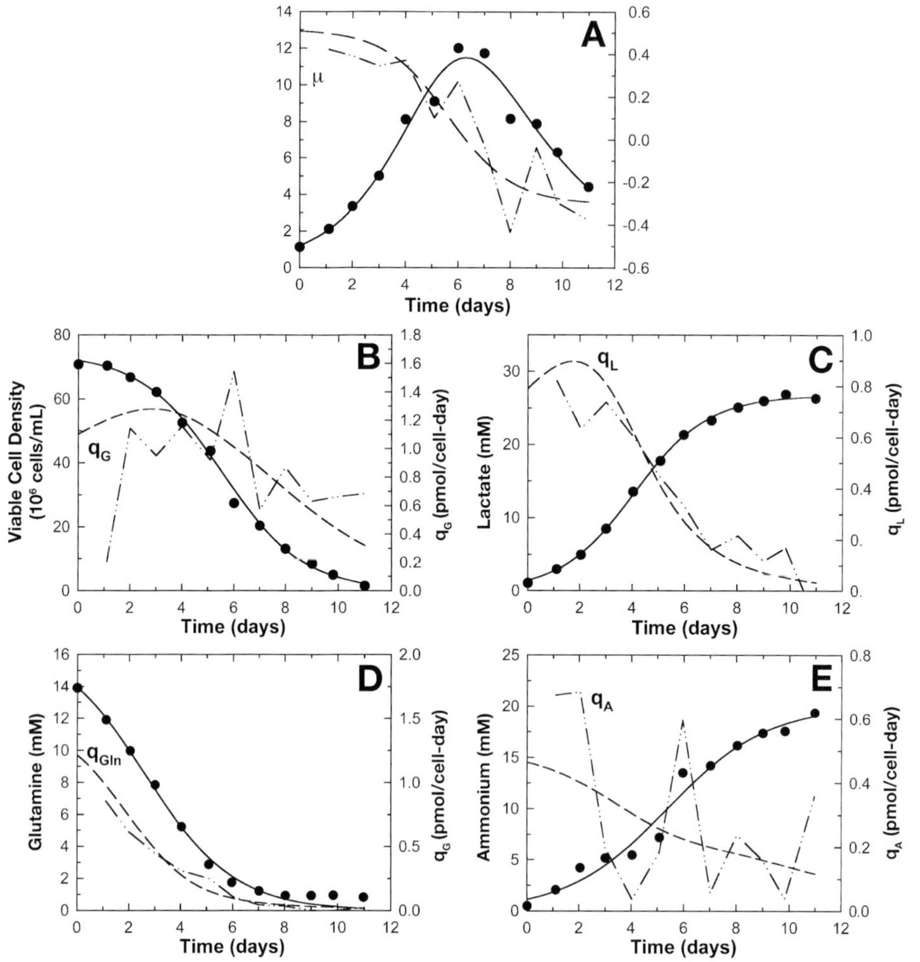

Fig. 1. Time profiles of cell density, nutrient, and metabolite concentrations for Chinese hamster ovary cells in 15 L batch culture. Experimental data (•); logistic (generalized logistic equation for cell density, logistic decline equation for glucose and glutamine, and logistic growth equation for lactate and ammonium) fit (——); logistic specific rate (------); discrete derivative specific rate (— ·· — ··). (From **ref. *32***.)

(v_{m3}) make up the measured extracellular rates. The network has a total of these four measured extracellular rates and six unknown intracellular fluxes to be estimated (v_{c1}–v_{c6}), including the net ATP production rate.

3.4.1. Flux Estimation Procedure

1. The stoichiometic matrix, $\boldsymbol{G}^T$, is first derived from the bioreaction network. Choosing glucose, lactate, CO_2, O_2, pyruvate, NADH, $FADH_2$, and ATP as the

Table 1
Specific Rate Expressions for a Perfusion System

Parameter	Expression[a,b]
Specific growth rate (μ', 1/d)	$\mu' = \frac{F_d}{V} + \frac{F_h}{V}\left(\frac{X_V^H}{X_V^B}\right) + \frac{1}{X_V^B}\left(\frac{dX_V^B}{dt}\right)$
Specific productivity (q_P, µg/cell/d)	$q_P = \frac{1}{X_V^B}\left(\frac{F_m P}{V} + \frac{dP}{dt}\right)$
Specific glucose consumption rate (q_G, pmol/cell/d)	$q_G = \frac{1}{X_V^B}\left(\frac{F_m(G_m - G)}{V} - \frac{dG}{dt}\right)$
Specific glutamine consumption rate (q_{Gln}, pmol/cell/d)	$q_{Gln} = \frac{1}{X_V^B}\left(\frac{F_m(Gln_m - Gln)}{V} - \frac{dGln}{dt} - k_{Gln}Gln\right)$
Specific lactate production rate (q_L, pmol/cell/d)	$q_L = \frac{1}{X_V^B}\left(\frac{F_m L}{V} + \frac{dL}{dt}\right)$
Specific ammonium production rate (q_A, pmol/cell/d)	$q_A = \frac{1}{X_V^B}\left(\frac{F_m A}{V} + \frac{dA}{dt} - k_{Gln}Gln\right)$
Specific oxygen uptake rate (OUR, pmol/cell/d)	$q_{O_2} = \frac{1}{X_V^B}\left(\frac{F_{total}\left(O_{2_{in}} - O_{2_{out}}\right)}{V}\right)$

F_d = cell discard rate (L/d). V = bioreactor volume (L).
F_h = harvest rate (L/d). X_V^B = bioreactor viable cell density (10^9 cells/L).
X_V^H = harvest viable cell density (10^9 cells/L).
t = time (d). F_m = medium feed rate (L/d).
G = glucose concentration (m*M*). P = product concentration (mg/L).
Gln = glutamine concentration (m*M*). G_m = medium glucose concentration.
k_{Gln} = first-order rate constant for glutamine degradation (1/d).
L = lactate concentration (m*M*). Gln_m = medium glutamine concentration (m*M*).
F_{total} = total gas flow rate (L/d). A = ammonium concentration (m*M*).
O_{2out} = outlet oxygen concentration (m*M*). O_{2in} = inlet oxygen concentration (m*M*).
[a]Expressions for amino acid catabolism are analogous to that for glucose.
[b]Estimating carbon dioxide evolution rate is complicated, and detailed information is available (*see* **Note 6**).

pathway intermediates results in an 8 × 10 matrix (number of rows corresponds to the number of pathway intermediates and the number of columns to the sum of known and unknown rates). The mass balances for these pathway intermediates, $\mathbf{G}^T\mathbf{v} = 0$, can be written as:

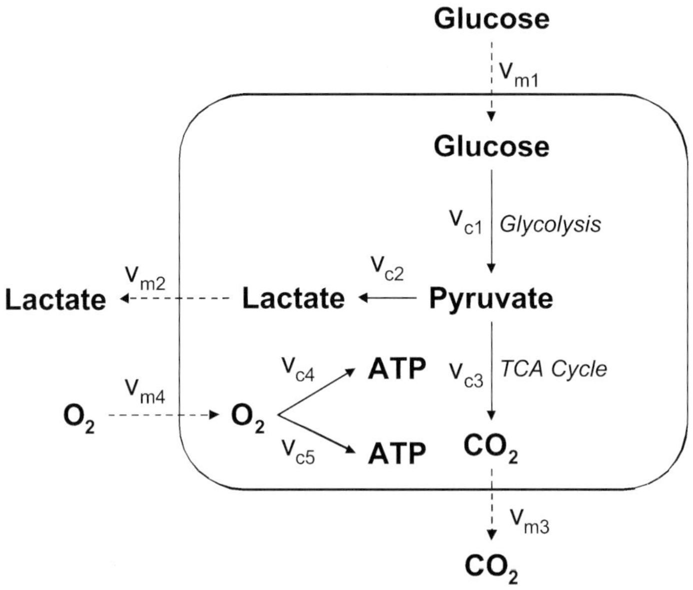

Fig. 2. A simplified network for mammalian cell metabolism with lumped reactions for glycolysis and tricarboxylic acid (TCA) cycle and those for lactate production and oxidative phosphorylation ***(33)***. The network consists of five unknown intracellular fluxes (v_{c1}–v_{c5}) and four extracellular measured rates (v_{m1}–v_{m4}).

$$\begin{pmatrix} -1 & 0 & 0 & 0 & -1 & 0 & 0 & 0 & 0 & 0 \\ 0 & -1 & 0 & 0 & 0 & 1 & 0 & 0 & 0 & 0 \\ 0 & 0 & -1 & 0 & 0 & 0 & 3 & 0 & 0 & 0 \\ 0 & 0 & 0 & -1 & 0 & 0 & 0 & -0.5 & -0.5 & 0 \\ 0 & 0 & 0 & 0 & 2 & -1 & -1 & 0 & 0 & 0 \\ 0 & 0 & 0 & 0 & 2 & -1 & 4 & -1 & 0 & 0 \\ 0 & 0 & 0 & 0 & 0 & 0 & 1 & 0 & -1 & 0 \\ 0 & 0 & 0 & 0 & 2 & 0 & 1 & 2.5 & 1.5 & -1 \end{pmatrix} \begin{pmatrix} v_{m1} \\ v_{m2} \\ v_{m3} \\ v_{m4} \\ v_{c1} \\ v_{c2} \\ v_{c3} \\ v_{c4} \\ v_{c5} \\ v_{c6} \end{pmatrix} = \begin{pmatrix} 0 \\ 0 \\ 0 \\ 0 \\ 0 \\ 0 \\ 0 \\ 0 \end{pmatrix} \quad (4)$$

Multiplying the first row of $\mathbf{G}^T$ with $\mathbf{v}$ results in $-v_{m1} - v_{c1} = 0$, the mass balance for glucose, and subsequent multiplications result in the other seven mass balance expressions. Matrix $\mathbf{G}^T$ is of full rank (rank = 8), indicating that all eight metabolite mass balances are independent of each other. It also has a low condition number of 7.6, suggesting low sensitivity of the calculated fluxes to the measured rates **(Note 8)**. This system is overdetermined since it has 2 degrees of freedom (degrees of freedom = number of reaction rates – rank of $\mathbf{G}^T$).

2. The rate vector, $\mathbf{v}$, is then split into the measured and calculated components ($\mathbf{v}_m$ and $\mathbf{v}_c$). If $\mathbf{G}_m^T$ and $\mathbf{G}_c^T$ are the corresponding splits in $\mathbf{G}^T$, the metabolite mass balance expression, $\mathbf{G}^T\mathbf{v} = 0$, becomes:

$$\mathbf{G}_m^T \mathbf{v}_m + \mathbf{G}_c^T \mathbf{v}_c = 0 \tag{5}$$

$$\begin{pmatrix} -1 & 0 & 0 & 0 \\ 0 & -1 & 0 & 0 \\ 0 & 0 & -1 & 0 \\ 0 & 0 & 0 & -1 \\ 0 & 0 & 0 & 0 \\ 0 & 0 & 0 & 0 \\ 0 & 0 & 0 & 0 \\ 0 & 0 & 0 & 0 \end{pmatrix} \begin{pmatrix} v_{m1} \\ v_{m2} \\ v_{m3} \\ v_{m4} \end{pmatrix} + \begin{pmatrix} -1 & 0 & 0 & 0 & 0 & 0 \\ 0 & 1 & 0 & 0 & 0 & 0 \\ 0 & 0 & 3 & 0 & 0 & 0 \\ 0 & 0 & 0 & -0.5 & -0.5 & 0 \\ 2 & -1 & -1 & 0 & 0 & 0 \\ 2 & -1 & 4 & -1 & 0 & 0 \\ 0 & 0 & 1 & 0 & -1 & 0 \\ 2 & 0 & 1 & 2.5 & 1.5 & -1 \end{pmatrix} \begin{pmatrix} v_{c1} \\ v_{c2} \\ v_{c3} \\ v_{c4} \\ v_{c5} \\ v_{c6} \end{pmatrix} = \begin{pmatrix} 0 \\ 0 \\ 0 \\ 0 \\ 0 \\ 0 \\ 0 \\ 0 \end{pmatrix} \tag{6}$$

3. The vector of unknown fluxes, $\mathbf{v}_c$, can now be computed from Eq. 5:

$$\mathbf{v}_c = -\left(\mathbf{G}_c^T\right)^{\#} \mathbf{G}_m^T \mathbf{v}_m \tag{7}$$

$$\mathbf{v}_c = -\begin{pmatrix} -0.3172 & 0.3414 & 0.0103 & -0.1034 & 0.2897 & 0.0517 & 0.0517 & 0 \\ -0.3414 & 0.8293 & -0.0052 & 0.0517 & -0.1448 & -0.0259 & -0.0259 & 0 \\ -0.0034 & -0.0017 & 0.2121 & -0.1207 & -0.0621 & 0.0603 & 0.0603 & 0 \\ -0.2552 & -0.1276 & 0.6931 & -0.9310 & 0.4069 & -0.5345 & 0.4655 & 0 \\ 0.0483 & -0.0241 & 0.0310 & -0.3103 & -0.1310 & 0.1552 & -0.8448 & 0 \\ -1.2034 & 0.3983 & 2.0121 & -3.1207 & 1.3379 & -0.9397 & 0.0603 & -1 \end{pmatrix}$$

$$\begin{pmatrix} -1 & 0 & 0 & 0 \\ 0 & -1 & 0 & 0 \\ 0 & 0 & -1 & 0 \\ 0 & 0 & 0 & -1 \\ 0 & 0 & 0 & 0 \\ 0 & 0 & 0 & 0 \\ 0 & 0 & 0 & 0 \\ 0 & 0 & 0 & 0 \end{pmatrix} \begin{pmatrix} -1.4788 \\ 1.7293 \\ 5.8333 \\ -5.1369 \end{pmatrix} = \begin{pmatrix} 1.6512 \\ 1.6431 \\ 1.8592 \\ 8.9824 \\ 1.7456 \\ 30.2361 \end{pmatrix} \tag{8}$$

where $(\mathbf{G}_c^T)^{\#}$ is the pseudoinverse of $\mathbf{G}_c^T$ **(Note 9)** and the listed $\mathbf{v}_m$ is from CHO cells in perfusion culture at a cell density of 20×10^6 cell/mL. Flux estimation is essentially a three-step process that involves formulation of the stoichiometric matrix (step 1) followed by splitting the rate vector into measured and nonmeasured

Table 2
Reactions for the Simplified Biochemical Network in Fig. 2 ***(33)***

Flux	Reaction
V_{c1}	Glc + 2 NAD^+ + 2 ADP + P_i → 2 Pyr + 2 NADH + 2 ATP + 2 H_2O + 2 H^+
V_{c2}	Pyr + NADH + H^+ → Lac + NAD^+
V_{c3}	Pyr + 4 NAD^+ + FAD + ADP + 3 H_2O + P_i → 3 CO_2 + 4 NADH + $FADH_2$+ ATP + 4 H^+
V_{c4}	0.5 O_2 + 2.5 ADP + 2.5 P_i + NADH + 3.5 H^+ → 2.5 ATP + NAD^+ + 3.5 H_2O
V_{c5}	0.5 O_2 + 1.5 ADP + 1.5 P_i + $FADH_2$ + 1.5 H^+ → 1.5 ATP + FAD + 2.5 H_2O
V_{c6}	Net rate of ATP Production

components (step 2) and subsequent flux calculations (step 3). No further calculations are necessary for a determined system. For overdetermined systems, gross error detection and consistency analysis can be performed as shown below.

4. The redundancy matrix, **R**, is first computed as $\mathbf{R} = \mathbf{G}_m^T - \mathbf{G}_c^T\left(\mathbf{G}_c^T\right)^{\#}\mathbf{G}_m^T$

$$\mathbf{R} = \begin{pmatrix} -0.6828 & -0.3414 & -0.0103 & 0.1034 \\ -0.3414 & -0.1707 & -0.0052 & 0.0517 \\ -0.0103 & -0.0052 & -0.3638 & -0.3621 \\ 0.1034 & 0.0517 & -0.3621 & -0.3793 \\ -0.2897 & -0.1448 & -0.1862 & -0.1379 \\ -0.0517 & -0.0259 & 0.1810 & 0.1897 \\ -0.0517 & -0.0259 & 0.1810 & 0.1897 \\ 0 & 0 & 0 & 0 \end{pmatrix} \quad (9)$$

It is clear that **R** has dependent rows. A reduced redundancy matrix, $\mathbf{R_r}$, is computed from **R** by singular value decomposition (SVD) and includes only the independent rows:

$$\mathbf{R_r} = \begin{pmatrix} 0.8099 & 0.4049 & -0.2250 & -0.3599 \\ -0.3679 & -0.1839 & -0.6745 & -0.6131 \end{pmatrix} \quad (10)$$

5. The measurement error vector, $\boldsymbol{\delta}$, contains information on the error in measured rates, and, assuming 10% error in all the elements of $\mathbf{v}_m$ **(Note 10)**, it can be written as:

$$\delta = \begin{pmatrix} -0.1479 \\ 0.1729 \\ 0.5833 \\ -0.5137 \end{pmatrix} \quad (11)$$

from which the residual vector, $\boldsymbol{\varepsilon}$, and the variance covariance matrix of the measured rates, **F**, are computed (**Note 11**):

$$\varepsilon = \mathbf{R_r v}_m = \begin{pmatrix} 0.0388 \\ -0.5591 \end{pmatrix} \tag{12}$$

$$\mathbf{F} = E(\delta\delta^T) = \begin{pmatrix} 0.0219 & 0 & 0 & 0 \\ 0 & 0.0299 & 0 & 0 \\ 0 & 0 & 0.3403 & 0 \\ 0 & 0 & 0 & 0.2639 \end{pmatrix} \tag{13}$$

6. Once **F** is known, the covariance matrix of the the residuals, φ, is computed as:

$$\varphi = \mathbf{R_r F R_r}^{\mathbf{T}} = \begin{pmatrix} 0.0707 & 0.1011 \\ 0.1011 & 0.2580 \end{pmatrix} \tag{14}$$

from which the consistency index, h, is determined:

$$h = \varepsilon\varphi^{-1}\varepsilon^T = 3.36 \tag{15}$$

It has been shown that h follows a χ^2 distribution with the degrees of freedom equal to the rank of **R** *(34,35)*. If $h < \chi^2$ at the appropriate degree of freedom, no gross measurement errors are present and the bioreaction network chosen is a good representation. From **Table 3**, the h value of 3.36 (Eq. 15) is less than the χ^2 distribution at the 90% confidence level for two degrees of freedom, indicating that the data and model quality are acceptable. This essentially completes the error-diagnosis process.

7. If desired, improved estimates of the measured rates, $\hat{\mathbf{v}}_m$ can be obtained as:

$$\hat{\mathbf{v}}_m = \mathbf{I} - \mathbf{F R}_r{}^T \varphi^{-1} \mathbf{R}_r \mathbf{v}_m = \begin{pmatrix} -1.6698 \\ 1.5987 \\ 5.2243 \\ -5.2251 \end{pmatrix} \tag{16}$$

where **I** is an indentity matrix. The unknown flux vector, $\hat{\mathbf{v}}_c$ corresponding to $\hat{\mathbf{v}}_m$ is calculated from Eq. 7:

$$\hat{\mathbf{v}}_c = \begin{pmatrix} 1.6701 \\ 1.5986 \\ 1.7415 \\ 8.7079 \\ 1.7417 \\ 29.4639 \end{pmatrix} \tag{17}$$

Table 3
Values of χ^2 Distribution at Varying Degrees of Freedom and Confidence Levels

	Confidence level		
Degrees of freedom	90%	95%	99%
1	2.71	3.84	6.63
2	4.61	5.99	9.21
3	6.25	7.81	11.3
4	7.78	9.49	13.3

Flux estimates from Eq. 8 and Eq. 17 are very similar, but this may not always be the case.

8. If $h > \chi^2$, then either the experimental data contain gross errors or the model chosen is not appropriate. Error diagnosis is possible for overdetermined systems with at least 2 degrees of freedom ***(35)***. Let us assume that because of a measurement error, the CER has been inaccurately determined to be 7.2916 (i.e., 25% error; actual value = 5.8333) and the other measurements are unaffected. The unknown flux vector can be calculated from Eq. 7 as:

$$\mathbf{v}_c = \begin{pmatrix} 1.6663 \\ 1.6355 \\ 2.1684 \\ 9.9932 \\ 1.7902 \\ 33.1703 \end{pmatrix} \tag{18}$$

and has an h value of 9.64 that is higher than the χ^2 distribution, even at the 99% confidence level **(Table 3)**. To determine if a gross error in any of the measured rates is responsible for the high h value, they can be sequentially eliminated and the h value computed each time. Elimination of CER resulted in a dramatic reduction in the h value **(Table 4)**, correctly indicating an error in its measurement **(Note 12)**. A summary of the above flux estimation and error diagnosis procedure is shown in **Fig. 3**.

3.4.2. Software for Flux Estimation

Computer programs for flux estimation by the metabolite-balancing approach are typically available free of charge for academic use ***(36,37)***. These programs provide a graphical interface for specifying the reactions in the biochemical network and for inputting experimental data. Derivation of the stoichiometric matrix and the rate vector are automated, as are flux estimation and error diagnosis. These calculations can also be readily performed in computing environments such as MATLAB® (The Mathworks, Natick, MA) and Mathematica®

Table 4
Values of Consistency Index, *h*, After Sequential Elimination of Specific Rates

Measurement eliminated	h
None	9.64
Glucose uptake rate	5.87
Lactate production rate	5.87
CO_2 production rate (CER)	1.59
O_2 consumption rate (OUR)	8.21

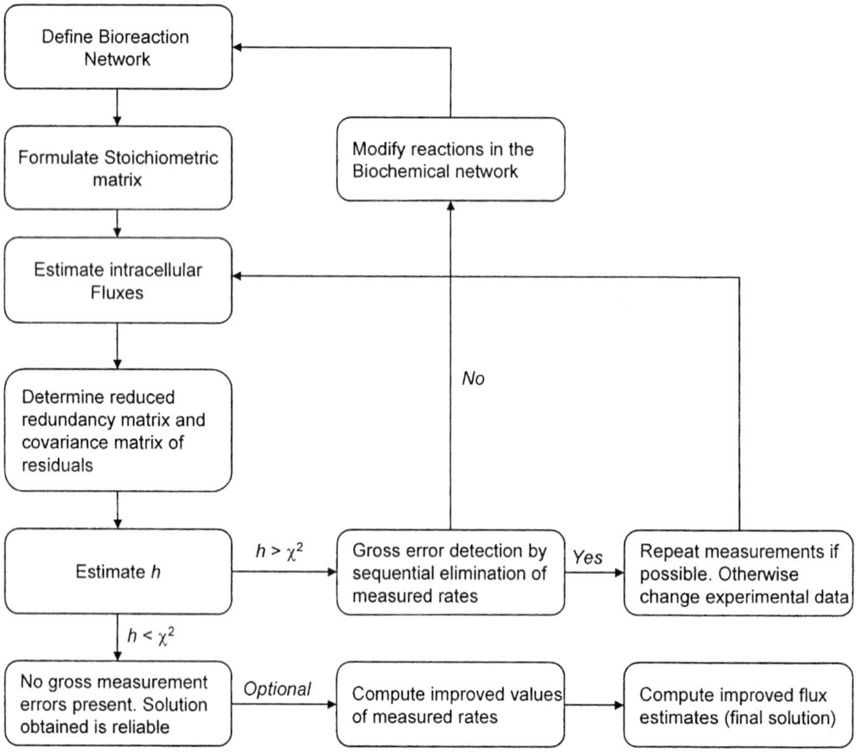

Fig. 3. Overview of the metabolite-balancing approach for flux estimation in overdetermined systems.

(Wolfram Research Inc., Champaign, IL). **Figure 4** shows a MATLAB script that performs some of the calculations presented in **Subheading 3.4.1**. Refinements to the program such as automatic stoichiometric matrix generation and error diagnosis **(Subheading 3.4.1., step 8)** can be readily made but have not been included in **Fig. 4**.

```
% mfa_balancing.m Computes Metabolic Fluxes for Mammalian Cells
% Reaction Network is from Balcarcel and Clark % (B. Prog., 19, 98-108 2003)
% Measurements include glucose, lactate, O2, and CO2
% References to Equation Numbers in text are included
clear all

% 1. MATRIX DEFINITIONS
% Stoichiometirc Matrix, GT (Eq.4)
GT = [-1 0 0 0 -1 0 0 0 0 0; 0 -1 0 0 0 1 0 0 0 0;...
    0 0 -1 0 0 0 3 0 0 0; 0 0 0 -1 0 0 0 -0.5 -0.5 0;...
    0 0 0 0 2 -1 -1 0 0 0; 0 0 0 0 2 -1 4 -1 0 0;...
    0 0 0 0 0 0 1 0 -1 0; 0 0 0 0 2 0 1 2.5 1.5 -1];

% rank and condition number of GT
rankGT = rank(GT); condGT = cond(GT);

% Rate Vector, v (Eq.4)
v = [-1.4788 1.7293 5.8333 -5.1369 0 0 0 0 0 0]';

%2. FLUX ESTIMATION
% Separation of G into Gm and Gc (Eqs.5&6)
GmT = GT(:,1:4); GcT = GT(:,5:10);

% Separation of v into vm and vc (Eqs.5&6)
vm = v(1:4,1); vc = v(5:10,1);

% Unknown Rate Vector (vc) Estimation (Eqs.7&8)
vc = -pinv(GcT)*GmT*vm;

% 3. ERROR DIAGNOSIS
% Redundancy Matrix (Eq.9)
R  = GmT - GcT*pinv(GcT)*GmT; rankR = rank(R);

% Reduced redundancy matric (Eq.10)
Rr=[0.8099    0.4049   -0.2250   -0.3599
   -0.3679   -0.1839   -0.6745   -0.6131];

% Delta, e and F estimation
measurement_error = 0.1*ones(1,4)';
delta      = vm.*measurement_error; %(Eq.11)
e1         = Rr*vm;             % (Eq.12))
delta1     = delta*delta';
F          = zeros(length(vm),length(vm));
for i=1:length(vm) F(i,i) = delta1(i,i); end %(Eq.13)

% Var-Cov Matrix of residuals and h estimation
tsi = Rr*F*Rr'           % (Eq.14)
h   = e1'*pinv(tsi)*e1   % (Eq.15)

% Improved vm and vc estimates
vm_new = (eye(4) - F*Rr'*inv(tsi)*Rr)*vm %(Eq.16)
vc_new = -pinv(GcT)*GmT*vm_new           %(Eq.17)
```

Fig. 4. A MATLAB® script that performs the calculations of **Subheading 3.4.1**. References to the equations in the text are included in the program.

4. Notes

1. Serum-containing medium has other nutrients, the uptake and production rates of which cannot be accurately determined, resulting in nonrepresentative flux estimates.
2. The recombinant protein production rate is often only a small fraction of the biomass formation rate (0.4% for human IFN-γ-producing CHO cells *[15]*) and can often be neglected without introducing substantial error into the flux estimates.

3. Fluxes in cyclical pathways (e.g., pentose phosphate pathway) and bidirectional reactions cannot be estimated using the balancing approach. Selection of a large number of reactions will result in a highly underdetermined system unless adequate experimental measurements are made.
4. Intracellular metabolite concentration changes of <10 m*M*/h suggest that pseudo-steady-state conditions prevail ***(1)***. For most mammalian cell cultures, the rate of change is much lower than 10 m*M*/h ***(10)***.
5. Once biomass composition is known, literature data can be used to quantify the amino acid requirement for biomass synthesis ***(38)***. Detailed information on the calculation procedure is available ***(10)***. The equation for recombinant protein production can be derived in a similar fashion.
6. The concentrations of O_2 and CO_2 in the inlet gas stream are also necessary to close their mass balances, and these are readily obtained when the composition of the inlet gas is known. CER estimation in a mammalian cell culture is not straightforward given the complex CO_2 dynamics in solution ***(39,40)***.
7. Determined systems are a special case of overdetermined systems, and the flux estimation procedure is identical. However, error and consistency testing is not possible in determined systems. Underdetermined systems have fewer experimental measurements than necessary and are formulated as linear programming problems ***(6,41–46)***. They do not have a unique solution and are influenced by the choice of objective function. Growth/product maximization and ATP/nutrient uptake minimization are examples of objective functions that have been used to solve underdetermined systems.
8. Rank and condition number of $\mathbf{G}^T$ must be checked, and this is easily done in most computing environments (rank (GT) and cond (GT) in MATLAB®). High condition numbers indicate that flux estimates are overly sensitive to measured rates, and condition numbers < 100 have been considered acceptable for flux analysis ***(15)***.
9. For overdetermined systems, $\mathbf{G}_c^T$ is nonsquare, and the pseudoinverse calculation is necessary (pinv (GcT) in MATLAB®).
10. Error in measured rates is influenced by both error in prime variable measurement and computational error, especially if derivatives are involved. For perfusion cultivations, measured rate errors are in typically in the 5–25% range.
11. The covariance matrix, **F**, is diagonal, reflective of the assumption that measurement errors are uncorrelated. This assumption may not be valid under all experimental conditions (for instance, bioreactor viable cell density, X_V^B, is present in all **Table 1** specific rate expressions, and errors in X_V^B measurement will impact all specific rates estimates, implying indirect correlation among specific rate errors). Methods to obtain a representative **F** when the errors are correlated are available ***(1,47)***. However, computation of **F** in this fashion is difficult because error information for all prime variables is seldom available. Using a diagonal **F** and selecting reasonable measured rate errors as in Eq. 11 should be a reasonable compromise for most experimental systems.
12. If no reduction in h value is seen after sequential elimination of all measured rates, it is likely that the bioreaction network chosen is not appropriate and should be modified.

References

1. Stephanopoulos, G., Aristodou, A., and Nielsen, J. (1998) *Metabolic Engineering. Principles and Methodologies*, Academic Press, San Diego.
2. Stephanopoulos, G. (2002) Metabolic engineering: perspective of a chemical engineer. *AIChE J.* **48**, 920–926.
3. Stephanopoulos, G. and Stafford, D. E. (2002) Metabolic engineering: a new frontier of chemical reaction engineering. *Chem. Eng. Sci.* **57**, 2595–2602.
4. Bonarius, H. P., Hatzimanikatis, V., Meesters, K. P. H., de Gooijer, C. D., Schmid, G., and Tramper, J. (1996) Metabolic flux analysis of hybridoma cells in different culture media using mass balances. *Biotechnol. Bioeng.* **50**, 229–318.
5. Follstad, B. D., Balcarcel, R. R., Stephanopoulos, G., and Wang, D. I. (1999) Metabolic flux analysis of hybridoma continuous culture steady state multiplicity. *Biotechnol. Bioeng.* **63**, 675–683.
6. Varma, A. and Palsson, B. O. (1994) Metabolic flux balancing: Basic concepts, scientific and practical use. *Bio/Technology* **12**, 994–998.
7. Bonarius, H. P., Ozemere, A., Timmerarends, T., Skrabal, P., Tramper, J., Schmid, G., and Heinzle, E. (2001) Metabolic-flux analysis of continuously cultured hybridoma cells using $^{13}CO_2$ mass spectrometry in combination with ^{13}C-Lactate nuclear magnetic resonance spectroscoopy and metabolite balancing. *Biotechnol. Bioeng.* **74**, 528–538.
8. Forbes, N. S., Clark, D. S., and Blanch, H. W. (2001) Using isotopomer path tracing to quantify metabolic fluxes in pathway models containing reversible reactions. *Biotechnol. Bioeng.* **74**, 196–211.
9. Wiechert, W. (2001) ^{13}C Metabolic flux analysis. *Metab. Eng.* **3**, 195–206.
10. Zupke, C. and Stephanopoulos, G. (1995) Intracellular flux analysis in hybridomas using mass balances and In Vitro ^{13}C NMR. *Biotechnol. Bioeng.* **45**, 292–303.
11. Szyperski, T., Bailey, J. E., and Wüthrich, K. (1996) Detecting and dissecting metabolic fluxes using biosynthetic fractional ^{13}C labeling and two-dimensional NMR spectroscopy. *Trends Biotechnol.* **14**, 453–458.
12. Cruz, H. J., Moreira, J. L., and Carrondo, M. J. (1999) Metabolic shifts by nutrient manipulation in continuous cultures of BHK cells. *Biotechnol. Bioeng.* **66**, 104–113.
13. Cruz, H. J., Ferreira, A. S., Freitas, C. M., Moreira, J. L., and Carrondo, M. J. (1999) Metabolic responses to different glucose and glutamine levels in baby hamster kidney cell culture. *Appl. Microbiol. Biotechnol.* **51**, 579–585.
14. Altamirano, C., Illanes, A., Casablancas, A., Gámez, X., Cairo, J. J., and GÒdia, C. (2001) Analysis of CHO cells metabolic redistribution in a glutamate-based defined medium in continuous culture. *Biotechnol. Prog.* **17**, 1032–1041.
15. Nyberg, G. B., Balcarcel, R. R., Follstad, B. D., Stephanopoulos, G., and Wang, D. I. (1999) Metabolism of peptide amino acids by Chinese hamster ovary cells grown in a complex medium, *Biotechnol. Bioeng.* **62**, 324–335.
16. Bonarius, H. P., Houtman, J. H., Schmid, G., de Gooijer, C. D., and Tramper, J. (2000) Metabolic-flux analysis of hybridoma cells under oxidative and reductive stress using mass balance. *Cytotechnology* **32**, 97–107.

17. Europa, A. F., Gambhir, A., Fu, P. C., and Hu, W. S. (2000) Multiple steady states with distinct cellular metabolism in continuous culture of mammalian cells. *Biotechnol. Bioeng.* **67**, 25–34.
18. Zupke, C., Sinskey, A. J., and Stephanopoulos, G. (1995) Intracellular flux analysis applied to the effect of dissolved oxygen on hybridomas. *Appl. Microbiol. Biotechnol.* **44**, 27–36.
19. Bonarius, H. P., Timmerarends, B., de Gooijer, C. D., and Tramper, J. (1998) Metabolite-balancing techniques vs. ^{13}C tracer experiments to determine metabolic fluxes in hybridoma cells. *Biotechnol. Bioeng.* **58**, 258–262.
20. Nadeau, I., Sabatié, J., Koehl, M., Perrier, M., and Kamen, A. (2000) Human 293 cell metabolism in low glutamine-supplied culture: Interpretation of metabolic changes through metabolic flux analysis. *Metab. Eng.* **2**, 277–292.
21. de Graaf, A. A., Mahle, M., Möllney, M., Wiechert, W., Stahmann, P., and Sahm, H. (2000) Determination of full ^{13}C isotopomer distributions for metabolic flux analysis using heteronuclear spin echo difference NMR spectroscopy. *J. Biotechnol.* **77**, 25–35.
22. Schmidt, K., Carlsen, M., Nielsen, J., and Villadsen, J. (1997) Modeling isotopomer distributions in biochemical networks using isotopomer mapping matrices. *Biotechnol. Bioeng.* **56**, 831–840.
23. van Winden, W. A., Heijnen, J. J., and Verheijen, P. J. T. (2002) Cumulative bandomers: A new concept in fluxanalysis from 2D[^{13}C, ^{1}H] COSY NMR data. *Biotechnol. Bioeng.* **80**, 731–745.
24. Wiechert, W., Möllney, M., Petersen, S., and de Graaf, A. A. (2001) A universal framework for ^{13}C metabolic flux analysis. *Metab. Eng.* **3**, 265–283.
25. Xie, L. and Wang, D. I. C. (1996) Material balance studies on animal cell metabolism using a stoichiometrically based reaction network. *Biotechnol. Bioeng.* **52**, 579–590.
26. Burgard, A. P. and Maranas, C. (2001) Review of the enzymes and metabolic pathways (EMP) database. *Metab. Eng.* **3**, 193–194.
27. Bree, M. A., Dhurjati, P., Geoghegan, R. F., and Robnett, B. (1988) Kinetic modeling of hybridoma cell growth and immunoglobulin production in a large-scale suspension culture. *Biotechnol. Bioeng.* **32**, 1067–1072.
28. Dalili, M., Sayles, G. D., and Ollis, D. F. (1990) Glutamine-limited batch hybridoma growth and antibody production: experiment and model. *Biotechnol. Bioeng.* **36**, 74–82.
29. Linz, M., Zeng, A. P., Wagner, R., and Deckwer, W. D. (1997) Stoichiometry, kinetics and regulation of glucose and amino acid metabolism of a recombinant BHK cell line in batch and continuous culture. *Biotechnol. Prog.* **13**, 453–463.
30. Altamirano, C., Paredes, C., Illanes, A., Cairó, J. J., and GÒdia, C. (2004) Strategies for fed-batch cultivation of t-PA producing CHO cells: substitution of glucose and glutamine and rational design of culture medium. *J. Biotechnol.* **110**, 171–179.
31. Zhou, W. C., Chen, C.-C., Buckland, B., and Aunins, J. G. (1997) Fed-batch culture of recombinant NSO myeloma cells with high monoclonal antibody production. *Biotechnol. Bioeng.* **55**, 783–792.

32. Goudar, C. T., Joeris, K., Konstantinov, K., and Piret, J. M. (2005) Logistic equations effectively model mammalian cell batch and fed-batch kinetics by logically constraining the fit. *Biotechnol. Prog.* **21**, 1109–1118.
33. Balcarcel, R. R. and Clark, L. M. (2003) Metabolic screening of mammalian cell cultures using well-plates. *Biotechnol. Prog.* **19**, 98–108.
34. van der Heijden, R. T. J. M., Romein, B., Heijnen, S., Hellinga, C., and Luyben, K. C. A. M. (1994) Linear constraint relations in biochemical reaction systems: II. Diagnosis and estimation gross errors. *Biotechnol. Bioeng.* **43**, 11–20.
35. Wang, N. S. and Stephanopoulos, G. (1983) Application of macroscopic balances to the identification of gross measurement errors. *Biotechnol. Bioeng.* **25**, 2177–2208.
36. Klamt, S., Schuster, S., and Gilles, E. D. (2002) Calculability analysis in underdetermined metabolic networks illustrated by a model of the central metabolism in purple nonsulfur bacteria. *Biotechnol. Bioeng.* **77**, 734–751.
37. Lee, D. Y., Hongsoek, Y., Park, S., and Lee, S. Y. (2003) MetaFluxNet: the management of metabolic reaction information and quantitative metabolic flux analysis. *Bioinformatics* **19**, 2144–2146.
38. Okayasu, T., Ikeda, M., Akimoto, K., and Sorimachi, K. (1997) The amino acid composition of mammalian and bacterial cells. *Amino Acids* **13**, 379–391.
39. Bonarius, H. P., de Gooijer, C. D., Tramper, J., and Schmid, G. (1995) Determination of the respiration quotient in mammalian cell culture in bicarbonate buffered media. *Biotechnol. Bioeng.* **45**, 524–535.
40. Frahm, B., Blank, H.-C., Cornand, P., et al. (2002) Determination of dissolved CO_2 concentration and CO_2 production rate of mammalian cell suspension culture based on off-gas measurement. *J. Biotechnol.* **99**, 133–148.
41. Fell, D. A. and Small, J. R. (1986) Fat synthesis is adipose tissue. An examination of stoichiometric constraints. *J. Biochem. (Tokyo)* **238**, 781–786.
42. Majewski, R. A. and Domach, M. M. (1990) Simple constrained optimization view of acetate overflow in *E. coli*. *Biotechnol. Bioeng.* **35**, 732–738.
43. Pramanik, J. and Keasling, J. D. (1997) A stoichiometric model of *Escherichia coli* metabolism. Incorporation of growth-rate-dependent biomass composition and mechanistic energy requirements. *Biotechnol. Bioeng.* **56**, 398–421.
44. Savinell, J. M. and Palsson, B. O. (1992) Network analysis of intermediary metabolism using linear optimization. I. Development of mathematical formulation. *J. Theor. Biol.* **154**, 421–454.
45. Savinell, J. M. and Palsson, B. O. (1992) Network analysis of intermediary metabolism using linear optimization. II. Interpretation of hybridoma cell metabolism. *J. Theor. Biol.* **154**, 455–473.
46. Varma, A., Boesch, B. W., and Palsson, B. O. (1993) Biochemical production capabilities of *Escherichia coli*. *Biotechnol. Bioeng.* **42**, 59–73.
47. Madron, F., Veverka, V., and Vanecek, V. (1977) Stastical analysis of material balance of a chemical reactor. *AIChE J.* **23**, 482–486.

IV

Special Cultivation Techniques

15

Disposable Bioreactors for Inoculum Production and Protein Expression

Regine Eibl and Dieter Eibl

Summary

Although glass-stirred or stainless-steel-stirred reactors dominate in the area of animal cell cultivation, users are increasingly trying to integrate disposable bioreactors wherever possible. Today applications range from inoculum to glycoprotein-production processes, which are most commonly performed in various membrane bioreactors and the Wave in small- and middle-scale production. For this reason we present protocols instructing the reader how to handle the CeLLine, the miniPerm, and the Wave. In our experience, the methods described for a Chinese hamster ovary (CHO) suspension cell line can also be successfully applied to other animal suspension cells, such as insect cells (Sf-9) and human embryogenic kidney cells (HEK-293 EBNA cells).

Key Words: Disposable bioreactor; bag bioreactor; Wave; membrane bioreactor; CeLLine; miniPerm; inoculum production; protein expression; Chinese hamster ovary cells.

1. Introduction

Disposable bioreactors consist of a sterile chamber or bag made of polymeric materials such as polyethylene, polystyrene, and polypropylene. They are partially filled with medium, inoculated with cells, and discarded after harvest. The single-use chambers or bags make cleaning and sterilization in place unnecessary and guarantee high flexibility as well as process security, with contamination levels below 1%. Their use results in minimized complex cleaning, sterilization, and validation and can improve process efficiency as well as reduce the time-to-market for new products. Therefore, disposable bioreactors are becoming standard for animal cell-based seed inoculum production, process development, and GMP (good manufacturing practice) manufacturing in small- and middle-scale production focused on r-proteins (recombinant proteins) and monoclonal antibodies for therapy and diagnostics ***(1–4)***.

From: *Methods in Biotechnology, Vol. 24: Animal Cell Biotechnology: Methods and Protocols, 2nd Ed.* Edited by: R. Pörtner © Humana Press Inc., Totowa, NJ

With respect to disposable bioreactors for animal cells as reported in the literature and their categorization, the basic structure and working principles of CeLLine (Integra Biosciences), miniPerm (Vivascience, Sartorius Group), and the Wave (Wave Biotech) are described in this chapter. These minibioreactors are widely used in fundamental research and for clinical applications. Furthermore, typical seed inoculum production and glycoprotein expression procedures for suspension of CHO XM 111-10 cells in the CL 1000, miniPerm (classic), and BioWave 20 SPS with 1-L culture volume are given and discussed.

1.1. Disposable Bioreactors for Animal Cells

Table 1 summarizes the disposable bioreactors available on the market today for animal cells and culture volumes from 2.5 mL up to 500 L. If traditional disposable t-flasks and minibag systems such as VectraCell *(5)* for stationary cultures are excluded, there are generally four main categories of disposable bioreactors for animal cells: tissue flask-like bioreactors, membrane bioreactors, oscillating bioreactors, and mechanically driven bioreactors. Until now, small-scale bioreactors equipped with a dialysis membrane ***(1–3,10–20)***, which guarantee high cell densities, and mechanically driven bag bioreactors of up to 500-L culture volume ***(1–4,13,24–27)*** represent the most frequently cited groups among the disposable bioreactor types. Here the CeLLine, the miniPerm, and the Wave have leading positions beside the different types of hollow-fiber reactors. It should be mentioned that the Optima, the OrbiCell, the AppliFlex, and the Mini-Tsunami bioreactors have working principles similar to that of the Wave.

1.2. Double-Membrane Bioreactors for mL Scale: CeLLine and miniPerm

As **Fig. 1** illustrates, CeLLine and miniPerm are double-membrane systems consisting of a cultivation or production compartment and a medium storage compartment. They work in a quasi-perfusion mode in which the cells are kept in the cell compartment and fed by the nutrients from the medium storage compartment, the nutrients passing over a semi-permeable membrane while desired proteins are secreted into the production compartment. Efficient gas transfer is ensured by the 0.1- to 0.2-mm-thick silicone membrane on the outside. While CeLLine is based on a static t-flask, miniPerm is a modified roller bottle apparatus.

Both reactors require a CO_2 incubator at a minimum relative humidity of 70% for optimum cultivation conditions. In the case of miniPerm, there is a bottle tuning device providing the necessary rotation movement of between 5 and 20 rpm. Depending on the growth and production characteristics of the cell lines cultivated, the user can select the most suitable CeLLine type (CL 350, CL 1000, AD 1000) and production module of the miniPerm (classic, HDC 5, SM).

Table 1
Disposable Bioreactors for Animal Cells

Main category	Subcategory	Bioreactor	Manufacturer/supplier	Ref.
Tissue flask-like bioreactors	Cell factories and further developments	Nunclon Cell Factory	NUNC, www.nuncbrand.com	***1,6,7***
		$Cyto^3$	Osmonics, www.osmonics.com	***1***
		CellCube	Corning, www.corning.com	***8,9***
Membrane bioreactors	Double-membrane systems	CeLLine	Integra Biosciences, www.integra-biosciences.com	***1–3,10–13***
		miniPerm	Vivascience, Sartorius Group,www.ivss.de	***1–3,14–16***
	Single-membrane systems or hollow fiber bioreactors	CellMax	Cell Culture Spectrum Labs, www.spectrapor.com	***17***
		FiberCell	FiberCellSystem, www.fibercellsystem.com	***18***
		ResCu-Primer HF, miniMax, Maximizer, Xcellerator	BioVest, www.biovest.com	***1–3,19,20***
Oscillating bioreactors	Bellow systems with microcarriers	FibraStage	New Brunswick Scientific's, www.nbsc.com	***4***
		BelloCell	Cesco Bioengineering, www.cescobio.com.tv	***21,22***
Mechanically driven bioreactors	Bag bioreactors	MantaRay	Wheaton Science Products, www.wheatonsci.com	***23***
		BioWave	Wave Biotech, CH, www.wavebiotech.ch	***1–3,13,***
		Wave	Wave Biotech, USA, www.wavebiotech.com	***24–27***
		Optima, OrbiCell	MetaBios, www.metabios.com	***3***
		AppliFlex	Applikon, www.applikonbio.com	***4***
		Mini-Tsunami	Mega Partners Int'I BV, www.megainternational.com.hk	***9***
	Stirred bioreactors	Spinner-flask systems	various suppliers	***3***
		XDR	Xcellerex, www.xcellerex.com	***28***
		SUB	Hyclone, www.hyclone.com	***3,29,30***
	Rotating bioreactors	Roller bottle systems	various suppliers	***3***
		RCCS-D	Synthecon, www.synthecon.com	***31–33***

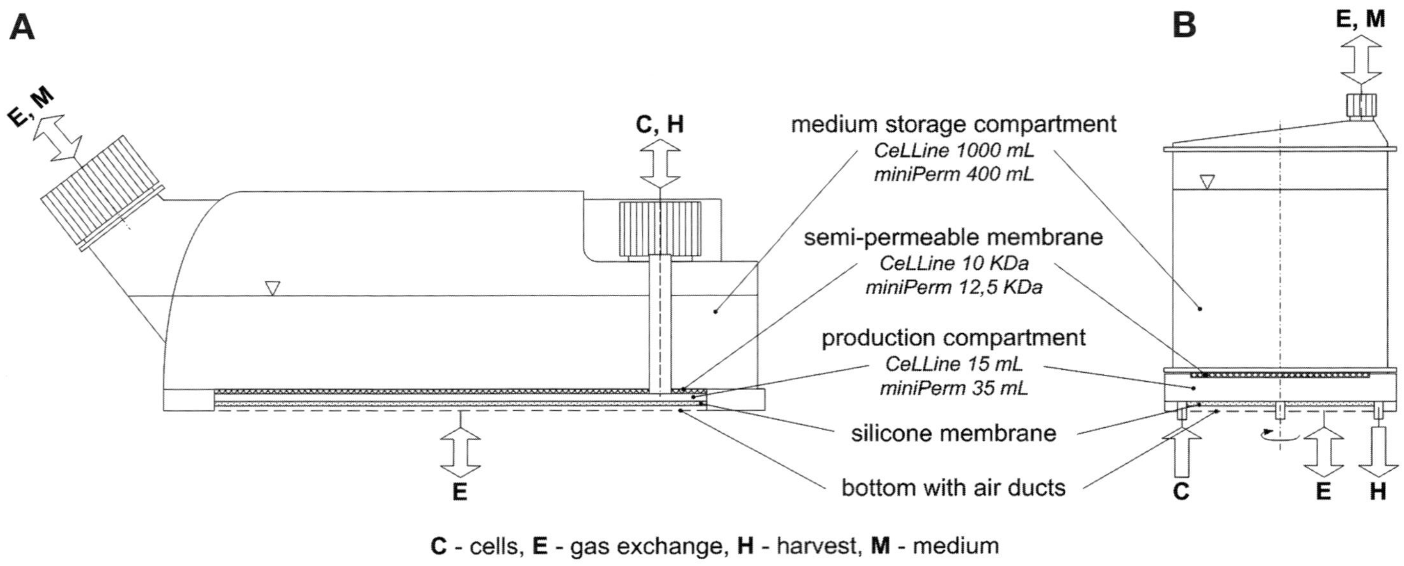

Fig. 1. Basic setup of **(A)** the CeLLine bioreactor, CeLLine CL 1000, and **(B)** the miniPerm bioreactor (classic system).

All these types guarantee cell densities of more than 10^7 cells/mL and protein concentrations in the milligram range. The maximum culture volume of 15 mL for CeLLine and 35 mL for miniPerm restricts the application of these bioreactors to research and production processes used to gain clinical samples.

1.3. Wave

The mechanically driven Wave is scalable up to 500 L culture volume. Wave reactors from benchtop to GMP manufacturing scale facilitate measurement and regulation of rocking angle (4–12°), rocking rate (5–42 rpm), temperature, aeration, and CO_2 rate. Optional monitoring and control of pH, dissolved oxygen (DO), and weight and flow rates in perfusion mode are also possible. The working principle of the Wave, shown in **Fig. 2**, is the wave-induced motion (WIM) produced by rocking the platform with the culture bag containing the medium and the cells. In this way, oxygenation and mixing with minimal shear forces result. The surface of the medium is continuously renewed and bubble-free surface aeration takes place. It should be noted that the culture bags available in sizes from 1 to 1000 L have varying bag geometries and are designed to operate with a minimum of 10% culture volume and a maximum of 50% culture volume.

Oxygen transfer and mixing time of the Wave are comparable to those of standard, bubble-free aerated stirred reactors for animal cell cultures. Residence time distributions in a continuously operating Wave can also be described by the ideally mixed stirred tank reactor model. Furthermore, the use of a modified Reynolds number, Re_{mod}, which depends on rocking rate, rocking angle, culture bag type, and culture volume, has been introduced to characterize the hydrodynamic environment in the Wave. Energy input modeling indicates a dependence on rocking rate, rocking angle, and culture volume of Wave reactors. All in all, these facts have contributed to a better understanding of cell behavior in the Wave and simplify the scale-up that can be facilitated by Re_{mod} or specific energy input.

The Wave provides cell densities between 4×10^6 and 1×10^7 cells/mL in dominant feeding mode as well as high cell densities in perfusion mode. The product achieves g-outputs *(3)*.

2. Materials

In addition to essential cell culture equipment (e.g., safety cabinet class II, centrifuge, water bath, magnetic stirrer, CO_2 incubator, centrifuge), 0.2-μm filters such as Midisart 2000 (Sartorius) or VacuCap 60 PF Filter Unit (Pall), tubes, sample vials, sterile bottles, serological pipets, and syringes, the materials and apparatus listed below are necessary for the experiments described in **Subheading 3**.

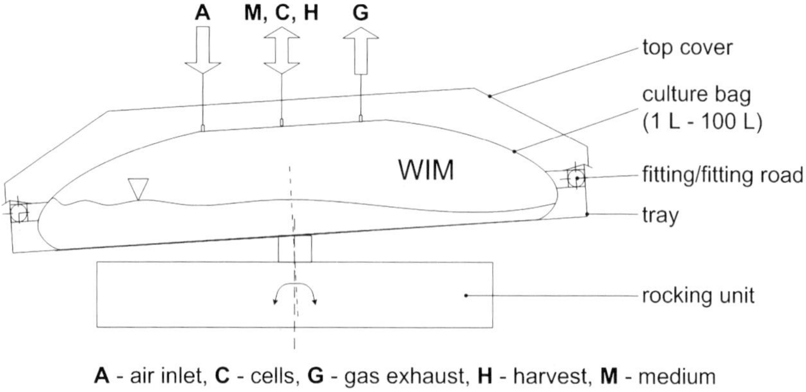

Fig. 2. Schematic diagram of the Wave.

2.1. Cell Cultivation and In-Process Control

1. Cell line CHO XM 111-10, obtained from Prof. Dr. Martin Fussenegger (Swiss Federal Institute of Technology in Zurich) (*see* **Note 1**).
2. Protein- and peptide-free, filter sterilized media ChoMaster HP-1 and ChoMaster HP-5 (Cell Culture Technologies) supplemented with D-glucose, sodium hydrogen carbonate, and L-asparagine monohydrate according to manufacturer's instructions, stored at 4°C in the dark, and used within 3 mo of preparation (*see* **Note 2**).
3. Additional medium supplements:
   a. Antibiotics (*see* **Note 2**): G418 sulfate, harmful (Alexis), puromycin dihydrochloride (Alexis), tetracycline hydrochloride (Sigma), sterile penicillin-streptomycin-fungizone solution (Bioconcept).
   b. Pluronic F68 solution (Sigma) (*see* **Note 3**).
4. Phosphate-buffered saline (PBS) solution (Fluka) (*see* **Note 6**).
5. Bioreactors (*see* **Notes 6–8**): BioWave 20 SPS (Wave Biotech, Switzerland) with Wave Bag 2 L D (Wave Biotech, Switzerland), CeLLine CL 1000 (Integra Biosciences), miniPerm classic with medium compartment and miniTuner (Vivascience, Sartorius Group).
6. Cedex cell counter (Innovatis) working with Trypan blue dye solution, toxic (Fluka) (*see* **Note 4**).
7. Nova BioProfile Analyzer 100 (Laborsysteme Flükiger) (*see* **Note 5**).

2.2. SEAP Analysis

1. Secreted alkaline phosphatase (SEAP) buffer solution (*see* **Note 9**): diethanolamines (Fluka), L-homoarginine hydrochloride (Sigma), magnesium chloride hexahydrate (Merck).
2. *p*-Nitrophenol phosphate solution (Sigma).
3. Flatbed plate shaker (Morwell Diagnostics).
4. Microtiter plate reader with a 405-nm filter (Dynex Technologies).

5. 96-Well plates (Costar).
6. Photometer Uvikon 710 (Kontron Instruments).
7. Heater (Witec).

3. Methods

Because CHO cells are easy to grow and scale up by a factor of 5 or 10 in serum-free and chemically defined media, their application is very popular today. The protocols we present for CHO XM 111-10 cell mass-propagation and resulting SEAP (human placental secreted alkaline phosphatase) expression involve preparation of cells and bioreactors, inoculation, operation, in-process control, and harvest. It should be noted that other cell lines may behave differently.

3.1. Seed Inoculum Production

The basic structure of the CeLLine and the Wave allows the feeding of medium and cells, the exchange of medium, and therefore the safe operation of these bioreactors in seed inoculum production. The CeLLine is preferable to t-flasks for long-term inoculum production where one CeLLine can be used over 3 mo. For one-time production of seed for larger bioreactors, we prefer the Wave because the maximal available culture volume is higher and spinner flasks are not required.

3.1.1. Long-Term Inoculum Production in the CeLLine CL 1000

1. In order to obtain a minimal inoculum concentration of about 1.5×10^6 viable cells/mL, harvest 2.5×10^7 viable cells from t-flasks, pool the cell pellets, and resuspend them in 15 mL fresh ChoMaster HP-5 growth medium without Pluronic (*see* **Notes 1** and **2**).
2. Sterilely equilibrate the semi-permeable membrane of the CL 1000 with warmed 25 mL culture medium (ChoMaster HP-5 growth medium; *see* **Note 2**). Place the CL 1000 in a CO_2 incubator operating at 37°C with humidified atmosphere of 10% CO_2 in air for at least 20 min.
3. Inoculate the CL 1000 in the safety cabinet (*see* **Note 7**).
   a. To prevent air lock, slightly loosen the front cap of the medium storage compartment.
   b. Aspirate the 15 mL cell suspension into a 25-mL serological pipet and open the production compartment by removing the production compartment cap.
   c. Inoculate the production compartment by inserting the pipet into the silicone cone.
4. Add 975 mL equilibrated ChoMaster HP-5 growth medium (*see* **Note 2**) to the medium storage department and completely tighten both caps.
5. Place the CL 1000 in the CO_2 incubator.
6. To assess the cell growth and metabolism (*see* **Notes 4** and **5**), take 0.5 mL sample from the production compartment and 1 mL from the medium storage compartment after 48 and 96 h, respectively (*see* **Note 7**).

7. Realize the harvest and splitting of the cells between 4 and 5 d after inoculation in the safety cabinet. Perform the cell growth and metabolite analyses using the Cedex (0.5 mL) and the BioProfile (1 mL) (*see* **Notes 4**, **5**, and **7**). When there is a viable cell density in the range of 4×10^7 cells/mL, the following steps should be taken:
   a. Remove and discard all medium from the medium storage compartment.
   b. Harvest all cell suspension from the production compartment with a serological pipet with removed production compartment cap.
   c. Take 3 mL of the cell suspension and add 12 mL of fresh ChoMaster HP-5 growth medium (*see* **Note 2**). Return the well-mixed cell suspension back to the cell compartment and tighten the production compartment cap (*see* **Note 7**).
   d. Add 1000 mL warm ChoMaster HP-5 growth medium (*see* **Note 2**) to the medium compartment and fully tighten the medium compartment cap.
8. Return the CL 1000 to the incubator until the next harvest and splitting, which will take place approx 3–4 d later.
9. Repeat **steps 7** and **8** until the interruption of the inoculum production.

3.1.2. Wave Inoculum Production Procedure (Culture Bag 2 L)

1. Put 200 mL PBS solution or culture medium in the bag before inoculation with cells and leave it for at least 24 h, having checked that the bag is tightly closed (*see* **Note 6**).
   a. Clamp the air filters and add 200 mL PBS solution or culture medium to the sterile 2-L culture bag in the safety cabinet.
   b. Fix the bag to the tray and inflate it with air (0.3 L/min) until it is tight (inlet air filter is opened and exhaust air filter is clamped).
   c. Clamp the inlet air filter, stop aeration, close the top cover, and start rocking at 6°, 14 rpm, and 37°.
   d. Stop rocking, remove the bag from the tray, and bring it into the safety cabinet.
2. Pool sufficient t-flasks to obtain an inoculum with a living cell count of 5×10^7 cells in 100 mL fresh, prewarmed ChoMaster HP1-growth medium (*see* **Notes 1** and **2**).
3. Replace 400 mL prewarmed ChoMaster HP1-growth medium containing 0.2% Pluronic F68 with 200 mL PBS solution or culture medium **(step 1)** from the culture bag by using sterile 50-mL syringes prior to inoculation (inlet air filter clamped, exhaust air filter opened).
4. Add the 100 mL cell suspension by inserting a 50-mL syringe into the culture bag's luer lock inoculation port, and close off the exhaust air filter.
5. Put the filled bag back on the tray, fix it in place, inflate the bag with 10% CO_2/air (0.4 L/min) until bag is tight (approx 10 min), open the exhaust filter, switch the aeration to 0.15 L/min, and start rocking and warming again.
6. Take the first 2-mL sample after 30 min of operation. Check the cell growth (1 mL sample for Cedex) and the medium (1 mL sample for BioProfile), paying particular attention to the pH. Use syringes and adjust the pH if necessary (*see* **Notes 4–6**).
7. After 24 h, take a sample as in **step 6**, add 250 mL warm ChoMaster HP-1 growth medium with 0.2% Pluronic F68, and switch to 18 rpm if possible (*see* **Notes 4–6**).

8. After 48 h of cultivation, take the next sample as in **step 6**, add 250 mL warm ChoMaster HP-5 growth medium with 0.2% Pluronic F68, and switch to 25 rpm if possible (*see* **Notes 4–6**).
9. Between day 3 and 4 after inoculation, middle cell densities of approx 5×10^6 living cells/mL are achieved. Switch off the Wave, close off the air filters, remove the bag from the tray, and transfer it to a laminar flow cabinet with integrated tripod.
10. Hang the bag on the tripod, open the exhaust filter (leave the inlet filter closed), and feed 200 mL fresh ChoMaster HP-5 growth medium with 0.2% Pluronic F68 (*see* **Notes 2** and **3**). Under gravity, allow the cells to settle on the bottom of the culture bag for about 3 h.
11. Remove the ChoMaster HP-5 growth medium via the tube of the fill/harvest port and feed 1 L fresh medium.

3.2. CHO Cell-Based SEAP Expression

The secreted product of the used CHO cell line is the model protein SEAP. Because SEAP expression is controlled by the tetracycline-responsive promoter PhCMV-1, cell growth behavior and glycoprotein production can be regulated by the addition or withdrawal of tetracycline *(34)*. Furthermore, a shift of cultivation temperature from 37 to 30°C causes a growth arrest in the G1 phase of the cell cycle, resulting in higher SEAP productivities *(35)*. Therefore, a two-stage process with proliferation phase (medium with tetracycline, 37°C) and SEAP production phase (medium without tetracycline, 30°C) is realized.

3.2.1. CelLine CL 1000

1. Carry out **steps 1–5** as described in **Subheading 3.1.1**.
2. On day 5 of cultivation put a 0.5-mL cell suspension sample and 1-mL medium sample in the safety cabinet (*see* **Notes 4** and **5**). If the viable cell number reaches approx 4×10^7 cells/mL, the viability decreases and/or the glucose level drops below 1 g/L, remove and discard all ChoMaster HP-5 growth medium from the medium storage compartment and feed fresh, preheated ChoMaster HP-5 production medium to the medium compartment (*see* **Note 2** and **7**).
3. For subsequent product analysis (*see* **Note 9**), take a 0.5-mL sample from the production compartment again and freeze it at –20°C. Check that the caps of the CL 1000 are tightly closed.
4. Place the CL 1000 in the CO_2 incubator and decrease the temperature to 30°C.
5. A 1-mL sample from the production compartment should be removed daily to monitor cell growth as well as product formation, and likewise a 1-mL sample from the medium storage compartment to observe cell metabolism (*see* **Notes 4, 5, 7**, and **9**).
6. Stop the process at cell viabilities around 30%. This usually takes 10 d.
7. Estimate the product activities of the collected samples (*see* **Note 9**). Expected maximum SEAP activities are in the range of 30–40 U/mL.

3.2.2. miniPerm (Classic Kit)

1. Resuspend 5.5 × 10^7 viable cells from a pooled t-flask preculture in 35 mL fresh, preheated ChoMaster HP-5 growth medium with 0.1% Pluronic F68. This results in the minimal necessary concentration of 1.5 × 10^6 viable cells/mL (*see* **Notes 1–2**) for miniPerm inoculation.
2. Prepare the miniPerm by assembling the production and the medium compartment and place it on the plastic rack in the safety cabinet. Ensure visually that the bioreactor is not damaged.
3. Equilibrate the semi-permeable membrane of the miniPerm with 25 mL culture medium (ChoMaster HP-5 growth medium without Pluronic; *see* **Note 2**) by adding the latter to the medium storage compartment. Subsequently, place the miniPerm in a CO_2 incubator (37°C, 10% CO_2) for at least 20 min. Ensure caps are closed.
4. Aspirate the 35 mL cell suspension into a 50-mL sterile syringe in the safety cabinet and inoculate the production module via one Luer septum (*see* also **Note 8**). Tighten all product compartment caps.
5. Add 400 mL of preheated ChoMaster HP-5 growth medium without Pluronic to the medium compartment (*see* **Notes 2**, **3**, and **8**).
6. Place the tightly closed miniPerm on the roller device in the CO_2 incubator operating at 37°C with humidified atmosphere of 10% CO_2 in air, and start the bioreactor at 7.5 rpm.
7. After 30 min of operation, stop the miniPerm and take the first 0.5-mL sterile sample from the production module using a 2-mL sterile syringe and a 1-mL sample from the medium storage compartment (5-mL pipet) (*see* **Notes 4**, **5**, **8**, and **9**).
8. Start up the miniPerm placed in the CO_2 incubator.
9. To make estimates of cell growth and metabolism, the sampling described above is repeated every 2 d and the rocking rate is increased in steps up to 15 rpm, depending on cell growth and viability (*see* **Notes 4**, **5**, **8**, qnd **9**).
10. When viable cell density reaches 4 × 10^7 cells/mL, the production phase is introduced.
    a. Replace the ChoMaster HP-5 growth medium in the medium compartment with ChoMaster HP-5 production medium.
    b. Operate the miniPerm at 15 rpm, switch the temperature to 30°C in the CO_2 incubator, and take a 0.5-mL sample for subsequent SEAP estimation from the production compartment after 30 min (*see* **Notes 8** and **9**).
11. Take a daily 1-mL sample from the production compartment and a daily 1-mL sample from the medium storage compartment (*see* **Notes 4**, **5**, **8**, and **9**).
12. Stop the process at cell viabilities around 30%.
13. Estimate the product activities from the collected samples (*see* **Note 9**). Expected maximum SEAP activities are in a similar range to those provided by the CeLLine.

3.2.3. Wave (Culture Bag 2 L)

1. Follow **steps 1–10** from **Subheading 3.1.2**.
2. Replace the ChoMaster HP-5 growth medium with 1 L ChoMaster HP-5 production medium containing 0.2% Pluronic F68 via the tube of the fill/harvest port (*see* **Subheading 3.1.2.**, **steps 9** and **10** for more details, and **Note 2**).

3. Keep the inflated bag at 6°, 25 rpm, 0.15 L/min and 30°C for 30 min before taking a 0.5-mL sample for subsequent SEAP estimation (*see* **Notes 4**, **5**, **6**, and **9**).
4. Continue to operate the Wave at the parameters described above.
5. Take a daily 2-mL sample to measure cell count as well as metabolite concentration and a daily 0.5-mL sample for the further observation of product formation (*see* **Notes 4**, **5**, and **9**).
6. Stop the process at cell viabilities around 30% (production time approx 10 d).
7. Analyze the product activities (*see* **Note 9**). Maximum SEAP activities around 10 U/mL will be achieved.

4. Notes

1. A rapidly growing, healthy cell line that is fully adapted to the suspension culture is required. Ensure that the CHO XM 111-10 cells (routinely maintained in 75- and 125-cm^2 t-flasks containing FMX-8 growth medium [Cell Culture Technologies] and subcultured when cell densities have reached values around 1×10^6 viable cells/mL) are in the logarithmic phase. Incubate at 37°C and 10% CO_2.
2. Chemically defined ChoMaster media allowing cell growth and r-protein production under serum-, protein-, and peptide-free culture conditions have to be sterile and conditioned (37°C, pH 7.2) before use. They vary in glucose and amino acid content. It is advisable to add sterilely important supplements to the medium immediately before use: 4 mL G418 sulfate (from frozen stock: 100 mg/mL), 1.3 mL puromycin dihydrochloride (from frozen stock: 5 mg/mL), 2.5 mL tetracycline hydrochloride (from frozen stock: 1 mg/mL), and 10 mL penicillin-streptomycin-fungizone solution. Note that there is a simple difference between growth and production media: While the growth medium contains the agent tetracycline, in the production medium tetracycline is not available. Because all cell culture media contain light-sensitive compounds, protect them against damage caused by incidence of light by storing and incubating the media in the dark. Consequently, apply the Wave bioreactor with a nontransparent top cover.
3. Pluronic F68 protects sensitive cells against shear forces in medium without serum. Thus, adding Pluronic in concentrations between 0.1 and 0.2% to the production compartment of mechanically driven as well as bubble-aerated bioreactors by applying a syringe filter is recommended.
4. The Cedex cell counter replaces time-consuming manual cell counting. Its standardized results are based on the Trypan blue exclusion method. Alternatively, comparable results to those achieved with Cedex for the suspension CHO XM 111-10 cell line are delivered by the NucleoCounter from ChemoMetec. Immediately after sampling from the production compartment or bag, the cell count should be measured as in the case of the metabolites. Refer to the manufacturer's manual.
5. Nova BioProfile 100 or its successor BioProfile 100+ provides an efficient, reliable measurement of glucose, lactate, glutamine, glutamate, and ammonium as well as the pH of the bags and of medium storage compartment samples. Replace the kits, sensors, and membranes according to the supplier's guidelines and recommendations.
6. Our experience shows that a rinsing (24–48 h) of the Wave culture bag using sterile PBS solution or medium shortens the lag phase und supports the growth of

different mammalian cell lines. Prevent overheating at low culture volumes when using the BioWave 20 SPS with an internal heater by checking the contact between the temperature probe and culture bag. In feeding mode, adjust the rpm in response to the increasing culture volume, cell growth, and oxygen demands of the culture. We observed a decrease in energy input when the Wave is operating at 6°, 20 rpm, and 50% filled 2-L culture bag. This alteration can be explained by the occurring phase shift of the Wave and causes a decrease in shear stress for the cells and higher oxygen transfer if rpm rises. It should be kept in mind that the cells in the Wave bag are exposed to the highest shear stress if the cultivation is started at low culture volumes. Excessive attachment of cells to the Wave culture bag indicates that the cells are under chemical or mechanical stress in the Wave bioreactor. Reduce the CO_2 content if pH value starts to drop in the Wave.

7. Prior to inoculation it is advantageous to equilibrate the membrane of the CeLLine bioreactor to ensure that the CeLLine membrane is compliant and prevent a possible membrane break if inoculation is realized. Consequently, the air trapped in the production compartment of CeLLine bioreactors is more easily removed. Remember that air bubbles and also shaking should be avoided for CeLLine bioreactors during inoculation and harvest (as membrane failure may result). In CeLLine bioreactors, although the mixing of cell suspension by slow pipetting up and down several times is essential in order to realize an optimal sampling or harvest of the cell suspension, do not pipet liquid from or into the production compartment with force. Remove air bubbles by carefully drawing them back into the pipet together with the fluid. Because of the small culture volume in CeLLine and miniPerm, for frequent sampling during long-term cultivations we recommend adding 0.5 mL of conditioned medium from the medium compartment to the cell compartment after every sample. This feeding amount corresponds to the sample amount.
8. In addition to membrane wetting, let superfluous air out and prevent air bubbles in the miniPerm bioreactor after inoculation and after every sample. This can be accomplished by the application of two syringes attached to opposite Luer sample ports of the cell compartment. Finally, adjust the rpm in accordance with cell growth.
9. The secreted model protein SEAP can be estimated indirectly by enzymatic reaction from *p*-nitrophenolphosphate to *p*-nitrophenol. The endogenous activity of phosphatase is stopped by warming the samples to 65°C and adding L-homoarginine. The SEAP activity is indicated in U/mL with reference to the reacted amount of *p*-nitrophenolphosphate. Every 0.5-mL frozen SEAP sample collected is incubated at 65°C after thawing and centrifuged at 9300 RCF for 3 min to separate cell debris and protein aggregates. The supernatant is transferred to a new Eppendorf tube. Forty µL of this supernatant as the case may be of a produced dilution are mixed with 100 µL 2x SEAP buffer (20 m*M* L-homoarginine, 2 *M* diethanolamine, 1 m*M* magnesium chloride hexahydrate, pH 9.8) and 40 µL entionized water, and then incubated at 37°C for 10 min. By adding preheated (37°C) 20 µL *p*-nitrophenol phosphate (120 m*M*) solution, the enzymatic reaction starts. The running hydrolysis can be monitored via extinction variation by using a microtiter plate reader operating

at 405 nm. The extinction is recorded every 3 min during a 30-min run. The SEAP activity can be calculated from the slope of the linear range estimated by plotting the measured extinctions over the time multiplied by the factor 2.4 (obtained for our equipment and procedure). Further dilutions of the SEAP samples have to be considered accordingly ***(36)***.

Acknowledgments

We wish to thank the Fussenegger group for engineering the cell line used in the experiments described and Lidija Lisica for her technical assistance.

References

1. Eibl, D. and Eibl, R. (2002) Entwicklungsstand und -trends in der Zellkulturreaktortechnologie, in *Technische Systeme für Biotechnologie und Umwelt-Biosensorik und Zellkulturtechnik* (Beckmann, D., Meister, M., Heiden, S., and Erb, R., eds.), Erich Schmidt Verlag, Berlin, pp. 255–263.
2. Eibl, D. and Eibl, R. (2004) Reaktorsysteme für die Herstellung von biopharmazeutischen Proteinwirkstoffen: Quo vadis? in *Proceedings zum 12. Heiligenstädter Kolloquium* (iba, ed.), Heiligenstadt, Germany, pp. 271–279.
3. Eibl, D. and Eibl, R. (2005) Einwegkultivierungstechnologie für biotechnische Pharma-Produktionen. *BioWorld* **3**, 1–2 (Supplement BioteCHnet).
4. Glaser, V. (2005) Disposable bioreactors become standard fare. *GEN* **25(14)**, 80–81.
5. BioVectra (2002) VectraCell-single use bioreactor system. *Protocol*, 1–6.
6. Chu, L. and Robinson, D. K. (2001) Industrial choices for protein production by large-scale cell culture. *Curr. Opin. Biotechnol.* **12(2)**, 180–187.
7. Schwander, E. and Rasmusen, H. (2005) Scalable, controlled growth of adherent cells in a disposable, multilayer format. *GEN* **25(8)**, 29.
8. Balsey, H., Isch, C., and Bernard, C. R. (1995). Cellcube: a new system for large scale growth of adherent cells. *Biotechnol. Tech.* **9(10)**, 725–728.
9. DePalma, A. (2005) Approaches for boosting cell productivity. *GEN* **25(13)**, 46.
10. Trebak, M., Chong, J. M., Herlyn, D., and Speicher, D. W. (1999) Efficient laboratory-scale production of monoclonal antibodies using membrane based high density cell culture technology. *J. Immunol. Meth.* **230**(1–2), 59–70.
11. Scott, L. E., Aggett, H., and Glencross, D. K. (2001) Manufacture of pure monoclonal antibodies by heterogeneous culture without downstream purification. *Biotechniques* **31(3)**, 666–668.
12. Bruce, M. P., Boyd, V., Duch, C., and White, J. R. (2002) Dialysis-based bioreactor systems for the production of monoclonal antibodies—alternatives to ascites production in mice. *J. Immunol. Meth.* **264**(1), 59–68.
13. Eibl, R., Rutschmann, K., Lisica, L., und Eibl, D. (2003) Kostenreduktion bei der Säugerzellkultivierung durch Einwegreaktoren? *BioWorld* **5**, 22–23.
14. Weichert, H., Behn, I., Falkenberg, F., Krane, M., and Nagels, H. O. (1994) Produktion monoklonaler Antikörper mit einem Minibioreaktor. *BIOforum* **17**, 356–361.

15. Vollmers, H. P., Zimmermann, U., Krenn, V., et al. (1998) Adjuvant therapy for gastric adenomacarcinoma with the apoptosis-inducing human monoclonal antibody SC-1: First clinical and histopathological results. *Oncol. Rep.* **5**(3), 549–552.
16. Müller, S. (2001) Kultivierung von HEK-U293 Zellen als Suspensionskultur im miniPerm Bioreaktor, *In Vitro News* **1**, 4.
17. Beck, T., Shueh, L., Sitlani, L., and Menaker, M. (1995) Monoclonal antibody production in the CELLMAX artificial capillary system. *Focus* **17(1)**, 2–5.
18. Isayeva, T., Kotova, O., Krasnykh, V., and Kotov, A. (2003) Advanced methods of adenovirus vector production for human gene therapy: Roller bottles, micro-carriers, and hollow fibres. *BioProc. J.* **2(4)**, 75–81.
19. Laub, R., Brecht, R., Dorsch, M., Valey, U., Wenk, K., and Emmerich, F. (2002) Antihuman CD4 induces peripheral tolerance in a human $CD4^+$, murine CD4, $HLA\text{-}DR^+$ advanced transgenic mouse model. *J. Immunol.* **169**, 2947–2955.
20. Schläpfer, B. S., Scheibler, M., Holtorf, A. D., Van Nguyen, H., and Pluschke, G. (1995) Development of optimized transfectoma cell lines for production of chimeric antibodies in hollow fiber cell culture systems. *Biotechnol. Bioeng.* **45(4)**, 310–319.
21. Hu, Y. C., Lu, J. T., and Chung, Y. C. (2003) High-density cultivation of insect cells and production of recombinant baculovirus using a novel oscillating bioreactor. *Cytotechnology* **42**(3), 145–153.
22. Ho, L., Greene, C. L., Schmidt, A. W., and Huang, L. H. (2004) Cultivation of HEK 293 cell line and production of a member of the superfamily of G-protein coupled receptors for drug discovery applications using a highly efficient novel bioreactor. *Cytotechnology* **45**(3), 117–123.
23. Wheaton Science Products (2003) MantaRay—a new single-use cell culture flask. *vwr-international* **6**, 6–7.
24. Singh, V. (1999) Disposable bioreactor for cell culture using wave-induced agitation *Cytotechnology* **30**(1–3), 149–158.
25. Weber, W., Weber, E., Geisse, S., and Memmert, K. (2002) Optimisation of protein expression and establishment of the Wave Bioreactor for Baculovirus/insect cell culture. *Cytotechnology* **38**(9), 77–85.
26. Pierce, L. N. and Sabraham, P. W. (2004) Scalability of a disposable bioreactor from 25 L-500 L run in perfusion mode with a CHO-based cell line. *Bioproc. J.* **3(4)**, 51–56.
27. Wernli, U., Eibl, R., and Eibl, D. (2006) CD 293 AGT medium for the cultivation of HEK-293 EBNA cells in small-scale bioreactors: an application report. *Quest*, in press.
28. Fox, S. (2005) The impact of disposable bioreactors on the CMO industry. *Contract Pharma* **6**, http://contractpharma.com/archive/2005/06/fature 2.php./21.8.05.
29. Kunas, K. T. (2005) Stirred tank single-use bioreactor-comparison to traditional stirred tank bioreactor. Presented at Biologic Europe 2005, Geneva, Switzerland.
30. Aldridge, S. (2005) New biomanufacturing opportunities & challenges. *GEN* **25(12)**, 1–16.

31. Joosten, C. E. and Shuler, M. L. (2003) Effect of culture conditions on the degree of sialylation of a recombinant glycoprotein expressed in insect cells. *Biotechnol. Prog.* **19**(3), 739–749.
32. Saarinen, M. A. and Murhammer D. W. (2000) Culture in the rotating-wall vessel affects recombinant protein production capability of two insect cell lines in different manners. *In Vitro Cell Dev. Biol. Anim.* **36**(6), 362–366.
33. Dutton, G. (2003) Enhancing cell culture production operations. *GEN* **23(8)**, 20–21.
34. Mazur, X., Fussenegger, M., Renner, W. A., and Bailey, J. E. (1998) Higher productivity of growth-arrested Chinese hamster ovary cells expressing the cyclin-dependent kinase inhibitor p27. *Biotechnol. Prog.* **14(5)**, 705–713.
35. Kaufmann, H., Mazur, X., Fussenegger, M., and Bailey, J. E. (1999) Influence of low temperature on productivity, proteome and protein phosphorylation of CHO cells. *Biotechnol. Bioeng.* **63(5)**, 573–582.
36. Rutschmann, K. (2001) Qualifizierung von Laborbioreaktoren für die Kultivierung von CHO-Zellen. Diploma thesis, University of Applied Sciences Wädenswil, not published.

16

Hollow Fiber Cell Culture

John M. Davis

Summary

The use of hollow fiber culture is a convenient method for making moderately large quantities of high molecular weight products secreted by human and animal cells at high concentrations and with a higher ratio of product to medium-derived impurities than is generally achieved using homogeneous (e.g. stirred-tank) culture. Methods and options for operating and optimising such systems are described in detail, with special reference to the Biovest AcuSyst systems.

Key Words: Hollow fiber; high density cell culture; secreted product; monoclonal antibody; lymphoblastoid cells; hybridoma; Chinese hamster ovary cells.

1. Introduction

Hollow fiber mammalian cell culture systems were first conceived *(1)* to mimic the in vivo cell environment. In tissues, cells exist immobilized at high density and are perfused via capillaries having semi-permeable walls. Fluid (blood) circulating within the capillaries brings oxygen and nutrients and removes CO_2 and other waste products. This description applies equally to hollow fiber culture systems but with culture medium in the place of blood and with capillaries made from ultrafiltration or microfiltration membranes.

Hollow fiber systems have a number of advantages over other culture systems. These include:

1. High product concentrations: Where a cell secretes a product of higher molecular weight than the cut-off of the fiber membrane, it accumulates in the cell-containing compartment (normally the extracapillary space [ECS]). As the vast majority of the medium circulates in the intracapillary space (ICS), the product is not diluted in this, as would be the case in a homogeneous system (e.g., a stirred tank). This also results in the next advantage.

From: *Methods in Biotechnology, Vol. 24: Animal Cell Biotechnology: Methods and Protocols, 2nd Ed.* Edited by: R. Pörtner © Humana Press Inc., Totowa, NJ

2. A higher ratio of product to culture medium-derived contaminants: This greatly facilitates the purification of the secreted product.
3. Reduced requirements for high molecular weight supplements: If the cells require supplements of higher molecular weight than the cut-off of the fiber, then they only need to be supplied to the small proportion of medium in the ECS. Often these supplements can be reduced or omitted entirely once the culture is well established, as the cells themselves may secrete factors sufficient to maintain their own viability.
4. A low shear environment: The main medium flow is separated from the cells by the capillary membrane. Oxygenation also takes place in the intracapillary (IC) circuit and/or uses a silicone membrane or similar gas exchange system, and thus the cells are not subjected to potentially damaging contact with bubbles.
5. Convenience: Compared to a stirred tank capable of producing an equivalent amount of material, a hollow fiber system requires only a supply of CO_2 and electricity (i.e., no O_2, N_2, compressed air, water, steam, or drainage). It is also much smaller, being a benchtop unit at pilot scale and the size of a large refrigerator–freezer at full production scale. Thus, hollow fiber equipment can be accommodated in a room with normal ceiling height. This contrasts with fermenters, where specially designed rooms with increased ceiling height will be required once the scale reaches hundreds of liters and where the largest vessels often occupy three stories.
6. Cost: Again, compared to a stirred tank of similar productivity, hollow fiber systems are much less expensive *(2)*.

For the above reasons, hollow fiber systems have been widely used for the production of monoclonal antibodies and other high molecular weight secreted products, particularly in the in vitro diagnostics industry, and this is the type of application that will be described here. These systems have also found favor for many other purposes, such as the extracorporeal expansion of tumor-infiltrating lymphocytes ***(3,4)*** and, in particular, as the basis for bioartificial organs such as liver ***(5–7)***, but such applications are beyond the scope of this chapter.

1.1. Principles

A basic hollow fiber system is shown diagrammatically in **Fig. 1**. Oxygenated medium at the appropriate pH for the cells is circulated through the thousands of capillaries within the hollow fiber cartridge before being returned to the reservoir and recirculated. In the simplest systems, oxygenation and pH control is achieved by having a gas-exchange surface (usually silicone tubing) in a CO_2 incubator. More complex systems have self-contained gas-exchange units with the mixture of CO_2 and air passing through it controlled by feedback from a sterile pH probe situated in the medium flow. (This implies that the medium has a CO_2/HCO_3^- buffering system, as is common in mammalian cell culture media.) The cells are located on the outside of the capillaries, in the ECS. In such a system, nutrient and waste product exchange occurs largely by diffusion,

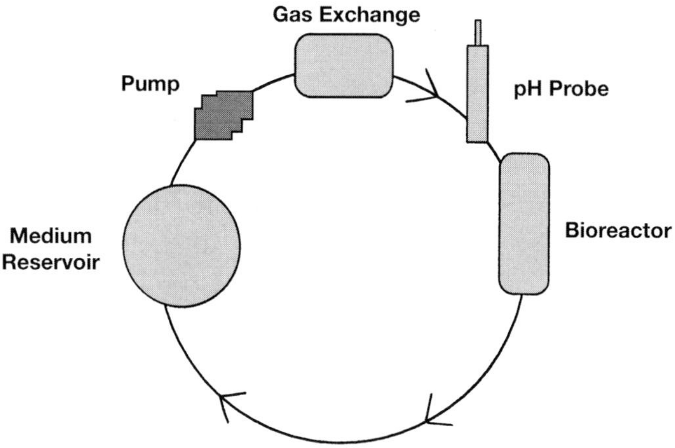

Fig. 1. Schematic diagram of a basic hollow fiber culture system. (From **ref. *16*.**)

supplemented by Starling flow *(8)*—flow out of the capillaries into the ECS at the upstream end of the fibers and back into the capillaries at the downstream end, resulting from the pressure drop along the length of the fibers. This can result in the formation of large axial and radial gradients of both nutrients and waste products within the cartridge, leading to nonuniform (and thus suboptimal) colonization of the ECS. This effect can be partially overcome by the periodic reversal of the direction of medium flow in the capillaries *(9)*, but the systems that will be used as the basis for the rest of this chapter, the benchtop Biovest AcuSyst-Jr (Biovest International Inc., Minneapolis, MN) and related Maximizer 500 and 1000, and the production-scale AcuSyst Xcell, overcome these problems by inducing mass flow of medium across the fiber walls. This is achieved by the application of a pressure differential to first push medium from the capillaries into the ECS and then, by reversing the pressure differential, back again into the capillaries ("cycling"). This system is shown diagrammatically in **Fig. 2**. The cycling is controlled by an ultrasonic detector in the base of the extracapillary (EC) expansion chamber.

Although what follows applies to the specific systems mentioned and is based on the use of hollow fibers with a 10,000 Da cutoff, the methods have been couched as far as possible in general terms and should be applicable, with greater or lesser modification, to most hollow fiber systems.

2. Materials

2.1. Materials for Growth Study in T-Flasks (see *Subheading 3.1.1.)*

1. Culture media and supplements.
2. 75-cm^2 tissue culture flasks.

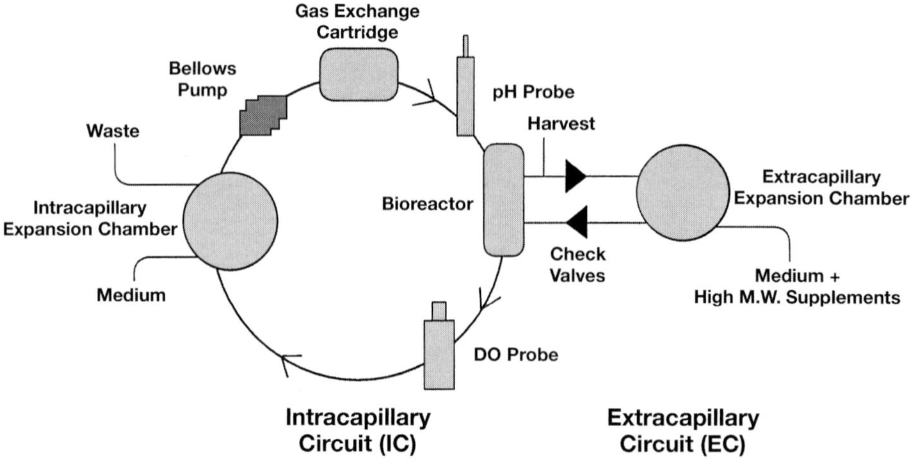

Fig. 2. Schematic diagram of an AcuSyst hollow fiber culture system. (From **ref.** ***16***.)

3. Carbon dioxide/air mixture(s).
4. Centrifuge.
5. 37°C incubator.
6. Sterile plugged Pasteur pipets.
7. pH meter.
8. Assays.

2.2. Materials for Flowpath Preparation (see Subheading 3.2.)

1. Autoclavable pH and dissolved oxygen (DO) probes and cases.
2. DO probe membranes and electrolyte.
3. Phosphate-buffered saline.
4. Autoclave (must be big enough to accept the pH and DO probes standing vertically).
5. Presterilized cultureware.
6. DO and pH probe leads.
7. Temperature probe.
8. Gas lines.
9. Multiport bottle caps with air vent filters (0.2 µm).
10. Sterile muslin gauze swabs.
11. 0.2% Hibitane (chlorhexidine gluconate) in 70% ethanol solution.
12. Sterile injection site septa.
13. Sterile containers with 3 L of sterile water, another with 10 L of basal medium, a third with medium supplemented with any high molecular weight substances required by the cells, and a fourth for waste. Bottles can be connected to the flowpath using **step 9** of **Subheading 3.2.**; other containers (e.g., sterile bags) should be equipped with female luer connectors.

2.3. Materials for Inoculation of Cells Into the Hollow Fiber Cartridge (see Subheading 3.3.)

1. pH meter with standards.
2. DO meter with standards.
3. A healthy culture of a sufficient number of appropriate cells in mid-logarithmic growth phase.
4. Sterile centrifuge tubes.
5. Centrifuge.
6. 80 mL of culture medium.
7. 60- and 10-mL syringes.

2.4. Materials for Growth Phase (see Subheading 3.4.)

1. pH meter with standards.
2. DO meter with standards.
3. Glucose and lactate assays.
4. Cell line characterisation/set point determination data (*see* **Subheading 3.1.**).

2.5. Materials for Monitoring and Maintenance During Production Phase (see Subheading 3.5.)

1. pH meter with standards.
2. DO meter with standards.
3. Assays for glucose and lactate (and ammonia and glutamine if required) and product.
4. Sterility assays.
5. Syringes for sucking cell debris from the flowpath.

2.6. Materials for Removal of Cells From the Hollow Fiber Cartridge at the End of a Run

1. Gate clamps.
2. 250 mL of basal (or supplemented) medium in a sterile bottle, fitted with tubing and adaptor to connect into the EC circuit between the lower nonreturn valve and the hollow fiber cartridge.
3. 250-mL (or larger) sterile bottle fitted with tubing and adaptor to fit EC bioreactor out/inoculation port.
4. Peristaltic pump, e.g., Watson-Marlow 500 series.

3. Methods

3.1. Cell Line Characterization/Set-Point Determination

Before attempting to culture any cell line in a hollow fiber system, it is best to characterize the cells' behavior in culture. In particular, it is important to ascertain the optimum pH for growth of the cell line, as the sooner after inoculation the cells multiply to the point where they have fully colonized all the available space within the ECS ("packed out" the ECS), the sooner high levels of product can be harvested. The cells' tolerance to lactate and requirements for glucose are also useful to know, as these will help in making decisions on the replacement rate of the basal medium (i.e. that circulated in the IC circuit). It is also useful to have some idea of the limits of pH at which the cells retain viability, as these will define the pH range within which the system can be run. Data on the cells' rate of production of the substance of interest at various pHs are also of use. Sadly, there is no simple way of quickly getting all of this information. However, a growth study in T-flasks **(Subheading 3.1.1.)** is a good starting point. Alternatively, the author has found from 17 yr of experience with hybridomas, heterohybridomas, lymphoblastoid cells, and Chinese hamster ovary (CHO) cells that many of them grow best at around pH 7.2 in hollow fiber culture, and this could be used as a starting point for studies in the actual hollow fiber system in use. Whichever approach is taken, optimization experiments will ultimately need to be performed in the hollow fiber system itself.

The AcuSyst-Jr or Maximizer is, in the author's experience, a good scaled-down model for the AcuSyst Xcell. However, further scale down necessitates the use of equipment (such as the Unisyn C100) that has no in-line pH control, no constant medium replenishment, no constant harvesting from the ECS, and different hollow fiber membranes, and thus the data obtained may at best be of only limited use and at worst may be positively misleading.

3.1.1. Growth Study in T-flasks

This method is intended for use with cells that grow in suspension and will require adaptation for cells that attach to the culture surface.

1. Decide on suitable culture media and supplement types to test, and adapt the cells to these media by multiple passage in tissue culture flasks.
2. Pregas three 75-cm^2 tissue culture flasks per test medium with carbon dioxide/air mixture(s) according to media requirements.
3. Prepare inocula by centrifuging mid-log-phase cell suspensions and resuspending the pellets in the minimum volume of the chosen media. Use cells at maximum possible viability.
4. Inoculate two 75-cm^2 flasks with between 1×10^5 and 2×10^5 viable cells per mL for each medium. The final culture volume should be 50 mL. The remaining flasks are blank controls with 50 mL of medium only.
5. Incubate the flasks at 37°C.
6. Aseptically remove a 3-mL sample from each flask 2 h after inoculation. Record this as time zero. The sample volume may be adjusted according to assay requirements (*see* **step 8** below).
7. Sample each flask daily for the following 10 d.

8. Measure pH and cell viability immediately after sampling. Also measure glucose, lactate, ammonia, glutamine, and product concentration. Control flasks are assayed for pH, ammonia, and glutamine only.
9. Display the data graphically. If the cultures have not reached a stationary/death phase by day 10, continue sampling every other day until day 14. If a stationary/death phase has still not been reached, repeat the procedure with double the inoculum cell density (*see* **step 4** above).
10. From the graph identify:
    - The pH at which maximum cell growth rate occurs. This is used to determine the growth-phase pH.
    - The pH at which maximum product secretion with minimum growth rate occurs. This is used to decide on the production-phase pH.
    - The glucose, lactate, and glutamine concentrations at and around maximum growth and maximum product secretion rates. These act as a guide in determining the glucose, lactate, and glutamine growth and production setpoints to aim for, and to identify limits during the culture.
    - The ammonia production profile. This indicates ammonia concentration limits during culture.

3.2. Flowpath Preparation

Different machines have flowpaths with different degrees of re-usability. However, in the AcuSyst systems, the pH and DO probes are the only major components of the flowpath that can be re-used; the rest comes as a unit that has been presterilized with ethylene oxide. Preparation involves sterilizing the probes and connecting them to the rest of the flowpath, adding any additional lines that may be necessary, installing the flowpath in the incubator/control module, pressure-testing the assembly, and flushing the system (to remove toxic ethylene oxide residuals and glycerin from the hollow fibers) and filling with medium.

In some cases (notably the AcuSyst Xcell), it may be worth securing those tubing connections subject to the highest pressures—usually those in the IC circuit—with self-locking cable ties to ensure that these connections do not come apart during use. If this is to be done, it is most easily performed before installing the flowpath in the incubator module.

In the method below, the details relate specifically to the AcuSyst-Jr. and will need to be adapted for use with any other system.

1. Remembrane the DO probe according to the manufacturer's instructions. Check operation. A 1% solution of Na_2SO_3 is useful in order to check a probe's ability to give a zero reading and the speed with which it will do so.
2. Assemble the pH and DO probes in their respective cases (these cases enable the probes to be plumbed into the flowpath such that medium in the IC circuit will flow past them continuously).

3. Place a small amount of phosphate-buffered saline in each probe case. Bag both probes in autoclave bags.
4. With the probes in an upright position (use a suitable rack or basket), autoclave the probes using a liquid cycle at 121°C for more than 30 min.
5. Remove the presterilized cultureware from its box. Spray and wipe down the covering with 70% ethanol in water and allow the package to dry in a laminar-flow hood or Class II (micro)biological safety cabinet. Check package for damage or punctures.
6. Open and discard the packaging and remove the cultureware. Untape and discard padding.
7. Immediately check and tighten all connections. The flowpath is a closed system with all possible routes into the system protected by filters, covers, or plugs. Ensure all of these are in place.
8. Close all clamps and unkink all tubes. Check that all lines are labeled correctly.
9. Aseptically connect the pH and DO probe assemblies following the instructions supplied with the flowpath. To facilitate aseptic manipulation of the connections, they may be wrapped for a minimum of 3 min in autoclaved muslin gauze swabs soaked in 0.2% Hibitane in 70% ethanol.
10. Aseptically attach sterile injection site septa to the sample ports and connect any other tubing lines as necessary.
11. Remove the flowpath from the hood or safety cabinet and gently slide it into the incubator chamber. Load the circulation pump into its holder and secure.
12. Plug the leads from the DO, pH, and temperature probes and IC and EC ultrasonic detectors into the appropriate sockets. Position the temperature probe in its holder in the flowpath.
13. Attach the gas lines to their correct ports and their corresponding positions in the flowpath.
14. Assemble the peristaltic pump heads and load all except the outflow pump tube segments.
15. Test the integrity of the flowpath following the steps detailed on the liquid crystal display (LCD) unit.
16. Reverse-load the outflow tubing into its pump head, aseptically connect 3 L of sterile water to this line and flush the system with it, again following the instructions on the LCD unit. Swap the vessel that contained the water for one containing sterile basal medium, and flush 3 L of this through the system.
17. Reload the outflow tubing, this time in its correct orientation, and connect to a waste vessel. Connect basal medium to the medium pump (which supplies the IC circuit) and medium containing any high molecular weight substances to the F3 pump (which supplies the EC circuit). If required, pump a suitable volume of this supplemented medium into the ECS to condition the outside of the fibers prior to inoculating with cells.
18. Set the medium pump at 25 mL/h and set the machine running (without cells) for 2 d. Check for the presence of any residual cytotoxic compounds in the flowpath by then taking a sample of the medium from both the IC and EC circuits and check that cells plated out at a low population density will grow in this medium. If not,

continue to run the machine, with the medium pump set at 25 mL/h, for several more days and try again. Once growth in both IC and EC samples is satisfactory, the flowpath is ready for inoculation.

3.3. Inoculation of Cells Into the Hollow Fiber Cartridge

Cells can grow to very high densities (1–2 × 10^8 per mL) in hollow fiber systems but must also be introduced in quite large numbers to ensure growth within the system. In our experience with antibody-producing cells, the number that should be inoculated to ensure growth is at least 2 × 10^6 per mL of extracapillary space in the hollow fiber cartridge, and it may be advantageous to add larger numbers, up to 8 × 10^6 per mL of extracapillary space. The number will, however, vary with the cell line. Fast-growing rodent cells may thrive from a smaller inoculum, while anchorage-dependent cells may need to be used at the upper end of this range.

1. Culture the cells so that they are growing at maximum rate and with maximum viability. It is often best to feed the culture with fresh medium 1–3 d before inoculation.
2. Calibrate the flowpath pH and DO probes.
3. In the control software, set the pH to the optimum for growth as previously ascertained (*see* **Subheading 3.1.**), set the basal medium feed to 25 mL/h/cartridge, and program in appropriate delays for cycling and for the supplemented medium and harvest pumps (suggested delays: 12 h to 7 d for cycling, 3–14 d for supplemented medium/harvest, depending on cell line).
4. Allow the pH to equilibrate to its setpoint.
5. Concentrate the cells by centrifuging under conditions that will maintain maximum viability (other strategies, such as settling out, are also possible). Resuspend the pellets in a total of 60 mL of medium, which may be either fresh or from the culture supernatant, depending on the preference of the cell line in use.
6. Transfer these cells to the 60-mL syringe. Add 10 mL of medium to the smaller syringe.
7. Connect the cell-containing syringe to the inoculation port (which connects with the ECS of the hollow fiber cartridge) and gently inject the cells. Swap syringes and flush the line through with the 10 mL of medium.

Increasingly, cells are being grown in protein-free medium. In this case it may not be necessary to concentrate the cells prior to inoculation into the hollow fiber system. The cells can be introduced into the ECS by applying gas pressure to the headspace of the vessel containing the cells (or by using a peristaltic pump, as long as this does not adversely affect cell viability), with the excess medium being removed by ultrafiltration through the pores in the hollow fiber membranes into the ICS. The main limitation on this approach is the time it may take to introduce the cells into the ECS, as during this period the cells are likely to be in an environment where neither the temperature nor the pH is controlled. The potential impact of this should be assessed for each cell type and hardware setup, and it may be appropriate to define a time limit for the transference of the cells into the ECS. With such a time limit defined, it may prove necessary to partially preconcentrate the cell suspension by centrifugation or other means.

3.4. Growth Phase

After inoculation there follows a period when the cells are actively multiplying to pack out the ECS of the hollow fiber cartridge (the growth phase). The conditions used at this time are generally those for optimal growth as ascertained in **Subheading 3.1**. In machines equipped with it, cycling is not started immediately in order to give the cells the opportunity of colonizing the fibers undisturbed by mass flow of medium. Similarly, the addition of supplemented medium to, and harvesting of product from, the ECS is delayed to allow the cells to condition the medium and grow to higher densities before medium removal starts. This has to be balanced against the possible build-up of any high molecular weight toxic compounds that a cell line might produce.

1. Once the temperature and pH have equilibrated after inoculation, check the calibration of the pH and DO probes. The probes will need to be recalibrated at regular intervals thereafter. (Recalibration every 3–7 d should be sufficient but may need to be done more frequently if significant drift is seen.)
2. Assay glucose and lactate levels in the medium in the IC circuit every 24–48 h.
3. When glucose and lactate assays indicate that glucose levels are starting to become limiting, or more commonly when lactate levels reach 50–75% of the maximum acceptable level defined previously, increase the medium pump rate. Increasing it from 25 to 50 mL/h will be adequate for most cell lines at this stage.
4. Repeat **steps 2** and **3**, increasing the medium pump rate as required until the conditions in **step 6** are fulfilled.
5. When addition of supplemented medium to the EC compartment and harvesting of product start, make sure the two pump rates are adjusted to provide the cells with the required concentration of supplements. The best way to do this is to supply the supplements at the required concentration and harvest product at the same rate.
6. When the medium pump rate reaches its maximum (or the maximum rate at which you are prepared to supply medium to the machine, whichever is lower), or the glucose and lactate (and product) levels reach a plateau, the growth phase is at an end.

3.5. Production Phase

Where the medium supply rate has been the deciding factor, you are now no longer able to cope with the metabolic demands of the cells at the growth pH, and it is necessary to change the pH with reference to the data obtained in **Subheading 3.1.**, or relevant published data, with the aim of reducing the growth and consequent metabolic activity of the cells while retaining product secretion. The hollow fiber cartridge may still not be completely packed out with cells at this stage. If, on the other hand, the glucose, lactate, and product levels have reached a plateau, then the cartridge has become completely packed out with cells without cellular metabolism placing inordinate demands on the medium supply. This usually only happens with slow-growing cell lines. Thus, although the growth phase has ended in terms of the cells multiplying freely to fill the cartridge, the culture may continue at this same pH or may be switched to another (often, but not always, lower) pH if it is advantageous in terms of product secretion rate.

The characteristic of the production phase is that the cells are held under more or less constant conditions, usually with a lower cell growth rate than in the growth phase. Thus, the frequency of monitoring assays can be reduced. However, a new requirement now arises because, as the hollow fiber cartridge is now filled with cells, excess cells and cellular debris are deposited in the bottom of the EC expansion chamber. If left to accumulate, this material will eventually affect the fluid flow and the ultrasonic detector in the chamber, which can lead to all medium being removed from the cells in the hollow fiber cartridge. Thus, regular removal of this material is essential.

Another consequence of the lower cellular growth rate (and hence lower metabolic rate) of the cells during the production phase is that it may now be possible to reduce the medium supply rate. Yet while the demands of the cells for nutrients may be considerably reduced, the requirement for buffering is still liable to be high, and this may be exacerbated if the production pH is higher than the growth pH. However, for economy of operation it is important to reduce the medium flow rate as far as possible, because medium is expensive. These two conflicting pressures may be accommodated by reducing the medium flow rate but adding additional buffering. This is most conveniently achieved by adding a feed of concentrated buffer solution. Buffers such as hydroxyethyl piperazine ethane sulfonate can be used but may interfere with pH control, which is achieved by adjusting the concentration of CO_2 supplied to the gas exchange unit and hence the pH of the CO_2/HCO_3^- buffering system in the medium. An alternative is to use a $NaHCO_3$ or Na_2CO_3 buffer solution: concentrations of around 22–30 g/L have proven suitable. One factor to be aware of, however, in adding such (alkaline) solutions to medium is that unless there is rapid mixing and dilution in the bulk of the medium, the pH may become very high and medium components may precipitate out. Thus, it may be important to add such solutions directly into the flow in the IC circuit, rather than into the fresh medium feed where flow, mixing, and dilution rates will be lower and tubing bores smaller and thus more prone to blockage by precipitates. If this is the case, it may be necessary to add a special line for this purpose to the IC circuit; this is best done in advance, during the initial setup of the flowpath (*see* **Subheading 3.2., step 10**).

Assuming no contamination, the length of the culture is determined by one of two factors. The culture may lose productivity over a period of months, usually because of increasing quantities of dead cells and cell debris in the hollow fiber cartridge. Alternatively, one or another part of the flowpath may become blocked, again because of the accumulation of dead cells and debris. This is usually first noticed in AcuSyst systems when they are unable to maintain their cycling time. In our experience, runs of 4–6 mo are normal.

3.5.1. Monitoring and Maintenance During the Production Phase

1. Inspect system daily for leaks, irregular cycling, adequate liquid levels in media feeds, harvest and waste containers, and generally for normal functioning. Keep a log of every action taken. Do not forget to check the CO_2 supply—running out can kill the culture.
2. Calibrate the pH and DO probes regularly (*see* **Subheading 3.4., step 1**). Recalibrate both probes immediately after any power interruption or (temporary) microprocessor failure.
3. During the growth phase, glucose and lactate (and, if required, ammonia and glutamine) levels were measured every 24–48 h. Once in production phase, these measurements may be made less frequently, but not less than twice a week, and additionally if a process control parameter (e.g., pH, media feed rate) is changed.
4. Measure product concentration regularly, i.e., on every batch of product, plus samples from the EC circuit if required.
5. Both IC and EC samples should be checked at least once a month for sterility.
6. Check for the accumulation of debris in the EC chamber. This should be removed at regular intervals that will depend on the cell line. Too much debris will lead either to blocking of the EC circuit or to malfunction of the ultrasonic detector.

3.6. Removal of Cells From the Hollow Fiber Cartridge at the End of a Run

Eventually a run will be terminated, either for one of the reasons mentioned in **Subheading 3.5.** or because enough product has been made. At this time it may be appropriate to remove a sample of the cultured cells from the hollow fiber cartridge, possibly for testing immediately, but more often to enable samples to be frozen down for future reference, i.e., for making a Postproduction Cell Bank. The method that follows, described specifically for the AcuSyst-Jr., has proven satisfactory for (among others) human lymphoblastoid cells, yielding (after 4 mo of hollow fiber culture) Postproduction Cell Banks indistinguishable by both single- and multi-locus DNA fingerprinting from the cells used for inoculation. Other approaches to extracting the cells can be taken, and in particular it may be necessary to use enzyme treatment to remove cells that are normally anchorage dependent.

1. Abort the process (this stops all pumps, cycling, etc.).
2. Clamp off IC circuit tubing into and out of the hollow fiber cartridge.
3. Clamp EC circuit tubing, upstream of lower nonreturn valve, and between the harvest removal line and the upper nonreturn valve.
4. Using aseptic technique, pull the EC circuit tubing off of the downstream end of the lower nonreturn valve, and connect to the bottle containing 250 mL of medium.
5. Connect the empty bottle to the EC bioreactor out/inoculate port.
6. Check that all clamps are open along the path that the medium will take from the bottle through the hollow fiber cartridge to the receiving bottle.
7. Load part of the tubing between the medium bottle and the flowpath into the peristaltic pump.
8. Pump the medium rapidly through the hollow fiber cartridge (but not so fast as to blow off the tubing connectors). This will flush a significant proportion (but not all) of the cells from the cartridge, along with dead cells and debris.
9. Separate the viable cells from the dead cells and debris either by repeatedly subculturing them or by using a physical separation technique.
10. If a Postproduction Cell Bank is to be made, culture the cells to obtain maximum viability and freeze according to standard procedures.

4. Notes

1. While the T-flask method for cell line characterization/setpoint determination described in **Subheading 3.1.1.** is easy to carry out, extrapolating the data from such a closed system, where cells experience constantly changing pH, nutrient, and waste levels, to the more open hollow fiber system, where most parameters are held constant by the supply of fresh nutrients/buffering agents and the constant removal of waste products, is difficult. In practice, the growth pH is usually easy to ascertain, while the production pH is usually a compromise between productivity and excessive metabolic demands of the cells. Using the data gained from culture in flasks the growth pH is usually accurate, but the pH that appears to be optimum for the production phase frequently proves to be too low when tried in a hollow fiber system. Err, therefore, on the side of caution, or you could kill your cells by using too low a pH. The question of whether, in fact, a pH change is indeed required between growth and production can only be answered by experimentation using a hollow fiber system.

   The method described will, by its very nature, identify a production pH that is lower than the growth pH. However, with some cells, e.g., CHO cells ***(10)***, productivity may be greater at a *higher* pH, in which case the optimum production pH will have to be ascertained by methods other than those described here, possibly by testing in the hollow fiber system itself. Note that, as with a decrease in pH, too large an increase in pH may kill the cells.

   Another approach that can be used in the production phase to limit the metabolic activity of the cells while retaining productivity is to change the incubation temperature. This appears from the literature to have been investigated mostly using CHO cells in non-hollow-fiber culture ***(11–15)***, but it can be applied to hollow fiber culture and may be worthy of investigation. Note, however, that if one wishes ultimately to use this approach in a system such as the AcuSyst Xcell, where two otherwise completely independent hollow fiber cultures can be run in a single incubator, it will not be possible to adjust the temperature of each culture independently. Consequently it will be necessary either to inoculate both flowpaths at the same time, using the same number of cells and the same growth phase conditions so that the switch from growth to production phase temperature is required at the same time by both cultures, or to use only one flowpath in the incubator (i.e., use only half the system's designed capacity).

2. The flowpath preparation procedure described in **Subheading 3.2.**, and all the subsequent steps in running a hollow fiber culture system, requires aseptic technique of the highest caliber. A large number of aseptic procedures are involved, particularly if one is to run the system for many months, and each carries with it a risk of contaminating the culture. Given good technique, it is possible to run a system in an ordinary laboratory, without the use of antibiotics, for 6 mo or more, as we have done on numerous occasions.
3. The condition of the cells used for inoculation **(Subheading 3.3.)** is of paramount importance to the subsequent success and productivity of the culture. Do whatever is necessary to get cells of the highest possible viability, and which are growing as fast as possible and are in mid-log phase.

   Some cell types are more sensitive to physical forces than others. Be careful not to damage the cells when concentrating them prior to inoculation. Similarly do not inject them into the hollow fiber system too fast, as they may then be damaged by high shear forces when passing through narrow-bore tubing.

   The approach described for possible use with cells that have been grown in protein-free medium, i.e., use of the ultrafiltration properties of the hollow fibers to remove large volumes of excess medium from the inoculum, is not recommended for use where the cell growth medium contains serum or other high molecular weight components. Not only will molecules larger than the membrane pore size become highly concentrated in the ECS, potentially affecting cell growth, but they will also tend to block the pores and prevent the bulk of the medium passing into the ICS, thus making it difficult to get the cell suspension into the ECS.
4. During the growth phase **(Subheading 3.4.)**, keep a close watch on the glucose and lactate levels, as these can change quite quickly and if adjustments are not made, for example, to the medium flow rate, they may kill the culture. We always measure the glucose and lactate levels every day for at least the first 10 d, and longer if necessary. My advice is to do the same and arrange that someone who is competent to make decisions regarding pump rates, etc., be available to take measurements over the weekend(s).
5. Barring contamination, the production phase **(Subheading 3.5.)** should be the vast majority of the culture period. It is less demanding on labor and assays than the growth phase, and weekend working should not be necessary on a routine basis.

   The key during monitoring and maintenance is just to keep a careful watch on everything, both the culture itself and the hardware containing it—and never relax the quality of your aseptic technique. If microbial contamination should occur it is usually obvious, characterized either by a rapid drop in dissolved oxygen levels in the medium or a disruption of the normal relationship between glucose utilization and lactate production rates. However, cryptic contaminations can occur, hence the recommendation to perform sterility tests on a regular basis.

6. It is useful to remove cells from the cartridge and make a Postproduction Cell Bank (*see* **Subheading 3.6.**) at the end of every run, as material is then always available for reference if any queries arise regarding a culture at a later date. Cells can also be extracted from the system during the course of a culture, and cell banks made if required, by removing medium from the harvest line prior to it reaching the peristaltic pump head. The medium will again contain a mixture of live cells, dead cells, and cell debris. It is usually necessary to remove a significant volume (e.g., 50 mL) in order to get sufficient cells for most purposes, such as to culture to make a cell bank, as in our experience the viable cell densities are well below 10^5 per mL. This will, however, vary with the stage of the culture and the cell line.

References

1. Knazek, R. A., Gullino, P. M., Kohler, P. O., and Dedrick, R. L. (1972) Cell culture on artificial capillaries: an approach to tissue growth *in vitro*. *Science* **178**, 65–67.
2. Hanak, J. A. J. and Davis, J. M. (1995) Hollow fiber bioreactors: the Endotronics Acusyst-Jr and Maximiser 500, in *Cell and Tissue Culture: Laboratory Procedures* (Doyle, A., Griffiths, J. B., and Newell, D. G., eds.), John Wiley & Sons, Chichester, Module 28D:3.
3. Freedman, R. S., Ioannides, C. G., Mathioudakis, G., and Platsoucas, C. D. (1992) Novel immunologic strategies in ovarian cancer. *Am. J. Obstet. Gynecol.* **167**, 1470–1478.
4. Hillman, G. G., Wolf, M. L., Montecillo, E., et al. (1994) Expansion of activated lymphocytes obtained from renal cell carcinoma in an automated hollow fiber bioreactor. *Cell Transplant.* **3**, 263–271.
5. Nyberg, S. L., Shatford, R. A., Peshwa, M. V., White, J. G., Cerra, F. B., and Hu, W.-S. (1993) Evaluation of a hepatocyte-entrapment hollow fiber bioreactor: a potential bioartificial liver. *Biotechnol. Bioeng.* **41**, 194–203.
6. Dixit, V. (1995) Life support systems. *Scand. J. Gastroenterol. Suppl.* **208**, 101–110.
7. Takeshita, K. Ishibashi, H., Suzuki, M., Yamamto, T., Akaike, T., and Kodama, M. (1995) High cell-density culture system of hepatocytes entrapped in a three-dimensional hollow fiber module with collagen gel. *Artificial Organs*, **19**, 191–193.
8. Starling, E. H. (1896). On the absorption of fluids from the connective tissue spaces. *J. Physiol.* **19**, 312–326.
9. Piret, J. M. and Cooney, C. L. (1990). Mammalian cell and protein distributions in ultrafiltration hollow fiber bioreactors. *Biotechnol. Bioeng.* **36**, 902–910.
10. Borys, M. C., Linzer, D. I. H., and Papoutsakis, T. (1993) Culture pH affects expression rates and glycosylation of recombinant mouse placental lactogen proteins by Chinese hamster ovary (CHO) cells. *Bio/technology* **11**, 720–724.

11. Furukawa, K. and Ohsuye, K. (1998) Effect of culture temperature on a recombinant CHO cell line producing a C-terminal α-amidating enzyme. *Cytotechnology* **26**, 153–164.
12. Furukawa, K. and Ohsuye, K. (1999) Enhancement of productivity of recombinant α-amidating enzyme by low temperature culture. *Cytotechnology* **31**, 85–94.
13. Hendrick, V., Vandeputte, O., Raschella, A., et al. (1999) Modulation of cell cycle for optimal recombinant protein production, in *Animal Cell Technology: Products from Cells, Cells as Products* (Bernard, A., et al., eds.), Kluwer, Dordrecht.
14. Kaufmann, H., Mazur, X., Fussenegger, M., and Bailey, J. E. (1999) Influence of low temperature on productivity, proteome and protein phosphorylation of CHO cells. *Biotechnol. Bioeng.* **63**, 573–582.
15. Rodriguez, J., Spearman, M., Huzel, N., and Butler, M. (2005) Enhanced production of monomeric interferon-beta by CHO cells through the control of culture conditions. *Biotechnol. Prog.* **21**, 22–30.
16. Davis, J. M. and Hanak, J. A. J. (1997) Hollow fiber cell culture, in *Basic Cell Culture Protocols*, 2nd ed. (Pollard, J. W. and Walker, J. M., eds.), Humana Press, Totowa, NJ, pp. 77–89.

17

Cultivation of Mammalian Cells in Fixed-Bed Reactors

Ralf Pörtner and Oscar B. J. Platas Barradas

Summary

Fixed-bed reactors for cultivation of mammalian cells can be used for a variety of applications such as the production of monoclonal antibodies with hybridoma cells or proteins for diagnostic or therapeutic use with recombinant, mostly anchorage-dependent animal cells, production of retroviral vectors, as well as for cultivation of tissue cells for artificial organs.

In **Subheading 2.**, suppliers for fixed-bed reactor systems and macroporous carriers as well as components of a lab-scale fixed-bed reactor and important mass balance equations are presented. **Subheading 3.,** covers topics such as selection of suitable carriers, procedures for cultivation in lab-scale fixed-bed reactors (setting up of the reactor, autoclaving, sterility test, preparation of seed culture and inoculation, start-up and process monitoring during culture, batch and continuous operation, activities after the culture), and additional requirements for larger reactor volumes (setup and sterilization, "inoculation during cultivation strategy").

Key Words: Fixed-bed reactor; macroporous carrier; radial flow; continuous cultivation; hybridoma; recombinant protein; retrovirus production; bioartificial liver.

1. Introduction

Fixed-bed reactors for cultivation of mammalian cells can be used for a variety of applications, such as the production of monoclonal antibodies with hybridoma cells or proteins for diagnostic or therapeutic use with recombinant, mostly anchorage-dependent animal cells, production of retroviral vectors, as well as for cultivation of tissue cells for artificial organs. They enable high-volume specific cell density and productivity under low-shear conditions as a result of the immobilization of cells within macroporous carriers. In principle, the fixed-bed reactor system (**Fig. 1**, left) consists of the fixed-bed unit containing macroporous carriers for cell immobilization and a vessel for conditioning of the medium. The oxygen-enriched medium is pumped from the conditioning vessel through the fixed bed and back. While the cells are retained in the

From: *Methods in Biotechnology, Vol. 24: Animal Cell Biotechnology: Methods and Protocols, 2nd Ed.* Edited by: R. Pörtner © Humana Press Inc., Totowa, NJ

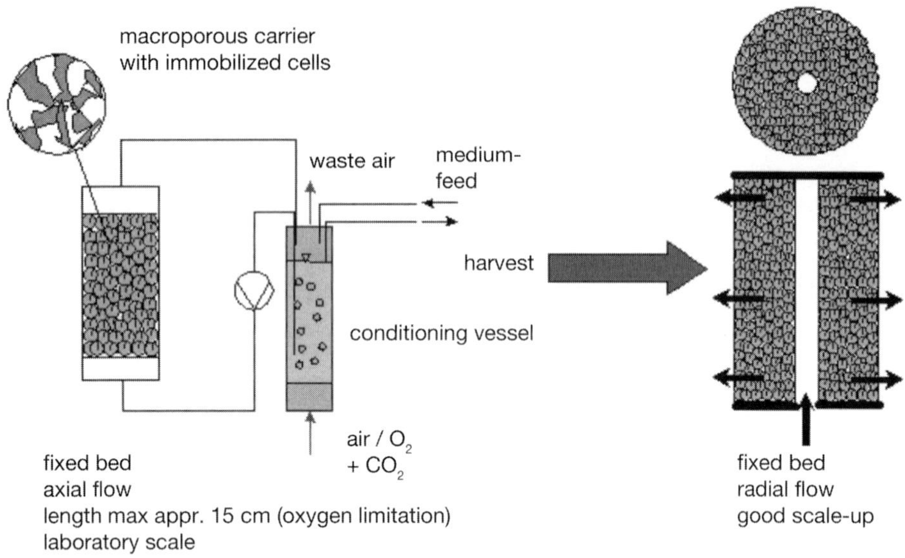

Fig. 1. Principle of a fixed-bed reactor system. The fixed bed is supplied constantly with nutrients and oxygen pumped from the conditioning vessel through it. The medium in the conditioning vessel is exchanged for fresh medium. (Left) Axial flow of medium through the fixed bed, used for bed lengths up to approx 15 cm. (Right) Radial flow—this avoids oxygen limitation for larger scales.

fixed-bed matrix, the spent medium with enriched product can be harvested either batchwise or continuously. In small-scale fixed-bed systems (length up to approx 15 cm), the medium flows axial through the matrix (**Fig. 1**, left). Larger fixed-bed volumes require the medium flow through the bed to be radial, e.g., from the center of the bed to the outer radius (**Fig. 1**, right), because of possible oxygen limitation ***(1,2)***.

These systems are characterized by high volumetric cell density and productivity, cell–cell contact (cell messaging over surface receptors), separation of cells from conditions of high shear stress caused by aeration and agitation, and the cell retention within the bioreactor, which allows an easier medium exchange and reduces the steps required for downstream processing. Furthermore, some adherent cell lines cannot be cultivated in suspension and need a suitable surface for attachment and growth.

Drawbacks of systems with cell immobilization are the inability to directly access cells during cultivation and the heterogeneity within the reactor, i.e., concentration and cell gradients over the reactor length or height and within the carriers or fibers. Determination of the fixed bed's cell density during the culture is difficult. For rough estimation of cell density, the volume-specific uptake rate for glucose can be used.

Table 1
Examples for Applications of Fixed-Bed Reactor Systems

Type of cell	Product/Purpose	ref.
Hybridoma	Monoclonal antibodies	***3–5,13–21***
Transfectoma	Chimeric fragmented antibodies	***8,22***
MRC-5	Inactivated HAV vaccine	***23***
rBHK	Glycosylated antibody–cytokine	***24***
CHO		***25–27***
rCHO	Rec. erythropoietin	***28***
	Rec. human placental alkaline phosphatase (SEAP)	***6***
Vero		***26***
L293	Rec. protein	Own data
HeLa	Vaccine	***29***
NIH3T3		Own data
Insect cells	Baculovirus	***30***
	Rec. protein	***31***
Packaging cell lines	Retrovirus production	***9,10,32–34***
Immortalized hepatocytes	Bioartificial liver	***11,35–40***
Primary hepatocytes	Bioartificial liver	***12,41,42***
Human kidney cells	Membrane vesicles	Own data
Human melanoma cells	Growth factors	Own data

HAV, hepatitis A virus; BHK, baby hamster kidney; CHO, Chinese hamster ovary; Rec., recombinant; SEAP, secreted alkaline phosphatase.

Results published for fixed-bed cultures underline the great potential of these systems (*see* **Table 1**). Cell densities up to approx 1.2×10^8 cells per mL of carrier were reached ***(2–5)***. Long-term cultures up to several months were performed without loss of cell activity, productivity, or cell stability ***(6,7)***. One of the main advantages of fixed-bed reactors is the low shear stress favoring the application of serum- or protein-free medium ***(8)***. Recent publications underline the potential of fixed-bed systems for production of retrovirus particles for gene therapy ***(9,10)***. A further application for fixed-bed reactor systems is the cultivation of tissue cells, e.g., as bioartificial liver ***(11,/12)***.

2. Materials

2.1. Suppliers of Fixed-Bed Reactors

Suppliers of fixed-bed reactor systems are shown in **Table 2**. **Figure 2** shows a series of fixed-bed reactor systems ranging from 10 mL to 5.6 L.

Table 2
Suppliers of Fixed-Bed Reactors

1. Able Co., 4-15 Higashigokencho, Shinjuku-ku, Tokyo 162-0813, Japan
   Radial-flow fixed-bed (5 mL, 30 mL)
2. Bellco Biotechnology, 340 Edrudo Road, Vineland, NJ 08360
   BellcoCell-500, bed volume 100 mL
3. Bioengineering AG, Sagenrainstraße 7, CH-8636 Wald, Switzerland
   Radial-flow fixed-bed (1 L, 5 L fixed-bed volume), instrumentation and control devices
4. Medorex e.K., Industriestraße 2, Tor 1. 37176 Nörten-Hardenberg, Germany
   Fixed-bed volumes of 10 and 100 mL (axial-flow), 1.5 L and 5 L (radial-flow), instrumentation and control devices
5. New Brunswick Scientific Col., PO Box 4005, Edison, NJ
   Celligen Cell Culture Systems (2.2 L, 5 L, 7.5 L, 14 L)
6. Zellwerk GmbH, Ziegeleistraße 7, D-16727, Eichstädt, Germany
   Rotating bed bioreactor consisting of rotating discs (0.5 L, 5 L total reactor volume)

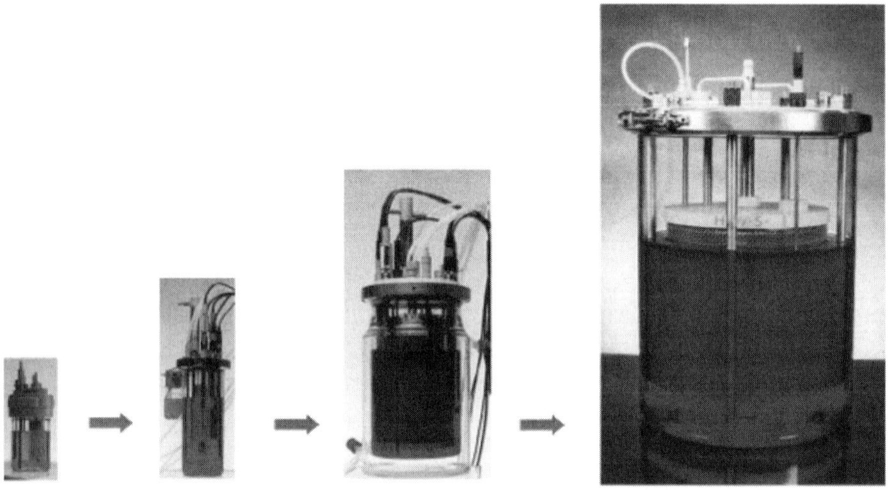

Fig. 2. Series of fixed-bed reactor systems in different scales.

2.2. Suppliers of Macroporous Carriers

A number of carriers for the immobilization of suspendable or anchorage-dependent cells are available **(Table 3)**. The modification of the surface is a critical parameter. Anchorage-dependent cells, especially primary or non-established cell

Table 3
Suppliers of Macroporous Carriers

1. ABLE Co., 4-15 Higashigokencho, Shinjuku-ku, Tokyo 162-0813, Japan Carrier: RS-001 (hydroxyl apatite), RS-002 (glass), RS-003/4 (PVA-block)
2. GE Healthcare. Amersham Biosciences. Munzingerstraße 9, 79111 Freiburg, Germany Carrier: Cytoline® (polystyrene with silica), Cytopore® (denatured cellulose)
3. Hiper Ceramics GmbH, Ziegeleistraße 7, D-16727 Eichstädt, Germany. Carrier: Sponceram® (zirconium oxide)
4. New Brunswick Scientific Co., Inc. PO Box 4005 Edison, NJ, USA Carrier: Fibra-Cel® (GMP certified)
5. QVF Engineering GmbH, Hattenbergstraße 36, D-55122 Mainz, Germany. Carrier: SIRAN® (borosilicate glass)

lines, have been shown to be much more sensitive to the carrier surface than established cell lines (e.g., hybridoma, Chinese hamster ovary [CHO], L293, NIH3T3). This has to be taken into account during the selection of a suitable carrier for each specific cell line. In general, the cell density of the immobilized cells in the carrier increases with decreasing carrier diameter. On the other hand, small carriers tend to block the free channels in the fixed-bed matrix. Therefore, diameters of 3–5 mm are an appropriate choice for fixed-bed reactors *(7)*.

2.3. Components of a Lab-Scale Fixed-Bed Reactor (< approx 1.5 L)

A fixed-bed bioreactor system consists mainly of a series of components **(Fig. 3)**. The components vary according to the scale of the process and the purpose of the cultivation to be carried out. The exact setup will depend on the specific model. In this section, the state-of-the-art technique for carrying out a process for the cultivation of animal cells in a 100-mL fixed-bed reactor (medorex) operated in a continuous mode is described **(Fig. 4)**:

1. The two bottles contain the media pumped into the conditioning vessel and back, defining with this the dilution rate D of the system controlled by a peristaltic pump. For cultivation in batch mode, it is necessary to empty the conditioning vessel completely. Therefore, a tube in the harvest line has to reach up to the bottom of the vessel. For continuous operation, the medium is preferably sucked off at the medium surface. The bottles for fresh and harvest medium should be connected to the reactor via sterile connections to enable an exchange of these bottles.
2. Another pump maintains the circulation of media to the fixed bed and back into the conditioning flask.
3. The aeration bottle allows for humidification of air, avoiding the loss of water from the medium (*see* **Note 2**). Temperature, dissolved oxygen, and pH (by addition of CO_2 into the air) are parameters to be controlled during the cultivation.

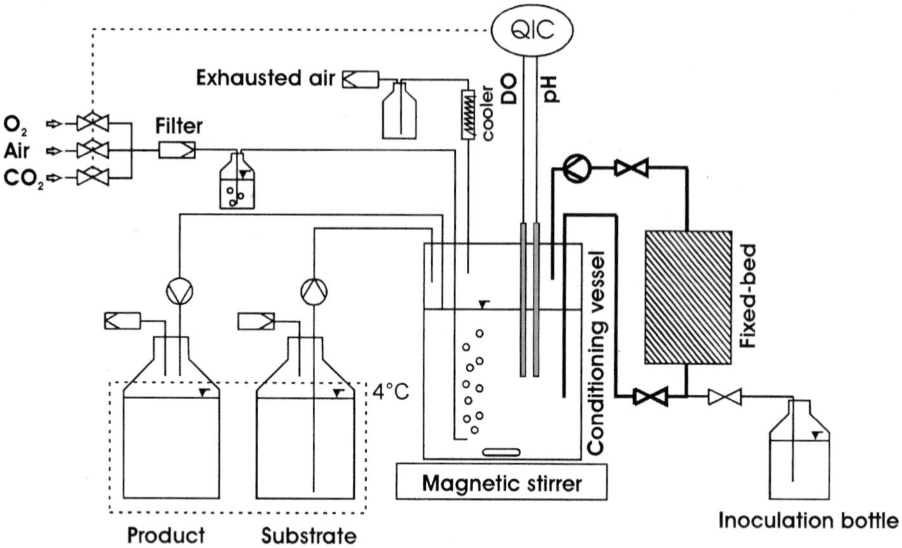

Fig. 3. Fixed-bed process for the lab scale. The medium is pumped through the fixed bed and back into the conditioning vessel. A second circuit constantly exchanges the medium of the conditioning vessel with fresh substrate, while the depleted, product-containing medium is removed. Different parameters like dissolved oxygen, pH, and temperature are measured and controlled.

4. Agitation with a magnetic stirrer is required to mix the medium homogeneously and to avoid sedimentation of cells and cell debris.
5. An inoculation bottle contains the cells to be pumped into the fixed bed for inoculation before starting the culture.
6. The sampling flask allows one to take a sample of medium with minimum contamination risks.
7. The parameters of the system (pH, DO, temperature) are displayed and can also be recorded.

2.4. Equations

The crucial parameter to fix the flow rate of the medium pumped through the fixed bed is the superficial flow velocity U. From this, the volumetric flow rate to be pumped to the fixed bed is defined by:

$$\dot{V} = U \cdot A_{FB} \tag{1}$$

with the cross-sectional area A_{FB} of the fixed bed.

The general mass balance of the fixed-bed reactor operated continuously is described by the following equation:

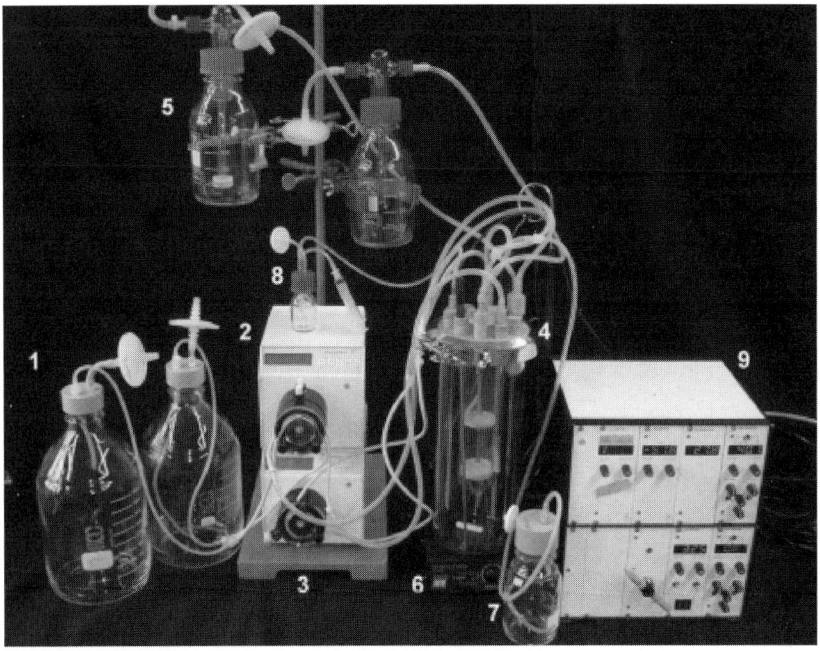

Fig. 4. Setup of a 100-mL fixed-bed reactor (Medorex, Germany) for continuous cultivation. The numbers on the figure indicate the components of the system: 1, inlet and outlet bottles; 2, peristaltic pump for medium inlet and outlet; 3, peristaltic pump for recirculation of media; 4, fixed-bed reactor with integrated fixed-bed in conditioning vessel; 5, aeration flasks; 6, magnetic stirrer; 7, inoculation bottle; 8, sampling flask; 9, unit for display and control of pH and dissolved oxygen.

$$\frac{dc_i}{dt} = D \cdot (c_{i0} - c_i) \pm q_i^* \qquad (2)$$

where D is the dilution rate of medium in the conditioning vessel, c_i the substrate or product concentration, c_{i0} the substrate or product feed concentration, and q^*_i indicates the substrate uptake (-) or metabolite production rate (+) per volume conditioning vessel. If in continuous mode the concentration c_i can be regarded as constant (steady state), $dc_i/dt = 0$. During batch or repeated batch mode, the substrate uptake and the metabolite production rate is defined by:

$$q_i^* = \frac{c_{1i} - c_{2i}}{t_2 - t_1} \qquad (3)$$

for the time interval t_1–t_2. The dilution rate D_{FB} per volume of fixed bed and the substrate consumption or metabolite production rates per volume fixed-bed

$q^*_{i,FB}$ are defined in Eqs. 4 and 5. V_C is the volume of the conditioning vessel and V_{FB} the volume of the fixed bed.

$$D_{FB} = \frac{V_C}{V_{FB}} \cdot D \tag{4}$$

$$q^*_{i,FB} = \frac{V_C}{V_{FB}} \cdot q^*_i \tag{5}$$

3. Methods

3.1. Selection of Suitable Carriers

3.1.1. Preselection (Proof of Suitability)

A preselection of an appropriate carrier can be performed in 12-well plates, especially for adherent cells. This test gives a first indication of whether cells can grow on the carrier, but the test is not suitable for estimation of the later performance of the fixed bed. A possible procedure is as follows:

1. Perform the test in 12-well plates (3 wells per carrier).
2. Pretreat the plate with silicone to prevent cell attachment and growth at the bottom is required.
3. Carriers (sterilized according to manufacturing manual) should cover the bottom of the plate, but should not form multiple layers.
4. Inoculate 3 mL of medium containing 2×10^5 cells/mL.
5. Cultivate for 12 d.
6. Exchange medium every third day.
7. Determine glucose concentration and glucose uptake rate (Eq. 3) after each medium exchange.
8. Store samples at –20°C for later analysis.
9. After experiment: cell counting by crystal violet, scanning electron microscopy (SEM).

3.1.2. Evaluation of Carrier Performance

Initial data on fixed-bed performance can be obtained with small-scale fixed-bed systems (5–10 mL, e.g., medorex, ABLE), which are usually run in a batch or repeated-batch mode. These systems are easy to handle, can be operated within an incubator, and do not require sophisticated devices for medium exchange or control. The procedure is as follows:

1. Take the most suitable carriers from the 12-well test.
2. Prepare the system according to the supplier or according to the procedure given below.
3. Run the system in repeated batch mode, as described below.

4. Adjust the intervals for medium exchange based on glucose measurement (minimal concentration approx 5 mmol/L). At the end of cultivation, several medium exchanges per day might be required.
5. Run the system for 10 d.
6. Calculate the glucose uptake rate according to Eq. 3.
7. Prepare samples of the carrier for cell count and SEM.

3.2. Lab-Scale Fixed-Bed Cultivation (< approx 1.5 L)

The procedures for setting up and running a fixed-bed reactor vary according to the specific model of the reactor. Below the main steps involved in operating a 100-mL axial-flow fixed bed (medorex) are summarized (*see* **Notes 1–3**).

3.2.1. Setting Up the Reactor

1. All the components should be washed with a mild soap and rinsed thoroughly with water. Components like silicon tubings, connectors, or some reactor pieces with dead ends where steam may not reach bacteria can be washed with a detergent solution (e.g., 1% Mucasol® in water, 20 min at 70ºC).
2. The carrier used should be rinsed twice with demineralized water to remove particles or remaining chemical substances. If the carrier can be re-used, a special cleaning procedure is required. A common procedure used for glass and ceramic carriers is as follows:
   a. Wash the carriers with a 1 *M* NaOH solution overnight (preferably at higher temperature, e.g., 50ºC).
   b. Neutralize the solution by carefully adding HCl. Always wear laboratory coat and protection glasses.
   c. Dispose the solution and rinse the carriers thoroughly with distilled water. Note any change in the pH value.
   d. Most carriers can be dried at 180ºC and stored for further use.
3. All o-rings have to be greased with silicon. It is important to ensure that the fixed bed is tight enough to avoid air escaping from it to the conditioning vessel. A leak test can be done by operating the fixed bed with water and observing the circulation from the bed to the conditioning flask and vice versa.
4. Peristaltic pumps as well as pH and oxygen probes must be calibrated.
   a. The peristaltic pumps are calibrated by measuring the volume pumped into a flask per time unit. It is important to use the same configuration (tubes, adapters, fittings) as in the final setup.
   b. pH sensors are calibrated by inserting the probe in standard solutions with known pH value.
   c. DO probe is calibrated, on the one hand, at 0% of dissolved oxygen by introducing the probe in a saturated sodium sulfite solution (Na_2SO_3) or special zeroing gel. On the other hand, the value for 100% dissolved oxygen concentration can be obtained by keeping the probe in a well-aerated solution (water or culture media) until the displayed value becomes constant.

3.2.2. Autoclaving

1. Once the reactor has been completely assembled, it can be sterilized (*see* **Notes 4** and **5**). All the bottle caps as well as the reactor's lid should be loosened to allow the steam to penetrate them. The closing metal ring of some reactor vessels should also be loosened.
2. The same applies for temperature sensors or heaters; this avoids pressure build-up inside of the system and allows the flow of steam through it.
3. The O_2 sensor, pH electrode, sampling tube, and filters should be covered with aluminum foil to protect them against humidity. Clips will be set to the sampling tube and to the medium flask.
4. After sterilization (standard conditions 121°C, 20 min), all the caps and sensors must be tightened quickly while the reactor is still in the autoclave.
5. Calibration of probes (especially pH and DO) should be checked after autoclaving.

3.2.3. Sterility Test

After sterilization it is recommended to perform a sterility test before inoculation of cells (*see* **Note 6**). For this the remaining water in the reactor is exchanged with fresh medium. Once all the medium is inside the reactor, the circulation to the fixed bed can be started. The reactor is aerated as during cultivation. The sterility test should run at least 3 d, the reactor should be observed, and samples should be taken for either consumption of glucose, change of turbidity, or color (if the medium contains phenol red as an indicator) of the medium.

3.2.4. Preparation of Seed Culture and Inoculation

At the start of the experiment a defined number of cells growing in the exponential phase is inoculated into the fixed bed. An inoculation density of $1–2 \times 10^5$ cells per mL of culture medium is appropriate *(1,7)*. The required amount of cells (depending on the amount of medium in the reservoir) is prepared in standard cell culture systems (T-flasks, roller bottles, etc.):

1. The cells must be removed from the culture flask. If working with adherent cells, prior treatment to remove the cells is necessary (e.g., trypsinization), the medium is removed and the flask walls are washed twice with sterile Ca^{+2}-and Mg^{+2}-free phosphate-buffered saline (PBS) and incubated for 2 min in a trypsin–ethylene diamine tetraacetic acid (EDTA) solution (0.5 g trypsin and 0.2 g EDTA per L of Hanks' balanced salt solution). The detached cells are centrifuged and resuspended in fresh medium.
2. The cell suspension is than filled into the sterile inoculation flask. The inoculation flask is connected to the fixed-bed reactor (this should be done in a clean bench).
3. The circulation loop between the conditioning vessel and the fixed bed is closed and the inoculation circuit opened. The cells are then pumped with a superficial velocity U of 0.15 mm/s through the fixed bed giving a flow rate according to Eq. 1 (*see* **Notes 7**).

Table 4
Recommended Superficial Velocities Through the Fixed Bed During Culture for Nonadherent Cells

Superficial velocity (mm/s)	When to use
0.2	After inoculation while the cells attach onto the carriers and the medium becomes clear (approx 2 d)
0.3	First 2 d of cultivation
0.4	Third and fourth days of cultivation
0.6	Fifth and sixth days of cultivation
0.8	Longer cultivation periods
1.0	Washout of hybridoma cells from the carriers

4. Observe the turbidity in the inoculation flask. If the medium becomes transparent, inoculation is completed.

3.2.5. Start-Up and Process Monitoring During Culture

1. Once the cells have been inoculated, the circulation to the fixed bed can be started (*see* **Notes 8–11**). For the first 2 h a minimum superficial velocity U is applied to let the cells attach to the carriers (*see* **Table 4**; the volumetric flow rate is defined by Eq. 1).
2. Light stirring should be applied during this time to avoid sedimentation of cells remaining in the medium (to avoid cellular damage due to the stirrer, shaking by hand can be used). After the medium has become clear, the aeration (5% CO_2 in air) can be started and the pumping velocity increased.
3. The superficial flow velocity is increased stepwise to avoid washout of cells from the fixed bed. For nonadherent cells grown on SIRAN carriers, recommendations are given in **Table 4**. For this type of cell above 1 mm/s a washout has to be expected. Adherent cells may tolerate higher flow velocities. This has to be tested individually.
4. Every day a sample should be taken and the residual glucose concentration as well as lactate concentration should be determined. With these values the glucose uptake rate, the lactate production rate, and the lactate to glucose yield can be determined.
5. Fur further analysis (glutamine, ammonia, amino acids, lactate dehydrogenase activity, etc.), the samples should be frozen (–20°C).

3.2.6. Operation

3.2.6.1. Batch Mode

After start-up the reactor is run in batch mode (feeding and harvest pump switched off). Here the substrate uptake and the metabolite production rate is defined by Eq. 3. When the glucose value approaches 5 mmol/L, the medium can be changed as follows:

1. The circulation pump is switched off, but the fixed bed should remain filled with medium.
2. The medium is pumped into the harvest bottle.
3. A new flask with sterile media is connected to the reactor, and the media is pumped to the conditioning flask.
4. The circulation pump is restarted.

3.2.6.2. Continuous Mode

1. Continuous feeding and harvesting can be started when the glucose values approach 5 mmol/L. The initial value for the dilution rate (Eq. 6) can be estimated from the glucose uptake rate of the system while it was working in batch mode (Eq. 5).
2. Glucose concentration should be measured every day to determine the glucose uptake rate. If the glucose uptake rate is almost constant for 3–4 d, double the perfusion rate.
3. Repeat this until there is no further increase in glucose uptake rate. At very high cell densities, perfusion rates up to 10 mL per mL fixed bed and day are realistic.
4. The system can be run in perfusion mode for several months (*see* **Note 12**).

3.2.7. Activities After Culture

1. Recovery of medium: The circulation pump is switched off, and the medium from the conditioning vessel is pumped to the harvest vessel. The media can be stored at 4°C for downstream purposes.
2. Final cell density: a representative sample of carriers (10–100) can be taken from different positions of the fixed bed and incubated for 1 h in 5 mL of a crystal violet solution (0.1% crystal violet in 0.1 *M* citric). The sample should be thoroughly shaken to remove the nuclei from the carriers, and the cell number for each position is determined by counting the nuclei in a hemocytometer.
3. SEM: a sample of carriers is kept for 1 d in a glutaraldehyde fixture (5% glutaraldehyde in PBS, pH 7.0). Afterwards, they are dehydrated by stepwise exchanging water with ethanol (20, 40, 60, 80, and 100% ethanol in PBS, pH 7.0, kept for 2 h in each dilution step) and stored for 1 d in 100% ethanol. Afterwards, ethanol is exchanged with amyl acetate (50, 100% *n*-amyl acetate in ethanol, kept for 2 h in the 50% dilution). The samples are then stored in 100% amyl acetate. Before carrying out the SEM, the samples are critical-point-dried and gold-sputter-coated.
4. Sterilization of the reactor: the reactor should be sterilized after the cultivation, washed with water and a mild soap, and rinsed with demineralized water. The carriers can be either disposed or prepared for re-use.

3.3. Larger Reactor Volumes (>1.5 L Fixed Bed)

3.3.1. Setup and Sterilization

These systems require some further considerations with respect to autoclaving. Always keep in mind the handling instructions of the suppliers (*see* **Notes 13** and **14**).

1. If the carriers endure higher temperatures, it is advisable to dry them for 2 h at 180°C (dry heat) to avoid survival of spores in the cavities of the carriers.
2. Larger volumes of carriers need a longer time to reach the sterilization temperature. Therefore, it is recommended to check this effect before by introducing a temperature sensor of the autoclave into a 2–L flask filled with dry carriers. This experiment gives an idea of how long it takes until the temperature in the fixed bed has reached the required value (121°C).
3. Because of the large fixed-bed volume, a high quantity of steam is required to penetrate the pores of the bed and the carriers. At least 300 mL of water should be poured into the reactor to increase the quantity of steam produced in the system during sterilization.
4. The sterilization time should be prolonged—if possible (stability of the carrier), up to 90 min.

3.3.2. Inoculation-During-Cultivation Strategy for Larger Fixed-Bed Volumes

Large-scale fixed beds need a high total cell number for inoculation. To reduce the required number of cells, a strategy suggested by Fassnacht ***(1)*** can be applied. This inoculation-during-cultivation strategy is possible if the fixed bed is introduced directly in the conditioning vessel (Bioengineering, medorex). At the start only a fraction of the fixed bed (e.g., one-third) is submerged and therefore active. Only this part of the fixed bed is then inoculated with a relatively low inoculation volume. After some days of cultivation, the cell density increases and the volume of media can be raised to the desired level. This activates the fresh part of the fixed bed, which is then inoculated by the low number of cells washed out of the fixed bed during cultivation. The maximal cell density within the fixed bed is finally reached a few days afterwards. Once the reactor has reached its maximum level of media, a batch or a continuous exchange of media can be performed.

4. Notes

1. Break-up of tubes (circulation loop, feeding, and harvesting) must be avoided in any case, as this usually means termination of the cultivation. Therefore, select a pump working at low rotating speed and a tube especially designed for use in peristaltic pumps (e.g., PharMed).
2. A condenser should be used to avoid the loss of water in the reactor during cultivation.
3. When autoclaving, the reactor, the aeration flasks, the sampling flask, as well as pH and DO sensors should be loosened. All the components are to be covered with aluminum foil. This works as a barrier against bacteria when air penetrates the system during cooling.
4. Water in all the vessels will create more steam during autoclaving. This is important to guarantee complete sterilization. Furthermore, the tips of probes for DO and pH should be covered with water.

5. The components inside the autoclave are still at high temperature when the sterilization process is finished. When the temperature is reduced, the volume of the air contained in the system is also reduced and air from the exterior flows into the reactor. Therefore, it is important to close all possible air inlets to the reactor except the two air filters at the aeration bottles. For higher reactor volumes, it could be advisable to let the autoclave reach the ambient temperature before opening it. This keeps a sterile environment inside while the reactor temperature decreases.
6. Connections and air leakage in the reactor should be tested before the sterility test. The recirculation to the fixed bed and back into the conditioning vessel should be checked.
7. Aeration of the system during inoculation of cells can cause cell death.
8. Stirring during cultivation keeps good homogeneity of the medium and reduces the amount of air required for aeration.
9. After a few days of cultivation the water in the flask for humidification will evaporate. A new flask with water should be autoclaved and exchanged under sterile conditions.
10. During cultivation, the inoculation loop should be closed, otherwise, the medium will flow in or out of the reactor due to gravity.
11. Glutamine has low stability at cultivation temperatures. It could be added to the reactor separately. The samples should be frozen to avoid glutamine degradation.
12. Oxygen concentrations above air saturation, which is usually toxic to mammalian cells, turned out to be nontoxic for cells immobilized in fixed-bed reactors and can improve the performance of fixed-bed reactors if oxygen limitation within the carriers is critical *(2)*. Therefore, it can be advisable to increase the oxygen concentration in the medium above air saturation. But this should be done after some time of cultivation (approx 2 wk).
13. Because the amount of medium required for larger reactor systems increases drastically (e.g., up to 50 L per day for a 5-L radial-flow reactor), medium bags of an appropriate size should be used for feed and harvest.
14. For larger fixed-bed volumes, the time required to heat the fixed bed up to 121°C must be determined to guarantee the appropriate temperature throughout the reactor.

References

1. Fassnacht, D. (2001). *Fixed-Bed Reactors for the Cultivation of Animal Cells,* Fortschritt-Berichte VDI, VDI Verlag GmbH, Düsseldorf, Germany.
2. Fassnacht, D. and Pörtner, R. (1999) Experimental and theoretical considerations on oxygen supply for animal cell growth in fixed-bed reactors. *J. Biotechnol.* **72**, 169–184.
3. Ong, C.P., Pörtner, R., Märkl, H., Yamazaki, Y., Yasuda, K., and Matsumura, M. (1994) High density cultivation of hybridoma in charged porous carriers. *J. Biotechnol.* **34**, 259–268 .
4. Looby, D., Racher, A. J., Griffiths, J. B., and Dowsett, A. B. (1990) The immobilization of animal cells in fixed bed and fluidized porous glass sphere reactors, in *Physiology of Immobilized Cells* (de Bont, J. A. M., Visser, J., Mattiasson, B., and Tramper, J., eds.) Elsevier Science Publishers B.V., Amsterdam, pp. 255–264.

5. Bohmann, A., Pörtner, R., and Märkl, H. (1995) Performance of a membrane dialysis bioreactor with radial-flow fixed bed for the cultivation of a hybridoma cell line. *Appl. Microbiol. Biotechnol.* **43**, 772–780.
6. Fussenegger, M., Fassnacht, D., Schwartz, R., et al. (2000) Regulated overexpression of the survival factor bcl-2 in CHO cells increases viable cell density in batch culture and decreases DNA release in extended fixed-bed cultivation. *Cytotechnology* **32**, 45–61.
7. Bohmann, A., Pörtner, R. Schmieding, J., Kasche, V., and Märkl, H. (1992) Integrated membrane dialysis bioreactor with radial-flow fixed bed—a new approach for continuous cultivation of animal cells. *Cytotechnology* **9**, 51–57.
8. Lüdemann, I., Pörtner, R., Schaefer, C., et al. (1996) Improvement of the culture stability of non-anchorage-dependent animal cells grown in serum-free media through immobilization. *Cytotechnology* **19**, 111–124.
9. Nehring, D., Gonzales, R., Pörtner, R., and Czermak, P. (2006) Experimental and modelling study of different process modes for retroviral production in a fixed-bed reactor. *J. Biotechnol.* **122**, 486–496.
10. Merten, O. W., Cruz, P. E., Rochette, C., et al. (2001) Comparison of different bioreactor systems for the production of high titer retroviral vectors. *Biotechnol. Prog.* **17**, 326–335.
11. Fassnacht, D., Rössing, S., Stange, J., and Pörtner, R. (1998): Long-term cultivation of immortalised mouse hepatocytes in a high cell density fixed-bed reactor. *Biotechnol. Techn.* **12**, 25–30.
12. Kurosawa, H., Yasumoto, K., Kimura, T., and Amano, Y. (2000) Polyurethane membrane as an efficient immobilization carrier for high-density culture of rat hepatocytes in the fixed-bed reactor. *Biotechnol. Bioeng.* **70**, 160–166.
13. Racher, A. J., Looby, D., and Griffiths, J. B. (1990) Studies on monoclonal antibody production by a hybridoma cell line (C1E3) immobilised in a fixed bed, porosphere culture system. *J. Biotechnol.* **15**,129–145.
14. Racher, A. J., Looby, D., and Griffiths, J. B. (1993) Influence of ammonium ion and glucose on mAb production in suspension and fixed bed hybridoma cultures. *J. Biotechnol.* **29**, 145–156.
15. Golmakany, N., Rasaee, M. J., Furouzandeh, M., Shojaosadati, S. A., Kashanian, S., and Omidfar, K. (2005) Continuous production of monoclonal antibody in a packed-bed bioreactor. *Biotechnol. Appl. Biochem.* **41**, 273–278.
16. Yang, S. T., Luo, J., and Chen, C. (2004) A fibrous-bed bioreactor for continuous production of monoclonal antibody by hybridoma. *Adv. Biochem. Eng. Biotechnol.* **87**, 61–96.
17. Wang, G., Zhang, W., Jacklin, C., Freedman, D., Eppstein, L., and Kadouri, A. (1992) Modified CelliGen-packed bed bioreactors for hybridoma cell cultures. *Cytotechnology* **9**, 41–49.
18. Bliem, R., Oakley, R., Matsuoka, K., Varecka, R., and Taiariol, V. (1990) Antibody production in packed bed reactors using serum-free and protein-free medium. *Cytotechnology* **4**, 279–283.
19. Pörtner, R., Lüdemann, I., and Märkl, H. (1997) Dialysis cultures with immobilized hybridoma cells for effective production of monoclonal antibodies. *Cytotechnology* **23**, 39–45.

20. Fassnacht, D., Rössing, S., Singh, R., Al-Rubeai, M., and Pörtner, R. (1999) Influence of BCL-2 Expression on antibody productivity in high cell density hybridoma culture systems. *Cytotechnology* **30**, 95–105.
21. Moro, A. M., Rodrigues, M. T., Gouvea, M. N., Silvestri, M. L., Kalil, J. E., and Raw, I. (1994) Multiparametric analyses of hybridoma growth on glass cylinders in a packed-bed bioreactor system with internal aeration. Serum-supplemented and serum-free media comparison for MAb production. *J. Immunol. Methods* **176**, 67–77.
22. Pörtner, R., Lüdemann, I., Reher, K., Neumaier, M., and Märkl, H. (1998) Fixed-bed dialysis culture of a transfectoma cell line producing chimeric Fab-fragments with "nutrient-split"-feeding strategy. *Biotechnol. Techn.* **12**, 501–505.
23. Aunins, J. G., Bader, B., Caola, A., et al. (2003) Fluid mechanics, cell distribution, and environment in CellCube bioreactors. *Biotechnol. Prog.* **19**, 2–8.
24. Burger, C., Carrondo, M. J., Cruz, H., et al. (1999) An integrated strategy for the process development of a recombinant antibody-cytokine fusion protein expressed in BHK cells. *Appl. Microbiol. Biotechnol.* **52**, 345–353.
25. Ducommun, P., Ruffieux, P. A., Kadouri, A., von Stockar, U., and Marison, I. W. (2002) Monitoring of temperature effects on animal cell metabolism in a packed bed process. *Biotechnol. Bioeng.* **77**, 838–842.
26. Matsushita, T., Ketayama, M., Kamihata, K., and Funatsu, K. (1990) Anchorage-dependent mammalian cell culture using polyurethane foam as a new substratum for cell attachment. *Appl. Microbiol. Biotechnol.* **33**, 287–290.
27. Thelwall, P. E., Anthony, M. L., Fassnacht, D., Pörtner, R., and Brindle, K. M. (1998) Analysis of cell growth in a fixed bed bioreactor using magnetic resonance spectroscopy and imaging, in *New Developments and New Applications in Animal Cell Technology* (Merten, O. W., Perrin, P., and Griffiths J. B., eds.), Kluwer Academic Publishers, Dordrecht, The Netherlands, pp. 627–633.
28. Deng, J., Yang, Q., Cheng, X., Li, L., and Zhou, J. (1997) Production of rhEPO with a serum-free medium in the packed bed bioreactor. *Chin. J. Biotechnol.* **13**, 247–252.
29. Hu, Y. C., Kaufman, J., Cho, M. W., Golding, H., and Shiloach, J. (2000) Production of HIV-1 gp120 in packed-bed bioreactor using the vaccinia virus/T7 expression system. *Biotechnol. Prog.* **16**, 744–750.
30. Lu, J. T., Chung, Y. C., Chan, Z. R., and Hu, Y. C. (2005) A novel oscillating bioreactor BelloCell: implications for insect cell culture and recombinant protein production. *Biotechnol. Lett.* **27**, 1059–1065.
31. Kwon, M. S., Kato, T., Dojima, T., and Park, E. Y. (2005) Application of a radial-flow bioreactor in the production of beta1,3-N-acetylglucosaminyltransferase-2 fused with GFPuv using stably transformed insect cell lines. *Biotechnol. Appl. Biochem.* **42**, 41–46.
32. McTaggart, S. and Al-Rubeai, M. (2000) Effects of culture parameters on the production of retroviral vectors by a human packaging cell line. *Biotechnol. Prog.* **16**, 859–865.
33. Sendresen, C., Fassnacht, D., Benati, C., and Pörtner, R. (2001) Possible strategies for the production of viral vector: the role of the engineering design, in *Animal Cell Technology: From Target to Market*: (Lindner-Olsson, E., Chatzissavidou, N., and Lüllau, E., eds.), Kluwer Academic Publishers, Dordrecht, Netherlands, pp. 538–540.

34. Forestell, S. P., Dando, J. S., Chen, J., de Vries, P., Bohnlein, E., and Rigg, R. J. (1997) Novel retroviral packaging cell lines: complementary tropisms and improved vector production for efficient gene transfer. *Gene Ther.* **4**, 600–610.
35. Kataoka, K., Nagao, Y., Nukui, T., et al. (2005) An organic-inorganic hybrid scaffold for the culture of HepG2 cells in a bioreactor. *Biomaterials* **26**, 2509–2516.
36. Hongo, T., Kajikawa, M., Ishida, S., et al. (2005) Three-dimensional high-density culture of HepG2 cells in a 5-ml radial-flow bioreactor for construction of artificial liver. *J. Biosci. Bioeng.* **99**, 237–244.
37. Iwahori, T., Matsuno, N., Johjima, Y., et al. (2005) Radial flow bioreactor for the creation of bioartificial liver and kidney. *Transplant Proc.* **37**, 212–214.
38. Akiyama, I., Tomiyama, K., Sakaguchi, M., et al. (2004) Expression of CYP3A4 by an immortalized human hepatocyte line in a three-dimensional culture using a radial-flow bioreactor. *Int. J. Mol. Med.* **14**, 663–668.
39. Kawada, M., Nagamori, S., Aizaki, H., et al. (1998) Massive culture of human liver cancer cells in a newly developed radial flow bioreactor system: ultrafine structure of functionally enhanced hepatocarcinoma cell lines. *In Vitro Cell Dev. Biol. Anim.* **34**, 109–115.
40. Ijima, H., Nakazawa, K., Mizumoto, H., Matsushita, T., and Funatsu, K. (1998) Formation of a spherical multicellular aggregate (spheroid) of animal cells in the pores of polyurethane foam as a cell culture substratum and its application to a hybrid artificial liver. *J. Biomater. Sci. Polym.* **9**, 765–778.
41. Yamashita, Y., Shimada, M., Tsujita, E., et al. (2001) Polyurethane foam/spheroid culture system using human hepatoblastoma cell line (Hep G2) as a possible new hybrid artificial liver. *Cell Transplant.* **10**, 717–722.
42. Gion, T., Shimada, M., Shirabe, K., et al. (1999) Evaluation of a hybrid artificial liver using a polyurethane foam packed-Bed culture system in dogs. *J. Surg. Res.* **82**, 131–136.

18

Configuration of Bioreactors

Dirk E. Martens and Evert Jan van den End

Summary

For basic research on animal cells and in process development and troubleshooting, lab-scale bioreactors (0.2–20 L) are used extensively. In this chapter setting up the bioreactor system for different configurations is described, and two special setups (acceleration stat and reactors in series) are treated. In addition, on-line measurement and control of bioreactor parameters is described, with special attention to controller settings (PID) and on-line measurement of oxygen consumption and carbon dioxide production. Finally, methods for determining the oxygen transfer coefficient are described.

Key Words: Bioreactor; measurement and control; oxygen transfer; K_lA; animal cell; carbon dioxide production; PID tuning.

1. Introduction

Laboratory-scale bioreactors are used extensively for fundamental cell culture research and have a working volume varying from about 0.5 to 20 L. The goal of cell culture experiments in bioreactors of this scale is to obtain a better understanding of the process and cell physiology in general, find critical process parameters, and optimize the production process. Furthermore, lab-scale bioreactors are used for troubleshooting purposes to understand and solve problems in the large-scale production process.

Bioreactors can be operated in different modes, such as batch, fed-batch, continuous, and continuous perfusion, where each mode asks for its own special configuration of the bioreactor.

In a batch culture no medium is added or removed from the culture during growth, meaning all nutrients are present from the start. This means nutrient concentrations are high at the start, which leads to overflow metabolism and the formation of toxic waste products like ammonia and lactate.

From: *Methods in Biotechnology, Vol. 24: Animal Cell Biotechnology: Methods and Protocols, 2nd Ed.* Edited by: R. Pörtner © Humana Press Inc., Totowa, NJ

In a fed-batch culture a concentrated feed is added continuously to the bioreactor. The aim is to keep nutrient levels and the growth rate at a constant predefined value and have no accumulation of waste metabolites. The growth rate is set by the rate with which the feed is added and can be derived from:

$$\frac{d(V_r \cdot C_v)}{dt} = (\mu - \mu_d) \cdot V_r \cdot C_v \tag{1}$$

where V_r Is the reactor volume (m^3), C_v is the viable cell concentration (cells/m^3), μ is the specific growth rate (per s), μ_d is the specific death rate (per s), and t is the time (s). To prevent accumulation of substrates, the medium should be well balanced, meaning that the ratio of the nutrients in the feed should match the ratio at which they are consumed. Waste metabolism can be minimized by keeping the concentrations of glutamine and glucose low. To be able to do this, a good on-line measurement of critical parameters and a proper control strategy of the feed addition is needed. Note that to establish a constant growth rate, an exponential feed rate in time is required.

Chemostats are often used to do physiological studies, because a steady state is reached (*see* **Note 1**) at a certain dilution rate and this steady state can in theory be maintained as long as is desired (*see* **Notes 2** and **3**). The specific growth rate and specific death rate can be derived from a balance over the viable cells and the dead cells respectively:

$$V_r \frac{dC_v}{dt} = F_m(C_{vi} - C_v) + V_r \mu C_v - V_r \mu_d C_v \Rightarrow \mu = D + \mu_d \tag{2}$$

$$V_r \frac{dC_d}{dt} = F_m(C_{di} - C_d) + V_r \mu_d C_v - V_r k_1 C_d \Rightarrow \mu_d = (D + k_1)\frac{C_d}{C_v} \tag{3}$$

where C_d is the dead cell concentration (cells/m^3), C_{vi} and C_{di} are, respectively, the viable and dead cell concentration in the incoming medium flow (cells/m^3), which are in this case zero, F_m is the medium flow (m^3/s), and k_1 is the first-order lysis rate constant of dead cells. Usually cell lysis is negligible. However, in some situations this is not the case, such as at low dilution rates where the viability of the culture is low and the residence time of the dead cells in the reactor is long or when shear forces are high. In this case it is assumed that viable cells first die and then lyse. In case of cell lysis, the specific death rate can be calculated from the release of the intracellular enzyme lactate dehydrogenase (LDH) ***(1)***:

$$\begin{aligned} &V_r \frac{dC_{LDH}}{dt} = F_m(C_{LDHi} - C_{LDH}) + V_r \mu_d C_v C_{LDHcell} - V_r k_{LDH} C_{LDH} \Rightarrow \\ &\mu_d = (D + k_{LDH})\frac{C_{LDH}}{C_v \cdot C_{LDHcell}} \end{aligned} \tag{4}$$

where C_{LDH} is the LDH activity in the medium (U/m^3), C_{LDHi} is the LDH activity in the incoming medium (U/m^3), which is zero in this case, k_{LDH} is the first-order inactivation constant of LDH in the spent medium at 37°C (per s), and $C_{LDHcell}$ is the intracellular LDH activity (U/cell) Since the intracellular LDH concentration may vary depending on culture conditions, on each sample point both the supernatant LDH activity and the LDH activity per viable cell has to be determined.

In a continuous perfusion system, viable and dead cells are retained in the system using a cell-separation device. Different retention systems are available. For a review, *see* Voisard et al. ***(2)*** and Woodside et al. ***(3)***. The medium flow going through the separation device is called the perfusion flow. It appears not possible to keep all cells in the reactor, and thus a small bleed is necessary of cell containing medium. The feed flow of fresh medium balances the sum of the perfusion and bleed. The growth rate can again be derived from the viable cell balance:

$$\frac{dC_v}{dt} = \mu \cdot C_v - B \cdot C_v - P \cdot C_{vp} - \mu_d \cdot C_v \Rightarrow \mu = B + \mu_d + P\frac{C_{vp}}{C_v} \quad (5)$$

where **B** is the specific bleed rate (per s), P is the specific perfusion rate (per s), and C_{vp} is the concentration of viable cells that passes through the separation device (cells/m^3). The last term is negligible at high separation efficiencies. Bleed rates are usually low, meaning that cells reside in the reactor for long times. In addition cell concentrations are high and some form of sparging is required to supply the system with sufficient oxygen. Because of the long residence times and shear through sparging, cell lysis is not negligible and the specific death rate should be calculated from, for example, the LDH balance ***(4)***:

$$\frac{dC_{LDH}}{dt} = \mu_d \cdot C_v \cdot C_{LDHcell} - F \cdot C_{LDH} - k_{LDH} \cdot C_{LDH} \Rightarrow \mu_d = (F + k_{LDH})\frac{C_{LDH}}{C_v \cdot C_{LDHcell}} \quad (6)$$

where F is the specific feed rate (per s), which is the sum of the specific perfusion and the specific bleed rate.

Proper control of fermentation parameters like pH, dissolved oxygen (DO), and temperature is essential. This is important not only to provide the proper conditions for cell culture, but also for the accurate on-line prediction of, for example, the oxygen uptake rate. The controllers for these parameters are generally of the PID type, where P stands for proportional, I for integrating, and D for the derivative action of the controller. The output of the controller is based on the difference between the actual value and the set point of the controller, the error ε, and the P, I, and D parameters ***(5)***. The error is defined as:

$$\varepsilon = sp - mv \tag{7}$$

where ε is the error, *mv* is the measured value, and *sp* is the set point. The P action works as a gain on the error and is used to amplify the error signal.

$$CO = P \times \varepsilon \tag{8}$$

where *CO* is the controller output and *P* is the proportional gain. The P value determines how vigorous the controller responds to the control error. Using only the P action for control will result in a process that never reaches its set point meaning an offset remains between the measured value and the set point.

The I action is used to compensate for the offset that originates from the P action. The I value can be seen as a time-span used by the controller to integrate the control error in time. By doing this, the controller incorporates errors made in the past.

$$CO = P \times \frac{1}{I}\int_0^t \varepsilon dt \tag{9}$$

where *I* is the integration time constant (s). When the controller does not reach its set point, as occurs with a P action only, this integrated error can change the controller output enough to get the controlled parameter at the set point. The I action can make the controller action slow with respect to sudden changes in the process.

The D action of the controller is meant to give the controller a shorter response time to sudden changes in the process. The D value acts as a gain factor. A disturbance could, for instance, be a change of set point.

$$CO = P \times D\frac{d\varepsilon}{dt} \tag{10}$$

where *D* is the derivation time constant (s). In Eq. 11, all the terms of a PID controller are incorporated to give a complete mathematical description of the controller:

$$CO = P \times \left(\varepsilon + \frac{1}{I}\int_0^t \varepsilon dt + D\frac{d\varepsilon}{dt}\right) \tag{11}$$

A PID controller can be used as P, PI, PD, and PID controller. For the three major control loops involved in culturing cells, the preferred type of control for pH is often only P control and sometimes PI; for control of DO and T it is PI control (*see* **Note 4**).

In **Subheading 2.**, the components of the bioreactor system are described, including equipment to measure on-line. In **Subheading 3.**, preparation of the

Table 1
Suppliers of Lab-Scale Fermentation Equipment

Applikon Biotechnology B.V.	De Brauwweg 13, P.O. Box 149, 3100 AC Schiedam, The Netherlands
Bioengineering	Sagenrainstraße 7, CH-8636 Wald, Switzerland
Dasgip	Rudolf Schulten Strasse 5, 52428 Jülich, Germany
New Brunswick Scientific Co.	PO Box 4005, Edison, NJ
Pierre Guerin Sas Devision Biolafitte	Grand-Rue 179, B.P.12 Mauze-Sur-Le-Mignon, France
Sartorius BBI Systems GMBH	P.O. Box 1363, 34203 Melsungen, Germany

bioreactor for the cultivation of animal cells is first described. Next, two special reactor setups, the acceleration stat (A-stat) and reactors in series setup, are described. Then, methods to find the proper PID settings for the most important controllers are presented. Finally, methods to determine on-line the oxygen consumption rate and carbon evolution rate are given, including different methods to determine the oxygen transfer coefficient.

2. Materials

2.1. Components of a Lab-Scale Bioreactor

1. Lab-scale bioreactors are usually made out of glass with a stainless steel top-plate. For suppliers of lab-scale bioreactor equipment, *see* **Table 1**.
2. For mixing a marine impeller is commonly used, which causes a gentle vertical circulating flow pattern and is well suited for enhancing oxygen transfer through the surface and mixing with a minimal amount of shear (*see* **Note 5**).
3. The reactor top-plate is clamped to the reactor with screws. Between separate solid parts of the reactor, greased rubber rings are present to make sure the connection between the parts is airtight and to compensate for differences in expansion of the individual parts during sterilization. The top-plate has a number of holes, which can be used for putting probes and pipes into the reactor. To make sure the connections are airtight, rubber or Viton rings are used, which are pressed tightly between the top-plate and the probe using nipples. Holes in the top-plate that are not used are closed with blind plugs.
4. Standard a temperature, pH and DO probe are present.
5. For temperature control, a water bath or thermo-circulator is used (*see* **Note 6**) and sometimes an electrical heating blanket.
6. The pH in mammalian cell culture systems is usually controlled by a combination of carbon dioxide in the gas phase and addition of a solution of 0.1–0.5 *M* NaOH or $NaHCO_3$ (*see* **Note 7**).
7. The DO concentration in mammalian cell culture is usually controlled by adjusting the percentage of oxygen and nitrogen in the headspace. Gasses are supplied from gas cylinders.

8. To control the rate of oxygen, nitrogen, and carbon dioxide supply through the headspace or sparger, mass-flow controllers are generally used. The mass-flow controllers must be calibrated for the gas for which they are used.
9. For high cell density and/or high volume-to-surface ratios, surface aeration is not sufficient and sparging is necessary. Sparging can be done with spargers generating normal bubbles (1–6 mm diameter) or microbubbles (<1 mm diameter).
10. The outlet gas should first be transported through a condenser to prevent excessive evaporation of the liquid medium.
11. Controllers are required to keep pH, DO, and temperature constant. Usually the controllers are integrated into one device. For some applications the control device also controls pump rates and the working volume of the reactor.

2.2. On-Line Measurement Equipment

1. The oxygen concentrations in the gas phase can be measured on-line using a mass spectrometer (MS). The accuracy of the MS is high, and it can determine the CO_2 concentration at the same time. However, a MS is quite expensive both to buy and to maintain. An alternative is the use of the cheaper paramagnetic gas analyzers. However, they are in general less sensitive and can only measure a single component. Both devices require precise calibration (*see* **Note 8**).
2. The concentration of CO_2 can be measured in the off-gas using a mass spectrophotometer or the cheaper infrared analyzers (*see* **Note 8**), which are less accurate.
3. On-line measurement of viable biomass *(6)* is of great value for controlling cell culture processes. An important difference between application of biomass monitors in microbial and animal cell culture is that for animal cell culture the measurement should discriminate between viable and dead cells. Direct and indirect measurement methods are being used *(7)*. Direct measurements are based on either optical measuring principles or dielectric properties of the cells. Indirect methods are based on measuring an overall metabolic activity of the cells. Examples of these are overall uptake of oxygen or glucose and heat generation. They are applied in situations where the cell concentration cannot be determined by counting, as, for example, in hollow fiber reactors or when porous carriers are used.
   a. Optical density (OD) sensors use the transmission or scatter of light or both to determine the biomass concentration. Since the amount of dry weight per OD unit may vary depending on cell line, medium, and process conditions used, proper calibration is of utmost importance. However, OD properties of cells may even change during a process. Usually discrimination between dead and viable cells is poor and the measurement can be disturbed by air bubbles.
   b. Probes based on dielectric properties are based on the fact that cells are liquid elements surrounded by a membrane that can be polarized *(8)*. Thus, depending on the cell concentration, the capacitance of the liquid varies. This method is also able to measure median cell diameter and viability.
   c. The method using heat release is based on the fact that overall cell metabolism is exothermic *(9)*. By measuring the total amount of heat released in the bioreactor, one can calculate the concentration of metabolically active cells. However,

this requires an accurate knowledge of all heat losses and gains in the system and their variation in time, or, in other words, a well-defined system in terms of heat transfer is required. As cultivation systems become more complex and/or larger, this may become difficult. On the other hand, at larger scales the heat signal is easier to measure.

d. Other indirect methods are measurement of consumption or production rates of metabolites, such as oxygen ***(10)***, carbon dioxide, and glucose ***(11)***. Usually indirect measurements are combined with a software sensor that estimates biomass concentrations and growth rate based on one or more of the above-mentioned parameters.

4. Besides DO and pH, CO_2 can also be measured in the liquid phase using an electrode.
5. With flow injection analysis, automatic sterile sampling is done and the sample is analyzed directly. A variety of analytical methods can be used, which makes it a very flexible method that is able to measure on-line a whole range of variables. The main drawback is the unreliability of the system and lack of systems for proper on-line fault detection and correction. For more information *see* **refs. *12–14***.
6. Infrared spectroscopy is based on the absorption of electromagnetic energy in the region from 500 to 2500 nm, caused by vibrational transitions at the molecular level ***(15–21)***. Advantages of this technique are that it is noninvasive, measures several analytes simultaneously, can be in situ sterilized, is inexpensive, and has low maintenance requirements and robust instrumentation. However, spectral bands are broad, with severe overlaps and poor baseline resolution, making spectral interpretation difficult. In addition, a typical compound absorbs at more wavelengths and absorbance at each wavelength may have contributions from more compounds. Spectral data can be used for quantification of particular compounds of interest. This requires the construction of calibration models correlating spectral data to chemical data obtained by reference analytical methods. In order to extract the relevant information from the spectra and develop the required calibration models, multivariate regression techniques must be used. Alternatively, the information of the complete spectrum can also be used without quantification of individual compounds. The development of the spectrum in time is then regarded as a kind of fingerprint of the process. In order to do this, a number of reference processes have to be done and spectra of these reference processes then serve as a standard.

3. Methods

3.1. Preparation of the Bioreactor

1. Before use, check that the glass vessel is not damaged. In order for the reactor to be airtight, check if all rubber rings are still in good shape and, if necessary, grease them.
2. If required, a sparger can be placed below the impeller.
3. Wash all glass ware thoroughly with soap and then rinse with double distilled water.
4. Treat glassware with 0.1 *M* NaOH (in double distilled water) overnight to remove soap residues and then rinse three times with bidest.

5. To prevent attachment of cells and especially microcarriers to the glass surface of the reactor vessel, it is important to treat the surface with silicon oil. The inside of the reactor is wetted with a small volume of siliconizing fluid and then allowed to dry. Next the vessel is rinsed three times with double distilled water.
6. The pH probe is calibrated using standard buffer solutions of pH 4 and 7 (*see* **Note 9**).
7. Calibrate the DO electrode by putting it into deaerated water (DO = 0%) and water saturated with air (DO = 100%), respectively (*see* **Note 10**).
8. Check the response time of the DO electrode by transferring the electrode from a deaerated liquid to an air-saturated liquid instantaneously. The response time is the time needed to reach a DO concentration of about 67%, or (1/e) of the saturation level (*see* **Note 11**).
9. Calibrate the temperature probe (Pt100) in melting ice (0°C) and boiling water (100°C). Usually this only needs to be done once and not for each new run.
10. Silicone tubing should be connected to the inlets on the top plate. If a welding device is used for making connections, it is sufficient to close the tubing with clamps. Otherwise, for all tubing through which liquid has to be transported, connectors have to be used, which allow for making sterile connections (*see* **Note 12**). Tie raps should be used to secure the tube connections. Tubing for inlet and outlet gas should be closed with a 0.2-μ air filter.
11. Fill the reactor with double distilled water so that at least the pH probe is submerged.
12. Check whether the reactor is airtight. Close all connections and connect a pressure meter to the headspace. Bring a slight overpressure in the headspace, and check whether the overpressure can be maintained (*see* **Notes 13** and **14**).
13. For sterilization of the reactor, first the screws of the top-plate should be slightly loosened to prevent too much tension between the steel and the glass. All tubes that are submerged in the liquid should be tightly closed. There should be preferably at least two connections between the headspace and outer environment opened to allow for pressure equilibration during sterilization. Next the reactor can be sterilized (*see* **Notes 15** and **16**). After sterilization, let the autoclave cool down to 80°C and then tighten the screws on the top-plate before taking the reactor from the autoclave.
14. Next all necessary probes and vessels can be connected. Water can be pumped out and medium can be pumped in (*see* **Note 17**). Connect pressurized air to the gas inlet for the 100% calibration of the DO probe.
15. After sterilization the pH and DO probe will have drifted. The pH probe should be recalibrated using a one-point calibration. The DO probe can be recalibrated by deaerating and saturating the medium inside with nitrogen and air, respectively (*see* **Note 18**).
16. Activate the controls and let the medium equilibrate until the desired set-points for temperature pH and DO are reached.
17. For fed-batch systems an extra port is required for the addition of the feed. Furthermore, to add the concentrated feed a peristaltic pump is needed that can operate at low pump rates.
18. For chemostats there is a constant input of fresh medium and removal of spent medium with cells (*see* **Note 19**) at the same rate in order to keep the volume

constant. Thus, an extra port for removing spent medium has to be present. There are two ways to attain a constant dilution rate and reactor volume. First, the medium inflow rate can be controlled at a certain value, while spent medium is removed using a tube that removes medium from the surface (chemostat tube). Second, the removal rate of spent medium can be set to a certain value, and medium is added as soon as the volume drops below a certain value. This can be done by installing a level sensor at the desired liquid height in the reactor or by measuring and controlling the weight of the complete bioreactor system.

19. For perfusion cultures, usually the perfusion and bleed flow are controlled at a predefined value and the feed flow is next controlled by a level sensor or by reactor weight.
20. In perfusion culture during the growth phase, the perfusion rate is gradually increased, keeping the perfusion rate per cell constant.
21. Liquid flows in general can be controlled using pumps. Most often these are peristaltic pumps. These pumps must be calibrated in advance with the tubing that will be used (*see* **Note 20**). An alternative way to control the medium flow is by measuring the change in weight of the medium container with time. However, care should be taken that the weight is not influenced by the position of tubing or condensation of water if the container is in the refrigerator.

3.1.1. Acceleration Stat

Often one is interested in the relationship between growth rate, productivity, nutrient concentrations, and metabolism. This can be studied in a series of chemostats at different dilution rates. However, for a single chemostat to reach steady state, it can take up to 30 d or longer. Thus, to study a range of growth rates is time consuming. To speed up this type of study, Paalme et al. *(22)* invented the A-stat. In the A-stat the dilution rate is gradually increased starting from a steady state at low dilution rate. The increase in dilution rate is so slow that the system remains in steady state. In this way, information is obtained on a whole range of dilution rates in about the same amount of time as about five chemostats. In addition, for animal cells it is usually very difficult to reach steady state at low dilution rates, which probably is caused by the low viability at low dilution rates. Therefore, for animal cells it is better to start at a steady state at a high dilution rate and then slowly decrease the dilution rate. A major problem with this cultivation method is to choose the right acceleration rate. The acceleration should not be too fast, since this will lead to a non-steady-state situation. On the other hand, slow acceleration rates lead to unnecessary long cultivation times. Values of 0.001–0.010 of the maximum growth rate seem to give the optimal balance between steady-state operation and length of the run *(23–25)*. To check whether the A-stat is still in steady state, the acceleration can be stopped at discrete points. All culture variables should then immediately remain constant. Since the A-stat measures at a continuous range of dilution rates and growth rates, dilution rates at which sudden switches in metabolism occur can be exactly pinpointed. However, if the switch causes sudden changes in concentrations of metabolites, one should realize that the culture is no longer in steady state. To do an A-stat a precise control of the medium flow is necessary. This can be achieved by on-line measuring the change in weight of the medium vessel, based on which the feed pump is controlled. It can also be done by directly controlling a calibrated pump. However, culture time is still long, and wearing of the pump tube may cause deviations from the calibration curve. Therefore, regular checks of the medium flow are required. For calculation of specific rates, the same relations as for a chemostat can be used with the exception that the accumulation term is no longer zero in all cases and thus must be estimated.

3.1.2. Reactors in Series

An important characteristic of mammalian cells is that they rapidly go into apoptosis and die, for instance, due to nutrient depletion and accumulation of toxic waste metabolites. A consequence of this is that culture viability rapidly drops at the end of batch and fed-batch cultures and is low in perfusion cultures. Preventing or delaying apoptotic cell death can significantly enhance volumetric productivity. One way to study apoptotic cell death at steady-state conditions is to use two chemostats in series ***(26)***. In the first chemostat nutrients are depleted and toxic waste metabolites are accumulated. The outflow of the first reactor is pumped to the second reactor. Because of the depletion of nutrients and accumulation of waste metabolites, cells massively go into apoptosis in this second reactor. In fact, conditions in the second reactor much resemble the conditions at the end of a (fed-) batch culture. Despite massive cell death in the second reactor, a steady state is still reached because of the continuous inflow of new viable cells from the first reactor. Therefore, this system makes it possible to study apoptosis and ways to inhibit apoptosis under steady-state conditions. The calculation of growth and death rates in the second reactor is comparable to that in one chemostat with the exception that the incoming concentration of dead and viable cells is no longer zero but equal to the steady-state concentration in the first reactor (*see* **Note 19**).

3.2. Measurement and Control

*3.2.1. PID Settings **(5)***

Sometimes it is necessary to readjust the P, I, and D values if the control of the process is not satisfactory, as in situations B and C in **Fig. 1**. Finding the optimal values for P, I and D is difficult, but with some understanding of how the controller works and with the practical methods stated below, adequate values can be found. (*see* **Notes 21** and **22**) Usually the manufacturer of the control equipment can give some starting values for the P and I values that can be optimized. In essence three methods are available to find the correct PID settings: rules of thumb, the method of Ziegler and Nichols, and autotuning procedures (not always available in the control device).

All three methods involve a disturbance of the process by either changing the set point or putting the controller in manual mode and changing the controller output with a step (*see* **Note 23**). The latter method gives the best results because the controller, when it is working, has knowledge of the past by means of the I action. Obviously the measured value of the controller that is going to be modified needs to be recorded.

3.2.1.1. Rules of Thumb

1. When possible, a disturbance is made by changing the set point and studying the behavior of the controller. A good time to do this would be before inoculation of the bioreactor or at the end of the run, as long as culture medium is used.
2. If the response is like line B in **Fig. 1**, it is satisfactory as long as the overshoot of the measured value does not have a negative effect on cell physiology. In case of, for instance, pH control, an overshoot of 0.1 pH units is usually acceptable. Besides the overshoot, the wavering of the measured value around the set point should weaken in two to three oscillations. To weaken the overshoot and the oscillating behavior, the P value should be decreased and the I value increased. Control is in this case relatively fast because the measured value is approaching the set point rapidly.
3. If the measured value has an offset from the set point, the I value should be altered somewhat. Adjustments should be done in small steps, and time should be given to the controller to stabilize.
4. If the measured value behaves like line C in **Fig. 1**, then the controller is oscillating around the set point. This oscillation is caused by too much amplification of the error. In this case the P value should be decreased until the response is stable. Increasing the I value will further reduce the oscillatory behavior (*see* **Note 24**).
5. Line A shows the response of a controller with a low P value and a high I value. The response is slow but acceptable and will show stable behavior during the fermentation. Sample taking will not have much influence on the output of the controller. Increasing the P value and decreasing the I value can speed up the controller, and its behavior can be shifted toward the behavior of line B. The ideal response of the controller should look like a response that lies in between line A and line B.

3.2.1.2. The Method of Ziegler and Nichols

This method involves the recording of the change of the measured value after a step change of the controller output.

1. Put the controller in manual mode and let the measured value stabilize.
2. Make a step change in the output of the controller. The moment that the step change is done is the beginning of the measurement.

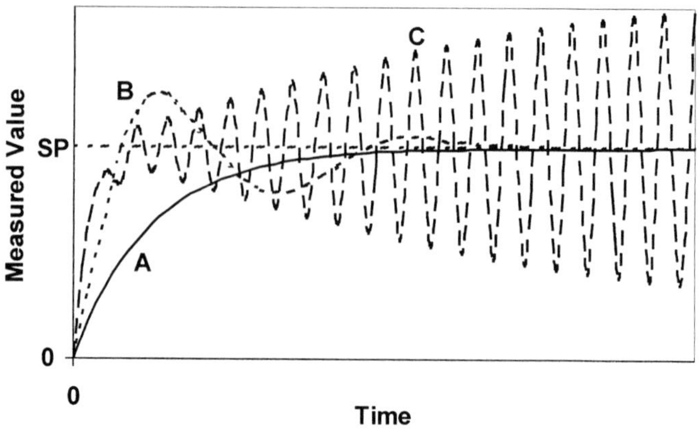

Fig. 1. Response from a controlled process to a change of the set point (SP) at $t = 0$. The different lines show the behavior of the process with different settings of the P and I values. Line A shows an acceptable response. Line B is for a controller that is close to becoming unstable, but the response is satisfactory. Line C is for a controller that is instable. The response of the controller to the change in set point is too strong.

3. **Figure 2** shows the response of the DO concentration in a headspace-aerated bioreactor. At $t = 0$ the liquid phase was in equilibrium with nitrogen gas; at that moment the gas supply was switched from nitrogen to air. In fact, this is a $K_{ol}A$ measurement, which is explained further on. The change of the DO concentration in time is described by an S-shaped curve. In the first part of the curve until the inflection point (IP), the nitrogen in the headspace is flushed out by the air. After the IP the headspace is considered to exist of air only. If the DO electrode is calibrated properly, the electrode will read 100% of saturation when the headspace is in equilibrium with the medium.
4. The method of Ziegler and Nichols uses this response curve to determine tv and $t1$. tv is the time delay of the response of the measured value after a change of the controller output, and $t1$ is the time at which the tangent line crosses the maximum measured value. The tangent line is drawn through the IP of the curve.
5. When the values for tv and $t1$ are known values for P, I and D (if applicable) are calculated with the help of **Table 2**.

The determination of tv should be done accurately since this value has a large influence on the calculation of the P, I, and D values. These values will need some readjustment when they are set in the controller (*see* **Note 21**).

3.2.1.3. Use Autotuning, If Available

Some controllers are equipped with an automatic "tuning" procedure. This procedure generally implies disturbances inflicted by the controller (*see* **Note 25**). The controller measures the response from the process and determines the

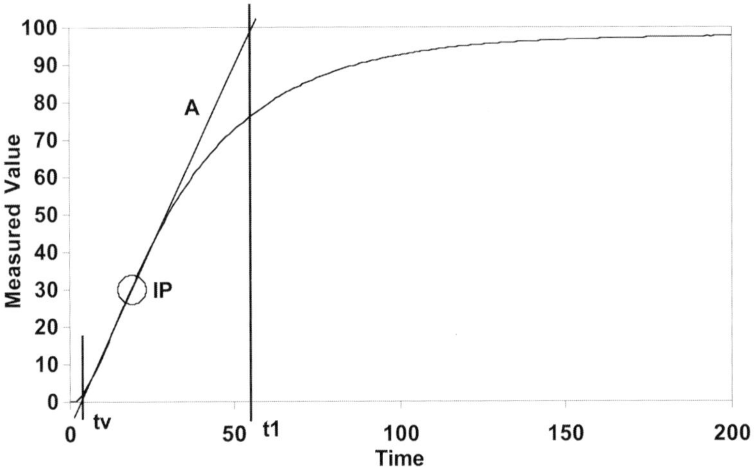

Fig. 2. Process response to a step change of the controller output from 0 to 100% at $t = 0$. In this case the measured value reaches a maximum of 100%. *tv* is the delay time, *t*1 is the time at which the tangent line crosses the maximum measured value, *A* is the tangent line, and IP is the inflection point.

best values for P, I, and D. The values for P, I, and D found by the controller form a good starting point for manual "tuning" of the controller. The D action is usually not wanted and switched off. When the D action of the controller is switched off, the values for P and I should be adjusted. The P value should be adjusted to 75% of its original value. The I value should be increased to 165% of its original value. This can be seen in **Table 2**. Thereafter it is best to observe how the controller behaves during the process and adjust the P and I values manually.

3.2.2. On-Line Measurements

3.2.2.1. Oxygen Consumption

In essence, three methods exist to measure the oxygen consumption rate on-line:

1. From an oxygen balance over the reactor as a whole *(27)*: By measuring the amount of oxygen entering and leaving the reactor the oxygen uptake rate, OUR ($mol/m^3/s$) can be calculated from the following equation:

$$OUR = \frac{F_{gi}C_{ogi} + F_{gs}C_{ogs} - F_{go}C_{ogo}}{V_r} \qquad (12)$$

where F_{gi} and F_{go} are the gas flows in and out of the headspace (m^3/s), F_{gs} is the gas flow through the sparger (m^3/s), C_{ogi} and C_{ogo} are, respectively, the oxygen

Table 2
Equations for Calculation of Controller Settings

Control type	P value	I value	D value
P	$\frac{t1-tv}{tv}$		
PI	$0.9\times\frac{t1-tv}{tv}$	$3.3\times tv$	
PID	$1.2\times\frac{t1-tv}{tv}$	$2\times tv$	$0.5\times tv$

concentration in the gas flow entering and leaving the headspace (mol/m^3), C_{ogs} is the oxygen concentration in the gas entering through the sparger (mol/m^3), and V_r is the reactor volume (m^3). For small-scale bioreactors the difference in ingoing and outgoing oxygen concentration is usually very small, which means they have to be measured with great accuracy in order to determine the OUR. The difference in oxygen concentration can in principle be increased by decreasing the gas flow rate. However, this may make control of the DO concentration more difficult and can lead to accumulation of CO_2.

2. Decrease of oxygen tension (*see* **Note 26**). During the fermentation oxygen control can be stopped and the gas phase can be flushed with gas being in equilibrium with 50% DO. The DO will now decrease according to:

$$\frac{dC_{ol}}{dt} = k_{ol}A(\frac{C_{og}}{m} - C_{ol}) - q_o \cdot C_v \tag{13}$$

where C_{ol} and C_{og} are the oxygen concentration in the liquid and gas phase, respectively (mol/m^3), k_{ol} is the mass transfer coefficient for oxygen (m/s), A is the specific surface area (m), m is the partition coefficient of oxygen for medium (mol/mol), and q_o is the specific oxygen-consumption rate (mol/cell/s). By decreasing the gas concentration in the gas and, if applicable, stopping sparging, the $k_{ol}A$ term will be negligible to the consumption term. And the oxygen consumption can be calculated in a straightforward manner. However, at low cell densities this will generally not be the case and the $k_{ol}A$ term will have to be included, making this approach rather inaccurate. Another option is to take a sample and put it in a biological oxygen meter. This is a small vessel that is completely filled so no gas exchange with the environment can occur. However, this requires sampling, resulting in a short change of conditions such as temperature, which may affect oxygen consumption by the cells ***(27)***.

3. Transfer method ***(28–30)***: This method is based on the fact that the oxygen uptake rate equals the oxygen transfer rate (OTR) (mol/m^3/s), which in turn is given by:

$$OUR = OTR = k_{ol}A(\frac{C_{og}}{m} - C_{ol}) \tag{14}$$

The concentration of oxygen in the gas can be calculated from a balance over the gas phase:

$$V_g \frac{dC_{og}}{dt} = F_{gi}C_{ogi} - F_{go}C_{og} - V_r k_{ol} A(\frac{C_{og}}{m} - C_{ol}) \quad (15)$$

where V_g is the volume of the gas phase (m^3) and the accumulation term on the left is usually negligible. (*see* **Note 27**). It is assumed that the gas phase is ideally mixed, or $C_{og} = C_{ogo}$.

3.2.2.2. Carbon Dioxide

The carbon dioxide measurement is important for closing the carbon balance. Furthermore, too high CO_2 concentrations have an adverse effect on growth and production ***(31,32)***. Finally, the respiration quotient, RQ—the carbon dioxide evolution rate (CER) divided by the oxygen uptake rate—gives information on the metabolic state of the cell and the substrates used ***(33)***. In contrast with oxygen, the CO_2 production rate cannot be directly calculated from the gas flow and gas CO_2 concentration. Carbon dioxide readily dissolves in medium, where it is part of the following equilibrium:

$$CO_2(g) \leftrightarrow CO_2(L) + H_2O \Leftrightarrow H_2CO_3 \Leftrightarrow H^+ + HCO_3^- \Leftrightarrow 2H^+ + CO_3^{2-}$$

Because of this the CO_2 in the medium cannot be ignored. For batch and fed-batch systems, accumulation in the medium may occur, while for perfusion cultures and chemostats significant amounts of CO_2 may be transported with the liquid. Finally, mammalian cell culture media are often buffered using a bicarbonate buffer. This means that part of the CO_2 in the gas phase may be coming from the medium and not from the cells. Thus, to estimate the carbon evolution rates, the gas phase/liquid phase balance together with the pH-dependent chemical equilibrium must be included ***(34–36)***:

$$V_r \frac{dC_A}{dt} = V_r \cdot CER - k_{cl} A(x^0_{CO_2}(t)\frac{P}{H^{CO_2}} - C_{CO_2})V_r - F_m[(1+\frac{K_l}{10^{-pH}})C_{CO_2}(t)) - C^i_A] \quad (16)$$

$$\frac{p}{RT}\frac{dx^0_{CO_2}(t)V_g}{dt} = -(x^0_{CO_2}(t)\cdot F^o_g - x^i_{CO_2}(t)\cdot F^i_g) + k_{cl}A(x^0_{CO_2}(t)\frac{P}{H^{CO_2}} - C_{CO_2})V_r \quad (17)$$

where C_A equals the sum of all CO_2-containing compounds ($CO_2 + H_2CO_3 + HCO_3^- + CO_3^{2-}$) (mol/m^3), x^o_{CO2} and x^i_{CO2} are the outgoing and incoming molar fractions of CO_2 (–), respectively, K_l is the equilibrium constant for the reaction of CO_2 to HCO_3^- and H^+ (mol/m^3), $k_{cl}A$ is the transfer coefficient for CO_2 (pr s), H^{CO2} is the Henry constant for CO_2 (Pa.m^3/mol), R is the molar gas constant (J/mol/K), T is the temperature in Kelvin, and C_{CO2} is the

CO_2 concentration in the liquid (mol/m^3). From the gas balance, C_{CO2} can be calculated. Using the liquid balance and this value, the CER can next be calculated. For the total amount of carbon dioxide in the liquid, C_A(CO_2 + H_2CO_3 + HCO_3^- + CO_3^{2-}), the amounts of H_2CO_3 and CO_3^{2-} at pH 7–7.2 are less than 1% of C_A and are disregarded. It is assumed that the incoming concentration C_{Ai} is known and constant. However, because of stripping of CO_2 during filter sterilization, storage, and pumping, this may not be the case. This can be solved experimentally or corrected for mathematically. In case CO_2 is also measured in the liquid, the transfer between liquid and gas need not be calculated and the CER can be directly calculated from the gas and liquid balance.

3.3. Determination of $k_{ol}A$

In order to be able to measure oxygen uptake and carbon dioxide production, the $k_{ol}A$ (*see* **Note 28**) must be determined. Again, different methods exist to measure this coefficient.

3.3.1. Dynamic Method (27,37)

1. The medium in the reactor is deaerated using nitrogen gas.
2. Next, air is blown over the surface and through the sparger and the change in oxygen concentration is followed in time (*see* **Notes 29** and **30**).
3. The change in time is given by:

$$\frac{dC_{ol}}{dt} = k_{ol}A(\frac{C_{og}}{m} - C_{ol}) \tag{18}$$

Integrating this equation gives:

$$C_{ol}(t) = C_{ol}(\infty) - (C_{ol}(\infty) - C_{ol}(0)) \cdot e^{-k_{ol}At} \tag{19}$$

where $C_{ol}(0)$ and $C_{ol}(\infty)$ are the DO concentrations (mol/m^3) at times zero and infinity, respectively. Using curve-fitting software, this equation can be fitted on the measured data with $C_{ol}(\infty)$, $C_{ol}(0)$, and $k_{ol}A$ as parameters. The equations assume a constant oxygen concentration in the gas phase (*see* **Note 31**).

3.3.2. Steady-State Method (28,38)

1. Connect a second deaeration reactor to the fermentation vessel, which is continuously sparged with nitrogen to keep it at zero DO concentration.
2. Oxygen-containing medium from the fermentation vessel is pumped to the deaeration vessel and deaerated medium is pumped back to the fermentation vessel at the same rate (*see* **Note 32**). The $k_{ol}A$ can now be calculated from a mass balance over the medium in the fermentation vessel:

$$V_r \frac{dC_{ol}}{dt} = F_p(C_{ol2} - C_{ol}) + k_{ol}A(\frac{C_{og}}{m} - C_{ol}) \tag{20}$$

where F_p is the liquid flow between both reactors (m^3/s) and C_{ol2} is the oxygen concentration in the second reactor (mol/m^3), which is usually zero.

3. After some time, steady state is reached and the accumulation term on the left can be disregarded. Also, this method becomes difficult to use at high $k_{ol}A$ values, because high liquid pumping rates are required to create sufficient driving force to accurately determine the $k_{ol}A$. These high liquid pumping rates and concomitantly short liquid residence times may affect the $k_{ol}A$ value itself by interfering with the fluid dynamics in the bioreactor.

3.3.3. Sulfite Method *(39)*

This method is one of the classical steady-state methods for volumetric oxygen transfer coefficient ($k_{ol}A$) determination using sodium sulfite oxidation. It is based on steady-state absorption of oxygen from the gas phase with simultaneous oxygen removal from the liquid phase through a chemical reaction. The reaction between sulfite ions and oxygen catalyzed by Cu^{2+} or Co^{2+} is instantaneous:

$$Na_2SO_3 + 0.5O_2 \xrightarrow{Cu^{2+}} Na_2SO_4$$

1. A known amount of sodium sulfite is added to air-saturated medium in the reactor. As a consequence, the DO concentration rapidly drops to zero.
2. If all sulfite has reacted with oxygen, the DO concentration rises again.
3. The $k_{ol}A$ can now be calculated according to:

$$k_{ol}A = \frac{0.5 \cdot n_s}{(\frac{C_{og}}{m} - C_{ol}) \cdot \Delta t} \quad (21)$$

where n_s is the amount of sulfite added (mol/m^3) For Δt the time between the points where the DO passes through the value of 50% air saturation is taken. Although it is well known that the values obtained are higher than those obtained from dynamic or gas balance methods, the sulfite oxidation method has been widely used for determining mass transfer characteristics in bioreactors and can also be used for high $k_{ol}A$ values.

4. Notes

1. After three volume replacements 97% of the volume is replaced (99% after five volume replacements). This is usually the minimum time necessary to reach steady state. Animal cells often require more time to reach steady state. Steady state is usually assumed if a number of parameters such as biomass, consumption rate of oxygen, glucose, glutamine, and lactate, are constant for five consecutive days.
2. Because continuous selection of cells occurs in chemostats, the cell population may change in time. Therefore, chemostats should preferably be run in parallel and not consecutive in one run. If done in one run, one should always return to the starting conditions to check whether the cells have remained the same *(40,41)*.

3. One should be aware that the steady-state situation reached may depend on the way the steady state is reached. In other words, multiple steady-state situations may occur at a certain dilution rate ***(42,43)***.
4. Usually PI control is sufficient, because biotechnological processes are slow and stable and a D action is not necessary. D makes the controller output noisy because of quick responses to small errors. Calculations done on the basis of the controller outputs will be difficult and less accurate. Also, the control equipment, such as pumps and valves, wears out faster.
5. Often an extra turbine impeller is present in the headspace for mixing the gas phase.
6. Water baths have to be regularly refilled with water. Thermocirculators are closed systems and do not have this disadvantage.
7. Higher concentrations of base may result in cell death because of local high osmolarity and pH value at the point of addition.
8. The off-gas must be dried before measuring the oxygen and carbon dioxide concentration, because water changes the molar fraction of the other gas components. If the water fraction is known, one can correct for it and drying is not required. However, for infrared meters water vapor interferes with the measurement and drying is an absolute requirement.
9. When the pH controller is temperature compensated, calibration can be done at room temperature. Otherwise it should be done at the fermentation temperature.
10. With regard to the DO probe, it should first be checked whether the membrane is still intact.
11. If calibration is difficult or the electrode is significantly slower than factory specifications, it may be sufficient to replace the electrolyte solution and membrane.
12. For continuous setups, special Maprene tubing should be used for insertion into pumps, because silicone tubing will break down when used for longer times in pumps as a result of the constant mechanical stress.
13. Reactor leaks can be found using a soap solution while keeping a slight overpressure on the reactor.
14. Be careful with overpressure. If the glass is damaged, even a slight overpressure may cause severe explosions as a result of the large surface area of the glass.
15. Be sure that all parts of the setup can be reached by steam. If certain parts cannot be reached by water, a small amount of water should be placed there.
16. When sterilizing keep in mind that the temperature of the autoclave is at the start higher than the temperature of the reactor liquid. Certainly at higher volumes the temperature of the reactor will seriously lag behind the autoclave temperature and sterilization times must be longer. The user manual of the autoclave will usually contain information on sterilization duration for different volumes.
17. If some kind of heating is used to make connections, be sure that the connection is cooled down sufficiently before passing medium or cells through.
18. Polarographic DO probes need up to 6 h after connection to the controller to become stable.
19. It is essential to check that the medium leaving the reactor has the same cell concentration as the medium in the reactor. Certainly at low dilution rates cells may settle

in the tubes, resulting in a perfusion type of culture. This is usually solved by having intermittent pumping at discrete time intervals at higher rates.

20. During the cultivation, tubing (even Maprene tubing) may wear out as a result of the constant mechanical stress, meaning that it is essential to check the liquid flows during the run. This can be done by checking the weight of the medium container or incorporating a pipet in the tubing connecting the medium vessel with the reactor.
21. Optimized PID values are in principle only applicable to the situation for which they were derived. Often it is better to choose a lower P value and a higher I value. By doing so control might be less optimal, but the process will be more stable with respect to fluctuations of the controlled parameters after taking a sample.
22. Make sure that the calculated units for P, I, and D correspond with the units of the controller. The P value is dimensionless. Some controllers use the P value a bit different. Check the manual of the controller. The I and D values are expressed in time units. **Figure 2** has minutes as the time unit, so the calculated values also have minutes as the time unit.
23. If it is not possible to make such a disturbance, controller behavior has to be deduced from the recent past.
24. It is advisable to stop and restart the controller after adjusting the P-value. This will help eliminate oscillations.
25. The disturbances inflicted by the controller might be large. If this is done in the presence of cells, it may lead to unwanted values for the parameter that is controlled. Care should be taken with critical parameters like the pH.
26. If the decrease is faster than the response time of the electrode, this method cannot be used because the electrode cannot follow the decrease.
27. In case the residence time of the gas is very short, meaning $F_{gi}C_{ogi}$ is very high compared to $k_{ol}A$, the concentration of oxygen in the gas is equal to the concentration of oxygen in the incoming gas.
28. It is difficult to calculate the $k_{cl}A$ because of the poor response times of the CO_2 electrodes and the need for pH control, which may influence transfer kinetics. Therefore, the $k_{cl}A$ is usually estimated using the fact that the ratio of K_lA values for carbon dioxide and oxygen is proportional to their liquid diffusivities resulting in $k_{cl}A = 0.89\ k_{ol}A$.
29. An important assumption is that the concentration in the gas phase is constant. However, when switching from nitrogen to air, it will take some time before the nitrogen in the gas phase is replaced by air. Therefore, the first data points usually fall below the curve described in Eq. 19. After three volume replacements of the gas phase, 97% of the gas phase is replaced. Data points acquired during this initial phase should not be used for parameter estimation.
30. Sparging with microbubbles may result in incorrect $k_{ol}A$ determinations. The bubbles have a small volume and a high surface-to-volume ratio and have long residence times in the reactor. This leads to a change in oxygen concentration in the gas phase during measurement. Consequently, the oxygen concentration is not constant and

the $k_{ol}A$ becomes dependent on the oxygen concentration used and is not constant during the measurement.

31. An important point is that the rate of oxygen transfer should be slower that the response time of the electrode, t_r (s), or in other words $(k_{ol}A)^{-1} > t_r$. If they have about the same value, the dynamic $k_{ol}A$ measurement can be corrected for the slow response of the electrode using a first-order approach ***(44)***. However, if the oxygen transfer is faster than the response of the electrode, the dynamic method can no longer be used.
32. When deaerated medium is pumped back to the measurement vessel, no nitrogen bubbles should be present, because these will influence the measurement and result in underestimation of the $k_{ol}A$ values. To prevent this, a bubble trap can be placed in the liquid flow line.

References

1. Goergen, J. L., Marc, A., and Engasser, J. M. (1993) Determination of cell lysis and death kinetics in continuous hybridoma cultures from the measurement of lactate dehydrogenase release. *Cytotechnol.* **11**, 189–195.
2. Voisard, D., Meuwly, F., Ruffieux, P. A., Baer, G., and Kadouri, A. (2003) Potential of cell retention techniques for large-scale high-density perfusion culture of suspended mammalian cells. *Biotechnol. Bioeng.* **82**, 751–765.
3. Woodside, S. M., Bowen, B. D., and Piret, J. M. (1998) Mammalian cell retention devices for stirred perfusion bioreactors. *Cytotechnol.* **28**, 163–175.
4. Dalm, M. C. F., Cuijten, S. M. R., van Grunsven, W. M. J., Tramper, J., and Martens, D. E. (2004) Effect of feed and bleed rate on hybridoma cells in an acoustic perfusion bioreactor: Part I. Cell density, viability, and cell-cycle distribution. *Biotechnol. Bioeng.* **88**, 547–557.
5. Cool, J. C., Schijff, F. J., and Viersma, T. J., Regeltechniek, Elsevier Nederland B.V., Amsterdam, 275–277.
6. Chattaway, T., Demain, A. L., and Stephanopoulos, G. (1992) Use of various measurements for biomass estimation. *Biotechnol. Prog.* **8**, 81–84.
7. Olsson, L. and Nielsen, J. (1997) On-line and in situ monitoring of biomass in submerged cultivations. *TibTech.* **15**, 517–522.
8. Cannizzaro, C., Gugerli, R., Marison, I., and von Stockar, U. (2003) On-line biomass monitoring of CHO perfusion culture with scanning dielectric spectroscopy. *Biotechnol. Bioeng.* **84**, 597–610.
9. Kemp, R. B. and Guan, Y. (1997) Heat flux and the calorimetric-respirometric ratio as a measure of catabolic flux in mammalian cells. *Thermochim. Acta* **300**, 199–211.
10. Dorresteijn, R. C., Numan, K. H., Degooijer, C. D., Tramper, J., and Beuvery, E. C. (1996) On-line estimation of the biomass activity during animal- cell cultivations. *Biotechnol. Bioeng.* **51**, 206–214.
11. Ducommun, P., Bolzonella, I., Rhiel, M., Pugeaud, P., von, S. U., and Marison, I. W. (2001) On-line determination of animal cell concentration. *Biotechnol. Bioeng.* **72**, 515–522.

12. Becker, T., Schuhmann, W., Betken, R., Schmidt, H.-L., Leible, M., and Albrecht, A. (1993) An automatic dehydrogenase-based flow-injection system: application for the continuous determination of glucose and lactate in mammalian cell-cultures. *J. Chem. Techn. Biotechnol.* **58**, 183–190.
13. Schugerl, K. (2001) Progress in monitoring, modeling and control of bioprocesses during the last 20 years. *J. Biotechnol.* **85**, 149–173.
14. Stoll, T. S., Ruffieux, P. A., Schneider, M., Vonstockar, U., and Marison, I. W. (1996) On-line simultaneous monitoring of ammonia and glutamine in a hollow-fiber reactor using flow injection analysis. *J. Biotechnol.* **51**, 27–35.
15. Ferreira, A. P., Alves, T. P., and Menezes, J. C. (2005) Monitoring complex media fermentations with near-infrared spectroscopy: Comparison of different variable selection methods. *Biotechnol. Bioeng.* **91**, 474–481.
16. Hisiger, S. and Jolicoeur, M. (2005) A multi-wavelength fluorescence probe: Is one probe capable for on-line monitoring of recombinant protein production and biomass activity? *J. Biotechnol.* **117**, 325–336.
17. Kornmann, H., Valentinotti, S., Marison, I., and von Stockar, U. (2004) Real-time update of calibration model for better monitoring of batch processes using spectroscopy. *Biotechnol. Bioeng.* **87**, 593–601.
18. Rhiel, M., Cohen, M. B., Murhammer, D. W., and Arnold, M. A. (2002) Nondestructive near-infrared spectroscopic measurement of multiple analytes in undiluted samples of serum-based cell culture media. *Biotechnol. Bioeng.* **77**, 73–82.
19. Rhiel, M. H., Cohen, M. B., Arnold, M. A., and Murhammer, D. W. (2004) On-line monitoring of human prostate cancer cells in a perfusion rotating wall vessel by near-infrared spectroscopy. *Biotechnol. Bioeng.* **86**, 852–861.
20. Riley, M. R., Crider, H. M., Nite, M. E., Garcia, R. A., Woo, J., and Wegge, R. M. (2001) Simultaneous measurement of 19 components in serum-containing animal cell culture media by Fourier transform near-infrared spectroscopy. *Biotechnol. Prog.* **17**, 376–378.
21. von Stockar, U., Valentinotti, S., Marison, I., Cannizzaro, C., and Herwig, C. (2003) Know-how and know-why in biochemical engineering. *Biotechnol. Adv.* **21**, 417–430.
22. Paalme, T., Kahru, A., Elken, R., Vanatalu, K., Tiisma, K., and Vilu, R. (1995) The computer-controlled continuous culture of Escherichia coli with smooth change of dilution rate (A-stat). *J. Microbiol. Meth.* **24**, 145–153.
23. Barbosa, M. J., Zijffers, J. W., Nisworo, A., Vaes, W., van Schoonhoven, J., and Wijffels, R. H. (2005) Optimization of biomass, vitamins, and carotenoid yield on light energy in a flat-panel reactor using the A-stat technique. *Biotechnol. Bioeng.* **89**, 233–242.
24. Kasemets, K., Drews, M., Nisamedtinov, I., and Aadamberg, K. P. T. (2003) Modification of A-stat for the characterization of micro-organisms. *J. Microbiol. Methods* **55**, 187–200.
25. Sluis, C., Westerink, B., Dijkstal, M., et al. (2001). Estimation of steady-state culture characteristics during acceleration-stats with yeasts. *Biotechnol. Bioeng.* **75**, 267–275.

26. Bakker, W. A. M., Schaefer, T., Beeftink, H. H., Tramper, J., and Gooijer, C. D. de. (1996) Hybridomas in a bioreactor cascade: modeling and determination of growth and death kinetics. *Cytotechnology* **21**, 263–277.
27. Riet, K. van 't and Tramper, J. (1991) *Basic Bioreactor Design*, Marcel Dekker, New York.
28. Dorresteijn, R. C., Gooijer, C. D. d., Tramper, J., and Beuvery, E. C. (1994) A method for simultaneous determination of solubility and transfer coefficient of oxygen in aqueous media using off-gas spectrometry. *Biotechnol. Bioeng.* **43**, 159–154.
29. Eyer, K., Oeggerli, A., and Heinzle, E. (1995) On-line gas analysis in animal cell cultivation: II. Methods for oxygen uptake rate estimation and its application to controlled feeding of glutamine. *Biotechnol. Bioeng.* **45**, 54–62.
30. Singh, V. (1996) On-line measurement of oxygen uptake in cell culture using the dynamic method. *Biotechnol. Bioeng.* **52**, 443–448.
31. deZengotita, V. M., Schmelzer, A. E., and Miller, W. M. (2002) Characterization of hybridoma cell responses to elevated pCO(2) and osmolality: Intracellular pH, cell size, apoptosis, and metabolism. *Biotechnol. Bioeng.* **77**, 369–380.
32. Goyal, M. M. Z. A., Rank, D. L., Gupta, S. K., Boom, T. V., and Lee, S. S. (2005) Effects of elevated pCO(2) and osmolality on growth of CHO cells and production of antibody-fusion protein B1: a case study. *Biotechnol. Prog.* **21**, 70–77.
33. Neeleman, R., End, E. J. van den, and Boxtel, A. J. B. van (2000) Estimation of the respiration quotient in a bicarbonate buffered batch cell cultivation. *J. Biotechnol.* **80**, 85–95.
34. Bonarius, H. P. J., Gooijer, C. D. de, Tramper, J., and Schmid, G. (1995) Determination of the respiration quotient in mammalian cell culture in bicarbonate buffered media. *Biotechnol. Bioeng.* **45**, 524–535.
35. Frahm, B., Blank, H. C., Cornand, P., et al. (2002) Determination of dissolved CO2 concentration and CO(2) production rate of mammalian cell suspension culture based on off-gas measurement. *J. Biotechnol.* **99**, 133–148.
36. Wu, L., Lange, H. C., van Gulik, W. M., and Heijnen, J. J. (2003) Determination of in vivo oxygen uptake and carbon dioxide evolution rates from off-gas measurements under highly dynamic conditions. *Biotechnol. Bioeng.* **81**, 448–458.
37. Tribe, L. A., Briens, C. L., and Margaritis, A. (1995) Determination of the volumetric mass transfer coefficient (KlA) using the dynamic "gas out-gas in" method: Analysis of errors caused by dissolved oxygen probes. *Biotechnol. Bioeng.* **46**, 388–392.
38. Sonsbeek, H. M. v., Gielen, S. J., and Tramper, J. (1991) steady-state method for KA measurements in model systems. *Biotechnol. Techn.* **5**, 157–162.
39. Cooper, C. M., Fernstorm, G. A., and Miller, S. A. (1944). Performance of agitated gas liquid contractors. *Ind. Eng. Chem.* **36**, 504–509.
40. Coco-Martin, J. M., Martens, D. E., Velden-de Groot, T. A. M. v. d., and Beuvery, E. C. (1993) Cultivation of the hybridoma cell line MN12 in a homogeneous continuous culture system: effect of culture age. *Cytotechnology* **13**, 213–220.
41. Korke, R., Gatti, M. D., Lau, A. L. Y., et al. (2004) Large scale gene expression profiling of metabolic shift of mammalian cells in culture. *J. Biotechnol.* **107**, 1–17.

42. Europa, A. F., Gambhir, A., Fu, P. C., and Hu, W. S. (2000) Multiple steady states with distinct cellular metabolism in continuous culture of mammalian cells. *Biotechnol. Bioeng.* **67**, 25–34.
43. Follstad, B. D., Balcarcel, R. R., Stephanopoulos, G., and Wang, D. I. C. (1999) Metabolic flux analysis of hybridoma continuous culture steady state multiplicity. *Biotechnol. Bioeng.* **63**, 675–683.
44. Vandu, C. O. and Krishna, R. (2004) Influence of scale on the volumetric mass transfer coefficients in bubble columns. *Cheml Eng. Proc.* **43**, 575–579.

V

DOWNSTREAM TECHNIQUES

19

Membrane Filtration in Animal Cell Culture

Peter Czermak, Dirk Nehring, and Ranil Wickramasinghe

Summary

Membrane filtration is frequently used in animal cell culture for bioreactor harvesting, protein concentration, buffer exchange, virus filtration, and sterile filtration. A variety of membrane materials and pore sizes ranging from loose microfiltration membranes to tight ultrafiltration membranes, which reject small proteins, are frequently found in a purification train. While all of these operations make use of the same size-based separation principle, the actual methods of operation vary significantly.

Microfiltration is often the first of the unit operations in the purification train. Microfiltration membranes have pores in the micrometer size range. Microfiltration is used to remove cells and cell debris. This chapter begins by describing tangential flow microfiltration. A typical method of operation is included.

Concentration of the product and buffer exchange is often required toward the end of the purification train. Ultrafiltration membranes are used for both operations. The theory of tangential flow ultrafiltration is briefly described followed by a typical method of operation.

Today, large-pore ultrafiltration membranes (molecular weight cutoff 100–500 kDa) are finding increasing applications for virus filtration. Validation of virus cleaner is a major concern in the biopharmaceutical industry. At the same time, purification of virus particles for viral vaccines and applications in gene therapy is a major separations challenge. Consequently, the second part of this chapter focuses on these membrane filtration applications.

Key Words: Membrane filtration; microfiltration; ultrafiltration; tangential flow filtration; cell harvest; protein concentration; virus filtration; virus particle; murine leukemia virus; parvovirus.

1. Introduction

Membrane filtration is an essential part of biotechnology processes *(1)*. Standard practice in mammalian cell processes is the sterile filtration of media, purification buffers, and protein product pools. Also, the membrane filtration is used as a part of the overall strategy for viral clearance and retrovirus vector

From: *Methods in Biotechnology, Vol. 24: Animal Cell Biotechnology: Methods and Protocols, 2nd Ed.* Edited by: R. Pörtner © Humana Press Inc., Totowa, NJ

production. In addition, testable membrane filtration ***(2)*** is used for the harvest and clarification of mammalian feed streams in combination with depth filtration and centrifugation. For protein purification and improved protein–virus separation, membrane chromatography and high-performance tangential flow filtration will be of growing interest ***(1)***.

Selectivity for membrane separations is based on the retention of solutes larger than the pores through size exclusion and the permeation of smaller solutes. The driving force for flow through the membrane pores is the transmembrane pressure drop. Membrane-based operations can be conducted in either the tangential or the direct flow mode.

Tangential flow filtration (TFF) is traditionally used for cell culture clarification or protein concentration and diafiltration in the biotechnology industry. During cell culture clarification, cells, cell debris, and other insoluble particle matter typically 0.02–20 μm in diameter are removed from the suspending medium while the target protein is recovered in the permeate ***(3)***. During protein concentration/diafiltration (UF/DF) the target protein is retained by the membrane and recovered from the retentate side. A typical operation consists of a concentration step followed by diafiltration, where buffer is added to the feed tank at a rate equal to the filtration rate to keep the volume in the tank constant. The purpose of the diafiltration step is to recover the protein remaining in the feed at the end of the concentration process for cell culture clarification and to exchange buffer for UF/DF ***(4)***.

For TFF, high recovery of target protein at high filtration rates is desired. The economic viability of these operations depends on the permeate flux, the capacity of the module, and the yield for the target protein. High permeate fluxes and capacities imply short operations with small-surface-area systems.

During TFF only a small portion of the feed flow rate permeates through the membrane, with the remaining fluid flowing tangentially to the membrane surface to reduce the accumulation of retained solutes on the surface ***(5,6)***. The permeate flux is, however, limited by the formation of a concentration polarization boundary layer consisting of a high concentration of particles retained by the membrane. In addition, both the permeate flux and the module's capacity are compromised by the fouling of the membrane as a result of deposition of suspended and dissolved solutes within or on the surface of the membrane. Both polarization and fouling are caused by the net amount of species transported to the membrane. That amount is determined by the balance between the convection of particles toward the membrane and their backmigration away from the membrane because of mechanisms such as Brownian diffusion, shear-induced diffusion, and inertial lift forces. A number of theoretical models have been developed in order to predict the quasi-steady-state permeate flux during TFF. These models relate the permeate flux to variables such as the wall shear rate, the average particle diameter, the viscosity of the feed, the length of the hollow fibers, and the particle concentration *(7)*.

Fouling usually manifests itself as a rapid flux decline followed by an extended period of quasi-steady-state operation *(7)*. The flux decline is typically a result of various mechanisms such as pore plugging, pore constriction, and cake formation. Biological feeds are usually difficult to filter because of the presence of small particles causing pore plugging and species capable of forming highly compressible cake layers *(8)*. The filterability of the feed may depend on cell viability at the time of bioreactor harvesting. Low cell viability implies a large number of dead or lysed cells with smaller particles that can easily plug the pores.

Numerous investigators have attempted to maximize the permeate flux during TFF through a number of different approaches *(7)*. These approaches can be divided into three groups. Chemical methods involve modification of the surface chemistry of the membrane, thus increasing the electrostatic repulsion between the membrane and the particulate matter. This increased repulsion results in reduced deposition and fouling of the membrane. Hydrodynamic methods involve modifying the design of the module to induce turbulent flow. Physical methods involve pretreatment of the feed stream ***(5,6)***.

Given the numerous applications of membranes in animal cell biotechnology, a chapter like this is never complete. This chapter gives some general information about the established methods for cell harvesting, cell retention, protein concentration, and buffer exchange. For virus filtration and retroviral vector filtration, the chapter provides more detailed methods. The chapter does not consider important operations such as sterile filtration for the protection of any process against contamination (e.g., mycoplasm contamination) and laboratory concentration methods. These methods may be found in the literature.

1.1. Cell Harvest and Cell Retention

1.1.1. Cell Harvest

Mammalian cells are capable of expressing high molecular weight recombinant proteins that are properly folded and glycosylated. They can be cultured by a variety of methods including batch, fed-batch, and perfusion at volumes ranging from microliters to 20,000 L. The most commonly used cell lines are Chinese hamster ovary (CHO), murine myeloma (NSO), and human embryonic kidney (HEK) cells. Both cell density and productivity for cell cultures have increased significantly within the last 10 yr. Protein titers of 5 g/L are now commonly seen compared to 0.1 g/L seen in the early 1990s.

The removal of solids from cell culture is the first step in downstream processing ***(9–11)***. This clarification operation has become challenging because of the increase in cell densities and protein titers and the higher concentration of antifoam that is now being used. Different technologies such as centrifugation, tangential filtration, and depth filtration are used to achieve this difficult separation objective.

1.1.2. Cell Retention in Perfusion Systems

The most common cell culture operation modes are batch- (up to 7 d), fed-batch (7–15 d), and perfusion mode (more than 15 d). Continuous perfusion systems need an integrated system for cell retention to reach high cell densities ***(12–14)***. This allows high productivity. Perfusion usually results in cell viabilities under 50%, with the consequence of a high content of cell debris and solids and a very high turbidity (>1000 NTU) in the product stream.

1.2. Protein Concentration and Buffer Exchange

Target proteins from mammalian cell cultures are concentrated and diafiltered by tangential flow ultrafiltration ***(1,4,10,15–17)***. For buffer exchange of proteins at industrial scale, TFF competes with size exclusion chromatography and countercurrent dialysis ***(15)***.

Ultrafiltration is used to concentrate proteins, remove small molecules from the biological solution, and change the buffer environment of the protein. It is a pressure-driven process through an asymmetrical membrane where a thin skin layer provides for selectivity. Ultrafiltration membranes are characterized by a molecular weight cut-off, which is the molecular weight of a solute with a particular retention. Retention is defined as:

$$R = 1 - C_p / C_f$$

where R is the retention coefficient, C_p is the concentration of the solute in the permeate, and C_f is the concentration of solute in the feed. The pore size distribution and hence the molecular weight cutoff of ultrafiltration membranes is determined by challenging the membrane with a mixture of dextran molecules of different sizes and determining retention at each molecular size. Another important characteristic for these types of membranes is water permeability, which is defined as the flux divided by the pressure applied to the membrane.

Ultrafiltration is usually conducted in the tangential filtration mode, where the feed solution flows tangentially to the surface of the membrane, reducing the accumulation of retained solutes. A portion of the feed solution permeates through the membrane, forming the permeate stream, while the remaining flow is returned back to the feed tank.

Two important phenomena affect the performance of ultrafiltration operations. Polarization occurs when the flux reaches a critical value and does not increase with increasing transmembrane pressure. It is a result of the formation of a reversible concentration boundary layer near the surface of the membrane. Solutes that are retained by the membrane are convected to the surface by the permeate flow and start to accumulate while the concentration difference between the surface and the bulk causes a back diffusion. At steady state these two transport rates become equal and the flux can be expressed as a function of the mass transfer coefficient, the concentration at the surface of the membrane, and the bulk concentration:

$$J = k\ln(C_w/C_b)$$

where J is the flux, k is the mass transfer coefficient, and C_w and C_b are the surface and bulk concentrations of the solute, respectively. The mass transfer coefficient is a measure of the back transport. It is a function of the device hydrodynamics (increasing with shear rate), solution properties, and device geometry. Fouling is an irreversible decline in flux resulting from the formation of a cake layer, pore blocking, or pore constriction. For a fouled membrane, flux can only be recovered through cleaning ***(16)***. Typical cleaning solutions are sodium hydroxide, sodium hypochlorite, and enzymatic cleaners. The cleaning cycle can be performed at room or elevated temperatures (40–50°C) depending on the extent of fouling.

Batch or fed-batch systems can be used to conduct ultrafiltration. In batch systems the entire feed volume is within one tank. This can hinder the extent of concentration that can be achieved. In fed-batch systems an extra recycle tank is connected to the feed tank. The smaller size of the recycle tank enables higher concentration factors ***(16)***. The disadvantage of fed-batch systems is slower fluxes because of the higher concentrations in the recycle tank.

During concentration operations, permeate is diverted to the drain or a permeate tank. The removal of impurities and the product yield can be obtained from solving mass balances. The yield of the retained solute at a volumetric concentration factor X is:

$$Y = X^{R-1}$$

where Y is yield, X is the volumetric concentration factor (initial volume divided by final volume), and R is the retention for the solute. During diafiltration, buffer is introduced to the feed tank at a rate equal to the filtration rate, making the process a constant volume process. The concentration of a solute at the end of N diafiltration volumes can be obtained from mass balances as:

$$C_N = C_1 e^{-N(1-R)}$$

where C_N is the concentration after N diafiltration volumes and R is retention for the solute. The number of diafiltration volumes can be calculated by dividing the volume of the permeate by the constant feed volume.

1.3. Virus Filtration

Viruses have the potential to contaminate a biotherapeutic production process. They can enter the process via a number of different routes. Some cell lines carry the virus within their genomes. For example, endogenous retroviruses can be part of hybridoma cell lines, while retrovirus like particles can be part of CHO cell lines. Adventitious viruses such as parvoviruses can enter the process via the raw materials: growth media, chromatography beads, buffers, or through operator contact. A virus-control strategy that minimizes the risk of viral contamination in the final product consists of three different activities. First, the cell lines are tested for viral contamination and virus-free cell lines are selected together with the testing and screening of the raw materials. Second, the purified product is tested at different steps of the purification process for the presence of viruses. Finally, virus-removal and -inactivation steps are incorporated into the process. Virus-inactivation steps include exposure to solvent/detergent, low pH, temperature, or irradiation. Virus-removal steps include filtration, chromatography, and centrifugation. Ultrafiltration could be used not only to concentrate viruses, but also to remove virus from final product by retaining virus while allowing product to pass through the filter ***(17–20)***. The guidelines for viral safety recommend the use of at least two robust orthogonal processes with a viral titer reduction of 4 log for the removal/inactivation of viruses ***(21)***. At least one of these steps needs to be effective for the removal of nonenveloped viruses ***(21)***. A robust process is one where performance does not vary with changes in solution properties, such as pH and ionic strength. Orthogonal processes are processes relying on different mechanisms for the inactivation or removal of viruses.

Filtration removes viruses mostly by size exclusion with very little retention as a result of adsorption ***(22)***. It can be conducted in the tangential or direct flow modes, even though the majority of use in the recent years has been in the direct mode. Typical filters contain single or multilayers of membranes, having asymmetrical structures with a thin skin layer providing their retention characteristics. Available virus filters are made of polyvinylidene fluoride (PVDF), polyether sulfone (PES), or cuprammonium regenerated cellulose with pore size ratings of 15, 20, 35, 50, and 75 nm. They are manufactured in a way to minimize the presence of defects because very high levels of retention are expected of them. Filter manufacturers ensure the high retention of the filters by using a series of integrity tests during and after production. An integrity test is also conducted before and after processing a biological solution to have a high level of assurance on the required performance of the virus filter.

Integrity tests are usually correlated to the titer reduction of selected viruses ***(22)***. They can be based on different mechanisms, such as particle retention, liquid intrusion, or air diffusion through wetted cartridge. Air diffusion is based on the principle that when exposed to air, a wetted cartridge will not allow the flow of air through the pores until a pressure that provides for a force exceeding the capillary forces at the gas/liquid interface is exceeded. The pressure that allows the expulsion of liquid from the largest pores is called the bubble point. The bubble point is a function of pore diameter, surface tension of the gas/liquid system, and the contact angle. It can be expressed as:

$$P = K[(4\sigma\cos\theta)/d]$$

where P is the bubble point pressure, K is the correction factor that accounts for the tortuosity of the pores, σ is the surface tension of the wetting liquid, θ is the contact angle, and d is the diameter of the largest pore. The bubble point for virus filters is too high to be used for testing the integrity of such filters. The typical integrity tests instead rely on the use of pressures lower than the bubble point. If the filter is integral, the only flow is a result of the dissolution and diffusion of the gas through the liquid. This is called the diffusive flow *(2,22)*. For defective filters the defects will have much lower capillary forces leading to the expulsion of the liquid from the defect and to a convective gas flow that will be much higher than the diffusive flow through the integral membrane. The filter is considered as integral if the flow rate is lower than the manufacturer's specification. Liquid intrusion tests are based on the capillary forces created at the interface of two immiscible liquids. The membrane is first wetted with one of the liquids. A second liquid is then introduced and the flow rate at a particular pressure measured. If there are no defects, the second liquid will not be able to intrude the pores and hence no flow will be observed. Particle challenge tests are based on the concept of the transmission of virus size particles through the membrane. A commonly used test is the colloidal gold test. These types of tests are destructive and can only be performed post use.

In a typical virus-filtration operation the filter is autoclaved, flushed with water for injection (WFI), and integrity tested. Filters can also be steamed in place or sanitized by sodium hydroxide prior to the operation. The filtration is usually conducted in the constant pressure mode at pressures between 1.3 and 2 bar. At the end of the operation the filter is cleaned and a post use integrity test is performed. Virus filtration is typically conducted toward the end of the purification train since the volumes to be processed are smaller and solutions are purer *(20)*.

The important performance parameters for virus filtration are viral titer reduction, protein recovery, flux, and throughput. Experiments should be conducted with scale-down units to find the best filter operating pressure conditions. Virus filters also need to be validated to demonstrate their effectiveness in removing viruses. These validation studies are usually conducted on scale-down units with process streams spiked with a panel of viruses. The viruses in the panel need to be relevant to the starting material and they need to encompass large and small DNA and RNA viruses. The selection should also include enveloped and nonenveloped viruses.

Besides the optimization of production processes, a robust and efficient virus-reduction process is necessary. Focusing on the filtration, this chapter provides protocols for the production of, filtration to purify, and quantification of *Aedes aegypti* densonucleosisvirus (AeDNV) *(23–25)*. Such parvoviruses are similar in size to adeno-associated virus. The virus particles are easy to grow using a cell-culture-based system.

1.4. Filtration of Retroviral Vectors With Ceramic Membranes

Over the last decade, recombinant retroviruses such as murine leukemia virus (MLV) have been used in many clinical trials *(26)*. In order to produce sufficient amounts and quality of MLV, some of the current production tools and strategies of pharmaceutical vaccine and recombinant protein production are applicable for the concentration of viral vectors. However, the instability of retroviral pseudo-type vectors appears to require advanced filtration technologies *(26–30)*. Recently it has been reported that because of fast degradation and relatively low production rates, the concentration does not reach the appropriate level for successful application in gene therapy *(31,32)*. Besides the optimization of bioreactors and production processes, a robust and efficient filtration process is necessary. Focusing on filtration, this chapter provides protocols for the production, filtration, and quantification of a certain recombinant vector.

2. Materials
2.1. Cell Harvest and Cell Retention
2.1.1. Cell Harvest Techniques

1. Centrifugation in combination with depth filtration.
   a. Work with genetically modified organism (GMO): centrifugation, depth filtration, and final membrane filtration (with the possibility of integrity testing of the last step).
2. Direct flow filtration (dead end filtration).
3. Expanded bed chromatography.
4. TFF (flat membranes—cassettes, hollow fiber membranes, dynamic membrane systems, e.g., controlled horizontal oscillation) membranes with 0.1- or 0.65-µm pores.

The selection of cell harvest technique depends on the ease of filtration of the particular culture and the volume that needs to be processed. Volumes less than 1000 L can be processed by direct flow filtration, which typically is a series of depth filters of different grades followed by sterile filtration membranes. In general volumes larger than 1000 L can be processed by either TFF or centrifugation. Cell harvesting by membranes utilizes tangential flow (TFF) (also called cross-flow filtration, CFF) to reduce product concentration polarization and fouling. TFF can be conducted using 0.65-, 0.45-, or 0.2-µm membrane filters. The diversity of membrane types and configurations makes testing and evaluation of systems and/or relying on manufacturers' recommendations necessary.

Membrane characteristics, module and skid geometry, and operating conditions all have to be considered carefully for a successful TFF operation. Membrane requirements like low nonspecific protein binding, good compatibility with cleaning agents, good fouling resistance, and the availability of scalable filter formats are all important in the selection of the proper filter. In addition, there are TFF system requirements like low system shear forces (pump, module, pipework, etc.), good scale-up properties, low working volumes, complete drainage, and cleaning in place (CIP) and sanitation in place (SIP) capabilites *(1,9–11)*.

The key operating variables in TFF for cell harvest are the transmembrane pressure (TMP) = $[(P_F + P_R)/2] - P_P$ and the cross-flow flux.

During process development attention needs to be paid to come up with conditions that minimize cell damage and result in maximum target protein recovery. To meet these demands the following requirements pertain: low TMP, moderate cross-flow flux, shallow pressure gradient in feed channels, and an appropriate module configuration with regard to cell type, product viscosity, and transmission needs *(1,9,10)*.

2.1.2. Cell-Retention Techniques in Perfusion Systems *(12–14)*

1. Spin-filter.
2. Inclined settler.
3. Centrifugation.
4. Ultrasonic separation.
5. TFF (cross-flow filtration).
6. Integrated direct flow membrane filter.

Low protein-binding hollow fiber membranes with 0.1- or 0.65-µm pores (e.g., PVDF membranes) or cylindrical ceramic membranes *(14)* are used.

2.2. Membrane Modules for Protein Concentration and Buffer Exchange

Ultrafiltration modules are available in different configurations, such as cassettes (plate and frame), hollow fibers, and spirals. Most organic membranes available are regenerated cellulose, polysulfone, or PES. Inorganic ceramic membranes are also available.

2.3. Materials for Filtration of Retroviral Vectors

2.3.1. Cell Culture, MLV Vector Production (see also *Chapter 22*)

2.3.1.1. Production of Packaging and Target Cell Line in Culture Flasks

1. Cell culture flasks: 75, 150, and 300 cm^2 (Biochrom AG, Germany).
2. Cell culture wells, 6-well, 9.03 cm^2 (Costar, Germany).
3. Culture medium: Dulbecco's modified Eagle's medium (DMEM; Gibco/BRL, Eggenstein, Germany). supplemented with 5% (v/v) fetal bovine serum (PAA, Germany), 800 mg/L neomycin sulfate (Roth, Giessen, Germany), 5 mg/L blasticidin S (ICN-Flow, Meckenheim, Germany), and 2 m*M* glutamine (ICN-Flow, Meckenheim, Germany).
4. Phosphate-buffered saline (PBS): sodium chloride 8000 mg/L (Sigma, Germany), potassium chloride 200 mg/L (Sigma Germany), disodium hydrogen phosphate 710 mg/L (Sigma, Germany), and potassium dihydrogen phosphate 200 mg/L (Sigma, Germany) in sterile water.
5. Trypsin solution (0.2%) containing ethylenediamine tetraacetic acid (EDTA) 1 m*M* (Sigma, Germany).

2.3.1.2. Production of Retroviral Particles in a Fixed-Bed Reactor (*see also* Chapter 17)

1. Bioreactor and control unit Biostat B with 5-L conditioning vessel (Sartorius BBI Systems GmbH, Melsungen, Germany).
2. Fixed-bed unit, 200 mL (KGW Isotherm, Karlsruhe, Germany).
3. Cell culture carriers for adherent cells, Fibra-Cel (New Brunswick Scientific Inc., Edison, NJ).
4. One 2-L flask and three 1-L flasks (Schott, Mainz, Germany).
5. Four air filters, 0.2 μm (Millipore, Germany).
6. Silicon tubes, 3 and 5 mm inner diameter (Roth, Giessen, Germany).
7. Peristaltic pump Type Easy load, Model 7518-00 (Masterflex, Germany).

2.3.2. Filtration of Retroviral Particles

1. Ceramic aluminum oxide cross-flow filtration module, length 450 mm, outer diameter 25 mm, filtration surface 0.1 m^2 with 19 channels (Atech Innovations GmbH delivered by Amafilter GmbH, Hannover, Germany). Each channel has a diameter of 3 mm and a cutoff of 20 kDa specified by the manufacturer **(Fig. 1)**.
2. Stainless steel housing Type M05 (Amafilter GmbH, Hannover, Germany) for ceramic membrane filtration module with two connections at the permeate side and two connections at the retentate side of the membrane **(Fig. 1)**.
3. Two ½-in. connectors and two 1-in. connectors (Serto Jacob GmbH, Fuldabrück, Germany) to attached silicon tubes on both sides of the membrane.
4. Two 1-L flasks with four sterile tube connections (Schott, Mainz, Germany).
5. One peristaltic pump Type Easy load, Model 7518-00 (MasterFlex, Gelsenkirchen, Germany).

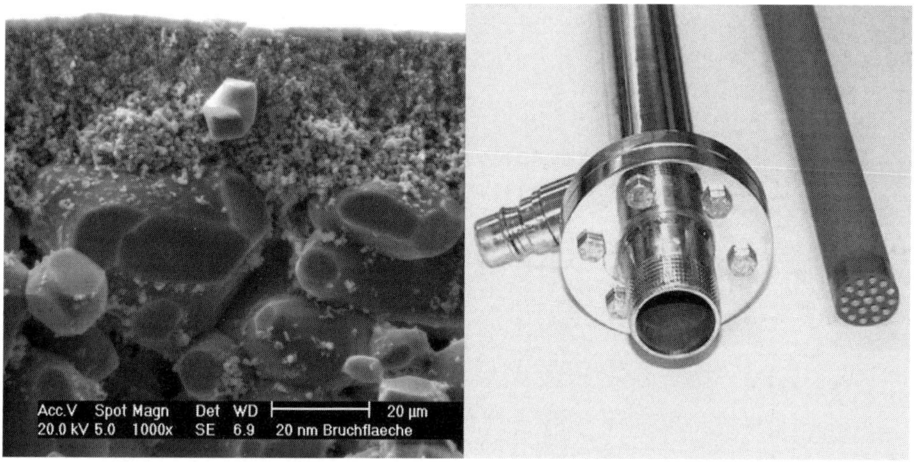

Fig. 1. (Left) Scanning electron microscopy of the cross section of the used ceramic membrane. The membrane consists of an Al_2O_3 support material and two asymmetrical layers. The first layer is built of zirconium oxide, while the second layer that gives the cutoff molecular size consists of TiO_2. The membrane cutoff was specified at 20 kDa by the manufacturer. (Right) Stainless steel module for the cylindrical multichannel ceramic membrane with an outer diameter of 25 mm and 19 inner tubes with a diameter of 3 mm.

6. Vacuum pump (N 035.3.AN.18, KNF Neuberger, Freiburg, Germany).
7. Dextrane blue 1000 kDa (Fluka, Buchs, Switzerland).

2.3.3. Retroviral Vector Quantification

1. Cell culture wells, 24-well, 1.91 cm^2 (Costar, Germany).
2. Culture medium DMEM as above, but without blasticidin.
3. X-gal solution: 1 mg/L 5-bromo-4-chloro-3-indolyl-β-D-galactopyranoside (Sigma, Deisenhofen, Germany) in dimethyl formamide (VWR, Darmstadt Germany).
4. Formaldehyde solution 2% (v/v) (VWR, Darmstadt, Germany).
5. Glutaraldehyde solution 2% (v/v) (Sigma, Deisenhofen, Germany).
6. $K_3Fe(CN)_6$ solution, 5 m*M* (Merck).
7. $K_4Fe(CN)_6$ solution, 5 m*M* (Merck).
8. $MgCl_2$ solution, 2 m*M* (Merck).

3. Methods

3.1. Membrane and Module Selection for Cell Harvest

Usually the choice of membrane and membrane module (design) should be based upon small-scale testing. The selection is based on a trade-off between the capacity of the membrane before it plugs, the turbidity of the filtrate, and the yield of the protein product. TFF is typically conducted in the constant flux mode by the use of a filtrate pump. The process is stopped when the transmembrane pressure reaches a critical value because the transmission of the target protein is lower when that critical value is exceeded, resulting in a low protein yield. Experiments are usually conducted with different-pore-size membranes—the capacity, yield, and filtrate turbidity are determined. The selection is then made based on process economics and the robustness of the operation (consistent filtration performance with changing feedstocks).

3.1.1. Membranes

For mammalian cell harvesting, low-protein-binding membranes made from PVDF, modified PES, or regenerated cellulose are available from several suppliers. In general, more hydrophilic membranes demonstrate better performance in the presence of antifoams. The membrane pore sizes are between 0.1 and 1.2 μm. The selection between the most commonly used membrane pore size of 0.2 or 0.65 μm for cell harvesting and clarification depends on particle size distribution in the feed solution (*see* **Note 1**).

3.1.2. Modules

For TFF flat membrane cassettes, hollow fibers or cylindrical multichannel tubular membranes are used (membrane area between 0.1 and 2.8 m^2). With an open-channel design of the membrane modules, it is possible to process feedstock solutions with low to medium cell densitiy and solids at high flux and at low transmembrane pressure.

A good portion of players in the industry have been using 0.65-μm rated membranes followed by depth filtration to remove the fines in the filtrate, and finally sterile filtration. As TFF devices, membrane cassettes with suspended or coarse screens and hollow fibers are used. If the solid content in the solution is too high, centrifugation becomes the preferred technology.

3.2. Membrane Techniques for Protein Concentration and Buffer Exchange by Ultrafiltration

During ultrafiltration operations, the following steps are typically conducted:

1. Install: The filters should be installed according to the manufacturer's instructions, which are usually packed with the filters. Cassettes need to be torqued to the manufacturer's specifications. Typical specifications are 35–45 Nm.
2. Sanitize: Flush the devices with WFI with the retentate and permeate valves open. Use approx 20 L of WFI per m^2 of membrane surface area. After the first 10 L/m^2 close the retentate valve to obtain a TMP of 1.3–1.7 bar. Place 40 L/m^2 of NaOH (0.1–0.5 *N* at 35–45°C) into the feed tank. Divert the first 10 L/m^2 to the drain. Recirculate the remaining solution for 30–45 min with a TMP of 1 bar. Drain the NaOH solution. Flush the system with WFI until both the retentate and permeate are neutral (20–40 L/m^2).

3. Water permeability test: Conduct the test with WFI. With the filtrate valve completely open, increase the TMP by adjusting the retentate valve to 0.3, 0.6, and 1.0 bar. At each pressure point, record the permeate and retentate flow rates and the temperature. Plot the water flux as a function of TMP. The slope of the line gives the water permeability. If the temperature is not 20 or 25°C, use the temperature correction factor supplied by the manufacturer to normalize the permeability.
4. Integrity test: Connect the feed side to an air supply with permeate and retentate valves open. Set the pressure regulator to the pressure recommended by the manufacturer (1.3–2.0 bar). Purge the liquid on the retentate side of the system. Close the retentate valve. Measure the airflow rate on the permeate line. If the airflow rate is below the manufacturer's specification, continue to the next step.
5. Buffer conditioning: Use the same buffer as for the protein solution to be processed. Pump approx 10 L/m^2 of buffer with a TMP of 1.0 bar and the retentate and permeate lines diverted to drain. Recirculate another 10 L/m^2 for 10 minutes with the retentate and permeate lines diverted to the feed tank. Drain the feed tank. Keep all piping, tubing, and the cassette assembly flooded with buffer.
6. Introduce protein to feed tank: Divert the retentate flow to the feed tank and the permeate to the a tank or drain with the permeate valve closed.
7. Concentrate: Slowly increase the cross-flow rate to the specified level. At that point slowly open the permeate valve. Keep track of the total amount of solution filtered and the amount on the retentate side. Stop when the desired concentration factor is achieved.
8. Diafilter: Turn on the diafiltration buffer pump and set it to a level equal to the permeate flow rate. Keep track of the amount of diafiltration buffer used. Stop the process when the desired number of diafiltration volumes has been processed.
9. Recover protein: Remove product as completely as possible from the system to maximize yield. If necessary, use nitrogen to blow down the remaining liquid on the retentate side. Buffer can also be used to displace the remaining liquid.
10. Buffer flush: Flush system with approx 20 L/m^2 of buffer. For the first 10 L/m^2 divert the retentate flow to drain with the permeate close. For the next 10 L/m^2 apply a TMP of 1.0 bar with the permeate valve open and the flow diverted to the drain.
11. Clean: Place 40 L/m^2 of NaOH (0.1–0.5 N at 35–45°C) into the feed tank. Divert the first 10 L/m^2 to the drain. Recirculate the remaining solution for 45–60 min with a TMP of 1.0 bar. Drain the NaOH solution. Flush the system with WFI until both the retentate and permeate are neutral (20–40 L/m^2).
12. Measure water permeability: Follow procedure outlined in **step 3**.
13. Storage: Introduce storage solution (0.05 or 0.1 N NaOH) into the device.

Certain rules need to be followed to scale up UF operations. The following are typically kept constant at different scales:

1. Membrane.
2. Device configuration (channel height, screen, channel length).
3. Solution.

4. Temperature.
5. Loading (ratio of solution volume to membrane surface area).

The operation is usually carried in the constant transmembrane pressure mode. During process development, experiments need to be conducted to find the optimum transmembrane pressure. In a typical experiment, flux is measured at different transmembrane pressures at a constant flow rate and protein concentration. The operation should be conducted at a pressure 10–20% lower than the pressure at the start of polarization. The cross-flow rate is usually recommended by the filter vendor. In certain cases users also conduct experiments to find an optimum cross-flow rate. During these experiments flux vs TMP data are generated at different cross-flow rates and the optimization is conducted via a tradeoff between the size of the pump and the high filtration rate. Membranes of molecular weight cutoffs of 1, 3, 5, 10, 30, 70, 100, 300, and 500 kDa are offered by different suppliers. The membrane molecular weight cutoff is selected to be at least three times (three to six times) tighter that the molecular weight of the target protein. Cassettes with fine or medium screens can be used according to the viscosity of the solution to be processed. Scale-up is conducted according to scaling rules where membrane, device configuration (channel height, screen, channel length), solution composition and temperature, and loading (ratio of solution volume to membrane surface area) are kept constant at different scales.

3.3. Methods for Virus Filtration

3.3.1. Production of AeDNV Particles in Serum and Serum-Free Medium

AeDNV particles are produced using the *A. albopictus* cell line C6/36 in either a serum medium or a serum-free medium. All experiments should be run under sterile conditions. Medium changes in culture flask and feed and harvest bottles of the bioreactor should be opened and closed under a laminar flow clean bench.

1. In the serum medium culturing process, grow the C6/36 cell line at 28°C in T-75 flasks containing 10 mL Leibovitz's L-15 medium supplemented with 10% fetal bovine serum and 1% penicillin-streptomycin (Invitrogen Co., Carlsbad, CA).
2. The pH of the medium is 7.4.
3. When the cells confluent reached 80%, transfect the *A. albopictus* cell line C6/36 with pUCA, an infectious clone containing the AeDNV genome, by using Qaigen effectine kit (Qiagen, Valencia, CA).
4. Change the media 8–18 h posttransfection to remove the pUCA plasmid.
5. Four days posttransfection, freeze the T-75 flasks (at −80°C) and thaw (in 37°C water bath) three times, then centrifuge at 3200g for 15 min at 4°C to remove cellular debris.

6. Filter the supernatant containing AeDNV particles using a 0.45-μm sterilized filter (Nalge Nunc International, Rochester, NY) and store the particles at −80°C for future use.

These AeDNV particles produced by this cell culture technique are referred to as virus in serum medium.

AeDNV particles could be also produced using the *A. albopictus* cell line C6/36 in a serum-free medium.

1. Grow the cells first in T-75 flasks containing 10 mL serum- and protein-free medium (SFPFM) (S-F900 II SFM, Invitrogen Corporation, Grand Island, NY) at 28°C.
2. When the cells confluent reached 80%, transfect the cells with pUCA by using Qaigen effectine kit.
3. Change the media 8–18 h posttransfection to remove the pUCA plasmid.
4. Transfer the transfected C6/36 cells from the T-75 flask to a 1000-mL spin flask (stirred bioreactor) (Wheaton Science Products, Millville, NJ) at a cell concentration of 5.5×10^5 cell/mL.
5. The total medium volume is 500 mL.
6. Stirr the bioreactor at 300 rpm/min at 28°C.
7. Five to 6 d later, when the cell concentration reaches around $5-6 \times 10^6$ cells/mL, remove 450 mL medium and add 450 mL fresh medium to the bioreactor for continuously growing the cells.
8. Freeze and thaw the collected cells three times, and then centrifuge at 3200g for 15 min at 4°C to remove cellular debris.
9. Filter the supernatant containing AeDNV particles using a 0.45-μm sterilized filter and store the supernatant at −80°C for future use.

These AeDNV particles produced by this cell culture technique are referred to as virus in serum-free medium.

3.3.2. Tangential Flow Filtration

1. Conduct the TFF using flat sheet Sartocon™ Slice 200 cassettes (Sartorius AG, Goettingen, Germany). **Figure 2** shows the experimental setup.
2. Use ultrafiltration membranes with molecular weight cutoff of 30, 50, 100 kDa for the tests.
3. All membranes are made of PES and have a nominal filtration surface area of 0.02 m^2. The Sartocon™ Slice 200 cassettes consist of four membranes of 15 cm in length. The height of the feed channel is 200 μm.
4. Run all experiments at a feed flow rate of 150 mL/min controlled by a peristaltic pump.
5. Permeate flow is not controlled.
6. Run all experiments at a concentration mode (the retantate is returned to the feed reservoir, but not the permeate).
7. Measure the permeate mass by an electrical balance (Mettler Toledo, Columbus, OH) and record with an online personal computer connected to the electrical balance.

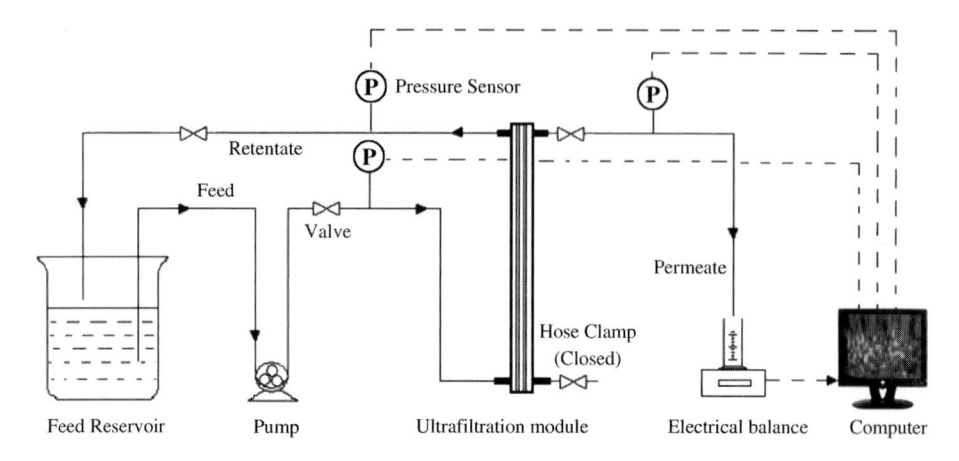

Fig. 2. Tangential flow filtration setup for virus reduction.

8. Measure the pressure in the feed side, retentate side, and permeate side by three microswitch sensing and control sensors (Honeywell International Inc., Morristown, NJ).
9. The calculated average transmembrane pressure has to be around 0.4 bar.
10. Prior to commencement of the virus filtration experiments, measure DI water fluxes at these operating conditions.
11. Add 500 mL virus-containing medium to a feed reservoir.
12. Cycle the feed through the membrane feed side for several minutes at the desired feed flow rate of 150 mL/min while the permeate outlet is closed.
13. Open the valve, record the permeate mass and the pressures, and take 1-mL samples at regular intervals from the feed, retentate, and permeate.
14. Continue this process until 400–450 mL of permeate has been collected.
15. After finishing the virus experiments, flush the membrane using DI water followed by 1 mol/L NaOH solution at 50°C for 1 hr.
16. Remeasure the DI water flux at the same condition. Usually more than 95% of the DI water flux can be recovered.
17. Store the membrane in a 0.1 mol/L NaOH solution supplemented with 20% alcohol by volume for future use.
18. Analyze samples of the feed, retentate, and permeate for virus titer.

(*See* **Notes 2** and **3**.)

3.3.3. Polymerase Chain Reaction Assay

1. Use a real-time reverse transcriptase polymerase chain reaction (RTPCR)-based assay to determine the virus titer in the infective solutions because AeDNV does not show cytopathic effects (CPE).
2. The primers and probe are designed within a conserved region of the viral NS1 gene. Use Primer Express™ oligo design software (Applied Biosystems, Foster

City, CA) to design forward primer: CAT ACT ACA CAT TCG TCC TCC ACA A; reverse primer: CTT GCT GAT TCT GGT TCT GAC TCT T; and TaqMan Probe: FAMCCA GGG CCA AGC AAG CGC CTAMRA.

3. Perform the reaction in 96-well format skirted v-bottomed polypropylene microplates (MJ Research, Inc., Waltham, MA) with optical caps (Applied Biosystems, Foster City, CA).
4. Use the Brilliant® Quantitative PCR core reagent kit (Stratagene, La Jolla, CA) as the RTPCR master mix. Add into each well 4 μL of unknown sample or standard control DNA pUCA plasmid, 10 μL of master mix, 2 μL of 0.05 mmol/L forward primer, 2 μL of 0.05 mmol/L reverse primer, and 2 μL of 5×10^{-3} mmol/L probe.
5. Choose the following thermal cycling conditions: stage 1, 50°C for 2 min; stage 2, 95°C for 10 min; stage 3, 95°C for 15 s; stage 4, 60°C for 1 min; repeat stages 3 and 4 39 times.
6. Perform all reactions in the Opticon 2 DNA Engine (MJ Research, Inc., Waltham, MA).
7. Analyze all samples three times and average the results.

(*See* **Notes 4** and **5**.)

3.4. Methods for Filtration of Retroviral Vectors

All experiments should be run under sterile conditions. Medium changes in culture flask and feed and harvest bottles of the bioreactor should be opened and closed under a laminar flow clean bench.

3.4.1. Cell Culture and Viral Vector Production

3.4.1.1. Production of Cells to Inoculate a Fixed-Bed Reactor *(30)* (*see also* Chapter 17)

1. In order to achieve vector-containing supernatant for the filtration process, human packaging cell lines can be used. In this method the packaging cell line TELCeB6/pTr712-K52s is used. This cell line derives from cell line TELCeB6 by transfection of the HIV-1 env gene with the plasmid pTr712 and produces MLV-based vectors.
2. After thawing, dilute the cells in culture medium and transfer them into a 75-cm^2 culture flask with 25 mL medium.
3. Incubate the cell culture flask at 37°C and 5% CO_2 atmosphere for 3 d.
4. Check under the microscope if cells are proliferating.
5. If the surface of the flask is covered completely by cells, split the cells by trypsinisation and transfer each half into a 150-cm^2 culture flask with 30–50 mL medium and incubate for 2 or 3 d.
6. Once the cells cover the surface of the 150-cm^2 flasks, transfer the cell content of each flask to two new 300-cm^2 flasks and incubate again with 50–75 mL.
7. If a confluent cell monolayer is reached, trypsinate all four 300-cm^2 flasks, fill each cell suspension into one 250-mL flask, and proceed with the inoculation of the fixed-bed reactor.

3.4.1.2. Production of Retroviral Supernatant in a Fixed-Bed Reactor

1. Connect all tubes of the harvest and feed vessel and attach sterile air filters to inlets and outlets of the conditioning vessel of the fixed-bed reactor.
2. Fill up the fixed bed with 200-mL cell culture carriers.
3. Fill up the all vessels with phosphate buffer and place the complete reactor setup, including the oxygen and pH sensors, into an autoclave. Cover the sterile filter opening with aluminum foil. Open the threats of all vessels so that steam will reach everywhere and operate the autoclave.
4. Build up the complete setup of the bioreactor, including the peristaltic pumps and the control unit.
5. Fill 2-L feed glass bottle with 1 L of medium.
6. Pump medium with a peristaltic pump of the control unit from the feed bottle into the conditioning vessel and check the operation of the circulation pump of the fixed.
7. Check the function of the oxygen and pH sensors and operate the aeration system.
8. When temperature is controlled at 37°C and the pH is stable at 7.2, inoculate the cell suspension.
9. During inoculation the bubble aeration should be shut off and the fixed-bed circulation pump should operate at relatively low medium speed of 1 mm/s in the fixed bed.
10. Observe the medium in the conditioning vessel. Once the medium is clear after 2–4 h, the aeration should be started.
11. Set the oxygen control to 90% of air saturation.
12. Take medium samples from the conditioning vessel and analyze the retroviral vector quantity and the glucose concentration.
13. Make sure that glucose concentration is greater than 5 mmol/L during the entire cultivation.
14. Once a glucose concentration of 5 mmol/L is reached, harvest the vector supernatant from the conditioning vessel and start with the filtration process.
15. Refill the conditioning vessel with fresh medium and perform another batch.

3.4.2. *Cross-Flow Filtration of Retroviral Vectors*

3.4.2.1. Checking the Cross-Flow Filtration Setup (Fig. 3)

1. Connect all the silicon tubes to the filtration module.
2. Make sure that the tube for the peristaltic pump is connected on the lower vertical connection of the module to provide the cross-flow to the inner channels of the module.
3. Connect permeate and retentate in 1-L bottles to both sides of the setup.
4. Connect the sterile filters to the inlet on the lid of each bottle.
5. Prepare 200 μL dextran blue in 1 L phosphate buffer solution.
6. Attach this bottle to the retentate side of the setup and connect a vacuum pump to the permeate bottle.
7. Operate the cross-flow with the peristaltic pump at a flow rate of 0.025 m^3/h.
8. Start the vacuum pump with a pressure of 0.6 bar absolute pressure and operate it for 1 h.

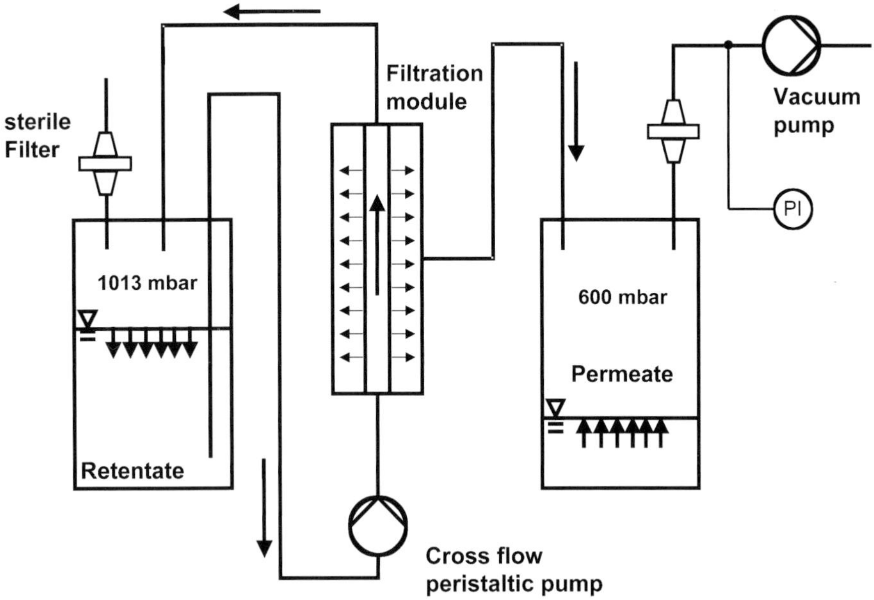

Fig. 3. Schematic filtration setup showing the batch operation of a cross-flow filtration using a vacuum pump to establish a transmembrane pressure.

9. Stop both pumps and take a sample of permeate in order to check the colorization.
10. Perform also a photometric check of the permeate. If blue dextrane is detected, a leakage of the system is very likely.
11. If no leakage can be detected, wash the membrane with phosphate buffer for 1 h.
12. Fill up both bottles with phosphate buffer and place the complete setup into an autoclave.
13. After sterilization for 40 min, start the vector filtration.

3.4.2.2. Cross-Flow Filtration of Vector Supernatant (**Figs. 3** and **4**)

1. Build up the sterilized filtration module and check the cross-flow of the peristaltic pump.
2. Remove the retentate bottle under a laminar flow clean bench and connect a bottle with 1 L vector containing supernatant from the fixed-bed cultivation.
3. Remove the permeate bottle and connect an empty sterile 1-L bottle.
4. Operate the cross-flow at 0.025 m^3/h.
5. Operate the vacuum pump at a pressure of 0.6 bar absolute pressure.
6. Observe the volume of filtrate with a scale.
7. Stop the filtration when 90% of the liquid weight has been filtrated.
8. Remove the filtrate and the permeate bottle and take samples.
9. Analyze the vector titer and store the retentate in a –80°C freezer or use the retentate for further purification steps.

(*See* **Notes 6–9**.)

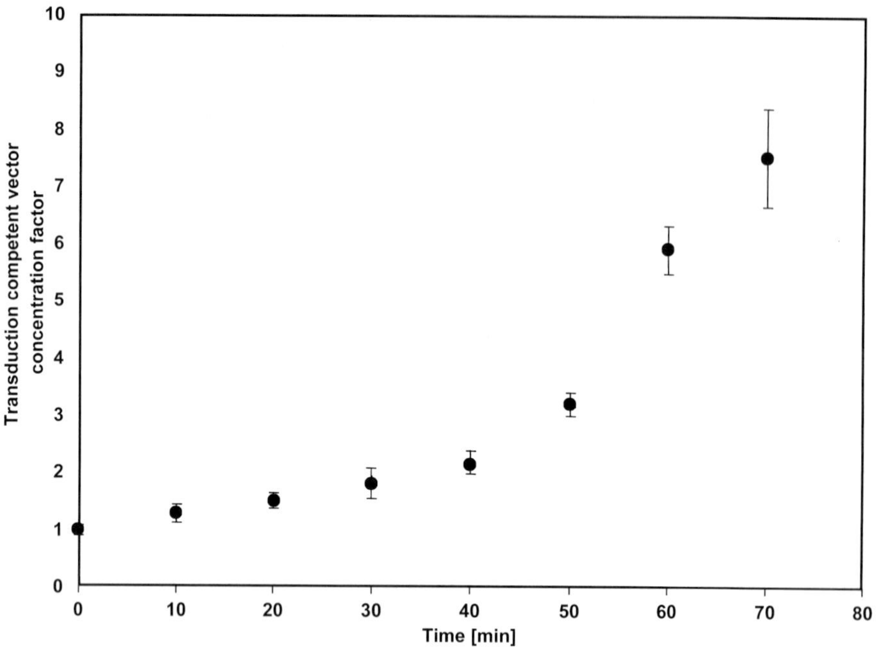

Fig. 4. Concentration factor in the retentate after cross-flow filtration with a ceramic membrane. Initial supernatant contained 2×10^5 IU/mL vector particles, and pressure on the filtrate side was kept at 600 mbar absolute pressure. Samples were taken from the retentate and analyzed with a cell-based assy.

3.4.3. *Retroviral Vector Quantification* ***(33,34)*** (see also *Chapter 22*)

1. In order to measure the transfusion efficiency and quantity of the active vector particles, a cell-based assay can be used. Prepare a target cell (HELA $CD4^+$) suspension with 2×10^5cells/mL.
2. Inoculate the cell suspension in a 24-well plate, 4.23×10^4 cells per well. Incubate the cells for 24 h in an incubator at 37°C and 5% CO_2.
3. Remove the medium from each well and wash with phosphate buffer solution.
4. Perform dilutions of 1:1, 1:10, 1:100, 1:1000 to the viral samples in a separate well plate.
5. Add 1 mL of viral dilution per well in triplicates. Incubate in an incubator at 37°C and 5% CO_2.
6. After 4 h of infection remove the supernatant wash two times with phosphate buffer solution.
7. Add 2 mL medium and incubate for another 48 h.
8. Prepare the fixing solution 0.2% (v/v) of glutaraldehyde and 2.0% of formaldehyde in 20 mL of PBS and store it at 4°C.

9. Prepare the staining solution for a 24-well plate: 600 μL of x-Gal (40 mg/mL), 120 μL of 1 M $K_3Fe(CN)_6$, 240 μL of 0.5 M $K_4Fe(CN)_6$, 48 μL of 1 M $MgCl_2$ in 24 mL of phosphate buffer solution.
10. Remove the medium from the plates and wash the plate with phosphate buffer solution.
11. Add 500 μL fixing solution per well; wait 10 min at room temperature and remove the fixing solution.
12. Wash the plates two times with phosphate buffer.
13. Add 250μL staining solution per well.
14. Incubate at least 12 h at 37°C.
15. Observe each viral dilution and count blue cells (infected cells expressing β-galactosidase) in the wells that have 10–90 blue cells per well.
16. The titer can be calculated as follows: titer = number of blue cells × dilution/1 mL.

3.5. Outlook

In the future, TFF will continue to be used for well-established operations such as clarification, protein concentration, and buffer exchange. Dead end and TFF will continue to be used for validation of virus clearance. In addition, TFF (microfiltration and ultrafiltration) will find greater applications in the purification of virus particles for viral vaccines and gene therapy applications. The use of ultrafiltration for fractionation of proteins with molecular weights less than an order of magnitude different may also be possible using techniques such as high-performance TFF and charge-assisted ultrafiltration.

4. Notes

1. Very often organic membrane modules are shipped wet in liquid containing a humectant and bactericidal storage solution. This solution consists of 15–20% glycerin and 0.05–0.1% sodium azide. The storage solution must be removed and the module flushed well with water prior to use to prevent product contamination. Membrane modules are also delivered with approx 0.3 N sodium hydroxide as the storage agent.
2. Usually parvovirus reduction is conducted with membranes with a pore size rating of 20 nm. Virus reduction is also possible with membranes with a nominal cutoff up to 300 kDa depending on the membrane material and production process of the membrane. Therefore, choice of the right membrane has to be tested under real conditions.
3. The use of other filtration devices (modules) and other organic or ceramic membranes is possible. Manufacturer of devices and organic membranes for that purpose include PALL, Sartorius, Millipore, Microdyn Nadir, and Asahi.
4. A RTPCR-based method was used for the quantification of AeDNV virus because more conventional biological assays are not straightforward *(23)*. The quantitative RTPCR assay is a rapid, sensitive, and efficient way to compare samples. Although

similar results could be obtained with naked viral genomic DNA, when batches of AeDNV prepared from cell culture or mosquito larvae as described are exposed to pancreatic DNase prior to RTPCR, there is little or no reduction in signal. Also, RTPCR on pellet fractions after ultracentrifugation under conditions that should pellet virus particles indicates that most of the DNA is pelleted. These results give confidence that we are measuring DNA from virus particles in these preparations rather than DNA from plasmid transfections or replicative forms.

5. The accuracy of the PCR assay was determined by analyzing 12 samples of the same infective solution and found to be within 0.5 log unit.
6. The membrane module was sterilized in the autoclave (30 min, 121°C) assembled completely. Use gaskets which are sterilizable. An additional possibility to get clean and pyrogen-free ceramic membranes and modules is to sterilize the membranes at 180°C in a hot air sterilizer (SL600, Memmert, Schwabach, Germany), rinse the assembled module with 1 M NaOH and 60% ethanol, and rinse it again with highly purified pyrogen-free water *(35)*. Then subject the whole test equipment to the same rinsing procedure.
7. Because of limited mechanical stability of the ceramic membrane, care is needed when assembling the membrane module.
8. The use of other ceramic or organic membranes is possible. Some additional manufacturers of ceramic membranes produced of aluminum oxide, titanium dioxide, or zirconium dioxide (or a combination of these materials) are PALL Exekia, Tami Industries, MemPRO Attaxx (manufacturer: Atech Innovations GmbH), and Rhodia Novasep. Manufacturers of organic membranes for that purpose are PALL, Sartorius, Millipore, Microdyn Nadir, and Asahi.
9. Choose a membrane with a cutoff three to four times smaller than the nominal cutoff given by the manufacturer.

References

1. van Reis, R. and Zydney, A. L. (2001) Membrane separations in biotechnology. *Curr. Opin. Biotechnol.* **12**, 208–211.
2. Czermak, P. and Catapano, G. (2003) *Accuracy of Automated Flow-Measuring Devices Used in the Pharmaceutical Industry for Testing Sterile Filter Integrity*, PDA J. Pharm. Sci. Technol. **57**, 277–286.
3. Davis. R. H. (1992) Microfiltration, in *Membrane Handbook* (Ho, W. S. W. and Sirkar, K. K., eds.), Springer, New York.
4. van Reis R. and Zydney, A. L. (1999) Protein ultrafiltration, in *Encyclopedia of Bioprocess Technology: Fermentation, Biocatalysis and Bioseparation* (Flickinger, M. C. and Drew, S. W. eds.), Wiley, New York, pp. 2197–2213.
5. Kim, J. S., Akeprathumchai, S., and Wickramasinghe, S. R. (2001) Flocculation to enhance microfiltration. *J. Membr. Sci.* **182**, 161–172.
6. Wickramasinghe, S. R., Han, B., Akeprathumchai, S., Chen, V., Neal, P., and Qian, X. (2004) Improved permete flux by flocculation of biological feeds: comparision between experiment and theory. *J. Membr. Sci.* **242**, 57–71.

7. Belfort, G., Davis, R. H., and Zydney, A. L. (1994) The behavior of suspensions and macromolecular solutions in crossflow microfiltration. *J. Membr. Sci.* **96**, 1–58.
8. Belter, P. A., Cussler, E. L., and Hu, W. S. (1998) *Bioseparations*. Wiley, New York.
9. van Reis R., et al. (1991) Industrial scale harvest of proteins from mammalian cell culture by tangential flow filtration. *Biotechnol. Bioeng.* **38**, 413–422.
10. Ng, P. K., et al. (2001) Filter applications in product recovery processes, *Membrane Seperations in Biotechnology* (Wang, W. K., ed.), Marcel Dekker, New York, pp. 205–224.
11. Russotti, G. and Göklen, K. E. (2001) Crossflow membrane filtration of fermentation broth, in W. K. Wang (ed.) *Membrane Separations in Biotechnology* (Wang, W. K., ed.), Marcel Dekker, New York, pp. 85–159.
12. Woodside, S. M., et al. (1998) Mammalian cell retention devices for stirred perfusion bioreactors. *Cytotechnology* **28**, 163–175.
13. Voisard D., Meuwly, F., Ruffieux, P.-A., Baer, G., and Kadouri, A. (2003) Potential of cell retention techniques for large-scale high-density perfusion culture of suspended mammalian cells. *Biotechol. Bioeng.* **82**, 751–765.
14. Dong, H., et al. (2005) A perfusion culture system using a stirred ceramic membrane reactor for hyperproduction of IgG_{2a} monoclonal antibody by hybridoma cells. *Biotechnol. Prog.* **21**, 140–147.
15. Kurnik, R.T., Yu, A.W., Blank, G. S., et al. (1995) Buffer exchange using size exclusion chromatography, countercurrent dialysis, and tangential flow filtration: models, development, and industrial application. *Biotechnol. Bioeng.* **45**, 149–157.
16. Zydney, A. L. and Kuriyel, R. (2000) Protein ultrafiltration, in *Downstream Protein Processing* (Desai, M., ed.), Humana Press, Totowa, NJ, pp. 35–46.
17. Ogle, K. F. and Azari, M. R. (2001) Virus removal by ultrafiltration, in *Membrane Separations in Biotechnology* (Wang, W. K. ed.), Marcel Dekker, New York, pp. 299–326.
18. Aranha-Creado, H. and Fennington, G. J. (1997) Cumultative viral titer reduction demonstrated by sequential challenge of a tangential flow membrane filtration system and a direct flow pleated filter cartridge. *J. Pharm. Sci. Technol.* **51**, 208–212.
19. DiLeo, A. J., Allegrezza, A. E., and Builder, S. E. (1992) High resolution removal of virus from protein solutions using a membrane of unique structure. *Bio/Technology* **10**, 182–188.
20. Huang, P.Y. and Peterson, J. (2001) Scale-up and virus clearance studies on virus filtration in monoclonal antibody manufacture, in *Membrane Separations in Biotechnology* (Wang, W. K., ed.), Marcel Dekker, New York, pp. 327–350.
21. Committee for Proprietary Medicinal Products (CPMP) Note for Guidance on Plasma Derived Products (CPMP/BWP/269/95).
22. Kuriyel, R. and Zydney, A. L. (2000) Sterile filtration and virus filtration, in *Downstream Protein Processing* (Desai M., ed.), Humana Press, Totowa, NJ, pp. 185–194.
23. Specht, R., Han, B., Wickramasinghe, S. R., et al. (2004) Densonucleosis virus purification by ion exchange membranes. *Biotechnol. Bioeng.* **88**(4), 465–473.

24. Han, B., Specht, R., Carlson, J. O., and Wickramasinghe, S. R. (2005) Virus purification using adsorptive membranes. *J. Chromatogr. A* **1092**, 114–124.
25. Grzenia, D. (2005) Virus removal from biological suspension using ultrafiltration, DA thesis, University of Applied Sciences, Giessen.
26. Andreadis, S. T., et al. (1999) Large scale processing of recombinant retroviruses for gene therapy. *Biotechnol. Prog.* **15**, 1–11.
27. Nehring, D., Gonzalez, R., Pörtner, R., and Czermak, P. (2004) Mathematical model of a filtration process using ceramic membranes to increase retroviral pseudotype vector titer. *J. Membrane Sci.* **237**(1–2), 25–38.
28. Cruz, P. E., Goncalves, D., Almeida, J., Moreira, J., and Carrondo, M. J. T. (2000) Modeling retrovirus production for gene therapy. 2. Integrated optimization of bioreaction and downstream processing. *Biotechnol. Prog.* **16**, 350–357.
29. Kuiper, M., Sanches, R. M., Walford, J. A., and Slater, N. K. H. (2002) Purification of a functional gene therapy vector derived from moloney murine leukaemia virus using membrane filtration and ceramic hydroxyapatite chromatography. *Biotechnol. Bioeng.* **80**, 445–453.
30. Nehring, D., Gonzalez, R., Pörtner, R., and Czermak, P. (2006) Experimental and modelling study of different process modes for retroviral production in a fixed bed reactor. *J. Biotechnol.* **122**, 239–253.
31. Clayton, T. M. (2000) Cell products-viral gene therapy vectors, in *Encyclopedia of Cell Technology* (Spier, R. E., ed.), Wiley and Sons, Chichester, UK, pp. 441–457.
32. Stitz, J., Mueller, P., Merget-Millitzer, H., and Cichutek, K. (1998) High-titer retroviral pseudotype vectors for specific targeting of human CD4-positive cells. *J. Biogenic. Amines.* **14**, 407–424.
33. Andreadis, S., Lavery, T., Davis, H. E., Le Doux, J. M., Yarmush, M. L., and Morgan, J. R. (2000) Toward a more accurate quantitation of the activity of recombinant retroviruses: alternatives to titer and multiplicity of infection. *J. Virol.* **74**, 1258–1266.
34. Cosset, F.-L., Takeuchi, Y., Battini, J.-L., Weiss, R. A., and Collins, M. K. L. (1995) High-titer packaging cells producing recombinant resistant retroviruses to human serum. *J. Virol.* **69**, 7430–7436.
35. Czermak P., Ebrahimi, M., and Catapano, G. (2005) New generation ceramic membranes have the potential of removing endotoxins from dialysis water and dialysate. *Int. J. Artif. Organs* **28**(7), 694–700.

20

Chromatographic Techniques in the Downstream Processing of (Recombinant) Proteins

Ruth Freitag

Summary

The purification of a (recombinant) protein produced by animal cell culture, the so-called downstream process (DSP), tends to be one of the most costly aspects of bioproduction. Chromatography is still the major tool on all levels of the DSP, from the first capture to the final polishing step. In this chapter we will first outline the commonly used methods and their setup, in particular ion exchange chromatography (IEX), hydrophobic interaction chromatography (HIC), affinity chromatography (AC), and gel filtration (gel permeation chromatography [GPC], size exclusion chromatography [SEC]), but also some lesser known alternatives. Then the rational design of a downstream process, which usually comprises three orthogonal chromatographic steps, is discussed. Finally, process variants deviating from the usual batch-column/gradient elution approach will be presented, including expanded bed, displacement, and continuous annular chromatography, as well as affinity precipitation.

Key Words: Affinity; capture; chromatography; downstream process; gel permeation chromatography; hydrophobic interaction chromatography; ion exchange chromatography; isolation; protein; purification; size exclusion chromatography.

1. Introduction

The isolation and purification of a recombinant protein produced by animal cells, i.e., the downstream process (DSP), remains one of the most costly and to some extent most difficult parts of a typical bioproduction process, even though recent developments on the cell culture side have rendered this step less cumbersome. Animal cells usually excrete the protein product into the culture media, which means that the amount of contaminating host cells proteins/components will be less pronounced than for most bacteria-derived products. The fact that the productivity of mammalian cells increased by more than an order of magnitude over the last decade *(1)* as well as the increased use of defined, serum- or even

From: *Methods in Biotechnology, Vol. 24: Animal Cell Biotechnology: Methods and Protocols, 2nd Ed.* Edited by: R. Pörtner © Humana Press Inc., Totowa, NJ

protein-free culture media has helped to make the DSP easier in terms of setup, handling, and standardization.

The working horse of the DSP is chromatography in its various forms and implementations ***(2,3)***. Chromatography includes a broad variety of techniques that are capable of separating mixtures of compounds on the basis of small differences in physicochemical parameters such as size, charge density, hydrophobicity, hydrophilicity (ability to form H-bridges), or mixtures thereof. Chromatography can be done under physiological conditions (aqueous solution, temperature 4–40°C, pressure <10 bar, neutral pH) and is presently the only cost-competitive, scalable, high-resolution method for protein isolation and separation. A typical setup for doing chromatography consists of one or more chromatographic columns, reservoirs for the mobile phases (buffers) and for the solutions needed for regenerating and cleaning/sanitizing the column, as well as the respective pumps and detectors, all controlled by a computer with suitable software. At small scale the sample is usually injected manually or automatically from a sample loop, whereas at larger scale the "feed" is pumped into the column.

The separation takes place in the chromatographic column. The steps involved tend to be similar in all cases. The compounds to be separated are injected into the mobile phase, which is pumped through column. This presents a form of convective or active mass transport. The column normally consists of a packed bed of stationary phase, typically porous particles with an average diameter of 10–500 μm. The surface of the stationary phase provides a certain affinity for the feed compounds. Most chromatographic methods are based on a form of non-covalent interaction (electrostatic, hydrophobic, H-bridges), although one method exists, namely gel filtration (also called size exclusion chromatography [SEC], or gel permeation chromatography [GPC]), that is based on differences is size. The major part of this interactive surface/accessible volume is located inside the porous space of the particles. In order to reach this space, and in particular the interactive surface, the target molecules need to overcome several mass transfer resistances. First they have to pass by diffusional transport through the stagnant layer of liquid around the outer surface of the particle. Then they have to move by diffusion (pore and/or surface diffusion) through the pore to reach the inner pore surface. Then they have to interact ("bind") with the surface. During elution the process is reversed, i.e., release from the surface, diffusion through the pore to the exit, diffusion through the stagnant layer around the particle, and finally convective transport by the mobile phase.

Transport by diffusion requires a concentration gradient. In the simplest form of chromatography, isocratic elution, the mobile phase composition is not changed during the separation and the sample mixture simply moves in peaks through the column, providing locally and temporarily a high concentration and driving force for diffusion into the pores, whereas an inverse concentration gradient—promoting desorption and diffusion out of the particle—establishes itself once the peak maximum has passed. A substance thus moves through the column by repeated interaction cycles. The diffusion speed, but also the strength of the surface interaction, will usually vary somewhat between molecules, and, given the large number of interaction cycles, significant differences in retention (or residence time) will ensue, leading to the separation of the substance mixture.

Because the diffusion speed depends on the size and shape of the molecules as well as on the viscosity of the liquid, diffusion is difficult to influence for a given system. The speed of convective transport, on the other hand, can easily be adjusted by the mobile phase flow rate (pumping speed). As a consequence, an increase in the pumping speed will increase the relative amount of time spent on diffusional transport and hence broaden the substance zone. Broader peaks reduce the efficiency of the separation. If one records the efficiency of the separation (approximated by the plate height or number; *see* **Note 1**) as a function of the mobile phase flow rate ("van Deemter curve"), an optimum is usually seen. Peaks will become broader at very low flow rates because of molecular diffusion in the mobile phase itself (this will rarely present a problem in the case of the fairly large proteins), but also at higher flow rates because of the necessity for diffusion in and out of the porous particles. In the case of protein chromatography the optimal flow rate for the highest column efficiency will often be too low for practical consideration.

Columns are thus typically operated below their highest possible efficiency. In fact, the problem of combining speed with capacity and efficiency has been called the "dilemma of protein chromatography." A high binding capacity calls for a large surface area, i.e., porous particles. The inside of these particles and therefore the larger part of the surface area can only be reached by diffusion, in case of the comparatively large proteins a rather slow process. An efficient use of the column capacity thus requires operation of the column at low speed. Efficient, high-speed-compatible columns can be packed from nonporous particles, albeit at the price of low capacity. Monolithic columns such as the UNO material from BioRad (Herkules, CA) may in certain cases present a possible solution to the dilemma. Monolithic columns consist of a single, porous polymer rod. Because in monolithic columns the mobile phase flows through the porous structure, the only diffusion still necessary is that through the stagnant layer of liquid covering the interactive surface. Such columns can be operated at high speed without measurable loss in efficiency. Unfortunately, the volume of such monolithic columns is at present restricted to a few milliliters, i.e., too small for most preparative applications. BioRad provides a UNO-type particular material for the packing of larger columns, which is supposed to retain the superior mass transfer properties of the monolithic columns even in this form.

While mixtures of small molecules can often be resolved by isocratic elution chromatography, proteins represent a large and very diverse class of molecules. It is therefore rare that a given protein mixture can be separated isocratically. Instead, an all-or-nothing type of binding behavior can usually be observed, where for any given mobile phase composition, one fraction of the protein mixture will not stick at all to the column, while another fraction will bind irreversibly, i.e., very strongly, and only a few closely related proteins will elute, albeit—because of their slow diffusion speed—in the form of rather broad and diluted peaks. The standard approach to protein chromatography is therefore gradient elution, a process whereby the composition of the mobile phase is changed either stepwise or continuously to increase the elution strength. Step gradients are especially popular at larger scale, where the homogeneous application of a continuous gradient over a large column may present a problem, or for well-characterized mixtures, where the method is tailored to the elution requirements of a given target molecule. A distinct advantage of gradient elution is the fact that the column capacity can be used efficiently and that the substances elute in concentrated, focused peaks.

It should be kept in mind that most theoretical concepts for interaction/ elution or transport in chromatography were developed for isocratic elution and cannot be transferred directly to gradient elution. It should also be kept in mind that proteins are dynamic structures that can change their mode/ strength of interaction, e.g., as a function of the salt content of the mobile phase or the protein surface concentration. A striking example is the increase in binding strength sometimes observed in ion exchange chromatography (IEX) at very high mobile phase salt concentration because of the manifestation of hydrophobic interactions at this point. Setting up an efficient protein purification strategy to this day remains therefore less of a theoretical exercise but rather calls for a judicious application of expert knowledge, design rules, and statistics ("chemometrics"). Instruments like the Äkta design chromatographic system (GE Healthcare, Life Sciences) come already provided with the necessary algorithms and can be a tremendous help in the quick design of an optimized separation strategy. However, even without such a support it will be possible to purify most proteins from a crude mixture to the desired purity in a few—usually not more than three—chromatographic steps.

2. Materials

2.1. Modes of Chromatography

Instruments, stationary phases, and prepacked columns for biochromatography are available from established suppliers such as GE Healthcare, BioRad, and Merck (*see* **Note 2**). Small, prepacked columns are preferable for method scouting and development. For financial reasons, loose material should be obtained for self-packing of larger columns, i.e., > 100 mL. In this case a packing apparatus adapted to the available chromatographic system/column type is necessary (*see* **Note 3**).

2.1.1. Ion Exchange Chromatography (see ***Note 4***)

1. Buffer A: 20 m*M* defined by the chosen pH (e.g., Na_2HPO_4 for neutral or Tris-HCl buffer for pH values around 8; *see* **Note 5**).
2. Buffer B: same as Buffer A, but containing in addition 1 *M* NaCl (*see* **Note 6**).
3. Solution for column cleaning/sanitization: 1 *M* NaOH (*see* **Note 7**).
4. Solution for column storage: 0.01 *M* NaOH (*see* **Note 7**).

2.1.2. Hydrophobic Interaction Chromatography (see ***Note 4***)

1. Buffer A: 50 m*M* Na_2HPO_4 buffer (neutral pH) containing in addition 1.5 *M* $(NH_4)_2SO_4$ (*see* **Note 8**).
2. Buffer B: 50 m*M* Na_2HPO_4 buffer (neutral pH) (*see* **Note 9**).
3. Solution for column cleaning/sanitization: 1 *M* NaOH (*see* **Note 7**).
4. Storage solution: 0.01 *M* NaOH (*see* **Note 7**).

2.1.3. Affinity Chromatography (see ***Note 4***)

1. Binding buffer: defined by the system. Often the clarified culture supernatant can be loaded directly onto the column, e.g., in Protein A affinity chromatography of antibodies.
2. Washing buffer: neutral (50 m*M*) phosphate buffers containing up to 1 *M* NaCl to prevent nonspecific electrostatic interaction are popular choices for binding and washing buffers.
3. Elution buffer: defined by the system, often characterized by a low pH. In the case of Protein A affinity chromatography for antibody purification, 100 m*M* glycine-HCl or citrate (pH 3) is often used (*see* **Note 10**).
4. Solution for column cleaning/sanitization, usually defined by the stability of the affinity ligand. Keep in mind that many biospecific ligands cannot withstand harsh cleaning/sanitization conditions and that the corresponding columns are therefore difficult to clean (always follow the manufacturer's instructions). Such columns should not be used for purifying different products to avoid cross-contamination (*see* **Note 11**).
5. Storage solution: defined by the system; 20% ethanol is often a good choice.

2.1.4. Gel Filtration (see ***Note 4***)

1. Mobile phase: can be chosen as desired (e.g., defined by the stability of the target molecule or the feed in general, but also by the requirements of an eventual subsequent purification/formulation step), since no interaction takes place.
2. Solution for column cleaning/sanitization, typically 0.1 *M* or 1.0 *M* NaOH (*see* **Note 7**).
3. Solution for column storage: typically 0.01 *M* NaOH (*see* **Note 7**).

2.1.5. *Reversed-Phase Chromatography* (see **Notes 4,12**)

1. Buffer A : H_2O (containing 0.1% trifluoroacetic acid [TFA]) (*see* **Note 13**).
2. Buffer B: acetonitrile (containing 0.1% TFA).

2.1.6. *Hydroxyapatite Chromatography* (see **Notes 4,14**)

1. Buffer A: 10 m*M* phosphate buffer pH 7.4 (*see* **Note 15**).
2. Buffer B: 500 m*M* phosphate buffer pH 7.4 (*see* **Note 15**).

2.2. *Alternatives*

2.2.1. *Expanded/Fluidized Bed Chromatography*

Buffers as well as solutions for column regeneration/cleaning/sanitization depend on the chromatographic mode; *see* **Subheading 2.1.** for details.

2.2.2. *Displacement Chromatography*

1. Mobile phase: defined by the system. Displacement chromatography is compatible with any chromatographic mode save gel filtration and carried out under binding conditions. Any of the above-mentioned (*see* **Subheading 2.1.**) binding buffers or buffers A would make a suitable mobile phase for the corresponding chromatographic separation in the displacement mode.
2. Displacer solution (often 20–30 mg/mL in mobile phase; *see* **Note 16**), defined by the system. A suitable displacer has a higher affinity to the stationary phase than the molecules of interest (*see* **Note 17**). In protein displacement chromatography often polymeric compounds are used as displacers such as polyacrylic acid in anion exchange displacement chromatography or polydiallyldimethylammonium chloride (PolyDADAMC) in cation exchange displacement chromatography (*see* **Note 18**).
3. Solution for column regeneration, defined by the system, similar to elution chromatography (*see* **Subheading 2.1.**).
4. Solution for column cleaning/sanitization, defined by the system, similar to elution chromatography (*see* **Subheading 2.1.**).
5. Solution for column storage similar to elution chromatography (*see* **Subheading 2.1.**).

2.2.3. *Continuous Annular Chromatography*

Buffers as well as solutions for column regeneration/cleaning/sanitization depend on the chromatographic mode; *see* **Subheading 2.1.** for details.

2.2.4. *Affinity Precipitation*

1. Binding buffer: defined by the system, usually identical to the corresponding binding buffer in affinity chromatography (*see* **Subheading 2.1.3.**; *see* **Note 19**).
2. Dissociation buffer: defined by the system, usually identical to the corresponding elution buffer in affinity chromatography (*see* **Subheading 2.1.3.**; *see* **Note 19**).

3. Methods

3.1. Modes of Protein Chromatography

Most chromatographic modes are based on a noncovalent interaction of the target molecule with the stationary phase. For reasons of simplicity, a single mode interaction is usually preferred. The major chromatographic modes used in protein chromatography are discussed here. The following steps apply in all cases:

1. Before you start, thoroughly familiarize yourself with the chromatographic system to be used, including software tools for running the system and data acquisition/export.
2. Prepare sufficient amounts of buffers A and B (mobile phases used for gradient formation) as well as the liquids to be used for column regeneration, column cleaning/sanitizing, and final system flushing.
3. Make sure the buffer passes through an in-line filter/guard column before entering the separation column.
4. Bring sample/feed to the same pH and ionic strength as the starting buffer.
5. Select/set the detection wavelength(s), typically 280 nm (*see* **Note 20**).
6. After the final experiment, clean the column and put it into proper storage; flush the entire chromatographic system with water to remove all residual buffers (*see* **Note 21**).

3.1.1. Ion Exchange Chromatography (IEX)

Most if not all proteins of interest to biotechnology bear charges. The distribution, accessibility, and density of these charges, and of course also the sign of the net charge, differ from protein to protein, and chromatographic methods based on electrostatic interaction (IEX) are therefore popular means for protein separation. Both cation and anion exchangers are used; in both cases strong and weak ion exchangers are known. Cation exchangers bear negative charges and bind positively charged molecules, while repelling negatively charged ones; anion exchangers bear positive charges and bind negatively charged molecules, while repelling positively charged ones. The surface charge of a strong ion exchanger is independent of the pH, while that of a weak ion exchanger varies with the pH. Often found interactive groups on the surface of ion exchangers in biochromatography are quarternary ammonium ions (Q materials, strong anion exchanger), diethylaminoethyl groups (DEAE materials, weak anion exchanger), sulfonic acid groups (S materials, strong cation exchanger), and carboxymethyl groups (CM materials, weak cation exchanger).

Because most proteins are negatively charged at neutral pH and somewhat pH sensitive; anion exchange chromatography on a strong anion exchanger material is the preferred form of IEX in protein chromatography (*see* **Note 22**). Elution in IEX is typically done in a gradient of increasing NaCl concentration; in the case of weak ion exchangers a pH gradient may also be used. The pertinent steps for method development are given below.

1. Run through **steps 1–5** in **Subheading 3.1**.
2. Select the type of ion exchanger to be used (*see* **Note 23**) based on the isoelectric point and the pH stability of the target molecule (*see* **Note 24**). The separation should be performed at least 0.5 pH units below (cation exchange chromatography) or above (anion exchange chromatography) the pI of the target molecule.
3. Choose the operating buffer and pH (*see* **Note 25**). The operating pH should be at least 0.5 pH units away from the isoelectric point of the target molecule. Because the pH dependency of the charge status of the protein impurities and sometimes even of the target molecule are often unknown, several (at least four) mobile phase pH values below and above the isoelectric point of the target molecule (or pH 7) should be investigated to optimize the separation (*see* **Note 24**).
4. Equilibrate the column with the first of the chosen starting buffers (at least 5 column volumes, CV); a stable detector signal/baseline should be seen at the end of the equilibration.
5. Adjust the pH/ionic strength of the sample/feed to that of Buffer A. If the salt content of the sample is too high (reduction of binding) a buffer exchange (or a dilution) may be necessary. A lower ionic strength in the sample than in Buffer A is permissible.
6. Load the clarified sample onto the column using no more than 20% of the dynamic protein-binding capacity during method development (*see* **Note 26**). A flow rate of 1 mL/min can be used with small (few mL) columns (*see* **Note 27**). The use of a guard column may be advisable in the case of complex sample (*see* **Note 28**).
7. Run a gradient from 0 to 100% Buffer B (corresponding to 0 to 1 *M* NaCl) in 10 CV at a flow rate of 1 mL/min.
8. Regenerate the column to remove remaining strongly bound compounds by flushing with up to 4 CV of Buffer B (*see* **Note 29**).
9. Reequilibrate the column with at least 2 CV of Buffer A (*see* **Note 30**).
10. Repeat the experiment with the next buffer until an optimized separation in terms of speed and separation quality is obtained.
11. Regenerate and clean the column, put it into proper storage, and clean the chromatographic system (*see* **Note 21**).
12. For optimizing the throughput of the method, try the following:
    - Increase the sample load, the steepness of the gradient, and/or the flow rate as much as possible while still maintaining sufficient resolution (*see* **Note 31**).
    - Investigate the possibility of using a step gradient instead of the linear one. At least three steps are required: one for the elution of the less well-binding impurities, one for the target molecule, and one for the strongly binding impurities. The composition of each step can be taken from the chromatogram recorded for the linear gradient.

3.1.2. Hydrophobic Interaction Chromatography (HIC)

The separation principle of hydrophobic interaction chromatography (HIC) relies on the hydrophobic effect, similar as seen in the salting out of proteins.

All proteins contain hydrophobic patches, which preferentially make contact with other hydrophobic surfaces rather than with water. The effect is enforced by high salt concentrations in the water (increase of surface tension), especially of salt with pronounced water-structure-enforcing ability such as ammonium sulfate (*see* **Note 32**). Binding in HIC hence takes place in a high-salt buffer, while elution is achieved by lowering the salt content of the mobile phase. HIC can thus be used very efficiently after steps such as IEX or ammonium sulfate precipitation. HIC uses stationary phases with distinct hydrophobic ligands linked to an overall hydrophilic matrix. Other than in the related reversed-phase chromatography (RPC), which uses strongly hydrophobic surfaces and organic solvents for elution, the native structure of proteins is usually well preserved in HIC. For method development, proceed as follows:

1. Run through **steps 1–5, Subheading 3.1**.
2. Select a column based on the hydrophobicity of the target molecule (*see* **Note 33**).
3. Choose Buffer A (e.g., 50 m*M* phosphate buffer pH 7, 1.5 *M* ammonium sulfate) and Buffer B (low salt) based on the stability and hydrophobicity of the target molecule (*see* **Note 34**).
4. Equilibrate the column with at least 5 CV of Buffer A; a stable detector signal/baseline should be seen at the end of the equilibration.
5. Adjust the pH of the sample/feed to that of Buffer A and adjust a fairly high ionic strength (4 *M* NaCl or 1.5 *M* $(NH_4)_2SO_4$).
6. Load the sample onto the column. A flow rate of 1 mL/min can be used with small (few mL) columns (*see* **Note 27**). The use of a guard column may be advisable in the case of complex sample (*see* **Note 28**). For a good separation, the total amount of protein loaded should not exceed 30% of the nominal dynamic binding capacity of the column as given by the manufacturer (*see* **Note 26**, which applies in similar manner to HIC).
7. Run a gradient from 0 to 100% Buffer B (i.e., 1.5 to 0 *M* $(NH_4)_2SO_4$) in 10 CV at a flow rate of 1 mL/min (*see* **Note 35**).
8. Regenerate the column to remove remaining strongly bound compounds by flushing with up to 4 CV of distilled water.
9. Repeat the experiment with the next buffer (*see* **Note 8**) until an optimized separation in terms of speed and separation quality is obtained.
10. Regenerate and clean the column, put it into proper storage, and clean the chromatographic system (*see* **Note 21**).
11. For optimizing the throughput of the method, try the following:
    - Increase the sample load, the steepness of the gradient, and/or the flow rate as much as possible while still maintaining sufficient resolution (*see* **Note 31**).
    - Investigate the possibility of using a step gradient instead of the linear one. At least three steps are required: one for the elution of the less well-binding impurities, one for the target molecule, and one for the strongly binding impurities.

The composition of each step can be taken from the chromatogram recorded for the linear gradient.

3.1.3. Affinity Chromatography (see **Note 36**)

The principle of affinity chromatography is very simple. A biospecific ligand that recognizes (binds) the target molecule is linked to the stationary phase. The mixture containing the target molecule is pumped through the column under binding conditions, the target molecule is retained on the column, and all other sample/feed components move through. Afterwards the target molecule is eluted from the column, most commonly by applying somewhat extreme and denaturing pH values in the elution buffer. Alternatively, changes in the ionic strength or the polarity of the mobile phase have been used for elution, as have agents that compete with the bound ligand for the target molecule ***(4)***. For method development, follow the following steps:

1. Run through **steps 1–5, Subheading 3.1**.
2. Find a suitable affinity ligand for the target molecule (*see* **Note 37**).
3. Prepare a stationary phase/affinity column. For some of the more common affinity ligands, prepacked columns can be bought. When a proprietary or specific ligand is to be used, this ligand may have to be coupled to a preactivated stationary phase (*see* **Note 38**).
4. Define the binding and eluting buffer. For most group-specific/commercially available ligands and affinity columns, instructions can be obtained from the manufacturer. Otherwise binding will usually take place under physiological conditions (especially in regard to the pH and ionic strength), while elution can usually be brought about by a pH shift, e.g., toward a pH of 3 (glycine or citrate buffer).
5. Equilibrate the column with the binding/washing buffer (at least 5 CV); a stable detector signal/baseline should be seen at the end of the equilibration.
6. Pass the feed through the column. A flow rate of 1 mL/min can be used with small (few mL) columns (*see* **Note 27**). Since most molecules pass through the affinity column, sample size does not matter as much as much as target molecule concentration in determining the loading volume. Usually up to 80% of the dynamic binding capacity of the column can be used even under process conditions before a significant breakthrough of the target molecule is observed.
7. Wash the column with 1–2 CV of binding/washing buffer after loading to remove residual sample matrix compounds.
8. Switch to the elution buffer (*see* **Note 39**).
9. Collect the eluting target molecule in a buffer that puts it quickly back into physiological conditions (such as pH 7, if acidic pH is used for elution) to reduce the risk of product loss through irreversible denaturing (*see* **Note 40**).
10. Regenerate and sanitize the column as recommended by the manufacturer, store it under proper conditions, and clean the chromatographic system.

3.1.4. Gel Filtration (SEC, GPC)

Gel filtration differs from most other chromatographic modes in that it does not require an interaction to achieve separation. Instead, the molecules are separated according to size. The mobile phase can be chosen exclusively according to the requirements of the protein, and GPC is therefore an extremely gentle form of separation. GPC columns are packed from porous particles that show a broad yet well-defined range of pore sizes. Small molecules have access to the entire internal porous space and elute late, whereas very large molecules will not be able to access any of the pores and elute early. Molecules of intermediate size will have access to a certain fraction of internal pore space determined by their size and thus will differ in column residence time, the basis for their separation. The drawback of GPC is that the sample zone is diluted considerably in the process and the columns have only a low separation capacity. Often the sample volume cannot exceed 1% of the column volume. GPC is therefore typically used late in most protein-purification schemes (polishing step), when the sample volume has already been considerably reduced (*see* **Note 41**). The pertinent steps for method development are as follows:

1. Run through **steps 1– 5, Subheading 3.1**.
2. Select a column according to the separation need. A wide variety of GPC columns is available covering large and small ranges of molecular sizes to be separated (*see* **Note 42**).
3. Choose a mobile phase. In principle any liquid may be used as mobile phase in GPC as the separation does not depend on it. Often the choice of mobile phase is governed by the needs of the subsequent step. However, keep in mind that some secondary interaction between the proteins and the matrix is always possible. A good choice is probably a neutral phosphate buffer containing some NaCl (e.g., 0.1 *M*) to reduce electrostatic interactions.
4. Equilibrate the column with at least 5 CV of the mobile phase; a stable detector signal/baseline should be seen at the end of the equilibration.
5. Load the sample onto the column. The sample volume should not exceed 1% of the column volume at this point.
6. Record the chromatogram at a flow rate of 1 mL/min (for small, few-mL columns, *see* **Note 27**).
7. After the separation has taken place, clean and sanitize the column according to the manufacturer's instructions, put it into proper storage, and clean the system (*see* **Note 21**).
8. For optimizing throughput of the method try the following (*see* **Note 43**):
   - Investigate the possibility of using higher flow rates without reducing the resolution too much.
   - Investigate the possibility of using a higher sample load (up to 15% of the column volume has occasionally been reported in GPC, although the maximum value for semi-preparative separations is probably 5%).

3.1.5. Alternative Methods

Affinity chromatography, HIC, IEX, and GPC are the four chromatographic methods most commonly used for protein purification. However, in certain cases alternative methods may be welcome, two of which are discussed below.

RPC, another popular chromatographic method, is usually not suitable for preparative protein separation because its hydroorganic mobile phases tend to denature proteins. However, RPC is a very powerful separation method and may occasionally constitute an interesting option, especially in the case of small proteins (large peptides). The downstream process for recombinant human insulin, for instance, contains an RPC step. The following considerations apply to the setup of an RPC protocol:

1. Stationary phases/prepacked columns for RPC can be obtained from a variety of suppliers (*see* **Note 12**).
2. Mobile phases in RPC contain organic solvents such as acetonitrile, methanol, or isopropanol (*see* **Note 44**).
3. Gradients are run from low content of organic solvent (<5%) to high content of organic solvent (>90%; *see* also **Note 45**).

Another chromatographic method that is usually not included in the standard repertoire of protein chromatography, but nevertheless constitutes an interesting option, is hydroxyapatite chromatography (HAC). Hydroxyapatite (HA) is a ceramic material that contains phosphate (P sites), calcium (C sites), and hydroxy groups. The surface bears a negative net charge; hence HA interacts with positively charged molecules like a cation exchanger. In addition, however, there is the possibility of an interaction of chelating groups, such as carboxylate residues on the protein surface, with the C sites. The interaction of HA with proteins is therefore of a mixed-mode type and difficult to predict, a fact that may have contributed to the relative reluctance to use the material. However, HAC has some distinct advantages. The native structure of the proteins is usually well preserved, and because of the mixed-mode interaction, very selective separations can be designed. The material is stable at elevated pH; hence sanitization with 1 *M* NaOH presents no problem. For setting up a HAC separation, proceed as follows:

1. Select a column. BioRad, for instance, provides a variety of ceramic hydroxyapatites that vary in surface composition (C site:P site ratio) and hence interaction potential. Although some recommendations are given as to which material should be used for which purpose/molecule class, in many cases it may be worthwhile to screen several materials.
2. Select Buffer A and Buffer B, typically based on 10 m*M* buffer for Buffer A and 500 m*M* phosphate for Buffer B (*see* **Note 46**).
3. Run a gradient from 10 to 500 m*M* phosphate buffer at pH 7.4 in 10 CV.

4. Adjust the pH. Since the interaction with hydroxyapatite depends on the charge status of the proteins, some pH screening should be done in the range defined by the stability of the HA toward lower pH (dissolution < 6.5) and the stability of the proteins toward higher pH values.
5. After the separation has been finished, regenerate the column with 1 *M* phosphate buffer and sanitize it with 1 *M* NaOH followed by reequilibration with Buffer A (*see* **Note 30**). HA materials may be stored in methanol to prevent bacterial growth. Follow the manufacturer's instructions in regard to regeneration, cleaning/sanitization, and storage whenever possible.
6. Clean the chromatographic system (*see* **Note 21**).
7. For improving the throughput of the separation, try the following:
   - Optimize the gradient volume.
   - Optimize the flow rate.

3.2. Setting Up an Orthogonal Purification Process

High purification factors can be obtained with a single, well-optimized chromatographic step. However, most protein purifications that aim at more than 90% purity will require more than one step. In order to keep product loss as low as possible, the required purity should be reached in the minimum number of steps, typically less than four. In order to design these steps efficiently, it is helpful to mentally divide the purification process into four distinct phases: (1) sample preparation (e.g., lysis, clarification, extraction), (2) capture (isolation, concentration, and stabilization of the product containing fraction), (3) intermediate purification (removal of most impurities), and (4) polishing (removal of remaining traces of impurities and closely related substances such as dimers).

Sample preparation should make the feed ready for entering the chromatographic sequence. In the case of packed bed columns, this usually means a clarification step, i.e., the removal of solids (danger of column clogging) from the product stream by filtration or centrifugation. In certain cases the sample matrix may not be suitable for binding the product to the stationary phase; in such cases a buffer exchange may be necessary. Speed and capacity are of the essence, especially in the capturing step, where the majority of the water is removed, the product is stabilized, and the most dangerous impurities (e.g., proteases) are removed. In intermediate purification and especially during polishing, resolution is more critical, while capacity becomes less critical as the sample size decreases with increasing purity.

Before actually developing the purification strategy, the following issues should be resolved:

1. Analytical procedures for quantification and characterization of the product and key impurities should be set up.
2. Suitable procedures for sample/feed preparation (i.e., phase 1 of the DSP) should be set up.

The following questions should be answered:

1. What is the final scale of the process (not all separation procedures scale well), and what is the required product purity/quality (overpurification leads to unnecessary costs and loss in product)?
2. What is the source of the feed, and what are the critical impurities/contaminants? In this context all available information about the properties of the target molecule and key impurities should be collected.
3. What is necessary/available in terms of time, personnel, apparatus, and financial support?

Finally, the following rules should be followed when setting up the purification process:

1. The developed procedure should be highly orthogonal, i.e., use a different separation principle at each step.
2. Potentially damaging contaminants (e.g., proteases) should be removed as early as possible (*see* **Note 47**).
3. The number of steps should be minimized.
4. The handling (e.g., buffer exchange) between separation steps should be minimized.
5. The use of additives should be minimized (*see* **Note 47**).

Extensive method and sequence scouting may require access to resources such as a ÄKTA design chromatographic system (GE Healthcare, Life Sciences). However, even without such a support the setup of a suitable separation method is normally possible. The following protein parameters are typically exploited for separation: charge (IEX), size (GPC), hydrophobicity (HIC), and capability of biospecific interaction (affinity chromatography [AC]). In order to minimize sample handling and the number of steps, the eluate of a given column should ideally already be suited for loading onto the following one. Typical sequences in protein isolation include IEX (elution in high-salt buffer)–HIC (column loading in high-salt buffer) or ammonium sulfate precipitation–HIC. GPC, on the other hand, can be used after any type of chromatographic separation, because it does not depend on a surface interaction. GPC can, in addition, combine a separation with a buffer exchange step.

In most cases the following sequence of chromatographic steps will lead to a suitable downstream process in the case of (recombinant) proteins: IEX (for capture)–HIC (for intermediate purification)–GPC (for polishing). For very dilute feed streams or for certain product classes (most notable antibodies), capture by AC, i.e., the sequence AC–IEX–GPC, may prove more efficient because the affinity step will be more selective towards the product molecule. However, keep in mind economical considerations such as the cost of the stationary phase and the danger of product loss because of the harsh elution conditions in AC when setting up such a strategy. Handling crude feeds may also require harsh column regenerating and sanitization protocols, which not all affinity ligands can take (*see* **Note 48**). If AC is used in the purification, a suitable assay for detecting any ligand leakage into the product must be set up (*see* **Note 49**). When IEX is to be used in the sequence, both anion and cation exchanger phases should be considered. Most proteins bear a negative net charge at neutral pH and hence bind to an anion exchanger. However, should the product bear a positive net charge (e.g., many antibodies), a very selective separation step can be designed on the basis of a cation exchanger step. RPC can be considered as alternative and extremely efficient polishing methods for such proteins that can withstand the hydroorganic phases used for this type of chromatography.

3.3. Alternatives to Elution Chromatography and Batch Column Approaches

Most chromatographic separations today are carried out in the discontinuous, gradient elution batch mode. While this format is certainly most suitable at small scale, it may contribute to the high operating costs of chromatography at larger scale. Some putative alternatives to this approach are discussed here.

3.3.1. Expanded/Fluidized Bed Chromatography

The "stationary" phase does not necessarily have to be in the form of a packed bed in chromatography; it can also consist of a fluidized or expanded bed with the solid adsorptive particles suspended in the upmoving liquid phase. Fluidized absorber beds mix freely in both the axial and radial directions. They behave therefore very much like simple one-stage adsorber tanks, where the solid phase particles are stirred into the product-containing solution and recovered—loaded with adsorbed product—afterwards by filtration. For the preparation of expanded adsorber beds (a concept commercialized by GE Healthcare, Life Sciences), particles that differ in density are fluidized, so that certain particles will show a tendency for remaining at the bottom of the bed while others will rise to the top. This assures a certain resolution in space. A special system is available from GE Healthcare, Life Sciences for Expanded bed chromatography **(Fig. 1)**.

The expanded bed approach has the advantage of speed and the fact that particulates in the sample (a huge problem for packed beds!) present no problem in this case. A typical operation sequence is shown in **Fig. 1**. In a first step the solid adsorber particles are expanded in equilibration buffer. For a given linear mobile phase flow rate, the expansion of the bed depends on the viscosity and the temperature of the expansion liquid. After equilibration the sample is applied (*see* **Note 50**) to the expanded bed followed by a washing step still in the expanded mode. Elution and column regeneration is then done in the packed

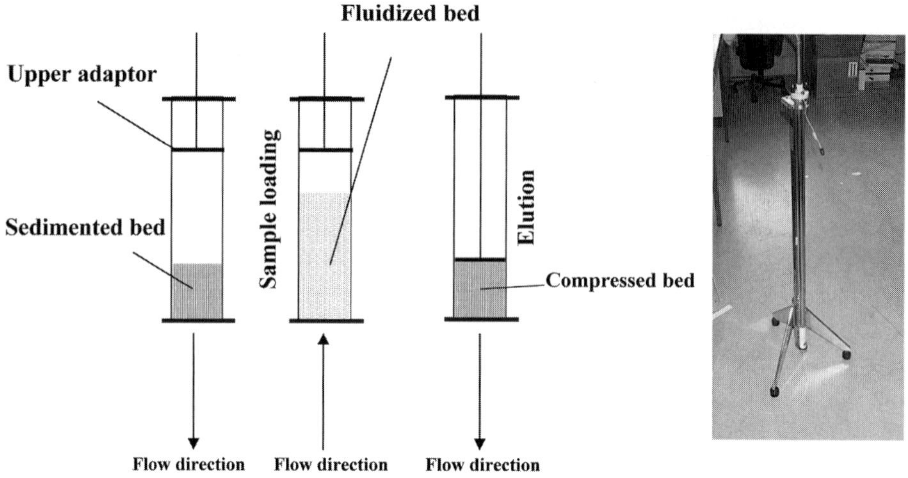

Fig. 1. Instrument and schematic presentation of a chromatographic separation in the expanded bed mode.

bed mode with the fluid phase flow downward, i.e., opposite the one used in the expanded mode.

Expanded beds have been used for direct product capture from unfiltered cell suspensions. The approach is best suited for affinity chromatography but has also been proposed for ion IEX and HAC. For method development, proceed as follows:

1. Develop the separation under standard packed bed column condition using a clarified feed (*see* **Note 51**) and keeping in mind that the stationary phase must be suitable for fluidization/expansion and that step gradients are more compatible with expanded/fluidized bed separations than continuous gradients. If necessary, adjust the sample composition (e.g., pH, ionic strength) to improve binding.
2. Transfer conditions to the fluidized/expanded bed approach using similar settled bed volumes and flow rates as with the batch column (*see* **Note 51**).
3. Further optimize binding, elution, and regeneration/cleaning steps for the small expanded bed.
4. Regenerate and clean the column as well as the system and put the solid phase materials into storage (*see* **Note 52**).
5. Once developed, scale up to process conditions of several hundred liters of stationary phase is possible (*see* **Note 53**).

3.3.2. Displacement Chromatography

Displacement chromatography is another way to intensify chromatographic separations *(5)*. In this mode the substance mixture is separated into consecutive, rectangular substance zones instead of peaks. **Figure 2** illustrates the principle

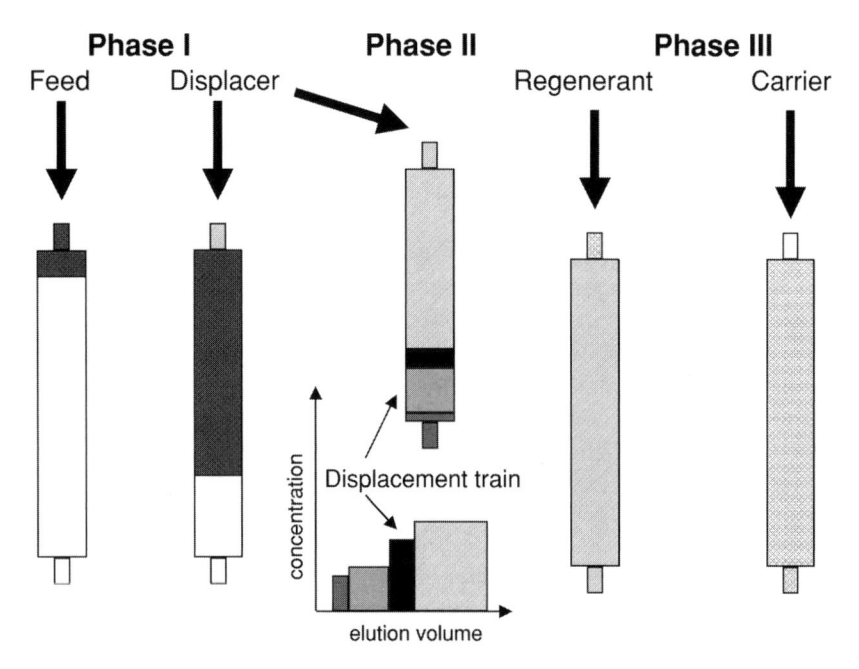

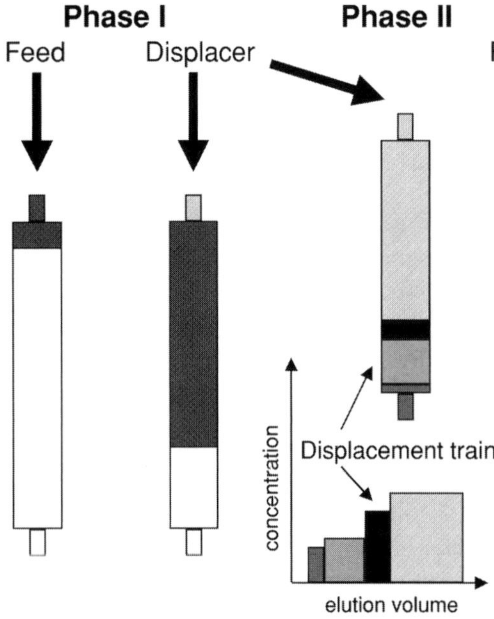

Fig. 2. Schematic presentation of a chromatographic separation in the displacement mode.

of a displacement separation. In a first step the column is equilibrated with the mobile phase (binding buffer). Then the feed is pumped through the column until exhaustion of the binding capacity. Then the displacer, i.e., a substance that binds even better than the feed components to the stationary phase, dissolved in the mobile phase, is pumped through the column. As a consequence, the displacement train develops, i.e., the substance mixture is resolved into consecutive zones of the pure substances followed by the displacer front. Finally, the column is regenerated (cleaned/sanitized if necessary) and reequilibrated with the mobile phase.

Displacement chromatography is an exclusively nonlinear form of chromatography using heavily overloaded columns. This causes a competition of the feed components for the binding sites already during column loading. The competition is enhanced as the displacer front moves through the column and the more strongly bound compounds displace the more weakly bound ones from the stationary phase surface. The result is the elution of all compounds from the column in the form of consecutive zones.

Advantages of displacement chromatography include (1) no need for gradients—a simple displacer step suffices, (2) efficient use of the stationary phase capacity, (3) concentration during separation (*see* **Note 54**), (4) control over the concentration in the substance zones (*see* **Note 54**), and (5) the displacer zone stays behind the separated feed components (*see* **Note 55**).

In order to develop a separation by displacement chromatography, proceed as follows:

1. Choose a stationary phase, i.e., for protein displacement chromatography typically an IEX material (*see* **Note 56**).
2. Choose a suitable mobile phase. Displacement chromatography requires strong binding of the compounds of interest to the column. A Buffer A used for elution chromatography on a given column will often be a suitable mobile phase for displacement chromatography.
3. Chose a suitable displacer. This substance should bind more strongly than any of the feed components to the stationary phase (*see* **Notes 54** and **55**). Information on possible displacers for various stationary phases can be found in **refs.** ***5*** and ***6***.
4. Record the displacer isotherm (e.g., as indicated in **ref.** ***7***), and if possible also the isotherm of the compound of interest under binding conditions, i.e., dissolved in the mobile phase.
5. Prepare the feed. If the feed composition is not suitable for strong binding of the compounds of interest, a buffer exchange may be required.
6. Load the column with feed and determine the breakthrough curve.
7. From this curve determine the dynamic capacity of the column for your separation (amount loaded up to the moment when 10% of the final concentration is breaking through).
8. Regenerate and reequilibrate the column. Regeneration buffers used in elution chromatography should also be suitable in the corresponding displacement approach.
9. Load to column with the feed mixture applying approx 80% of the binding capacity.
10. Switch to the displacer solution and pump the solution through the column at low flow rate (0.1 mL/min for small, few-mL columns) (*see* **Note 57**).
11. Collect fractions for future analysis (*see* **Note 58**).
12. After the displacer front has broken through (*see* **Note 59**), stop the flow and regenerate and reequilibrate the column.
13. Analyze the fractions, e.g., by RPC, to document the separation.
14. Optimization of the separation can be attempted by (1) changing the displacer concentration, (2) changing the mobile phase composition (*see* **Notes 54** and **55**), and (3) increasing the column length (*see* **Note 60**).

3.3.3. Continuous Annular Chromatography

Chromatography is mostly thought of as a typical batch process. However, continuous chromatography is possible. The most established approach to continuous chromatography is at present the simulated moving bed (SMB) approach. However, SMB is most easily applied to the separation of a two-component mixture, and protein purifications tend to be more complex. The continuous separation of a multicomponent mixture is possible by continuous annular chromatography (CAC) ***(8)***.

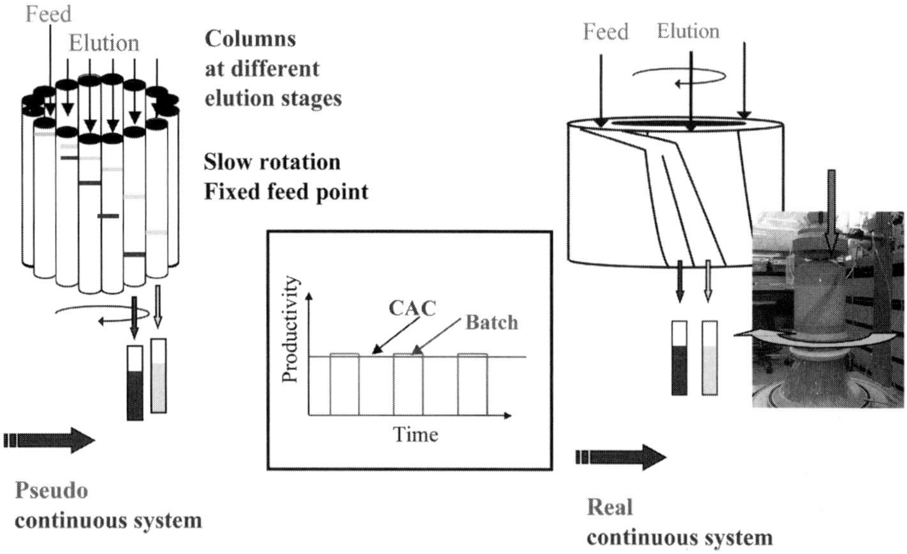

Fig. 3. Schematic principle of a chromatographic separation by continuous annular chromatography (CAC).

In CAC the stationary phase is packed into a slowly rotating hollow cylinder, which is continuously percolated from top to bottom by different liquids (e.g., mobile phase, feed, gradient steps, regeneration/cleaning solutions) (*see* **Fig. 3**). The resolution in time in the batch column approach thus becomes a resolution in space in CAC. Many chromatographic separations developed using a batch column can be directly transferred to CAC, including displacement chromatography and step gradients; only linear gradients are incompatible with CAC elution (*see* **Note 61**). In order to do so, follow these steps:

1. Set up and optimize the chromatographic separation using a small batch column packed with a similar stationary phase as intended for the CAC column, keeping the pressure limit of the CAC-system in mind.
2. Transfer the procedure to the CAC by using the loading factor concept ***(9)***, i.e., applying Eq. 1 to calculate the loading factor for the batch column and Eq. 2 for transfer to the CAC column (*see* **Note 62**):

$$LF[-]=\frac{Q_T\left[\mathrm{cm^3min^{-1}}\right]t_I[\mathrm{min}]}{H[\mathrm{cm}]S\left[\mathrm{cm}^2\right]} \quad (1)$$

$$LF[-]=\frac{Q_F\left[\mathrm{cm^3min^{-1}}\right]360[°]}{\mathrm{H[cm]S}\left[\mathrm{cm}^2\right]\omega\left[°\,\mathrm{min}^{-1}\right]} \quad (2)$$

where Q_T/Q_F is volumetric flow rate, t_I is time required for injection, H is bed height, S is column cross-subheadingal area (cylinder thickness in case of the CAC column), and ω is the speed of rotation for the CAC.

3.3.4. Affinity Precipitation

The use of chromatographic columns is less than ideal during the early capturing steps, where the feed is still comparatively complex. This is especially the case for the expensive affinity columns. Capturing by affinity, on the other hand, does not require the principal advantage of chromatography, namely the possibility of multistage interaction. In this context, affinity precipitation can present a viable alternative ***(10,11)***. In affinity precipitation the affinity ligand is linked to a stimuli-responsive, reversibly water-soluble polymer to create the affinity macroligand (AML). Upon the application of the stimulus (temperature increase), the formerly soluble AML precipitates abruptly. The process is fully reversible, and once the stimulus is removed (temperature lowered) the AML again becomes soluble. A separation by affinity precipitation proceeds as shown in **Fig. 4**. The AML is added to the feed under binding conditions and the affinity complexes form. By applying the stimulus, i.e., surpassing the critical solution temperature of the AML, the affinity complex is precipitated and recovered, e.g., by filtration or centrifugation. The product may be eluted directly from the precipitate or after dissolution of the complex in elution buffer (then typically followed by a precipitation of only the AML from the solution by application of the stimulus).

The exact conditions for the recovery of a given target molecule by affinity precipitation depend on the nature of the employed interaction. Here a general protocol is given using the recovery of a recombinant human antibody by a poly(*N*-isopropylacylamide) (PNIPAAm)–Protein A-AML (critical solution temperature = 34°C in pure water) as an example.

1. Choose suitable binding and dissociation buffers that support or interrupt the intended affinity interaction well (*see* **Note 63**). Indications can usually be taken from the corresponding affinity chromatography protocols. In the case of antibody recovery by PNIPAAm-Protein A-AML, capture can usually take place directly in the cell-free supernatant even in the presence of serum.
2. Stir the AML into the target molecule solution using a 10-fold molar excess (*see* **Note 64**). The temperature should be at least 5°C below the critical solution temperature of the AML in the binding environment during this process (binding temperature; *see* **Note 65**).
3. Stir at binding temperature for 30 min (*see* **Note 66**).
4. Raise the temperature at least 5°C above the critical solution temperature of the AML in the binding environment (precipitation temperature; *see* **Note 65**) and

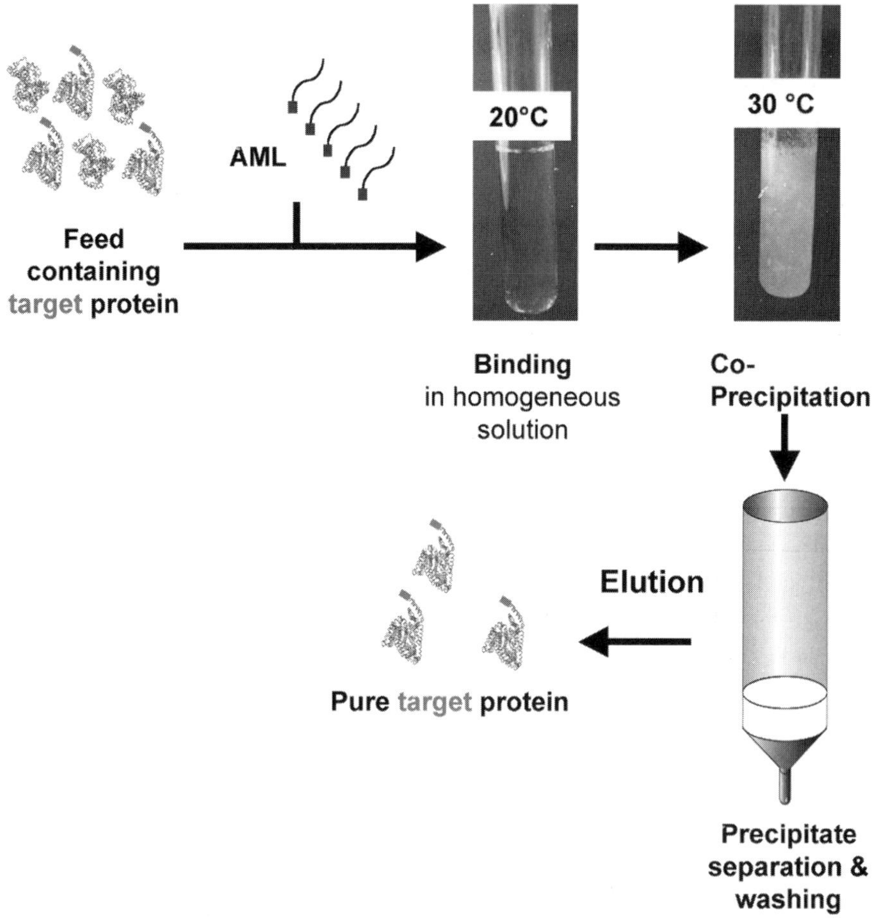

Fig. 4. Schematic presentation of a product capture by affinity precipitation. AML, affinity macroligand.

remove the precipitated affinity complex by centrifugation (10,000*g*, at precipitation temperature, 10 min).

5. Remove impurities entrapped in the precipitate by repeated thermoprecipitation in binding buffer (redissolution of the precipitate in 4°C cold binding buffer [*see* **Note 67**], precipitation at precipitation temperature, recovery of the precipitate at 10,000*g* for 10 min at precipitation temperature, followed by redissolution of the pellet in 4°C cold binding buffer, *see* **Note 68**).
6. For release of the target molecule, redissolve the pellet in 4°C cold dissociation buffer (*see* **Note 68**). In the case of antibody capture by PNIPAAm-Protein A-AML, a 0.1 *M* glycine-HCl buffer, pH 2.7, can be used as dissociation buffer.

7. Remove the AML from the purified product by thermoprecipitation followed by centrifugation (*see* **Notes 69** and **70**).

Compared to affinity chromatography, affinity precipitation has the following advantages: it is easily scalable and independent of the composition (e.g., viscosity) of the feed, it has superior ligand efficiency, and there exists the possibility of designing a "disposables-only" type of separation protocol. AML and instrumentation for affinity precipitation can be obtained from polyTag technology AG, Switzerland (http//www.polytag.ch).

4. Notes

1. Plate height (H) or plate number (N) denotes the efficiency of a column. Both depend strongly on the mobile phase flow rate. The plate number can be calculated by dividing the column length (L) by the plate height. The plate height at a given mobile phase flow can be calculated from the peak width at half height ($w_{1/2}$) and the retention time t_R as follows: $H = (w_{1/2}/t_r)^2(L/5.54)$.
2. Columns from one manufacturer can usually be used with the chromatographic system from another with suitable adapters. Biochromatography is usually performed at fairly low pressure (<10 bar) compared to high-performance liquid chromatography (HPLC) (up to 400 bar). The assembly of an instrument from different components is less common in biochromatography than in HPLC and can only be recommended when suitable low-pulse pumps are available. Certain types of biochromatography, i.e., capture by affinity chromatography, are also possible without an instrument. Instead, the sample is simply injected (by hand-held syringe) into the column, which is then washed followed by an elution step (injection of elution buffer by hand-held syringe).
3. Chromatographic stationary phases and columns are expensive high-tech materials. Detailed instructions on handling, including regenerating, cleaning, storing, and (re)packing can be obtained from the respective manufacturers. These instructions should be followed.
4. All liquids that pass through the column should be prepared with fresh, double distilled or deionized water (resistance >18 Ohm/cm). Afterwards liquids should be passed through a 0.45-μm filter to remove any particulates (1-μm filters for stationary phase larger than 50 μm) and finally be degassed, e.g., by applying vacuum (30 min). Such buffers should be stable without additives for several days. 0.02% sodium azide (NaN_3) can be added to prevent bacterial growth. Buffers can be kept longer if sterile (by filtration under sterile conditions through a sterile 0.2-μm filter, not by autoclaving!) and then should be stored at 4°C in the dark.
5. The buffer capacity should be high enough to keep the pH stable throughout the operation, i.e., also during sample application, but not much higher, since buffer ions also contribute to elution via charge screening in IEX. A buffer concentration of 20 mM is usually a good compromise. The pH should be at least 0.5 units

below (cation exchanger) or above (anion exchanger) the isoelectric point of the target molecule.

6. 1 *M* NaCl is usually sufficient to elute any given protein from an IEX column. Higher concentrations (up to 2.5 *M* NaCl) are occasionally used but may cause problems because of manifestation of hydrophobic effects (protein precipitation, interaction with the stationary phase because of hydrophobic rather than electrostatic interaction).
7. Stationary phases used for biochromatography are normally not operated at very high pressures; the use of silica-based materials (such as in HPLC) is therefore not necessary, and most of today's biochromatography materials are based on modified polysaccharides or organic polymers. Such materials can easily withstand up to 1 *M* NaOH. Since NaOH is a very efficient means for cleaning, it should be used whenever possible for that purpose and also (at a lower concentration) as medium for column storage. However, not all materials are compatible, and the manufacturer's instructions for cleaning and storage should be read carefully and followed in case of a diverging recommendation. Twenty percent ethanol can be used as an alternative (bacteriostatic) medium for column storage.
8. Buffers should have sufficient capacity to maintain the pH during the separation, including sample loading. In HIC the buffer strength is usually a little higher than in IEX (i.e., 50 rather than 20 m*M*) to provide some charge screening (i.e., suppression of electrostatic interaction) even at low salt concentration. The buffer pH is usually chosen according to the needs (stability, preservation of biological activity) of the target molecule. However, the pH also has an effect on the separation, which is not easy to predict for a given HIC separation. Some pH screening should be included in the method development. Salts are used in HIC to promote binding. $(NH_4)_2SO_4$ is a strong promoter of hydrophobic effects ***(12)*** and therefore a popular agent in HIC. The amount of $(NH_4)_2SO_4$ added to Buffer A should be great enough to assure good binding of the target molecule to the column; a concentration of 1.5 *M* is typical. However, certain proteins already precipitate at this concentration. In this case a lower amount has to be used. Na_2SO_4 can be used as alternative with similar salting-out potential. Occasionally, high amounts of NaCl (up to 5 *M*) are suggested, especially for HIC following an IEX step. However, adding $(NH_4)_2SO_4$ is recommended even in such cases.
9. Reducing the $(NH_4)_2SO_4$ content of the mobile phase should be sufficient for eluting most proteins from a HIC column. For a further increase of the elution strength up to 40% (v/v), ethylene glycol or isopropanol may be used as Buffer B (or C) instead.
10. A low-pH elution buffer in affinity chromatography serves also as virus-inactivation step.
11. Most recently a Protein A has come to the market (GE Healthcare, Life Sciences) that can withstand NaOH to some extent.
12. Because RPC is less common in protein chromatography, many of the suppliers for biochromatography materials do not provide a wide selection of such stationary

phase materials. On the other hand, suppliers such as Agilent may be a very good source of suitable columns and materials. Keep in mind that many RPC materials were originally designed for HPLC applications and therefore are based on silica. Such columns cannot be cleaned/sanitized with caustic solutions such as NaOH (rapid dissolution > pH 9). In addition, for many of these columns, a HPLC apparatus may be required because of the high backpressures of the columns.

13. Acetonitrile is a very common modifier in RPC of proteins. Methanol and isopropanol can be used instead and with some columns/applications are to be preferred. See the manufacturer's instructions (application notes) for hints in this regard.
14. Hydroxyapatite becomes unstable below a pH of 6 but can be treated (cleaned/sanitized) with 1 *M* NaOH. Recently a fluoroapatite material has become available for protein chromatography (BioRad) that can be used down to a pH of 5.
15. Hydroxyapatite (HA) shows a mixed-mode interaction with proteins, including both electrostatic interaction with positively charged substances and Ca-chelating interaction, e.g., with protein-bearing carboxylic acid groups on the surface. The buffer pH will influence the relative importance of these two types of interaction for a given protein. Buffers in HA chromatography usually have near-neutral pH, but values as low as 6 and as high as 9 (Tris-HCl buffer) have been reported. Elution in HA chromatography is usually done in a gradient of increasing phosphate concentration, which diminishes both types of interaction. Especially for proteins interacting predominately by electrostatic interaction, elution in a NaCl gradient (up to 1 *M*) is also possible.
16. The displacer concentration influences the concentration in the substance zones (*see* **Note 54**) and must be adjusted accordingly.
17. This often means that the displacer elutes after the proteins of interest under gradient elution conditions. However, this may not be the case for substances with crossing isotherms. Recording of the single substance isotherms (e.g., according to **ref.** ***7***) is therefore suggested for a full characterization of the system.
18. In addition, small molecules have been developed as displacer for a variety of chromatographic stationary phases and modes (*see*, e.g., **refs.** ***13,14***). In the case of hydroxyapatite displacement chromatography, ethylene-glycol-bis-(β-aminoethyl ether)-*N,N,N',N'* acetic acid (EGTA) has been suggested as displacer ***(15)***.
19. Affinity precipitation depends on the reversible precipitation of the AML/target protein affinity complex once a certain critical temperature has been passed. The critical temperature depends on the solution and is lowered by most salts. For any given buffer to be used in affinity precipitation, the critical solution temperature should therefore be determined. This can usually be done by eye as the precipitation manifests itself by a strong turbidity or in a photometer (*see* **Note 65**).
20. 280 nm detects the aromatic amino acids. Proteins that are poor in these amino acids give a comparatively weak signal at this wavelength. A detection wavelength of 220 nm (peptide bond) is more generic and often more (sometimes too) sensitive. Nucleic acids can be detected at 260 nm.

21. Column and system flushing is especially important in biochromatography, where typically aqueous, salt-containing mobile phases are used. Salt crystals forming after water evaporation can quickly damage pumps and columns.
22. Instead of binding the target molecule to the ion exchanger, i.e., work under binding conditions, it is also possible to bind the contaminants and impurities and let the product run through the column.
23. Even if the same type of surface group (i.e., a series of Q materials) is used, differences in protein adsorption behavior will be observed, e.g., because of secondary interactions with the stationary phase matrix. Making the best choice can therefore be difficult. However, often a standard column or column selection is already in use in the research group. This is a good starting point. If this is not the case, many established providers of chromatographic materials also sell sets of small, prepacked columns of their standard ion exchanger materials. A set of these may also provide a good basis for column scouting. If at all possible, prepacked columns should be used for scouting to avoid the challenge of column packing. If a stationary phase is only available in loose form, packing instructions can usually be obtained from the supplier. Empty columns for packing can be obtained from the supplier of the chromatographic system.
24. If the isoelectric point of the target molecule is not known, choose either a strong anion exchanger at pH 8 (Tris-HCl buffer) for binding IEX or a strong cation exchanger at pH 8 for nonbinding conditions to start the method development (*see* **Note 22**).
25. Keep in mind that the net charge of a protein represents the sum over all charges. Even a nominally negatively charged protein may still contain a number of positively charged amino acid residues. Moreover, most proteins dissolve least well at their isoelectric point. If the isoelectric point of the target molecule is not known, using a 20 m*M* Tris-HCl (pH 8) buffer together with a strong anion exchanger or a 20 m*M* phosphate buffer (pH 6) together with a strong cation exchanger is a good starting point in method development.
26. IEX of proteins normally uses strong binding conditions during loading. The sample volume is therefore less important than the total amount of protein contained in the sample. The dynamic protein-binding capacity is provided by the supplier of the column or can be determined by measuring the breakthrough curve. For method scouting, column loading should not be too heavy. In an optimized separation, up to 80% of the dynamic protein capacity can usually be used in gradient elution approaches.
27. By dividing the volumetric flow rate (mL/min) by the column's cross-sectional area, the linear flow rate (usually given in cm/min) is obtained. The linear flow rate is kept constant during column scale-up.
28. Any real-life protein feed, even after clarification (filtration), may still contain components that can harm the column. A guard column is a small column packed from the same stationary phase as the separation column and inserted into the flow line just ahead of the separation column. Most of the damage is then done in the guard column, which will routinely be thrown away after a few cycles.

29. The amount of regenerating solution can be reduced according to the sample behavior. Instead of regeneration with pure Buffer B, it may be advisable to regenerate with a high-salt buffer (e.g., 2 *M* NaCl in Buffer A).
30. A conductivity detector is a good means to see when the high-salt buffer has been completely displaced from the column during reequilibration.
31. When increasing the flow rate, keep the pressure limit of the column and other system components (detector cell) in mind.
32. The ability of a salt to enforce hydrophobic interaction is linked to its position in the Hoffmeister series (*see* **Note 8**). Sodium sulfate is a good alternative to ammonium sulfate. NaCl is a weak enforcer of hydrophobic interactions.
33. Column selection is not always straightforward in HIC because the hydrophobicity of a protein is difficult to predict. Very hydrophobic proteins need mildly hydrophobic stationary phases, otherwise the interaction becomes too strong (difficult elution) and vice versa. Binding strength will increase with ligands in the following order: ether, isopropyl, butyl, octyl, phenyl. Unless some prior knowledge of the behavior of the protein (mixture) is available, it will be a good idea to screen a set of HIC columns covering these ligands. Keep in mind that not only the ligand, but also the matrix may affect the separation. Hence two butyl columns may show completely different behavior.
34. The buffer concentration in HIC is generally a bit higher than in IEX (i.e., 50 m*M* instead of 20 m*M*) to reduce secondary electrostatic interactions. The pH as well as the temperature influences the hydrophobicity of most proteins. Generally hydrophobicity (and as a consequence the tendency for precipitation) is most pronounced at the isoelectric point of the protein. If the behavior of the target molecule is unknown, a 50 m*M* sodium phosphate buffer (neutral pH) is a good starting point for mobile phase development, but some pH scouting should be done to optimize the separation (*see* **Note 8**).
35. If desorption is not achieved in 1.5 *M* ammonium sulfate, organic solvents such as up to 40% of ethylene glycol or 30% isopropanol may be used (keep the stability of the target molecule in mind).
36. Because affinity chromatography is characterized by strong and specific binding of the target molecule in the binding buffer, affinity chromatography in its most simple form does not require a chromatographic system or even a column. Instead the adsorbent may just be stirred into the sample mixture for adsorption, recovered by filtration, and washed on the filter. Only the target molecule will bind, and it can be recovered by resuspending the adsorbent in elution buffer followed by filtration. Alternatively, the sample may be injected in a prepacked column by hand-held syringe, followed by washing and elution steps.
37. This is not a trivial task. For some molecule classes—most prominently antibodies—group-specific affinity ligands have been described (such as Protein A). The same is the case for certain peptide tags (such as the Histidin tag or the FLAG tag), which can be fused to recombinant proteins to facilitate the purification. In the case of enzymes, an inhibitor can sometimes serve as ligand. For most other proteins the possibility of producing a specific antibody exists. However, *in praxi* there will be

many proteins for which no affinity ligand can be identified and which are therefore not accessible through purification by affinity chromatography.

38. Preactivated stationary phases for affinity ligand coupling (carrying, e.g., epoxy or bromcyan groups for the coupling of molecules with amino groups *[16]*), including monolithic ones, can be obtained from commercial suppliers. Coupling should then be carried out according to the manufacturer's instruction. Afterwards the stationary phase has to be packed into a column. Protocols for column packing can be obtained from the manufacturer of the chromatographic system or the provider of the empty column. Because plate numbers are less significant in affinity chromatography, the packing of a serviceable affinity chromatography column is a comparatively easy task that should be accomplishable by any experimentally oriented investigator.
39. Because the binding tends follows the on/off-type in affinity chromatography, elution is always in a step gradient to the elution buffer. Especially when glycine buffers are used for elution, the elution buffer step will manifest itself in an increase in UV absorption of the mobile phase, as glycine also shows absorption in the UV range.
40. In this regard it is also important to increase the speed of the elution step as much as possible (high flow rate) in order to minimize the time, during which the target molecule finds itself under the (usually mildly denaturing) elution conditions.
41. Viscous fingering may manifest itself if the target molecule concentration is already very high in GPC.
42. While size (molar mass) is a good indicator for the behavior of a given molecule in GPC, the shape of the molecule may also be of influence. Especially, elongated rigid molecules may therefore appear larger than they are.
43. Because GPC does not use gradients, sufficient plate numbers are very important for good resolution. One means of improving the separation is therefore to increase the column length, e.g., by packing a longer column or by using two columns in a row.
44. The TFA that is often added to the mobile phase in RPC acts as an ion-pairing agent (increasing the hydrophobicity/retention) of the protein and to control the residual charges on the protein/stationary phase surface that may give rise to secondary electrostatic interactions. 1% phosphate acid or triethylammonium phosphate may be used alternatively.
45. Until recently, RPC stationary phases such as C18 columns could not withstand 100% water (irreversible collapse of the carbohydrate chains, together with pronounced loss in capacity/resolution). Refer to the manufacturer of the column if this applies in a given case.
46. Elution in HAC is usually done in a gradient of increasing phosphate concentration, since this will elute both the P-site-interacting components (because of shielding of the electrostatic interaction by the increase in ionic strength) as well as the C-site-interacting ones (because of a competing interaction of the phosphate ion with the C sites). However, the double gradient method has been proposed whereby the P-site-interacting components are first eluted in a salt gradient (0–1 *M* NaCl in 10 m*M*

phosphate buffer) followed by a second phosphate gradient for the selective elution of the C-site-interacting compounds.

47. Often impurities such as proteases are taken care of, e.g., via the addition of protease inhibitors. While this may present an elegant solution at small scale, keep in mind that such an approach means the addition of an active agent to the product stream, which later will have to be removed and which will add to the process costs. Even at small scale such additives may interfere with the sample or the chromatographic separation in a complex manner.
48. The incompatibility of most (protein/peptide-based) affinity ligands with caustic solutions presents a problem in AC. Alkaline-compatible Protein A has recently become available (GE Healthcare, Life Sciences). Columns that use small affinity ligands, most notably immobilized metal affinity chromatography (IMAC) columns bearing Ni^{2+} ligands (affinity to the Histidin tag), are more suited to such sanitization procedures, especially after the Ni^{2+} has been stripped (columns can be recharged with Ni^{2+} after cleaning).
49. Protein A is, for instance, a neurotoxin. Protein A traces in the final product would be a major problem. Protein A can be detected by enzyme-linked immunosorbent assay. The respective assays must be set up for each type of (recombinant) Protein A ligand; there is no generic assay. Moreover, the ligand immobilized on the column should be used in free form to determine the calibration curve. Nor all suppliers will provide samples of their ligand for this purpose.
50. The viscosity of the feed may differ from that of the equilibration buffer.
51. Ideally the stationary phase type and amount as well as the flow rate (transport kinetics) should be similar during method development and during method transfer to the expanded bed. The Streamline material (GE Healthcare, Life Sciences) can be obtained in 20-mL prepacked columns for method development and also be used directly in the expanded mode (20 mL settled bed volume). Such an approach will make method development and transfer more straightforward. Flow rates for stable expansions of the Streamline material are 200 and 400 cm/h at room temperature and for typical aqueous buffers. Under these conditions the bed expands to about two to three times the bed height.
52. The manufacturer of the adsorbent will usually suggest suitable liquids/conditions for column regeneration/cleaning/sanitization as well as column storage. Following this advice is strongly recommended to optimize column life and continued performance.
53. During scale-up parameters that determine the separation efficiency should be kept as constant as possible. These include the solid phase material, the height of the settled and the expanded bed, the linear flow rate, and the relative amounts of sample and buffers applied to the column (measured in CV of the respective settled beds).
54. For a given displacer concentration, the speed of the advancing displacer front, u_i, is given by:

$$u_i = \frac{u_o}{(1 + \phi \, {}^{\partial q_i}\!/\!_{\partial c_i})}$$

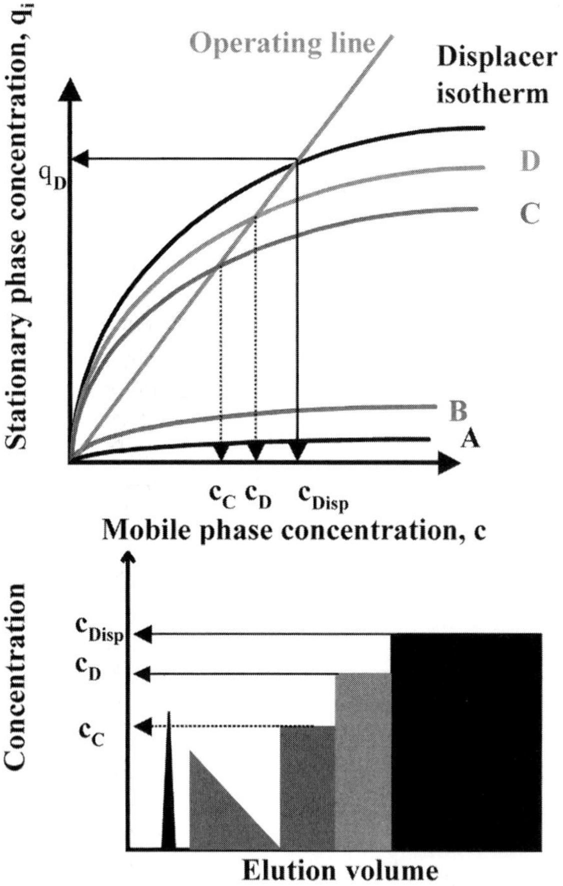

Fig. 5. Operation plot for displacement chromatography.

Since the mobile phase flow rate, u_o, and the phase ratio, ϕ, are constants for a given system and all substances in the displacement train move at the speed of the displacer front, the *q*:*c* ratio (amount adsorbed, *q*, for a given mobile phase concentration, *c*) must also be identical for all substances. In displacement chromatography the concentration in the substance zones thus depends only on the displacer concentration. Once this is fixed, an operating line can be drawn as indicated in **Fig. 5**, and the concentrations in the substance zones can be estimated from the intercept of the operating line with the single-component isotherm of the substance of interest. This treatment is based on some simplifications. Not all proteins show Langmuirian isotherms, and the protein microenvironment, e.g., in regard to local ionic strength, is not taken into account. A more comprehensive treatment of displacement chromatography has been proposed by Brooks and Cramer in the form of the steric mass action model *(17)*.

55. Instead of adjacent highly concentrated rectangular substance zones, which cannot exist in real life, displacement separations show abrupt but continuous changes in the concentration profile, and some zone overlap always occurs. Narrow overlaps require high-performance stationary phases, displacers of sufficient affinity (otherwise the final protein zone will tail into the displacer zone), and sufficiently fast adsorption/desorption kinetics.
56. The vast majority of protein separations by displacement chromatography have been performed in the ion exchange mode. HIC, AC, and HAC have been used occasionally ***(18)***. Peptides are often separated in the RPC mode.
57. Because of the requirement for adsorption equilibria and the high substance concentrations, the flow rate for a given column in the displacement mode will be roughly one order of magnitude lower than when the same column is used in the elution mode. This does not apply to monolithic columns, which can typically be used at similar flow rates in displacement and elution chromatography ***(19)***.
58. Since the substances in displacement chromatography are resolved into consecutive zones of high concentration, distinguishing between different substances, e.g., by UV detection, is impossible. Instead, fractions are collected in regular amounts and analyzed, e.g., by RPC or mass spectrometry.
59. The breakthrough of the displacer front in protein displacement chromatography often manifests itself by a diminution of the UV absorbance.
60. The column must be long enough to allow the full development of the displacement train.
61. CAC has been shown to be especially well suited for isocratic elutions, gel filtration, and refolding steps. While step changes in liquid composition can be transferred directly as indicated, continuous and especially linear gradients cannot be used in CAC.
62. In a batch column liquids can only be applied consecutively. In the CAC all parts of the column are continuously flushed with liquid. It is therefore possible that by applying the loading factor concept a section of the original mobile phases establishes itself, e.g., between two steps of the gradient. This is not a problem and is even necessary for a direct method transfer. Please note also that in CAC the volumetric flow rates as pumped into the column do not need to be (and in fact rarely are) equal for all liquids.
63. Just as in other affinity-based separations, the binding in affinity precipitation is relatively independent of the actual buffer composition (but *see* **Note 19**). Often the conditions in the raw feed solution will be suitable to affinity complex formation. In such cases the solid AML can be added directly to the feed in the desired concentration. Otherwise, buffer exchange must be performed via dialysis or ultrafiltration.
64. A 10-fold molar excess of binding places to target molecules usually works well. However, for fine-tuning of the process, e.g., in a large-scale bioseparation, it is usually worth the time to record an adsorption isotherm, i.e., determine for a given amount of AML the distribution of bound and unbound target molecules as a

function of the target molecule concentration over a wide concentration range from linear binding to complete AML saturation. This will give a very good idea of how much product will be recovered for a given amount of AML. In addition, the affinity constant can be determined from such a plot. Keep in mind that affinity precipitation is a one-stage process; there will always be an equilibrium between bound and unbound target molecules, although for a high-affinity AML this equilibrium may be well on the side of the affinity complex.

65. A binding temperature during affinity complex formation that is 5°C below the critical temperature or a precipitation temperature that is 5°C above this value usually works well in the case of PNIPAAm-based AML, because PNIPAAm shows a very sharp phase transition. However, in some solutions phase transition may occur over a broader temperature interval. High salt concentrations have been described not only to lower the critical solution temperature, but also to broaden the phase transition temperature interval. To be absolutely sure that a given temperature/solution will work, it is recommended to record the turbidity curve of the AML in that solution, i.e., measure the turbidity as a function of the temperature covering a temperature interval of 10–50°C. The information concerning the critical solution temperature as well as the broadness of the precipitation temperature interval will help to design a robust process. Also keep in mind that some PNIPAAm bioconjugates show a phase transition hysteresis, i.e., dissolution may occur at a lower temperature than precipitation.
66. A stirring of 30 min usually works well and allows enough time for establishing the binding equilibrium. However, the optimal duration of this process step is dictated by the kinetics of the affinity complex formation and may be much shorter. For fine-tuning of this parameter, the complex formation kinetics should be recorded, e.g., by determining the concentration of the unbound target molecule in the supernatant as a function of time.
67. Centrifugation works well especially for small samples. However, the recovered gelatinous pellet is often difficult to dissolve. Repeated thermoprecipitation under binding conditions for washing, i.e., the removal of physically entrapped impurities from the precipitate, may then become rather time-consuming. In such cases it may be helpful to add some fine cellulose fibers (e.g., Diacel 75 [75-μm fibers]; CFF GmbH & Co. KG, Thüringen, Germany) in a concentration of 50% ($w/w_{polymer}$) to the solution. The precipitate will then redissolve quickly. The cellulose does not interfere with the affinity interaction. At larger scale, filtration may be more useful than centrifugation for precipitate recovery. The interested reader is referred to the polyTag Technology web page (www.polytag.ch) for protocols and instrumentation to be used in such cases.
68. Usually the pellet dissolves more quickly at colder temperature. Especially for affinity complexes recovered by centrifugation in the absence of cellulose (*see* also **Note 67**), a dissolution temperature of 4°C is therefore recommended. For precipitates recovered in the presence of cellulose, a dissolution temperature of only a few degrees below the critical temperature (determination via turbidity curve in the

dissolution buffer; keep the possibility of a hysteresis in mind and record the cooling down curve as well in such cases; *see* also **Note 65**) may be more suitable.

69. It is also possible to recover the affinity complex via filtration in the presence of cellulose (*see* **Note 67**). Two methods are available for release of the target molecule. In the first case cold dissolution buffer is passed through the filter cake, thereby simultaneously redissolving the stimuli-responsive bioconjugate and releasing the target molecule from the complex. Alternatively, it is in this case also possible to pass a warm (temperature > critical solution temperature) dissociation buffer through the cake. In this case the target molecule is released, while the AML stays in the cake.
70. Depending on the intended use, but also on the biochemistry of the interaction and the release conditions, the AML may be recycled for reuse. Validation of this possibility similar to procedures used for affinity chromatography stationary phases is recommended.

References

1. Wurm, F. M. (2004) Production of recombinant protein therapeutics in cultivated mammalian cells. *Nat. Biotechnol.* **22**, 1393–1398.
2. Sofer, G. and Hagel, L. (1997) *Handbook of Process Chromatography: A Guide to Optimization, Scale-Up, and Validation.* Academic Press, New York.
3. Deutscher, M. P. (ed.) (1990) Guide to protein purification. *Methods Enzymol.* **182**, 1–818.
4. Turkova, J. (1993) *Bioaffinity Chromatography,* 2nd ed. Elsevier, Amsterdam.
5. Freitag, R. (1998) Displacement chromatography: application to downstream processing in biotechnology, in *Bioseparation and Bioprocessing,* Vol. 1: *Biochromatography* (Subramanian, G., ed.), VCH Verlagsgruppe, Weinheim, pp. 89–112.
6. Freitag, R. and Wandrey C. (2003) Synthetic displacers for preparative biochromatography, in *Synthetic Polymers in Biotechnology and Medicine* (Freitag, R., ed.), Landes Biosciences, Biotechnology Intelligence Unit 4, Landes Biosciences, Eurekah-Com.
7. Jacobson, J., Frenz, J., and Horvath, C. (1984) Measurement of adsorption isotherms by liquid chromatography. *J. Chromatogr.* **316**, 53.
8. Hilbrig F. and Freitag R. (2003) Continuous annular chromatography. *J. Chromatogr. B* **790**, 1–15.
9. Giovannini, R. and Freitag, R. (2002) Continuous isolation of plasmid DNA by annular chromatography. *Biotech. Bioeng.* **77**(4), 445–454.
10. Hilbrig, F. and Freitag, R. (2003) Affinity precipitation for protein purification. *J. Chromatogr. B* **790**, 79–90.
11. Freitag, R. and Hilbrig, F. (2005) Use of the avidin (imino)biotin system as a general approach to affinity precipitation, in (McMahon, R. J., ed.), Humana Press Inc., Totowa, NJ, submitted.
12. Graumann, K. and Ebenbichler, A. A. (2005) Development and scale up of preparative HIC for the purification of a recombinant therapeutic protein. *Chem. Eng. Technol.* **28**, 1398–1407.

13. Shukla, A. A., Barnthouse, K. A., Bae, S. S., Moore, J. A., and Cramer, S. M. (1998) Structural characteristics of low-molecular-mass displacers for cation-exchange chromatography. *J. Chromatogr.* **814**, 83.
14. Schmidt, B., Wandrey, C., and Freitag, R. (2002) Mass influence in the performance of oligomeric poly(diallyldimethylammonium chloride) as displacer for cation-exchange displacement chromatography of proteins. *J. Chromatogr. A* **944(1,2)**, 149–159.
15. Kasper, C., Vogt, S., Breier, J., and Freitag R. (1996) Protein displacement chromatography in hydroxy- and fluoroapatite columns. *Bioseparation* **6**, 247.
16. Hermanson, G. T., Mallia, A. K., and Smith, P. K. (1992) *Immobilised Affinity Ligand Techniques*. Academic Press, New York.
17. Brooks, C. A. and Cramer, S. M. (1992) Steric mass-action ion exchange: Displacement profiles and induced salt gradients. *AIChE J.* **38**, 1969–1978.
18. Freitag, R. (1999) Displacement chromatography of biomolecules, in *Analytical and Preparative Separation Methods for Biomolecules* (Aboul-Enein, H. Y., ed.), Marcel Dekker, New York, pp. 203–252.
19. Vogt S. and Freitag R. (1998) Displacement chromatography using the UNO™ continuous bed column as stationary phase. *Biotechnol. Prog.* **14(5)**, 742–748.

VI

SPECIAL APPLICATIONS

21

Vaccine Production

State of the Art and Future Needs in Upstream Processing

Yvonne Genzel and Udo Reichl

Summary

The production of viral vaccines in animal cell culture can be accomplished with primary, diploid, or continuous (transformed) cell lines. Each cell line, each virus type, and each vaccine definition requires a specific production and purification process. Media have to be selected as well as the production vessel, production conditions, and type of process. Here, we describe different issues that have to be considered during virus-production processes by discussing the influenza virus production in a microcarrier system in detail as an example. The use of serum-containing as well as serum-free media, but also the use of stirred tank bioreactors or wave bioreactors, is considered.

Key Words: Vaccine; virus; influenza; large-scale production; microcarrier; Madin-Darby canine kidney cell; adherent; virus harvest; virus quantification; wave bioreactor; stirred tank bioreactor.

1. Introduction

1.1. State of the Art

Most commercial viral vaccines for human or veterinary use comprise either attenuated (live) viruses, inactivated (killed) viruses, or purified viral components (subunit vaccines) *(1,2)*. In addition, new technologies are being developed such as viral vectors, vaccines consisting only of defined viral proteins (produced as recombinant surface proteins in yeast), virus-like particles (VLPs), or vaccines containing viral DNA bound to gold particles (DNA vaccines). For large-scale production viruses are typically produced in animals, fertilized eggs, cell lines, tissues, or by culture of genetically modified cells. Depending on the type of vaccine and the virus type, different production processes have to be used.

In animal cell culture viruses are produced either in primary, diploid, or continuous (transformed) cell lines *(3)*. Often adherent epithelial-like cells are

From: *Methods in Biotechnology, Vol. 24: Animal Cell Biotechnology: Methods and Protocols, 2nd Ed.* Edited by: R. Pörtner © Humana Press Inc., Totowa, NJ

used, which are first grown to high cell numbers and subsequently infected, resulting in a biphasic batch process. However, some viruses only replicate in actively growing and dividing cells; thus virus infection is in parallel to the inoculation of cells (*see* **Fig. 1**).

Safety is of utmost importance for any vaccine, whereas product specifications and production technologies partly differ between veterinary and human vaccines. For cell-culture-derived viruses, well-characterized cell substrates have to be used. Starting materials of animal origin should be avoided if possible, and media including any additives should preferably not contain any ingredients known to cause toxic, allergic, or other undesirable reactions in humans and animals. The titer per dose is defined, as well as the amount of remaining DNA, host cell protein, and endotoxin per dose after purification to reduce cross reactions. The product must be stable, conveniently applicable, and booster vaccinations possibly not be needed. Especially new vaccine types need to be cheap, and development time as well as the time needed for product-to-market must be short.

1.2. Future Needs

Because of new threats like fast-developing pandemics, newly emerging viruses, or bioterrorism, as well as new application fields for viral vaccines in gene or cancer therapy and low-cost vaccines for the third world, certain questions should be reconsidered.

- Is it economically feasible to optimize existing processes?
- Can this be done without new approval?
- Can existing cells lines be genetically modified to high-producer cells?
- Is it possible to design cells to express desired traits such as suspension growth in protein-free media with optimal virus replication?
- Is it possible to develop completely defined media?
- Will batch-to-batch consistency and reproducibility no longer be an issue?
- Are faster reaction times for adaptation of the process to new virus types possible?
- Will high-density cultures and perfusion systems allow higher productivity?

1.3. Influenza as an Example

At present not many details of large-scale vaccine-production processes can be found in the literature. Therefore, in this chapter the upstream processing of one viral vaccine production process, namely influenza, is described in detail.

Inactivated human influenza vaccines are still mostly produced in the allantoic cavity of embryonated hen's eggs. Veterinary influenza vaccines, however, have been produced in large-scale mammalian cell culture systems for many years ***(4,5)***. Pros and cons for both production systems can be cited ***(6–8)***.

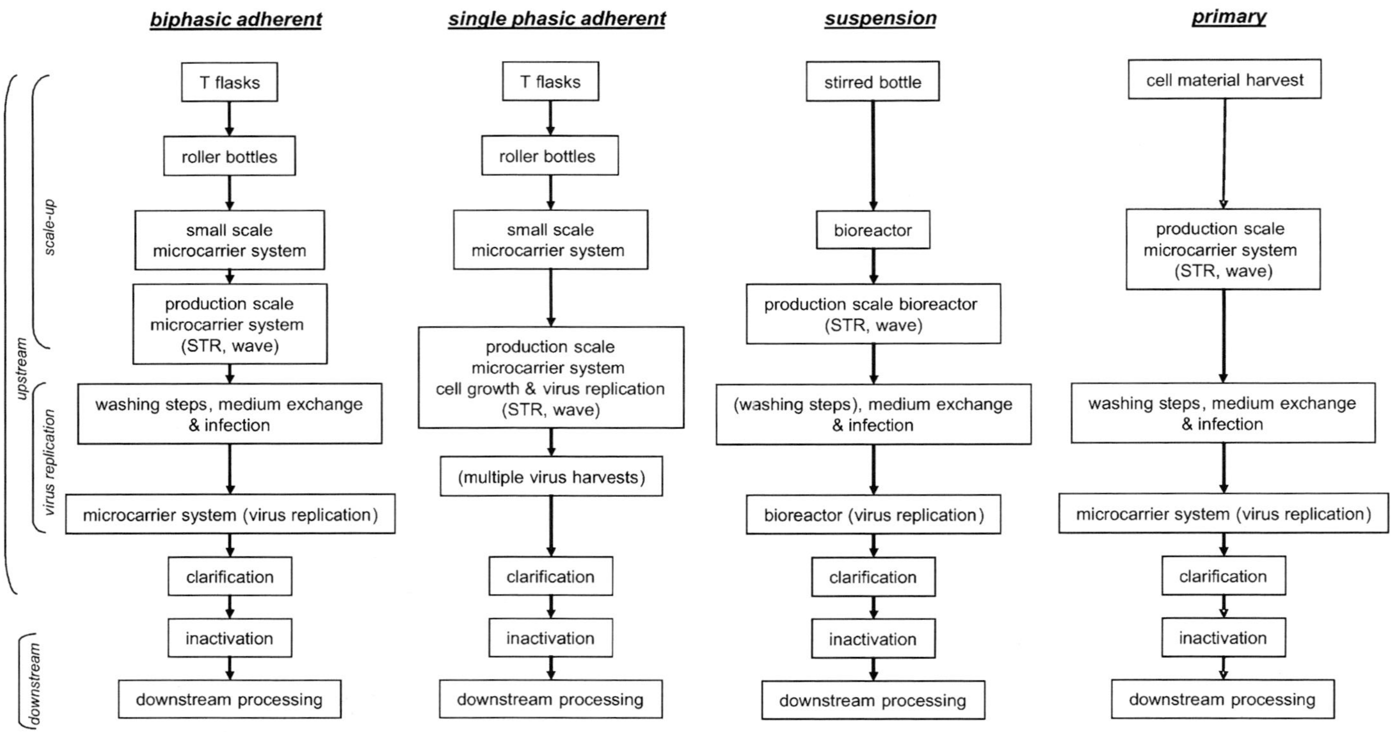

Fig. 1. Overview of inactivated cell culture derived viral vaccine production processes (continuous and diploid cell lines). For primary cell lines scale-up steps are omitted as cells are harvested from tissues directly into the production scale. STR, stirred tank bioreactor.

Not many continuous cell lines are good candidates as a host system. These cell lines have to fulfil the production requirements and be well characterized. Most cells used for virus production need immobilization on a growth surface, resulting in static systems (cell cubes, cell factories), roller bottle systems, or microcarrier cultures. Fetal calf serum (FCS) is still added to many cell lines for better growth and high virus yields, thus bringing animal-derived protein into the production system. FCS not only makes these processes expensive, but also involves the risk of prion contamination. Three cell lines, Madin-Darby canine kidney cells (MDCK) *(6,9–11)*, African green monkey kidney cells (Vero) *(12)*, and PER.C6 cells (human cell line derived from primary culture of human fetal retinoblast immortalized upon transfection with an E1 minigene of adenovirus type 5) *(13)*, have been cited in the literature for the production of influenza A and B viruses, which meet regulatory requirements. These three cell lines have all been adapted to grow in serum-free media, such as Ex-Cell MDCK for MDCK cells (JRH Biosciences technical bulletin) or EpiSerf (Gibco) *(6)*, Ex-Cell VPRO and Ex-Cell 525 *(13)* for PER.C6 cells, and Dulbecco's modified Eagle's medium for Vero cells *(12)*.

2. Materials
2.1. Cell Culture

1. Adherent MDCK cells from ECACC (Nr. 841211903) cultivated in serum-containing conditions (*see* **steps 2** and **3**). Adapted to growth in serum-free medium (SFM) (*see* **step 4**).
2. Serum-containing cell growth medium (CGM): GMEM (Gibco no. 22100-093, powder dissolved in tissue-culture Milli-Q water culture water) supplemented with glucose (final concentration 5.5 g/L) (Sigma no. G-8270), 10% FCS (Gibco no. 10270-106) and 2 g/L peptone (autoclaved 20% solution, International Diagnostics Group no. MC33) and 4.0 mg/mL NaHCO$_3$ (Merck, p.a.); pH adjusted to 6.8 with HCl; sterile filtered (0.22 µm); storage at 4°C (*see* **Notes 1** and **2**).
3. Virus maintenance medium (VMM): CGM without serum containing low levels of porcine trypsin (12.5 mg/L; Gibco no. 27250-018) to facilitate infection of cells; sterile filtered (0.22 µm); storage at 4°C (*see* **Notes 1** and **2**).
4. SFM: Ex-Cell MDCK (JRH Bioscience, no. 14580-1000M) supplemented with glutamine (2 m*M* final concentration; Sigma no. G3126), stock solution (200 m*M*, sterile filtered [0.22 µm], stable at 4°C for 3 months); for virus infection, addition of low levels of porcine trypsin (12.5 mg/L; Gibco no. 27250-018) (*see* **Notes 1** and **2**).
5. Trypsin solution for cell detachment: 0.5 g trypsin (1:250, powder; (Gibco no. 27250-018, 300 USP) and 0.2 g ethylene diamine tetraacetic acid (EDTA) (Sigma no. 101K0012) in 100 mL phosphate-buffered saline (PBS), filtered (0.22 µm) and stored at −20°C (stable for 1 yr) or at 4°C (stable for 3 mo) (*see* **Note 2**).

6. PBS solution: NaCl (8.00 g/L), KCl (0.20 g/L), KH$_2$PO$_4$ (0.20 g/L), Na$_2$HPO$_4$ (1.15 g/L) in tissue-culture Milli-Q water, autoclaved before storage at room temperature (*see* **Note 2**).

2.2. Cultivation Methods/Cell Growth

1. Static cultivation flasks and roller bottles (RB) 850 cm^2 (Greiner).
2. Microcarriers: Cytodex 1 (GE Healthcare).
3. Stirred tank bioreactor (STR): a 5-L bioreactor (Biostat C, B. Braun Biotech) with temperature, pO$_2$, and pH sensors as well as a dip tube; a PCS7 system (Siemens). Stirrer with two inclined paddle impellers (paddle: 8.5 cm length; 2 cm width; distance from bottom: 9 cm; distance between impellers: 4 cm), four baffles (30 cm in length).
4. Wave bioreactor (System 20P, Wave Biotech AG) with 19-in. instrument rack (Wave Biotech AG) and cellbags (CB2L, Wave Biotech AG, LDPE material).

2.3. Virus Infection

1. Equine influenza strain A/Equi 2 (H3N8) Newmarket 1/93 or human influenza A/PR/8/34 (H1N1) (NIBSC). Virus seed was stored at aliquots of 10 mL (2.1–2.4 log HA units/100 µL; equine influenza: 2.0–4.0 × 10^7 viruses/mL; human influenza: 3.2 × 10^7 viruses/mL from TCID$_{50}$) at −70°C. Influenza A viruses are classified as S2 pathogens and have to be handled under S2 biohazard laboratory conditions. All active virus should be handled under the laminar flow box. Virus can be inactivated under acidic conditions, by heat, or by corresponding disinfectants and inactivation agents.

2.4. Virus Harvest, Clarification, and Inactivation

1. Filters for harvesting: 5- and 1-µm Polyfil II depth filters (P05/P01, Porvair plc); 0.65-µm Flotrex AP depth filters (FAP10/FAP96, GE Infrastructure); 0.45-µm Memtrex AP membrane filter (MMP94, GE Infrastructure).
2. Binary ethyleneimine (BEI) inactivation: 2-bromoethylamine hydrobromide (Sigma, no. B-9258); hydroxyethyl piperazine ethane sulfonate (HEPES) buffer (pH 7.5; 0.5 *M*) (e.g., Sigma, no. H4034).
3. β-Propiolactone inactivation: β-propiolactone (Serva Electrophoresis, no. 33672.01); HEPES buffer (pH 7.5; 0.5 *M*) (e.g., Sigma, no. H4034).
4. Sterility test: casein peptone soy peptone medium (CASO): 30 g/L CASO (Fluka no. VM92175923) in tissue-culture Milli-Q water (25 mL per flask), autoclaved.
5. Innocuity assay: two 75-cm^2 T-flasks confluently grown with MDCK cells (CGM medium).

2.5. Virus Quantitation

1. Hemagglutination assay (HA): round-bottomed 96-well microtiter plates (with lid for active samples), a purified chicken erythrocyte solution (Chicken blood is diluted 1:2 with alsevers solution) (20.5 g/L glucose [Roth], 8.0 g/L sodium citrate

[Merck], 0.55 g/L citric acid [Merck], 4.2 g/L NaCl [Merck] in tissue-culture Milli-Q water), then washed three times with PBS and set to $1.9–2.1 \times 10^7$ red blood cells/mL with PBS), PBS, plate reader photometer (e.g., Rainbow Spectra, Tecan Instruments) (700 nm) (*see* **Note 3**).
2. Virus titration ($TCID_{50}$): VMM with addition of gentamycin (final concentration 0.1 g/L), PBS, 96-well plate (confluently grown with MDCK cells), ice-cold acetone-solution (80%), primary antibody (40 µL per well of a 1:5 dilution (filtered PBS) of pretreated primary antibody: equine influenza A anti-goat produced in goat (nanoTools). (As the primary antibody was obtained against equine influenza A plus MDCK cell debris a pretreatment by incubation of a 1:100 dilution [PBS] of the antibody solution with a confluent MDCK cell monolayer for 30 min at 37°C is needed.) Secondary antibody (40 µL per well of a 1:500 dilution with filtered PBS; Molecular Probes, # A-11015), fluorescecnce microscopy (*see* **Note 3**).

2.6. On-Line and Off-Line Measurements

1. Off-line cell counting: trypan blue solution (1:2 dilution of the stock solution [1.8 g NaCl and 1.0 g trypan blue in 100 mL Milli-Q-water, 0.22 µm filtered] with PBS), Fuchs-Rosenthal chamber.
2. Off-line measurement of basic metabolites: glucose, lactate, glutamine, ammonium, glutamate, sodium, and potassium using either a Bioprofile 100 Plus (Nova Biomedical) or a YSI model 2700 or YSI model 7100 Biochemistry Analyzer (Yellow Springs Instruments) and a Vitros DT60-II (Ortho Clinical Diagnostics) with corresponding consumables.
3. On-line measurements: pO_2 (InPro 6100/120/S/N, Mettler Toledo), pH (gel electrode 405-DPAS-SC-K8S/120, Mettler Toledo), and temperature (resistance thermometer PT100, JUMO MK Juchheim GmbH & Co.) sensors coupled to monitoring and control system (PCS7, Siemens).

3. Methods

Cell-culture-derived influenza vaccines are typically produced in a biphasic process, which comprises cell growth and virus replication. For adherent MDCK cells, the production process starts with the inoculation in T-flasks. After a scale-up into taller roller bottles, the cells are washed, trypsinized, harvested, and transferred into a vessel containing cell growth medium. The inoculation density is adjusted to cell numbers in the range $1–2 \times 10^5$ cells/mL at a concentration of microcarriers of about 2 g/L Cytodex 1. After a growth phase of 3–5 d, cells form a dense monolayer and cell numbers have increased four- to sixfold to $0.8–1.2 \times 10^6$ cells/mL.

In the next step spent growth medium is withdrawn and cells are washed several times with PBS before adding a serum-free medium containing low levels of trypsin to facilitate infection of cells. Virus seed is added at a low multiplicity of infection (MOI). The culture conditions are maintained at the same levels as before. Influenza is a lytic virus, and maximum cytopathic effect (CPE) occurs typically after 48–72 h with the detachment of the cells from the microcarriers and the release of virus into the medium. Eventually, more than 90% of the cells are in the supernatant and the demand for oxygen ceases, indicating cell death. For the manufacturing of dead vaccines, the bioreactor harvest is clarified into a new vessel and the virus is chemically inactivated by using chemicals such as β-propiolactone, binary ethyleneimine, or formaldehyde. Afterwards the harvest is transferred into a second inactivation vessel for a total inactivation time of 24–48 h at 37°C. The inactivated antigen is clarified through depth filters (optional step) and held at 4°C pending testing by quality control (QC). The final testing includes sterility, innocuity, typing, and a hemagglutination assay as a measure for virus yield. Eventually the inactivated harvest is concentrated and purified by several downstream processing methods before adjuvanting, blending, and filling takes place.

3.1. Cell Culture

1. Adherent MDCK cells are grown at 37°C in CGM in static T-flasks (5% CO_2) up to RBs (passaged every 4–7 d, when confluent). RB cultures (850 cm^2) are inoculated with approx 1.3×10^7 cells (1.5×10^4 cells/cm^2) and grown for 7 d in 250 mL CGM. When fully confluent ($1.0–1.4 \times 10^8$ cells or $1.2–1.6 \times 10^5$ cells/cm^2) the cells are washed three times with PBS (without Ca^{2+}/Mg^{2+}) and detached by exposure to 0.05% trypsin/0.02% EDTA (10 mL, about 30 min). The trypsin activity is stopped by addition of an equal volume of FCS to the trypsin/cell solution. The cell suspension is used to inoculate a 5-L bioreactor with microcarriers in CGM (*see* **Fig. 1**; *see* **Notes 1** and **4**).
2. For serum-free media, detachment with lower trypsin concentrations or only EDTA and shorter incubation times might be necessary. Trypsin activity is stopped by addition of medium or by adding PBS and washing of the cells.

3.2. Cultivation Methods/Cell Growth

1. Cultivation in roller bottles: As a typical split ratio from T175-flask to RB 1:2.4 is taken; 6–8 RBs are needed for a 5-L stirred tank bioreactor; a parallel T75-flask should be kept as reference and control for the microcarrier cultivation; RB rolling speed is typically set to 0.66 rpm (*see* **Note 5**).
2. Cultivation on microcarriers: The microcarriers are hydrated in PBS according to the manufacturer's instructions, autoclaved, and added to the medium. For serum-free conditions: additional preconditioning in serum-free medium at 4°C overnight before addition to the medium in the corresponding final concentration might be needed. Typical microcarrier concentrations are 2 g/L (*see* **Note 6**).
3. Bioreactor settings: 37°C, 50 rpm, pH control (1 *M* NaOH) at pH 7.3, aeration at 40% pO_2 by pulsed O_2-aeration through a sparger; sterilization, monitoring, and control by a digital control system (PCS7, Siemens). For the washing step before infection, a dip tube (6 mm inner diameter) is inserted into the reactor top to remove the medium above the settled microcarriers (*see* **Note 7**).

4. Wave bioreactor settings: 37°C, platform angle of 7°, rocking rate of 15 rocks per minute, and aeration with 2–5% CO_2 mixed with air at 0.1 NL/min. For cultivation in a 2-L wave bioreactor, either serum-containing medium or serum-free medium (1 L), microcarriers (2 g/L), and cells (start cell concentration: 2×10^5 cells/mL) are added in a feed bottle, transferred into the cellbag and incubated for 1 h without rocking at 37°C for better attachment. Then the rocking is started, and cells are grown to confluency on the microcarriers (*see* **Note 8**).

3.3. Virus Infection

Washing steps (typical dilution is about 1:2000 to remove spent cell growth medium containing serum) and medium exchange:

1. 5-L stirred tank bioreactor with dip tube: Remaining volume after removal of the supernatant is approx 400–500 mL; washing is done at least three times by addition of up to 4 L PBS. Virus and trypsin are added to 4.5 L VMM and transferred into the bioreactor via a steam sterilized addition port.
2. Wave bioreactor: The cellbag is taken under a laminar flow hood. After settling of the microcarriers, the medium is removed and the remaining suspension is washed three times with 1 L PBS. Virus and trypsin is added to 1 L VMM that is pumped into the cellbag.

Virus seed is added at a low MOI, typically in the range 0.001–0.01 based on $TCID_{50}$/mL, depending on optimal yield of the corresponding virus subtype (*see* **Notes 9–13**).

3.4. Virus Harvest, Clarification, and Inactivation

1. Harvesting, clarification: Cultivation broth is directly filtered through a 5-µm depth filter when coming from microcarrier cultivations. Then either a 1-µm Polyfil II depth filter or a 0.65-µm Flotrex AP depth filter is used, and the filtrate is inactivated. In an optional step the inactivated broth is filtered through a 0.45-µm polysulfone membrane filter (Memtrex AP, CMMP94, GE Infrastructure).
2. Binary ethyleneimine (BEI) inactivation: The filtrate is inactivated chemically by the addition of 1.5 mM BEI *(14)* (3.2 mL of a stock solution from 0.41 g 2-bromoethylamine hydrobromide in 4 mL NaOH is added together with 53 mL HEPES buffer to 1 L cultivation broth). After short incubation the reaction is transferred to a new vessel and incubated for 24 h at 37°C. (2-Bromoethylamine hydrobromide and BEI are very toxic; butyl rubber gloves must be used and waste has to be treated correspondingly; 2% citric acid should be used for neutralization.)
3. β-Propiolactone inactivation: The filtrate is inactivated chemically by the addition of 3 mM β-propiolactone. Addition of 6.4 mL from the stock solution (0.254 mL β-propiolactone and 7.75 mL PBS) to 53 mL 30 mM HEPES buffer and 1 L cultivation broth); the pH of the reaction is stabilized with HEPES buffer (pH 7.5 as recommended by Budowsky et al. *(15,16)*. After short incubation the reaction is transferred to a new vessel and incubated for 24 h at 37°C. (β-Propiolactone is very

toxic; butyl rubber gloves must be used, and waste has to be treated correspondingly; 2% citric acid can be used for neutralization.)
4. After clarification and inactivation, the filtrate is stored at 4°C until further processing.
5. Sterility test: From each sample two sterility tests are prepared by addition of about 2.5 mL per CASO medium flask and incubation at 37°C. Sterility is confirmed after a minimum of 14 d of incubation with negative result.
6. Innocuity assay: From the inactivated broth 1 mL is added to a T75 flask (confluently grown with MDCK cells) in VMM medium. After incubation for 3 d at 37°C the HA titer is determined and 1 mL of the supernatant of the first T75 flask is added to a second T75 flask (confluently grown with MDCK cells) in VMM medium. After incubation for 3 d at 37°C the HA is determined. The titer for the first flask must not exceed a HA corresponding to the dilution factor. For the second flask 0.0 log HA units per 100 µL indicate the successful inactivation.
7. For release of inactivated harvests for downstream processing, further testing might be required, for example, innocuity and typing, depending on the corresponding dossiers of the manufactures.

3.5. Virus Quantitation

1. HA: Titration of influenza virus by hemagglutination is based on the method described by Mahy and Kangro (17). Serial double dilutions of the test samples (100 µL) are made in round-bottomed 96-well microtiter plates containing 100 µL PBS. Each sample is measured in duplicate. When choosing the improved assay, two rows are needed for one sample as in the second row the sample is analyzed from a $1:2^{0.5}$ predilution for higher precision. For the standard assay only one row per sample is needed. To each well 100 µL of a chicken red blood cell solution (2×10^7 red blood cells/mL) is added and incubated for 60–90 min at room temperature. The last dilution showing complete hemagglutination is taken as the end point and is expressed as log HA units per test volume (100 µL). For photometric evaluation the plates are scanned with a plate photometer measuring extinction at 700 nm. A Boltzmann sigmoid is fitted to each data set (from one sample), and the dilution at the point of inflection (one of the parameters) is defined as the end point of the titration. The inverse of the dilution is defined as the volumetric HA activity with units 1 HAU (per 100 µL). An internal standard is used to compensate fluctuations caused by the varying quality of chicken erythrocytes (see **Fig. 2A**; see **Notes 14, 16,** and **17**).
2. For active virus titration ($TCID_{50}$) 10-fold serial dilutions of the culture supernatants are prepared in VMM with addition of gentamycin (1% v/v). Prior to inoculation, the cells are washed three times with 100 µL PBS per well. To each well of a 96-well plate (confluently grown MDCK cells), 100 µL of the diluted culture supernatants is added to inoculate (eight replicates per dilution). After 1 d at 37°C, 5% CO_2 100 µL of VMM with gentamycin is added to each well, and the plate is subsequently incubated for another day at 37°C, 5% CO_2. The plate is washed once with PBS and 100 µL of ice-cold acetone solution (80%) is added to each well for

Fig. 2. Virus titer during equine influenza virus production in Madin-Darby canine kidney cells (0–363 h) in roller bottles (MOI = 1). Cells have been cultivated in serum containing medium for 91 h before infection. Profiles of (**A**) virus titer in log HA units/100 µL (■), virus titer as TCID$_{50}$ (□); (**B**) virus titer in log HA units/100 µL (■), virus titer as TCID$_{50}$ (□), glutamate (●), and total cell numbers in supernatant (△).

fixation (30 min, 0°C). Then the plate is washed three times with filtered PBS before addition of the primary antibody (40 µL per well of a 1:5 dilution [filtered PBS] of the pretreated primary antibody). After 60-min incubation (37°C) with pretreated primary antibody, the plate is washed three times with filtered PBS and the secondary antibody (40 µL per well of a 1:500 dilution with filtered PBS) is added. The plate is washed three times with filtered PBS after 60-min incubation (37°C) and a final volume of 100 µL PBS is added before fluorescence microscopy.

The titers of infectivity are calculated from eight replicates according to the method of Spearman-Kärber (*see* **Fig. 2A**; *see* **Notes 15–17**).

3.6. On-Line and Off-Line Measurements

1. Cell counting is done according to standard procedures. For cell counting from microcarriers, it has to be checked if all carriers are empty after trypsinization. Especially, confluent microcarriers tend to agglomerate, which influences sampling. When counting virus-containing samples, all material has to be inactivated afterwards.
2. Basic metabolites like glucose, lactate, glutamine, ammonium, and glutamate can be measured using biosensors (e.g., a Bioprofile 100 Plus or a YSI model 2700 or YSI model). Depending on sample volume and concentration range, one of the two can be more advantageous (*see* **Table 1**; *see* **Note 18**). An on-line measurement of glutamate might be a fast means to follow virus production (*see* **Fig. 2B**). Samples can be measured directly or after storage at −70°C. Virus samples should be heat inactivated for 3 min at 80°C.
3. During the first part of the process (0–99 h, "cell growth") on-line data for a typical cultivation show an increased frequency of pure oxygen pulses because of an increasing number of MDCK cells attaching and actively growing on microcarriers (*see* **Fig. 4**, p. 471). Cells grow exponentially at the beginning. However, a linear increase in cell numbers is finally observed because of limitations of the available space on microcarriers resulting in contact inhibition. When fully confluent, a maximum cell number of 1.2×10^6 cells/mL is obtained. After washing and the addition of fresh VMM together with virus seed, the aeration pulse frequency slowly decreases and oxygen consumption finally stops completely at about 123 h when all cells are dead after virus replication (99–140 h, "virus production"). The increase of the pO_2 signal to about 165% at the end of the cultivation results from the diffusion of oxygen from the headspace of the bioreactor into the supernatant. (Because of sparging with pure oxygen, the partial pressure of oxygen in the headspace is higher than in air.)

4. Notes

1. A general overview on regulations, directives, and guidelines for the production of pharmaceuticals can be found in the *The Orange Guide (18)*. For the production of vaccines for human and veterinary use, the corresponding Pharmacopoeias such as EP or USP *(19,20)* are relevant. In addition, there are several documents giving guidance for the use of cell substrates for production of biologicals and viral safety, for example, by the Food and Drug Administration *(21,22)*. The main aspects of research and development of vaccines are thoroughly described by Gregersen *(23)*.
2. All media for production should preferably not contain any ingredients known to cause toxic, allergic, or other undesirable reactions in humans. Other compounds such as pH indicators or approved antibiotics should also be avoided or kept at the lowest effective concentration. Poorly defined components like peptone, serum, and soy hydrolysate should be thoroughly tested for product stability as well as for

Table 1
Comparison of Validation Results for Bioprofile 100 Plus and YSI 7200/Vitros

	Glucose		*Lactate*		*Glutamine*		*Glutamate*		*Ammonia*	
	Bp[c]	YSI	Bp[c]	YSI	Bp[c]	YSI	Bp[c]	YSI	Bp[c]	Vitros
Validated range (m*M*)	1.1–41.1	2.9–27.8	2.3–27.0	3.4–23.6	0.2–2.6	0.2–2.6	0.2–2.6	0.05–1.6	0.2–5.2	0.03–0.35
Linear?	Yes	Yes	Yes	Yes	No	Yes	Yes	Yes	Yes	Yes
XB[a]	2.72	0.89	2.13L	1.35	0.17	0.25	0.23	0.05	0.4	0.1
Slope	1.09	0.98	0.92	0.95	1.15	1.06	1.15	1	0.92	0.96
Medium	PBS	Water	GMEM	Water	GMEM	Water	GMEM	Water	GMEM	GMEM
SD[b]	1.3	0.6	1.5	1	1.2	1.7	1.6	1.8	1.5	5.3

[a]Limit of quantitation (m*M*) (definition: XB = 10 s_{xo}).

[b]Relative standard deviation of the method in %. Degree of freedom 7 (eight measurements for upper and lower limit), one-point detection for the calibration curve.

[c]Bioprofile 100 Plus.

PBS, Phosphate-buffered saline; GMEM, Glasgow minimum essential medium.

support of cell growth (especially on microcarriers) and virus yields. Additionally, the influence of culture medium ingredients such as antifoaming agents (pluronic) and detergents on downstream processing and analytics should be verified.

3. If possible, approved reference standards should be used, e.g., from the National Institute for Biological Standards and Control or the World Health Organization. For in-house standards, stability over time must be carefully monitored. Changing from one batch of reference standard to the next should be done with a time overlap for thorough parallel testing.
4. When using suspension cells, scale-up into larger production volumes is more easily achieved (*see* **Fig. 1**). For primary cells, production scale depends on the cell material. Predominantly, bioreactors are started directly from the harvested cell suspension (*see* **Fig. 1**).
5. For RBs 0.25 rpm might be required for serum-free media with poor cell attachment. Especially for serum-free media, the type of RB and surface treatment, for example, plasma-treated polystyrol or polyethylene terephthalate, might be important for cell attachment. Therefore, different suppliers should be checked.
6. The initial number of viable cells and attachment phase to the microcarriers are crucial; otherwise, cell recovery is poor, resulting in reduced cell and virus yields. Especially in wave cultivation, attachment was improved by discontinuous rocking at the beginning of the process. Typically, the microcarriers differ slightly in size.
7. During the washing steps pH control and aeration has to be modified to avoid foaming. Air bubbles take microcarriers to the liquid surface, leading to losses in cell and virus yield.
8. CO_2 and pH control could be added to the wave bioreactor setting. For sampling the rocking has to be stopped.
9. Washing steps are needed to remove the serum from the CGM because the trypsin activity needed for infection is inhibited by high protein concentrations. However, these washing steps are time-consuming and complicated and involve additional contamination risks. The use of serum-free media might allow using the same medium for cell growth and virus replication. Washing steps could be avoided, and the process would be simplified from a two-step process to a one-step process. Prerequisite is that no substrate limitations or inhibitions by metabolites occur. Further development can go into the direction of perfusion systems and higher cell densities. Viruses that replicate only in dividing cells have to be added directly at the beginning of cell growth; this results in single phasic (adherent) processes (*see* **Fig. 1**).
10. Infection can be performed at different MOI (e.g., from 0.0001 to 1). Typically, all MOI result in the same maximum HA titer, only with a different delay between HA increase and maximum titer. As the production of virus working seeds is time-consuming and expensive for industrial processes, lower MOI are preferred over faster virus replication times as long as the same maximum virus yields are reached.
11. The time point of infection should be optimized. Usually, cells are infected toward the end of their growth phase when maximum cell numbers are reached. In addition, the concentration of potential inhibitors of cell growth and virus replication like

ammonia and lactate should not be too high. For processes without pH regulation, such as RBs or wave bioreactors, the decrease of the pH value resulting from lactate formation should not exceed 6.8.

12. Trypsin activity can vary from batch to batch. If trypsin activity in VMM is too low, virus infection can be incomplete, resulting in reduced yield. Also, a delay in virus replication and an extension of harvest time is possible.
13. Cell numbers in the supernatant after infection should typically increase with time as the cells go into apoptosis and detach. At the end of the infection phase, most cells detach from their growth surface. However, because of the different cultivation conditions in the RB, stirred tank bioreactor, or wave bioreactor, not all cells detached can be found in the supernatant. In the wave bioreactor cells seem to disintegrate faster than in RBs and are no longer countable.
14. The time required for one HA assay is about 2–3 h. The detection limit is 0.3 log HA units/100 µL, which corresponds to about 4.1×10^7 virions/mL, assuming that the number of erythrocytes is proportional to the number of virus particles. The assay was validated with a dilution error for standard HA test: ± 0.3 log HA units/100 µL. For an improved assay, a resolution of ± 0.15 log HA units/100 µL with an error of ± 0.06 (95% confidence interval) can be obtained. To compare different HA data from other labs, the concentration of the chicken red blood cell solution should be defined. We have tested different assay volumes and obtained similar results; thus comparison between laboratories is possible. The number of virus particles is correlated with the number of chicken red blood cells, with a 1:1 ratio. For accurate mass balances the number of chicken cells should be exactly determined. Maximum value depends on virus strain, cell line, and cultivation conditions.
15. $TCID_{50}$ assays take about 1 wk; one sample requires one microtiter plate; good reproducibility and quality of data can only be obtained when performed by highly trained personal; detection limit: $10^{2.5}$ viruses/mL, dilution error: ± 0.3 log. Maximum value depends on virus strain, cell line, and cultivation conditions.
16. Other virus-quantification methods can be used for further details on the process, such as plaque-forming units (pfu), egg infectious dose (EID_{50}), laser scanning and electron microscopy, laser light scattering, real-time polymerase chain reaction, flow cytometry, and enzyme-linked immunosorbent assay. All assays have to be adapted to the corresponding medium, cell, and virus type.
17. To determine the optimal time point of virus harvest during the process, neither pfu nor $TCID_{50}$ analysis are adequate because analysis time is too long. For routine monitoring, HA assay is the best choice. Glutamate concentration in the cell culture medium, cell numbers, or lactate dehydrogenase activity give additional on-process information about the infection status of cells (*see* **Fig. 2B**). Analysis on optimal harvest time for either live attenuated or inactive vaccines can be supported by mathematical models ***(24)*** and virus stability studies under process conditions (*see* **Figs. 2A** and **3**).
18. The Bioprofile biosensor was developed for blood sample analysis. Therefore, calibration curves are needed for cell culture media. The slopes in **Table 1** indicate where the Bioprofile biosensor overestimates compared to the YSI measurement.

Vaccine Production 471

Fig. 3. Dynamics of equine influenza A virus replication. Experimental data (♦ $TCID_{50}$; ▲ HA) and simulation data (—) according to a mathematical model of virus replication. Optimal time for harvesting of live virus is about 20 h postinfection when infectivity peaks. Optimal time for harvesting for inactivated vaccines is about 25–40 h postinfection when HA achieves its maximum. Note: Dynamics strongly depends on virus subtype and multiplicity of infection.

Fig. 4. On-line data from a typical run during influenza virus production in Madin-Darby canine kidney cells on microcarriers in a 5-L stirred bioreactor (medium exchange and virus addition at 99 h): control at 40% pO_2 value by pulsed O_2-aeration, process monitoring, and control by Siemens PCS7 system.

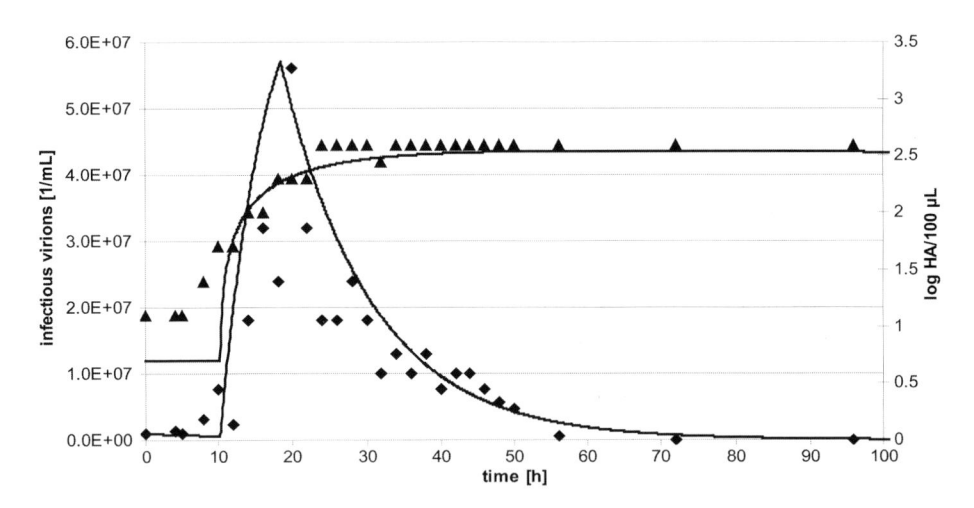

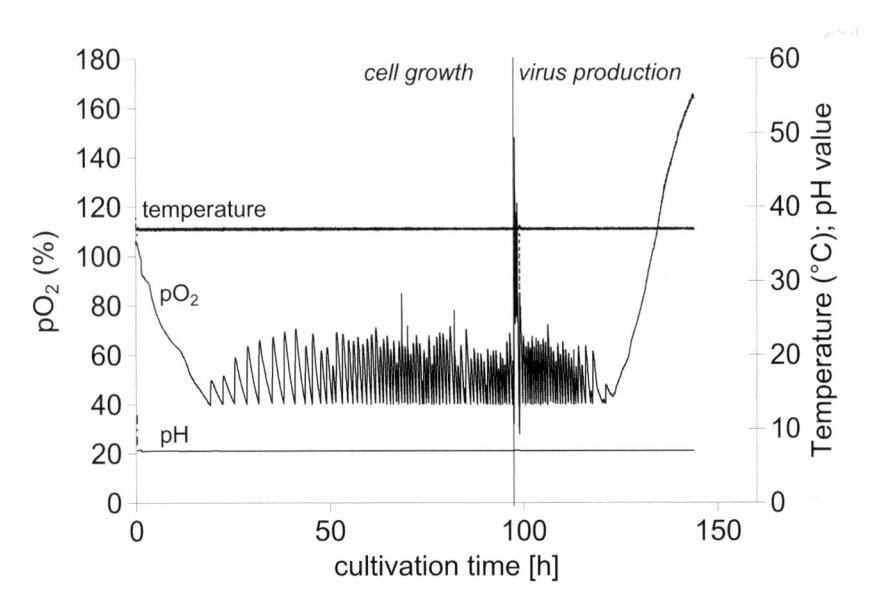

Each change of the service pack requires a new calibration curve. The enzyme membranes used in the biosensors vary in stability and quality, which must be monitored.

Acknowledgments

The authors thank N. Schlawin, C. Best, S. König, and I. Behrendt for their excellent technical assistance. Details on inactivation and the sensitive HA assay were provided by B. Kalbfuss, which is greatly appreciated. The authors would like to express their gratitude to D. Gade for critical reading of this manuscript.

References

1. Kaufmann, S. H. E. (ed.) (2004) *Novel Vaccination Strategies*. Wiley-VCH, Weinheim.
2. Huang, D. B., Wu, J. J., and Tyring, S. K. (2004) A review of licensed viral vaccines, some of their safety concerns, and the advances in the development of investigational viral vaccines. *J. Infect.* **49**, 179–209.
3. Aunins, J. G. (2000) Viral vaccine production in cell culture, in *Encyclopedia of Cell Technology* (Stier, R. E., ed.), Wiley & Sons, New York, pp. 1182–1217.
4. Reichl, U. (2000) ISCOM vaccines—antigen production and downstream processing, in *Proc. 4th Int. Congress on Biochemical Engineering*, Fraunhofer IRB Verlag, Stuttgart, pp. 314–318.
5. Genzel, Y., Voges, L., and Reichl, U. (2001) Development of bioprocess concepts on vaccine production: influenza virus as an example, in *Animal Cell Technology: From Target to Market* (Lindner-Olsson, Chatzissavidou, N., and Lüllau, E., eds.), Kluwer Academic Publishers, Dordrecht, The Netherlands, pp. 344–346.
6. Tree, J. A., Richardson, C., Fooks, A. R., Clegg, J. C., and Looby, D. (2001) Comparison of large-scale mammalian cell culture systems with egg culture for the production of influenza virus A vaccine strains. *Vaccine* **19**, 3444–3450.
7. Robertson, J. S., Cook, P., Attwell, A.-M., and Williams, S. P. (1995) Replicative advantage in tissue culture of egg-adapted influenza virus over tissue culture derived virus: implications for vaccine manufacture. *Vaccine* **13**(16), 1583–1588.
8. Govorkova, E. A., Kodihalli, S., Alymova, I. V., Fanget, B., and Webster, R. G. (1999) Growth and immunogenicity of influenza viruses cultivated in Vero or MDCK cells and in embryonated chicken eggs. *Dev. Biol. Stand.* **98**, 39–51.
9. Genzel, Y., Behrendt, I., König, S., Sann, H., and Reichl, U. (2004) Metabolism of MDCK Cells during cell growth and influenza virus production in large-scale microcarrier culture. *Vaccine* **22**(17–18), 2202–2208.
10. Genzel, Y., Ritter, J. B., König, S., Alt, R., and Reichl, U. (2005) Substitution of glutamine by pyruvate to reduce ammonia formation and growth inhibition of mammalian cells. *Biotechnol. Progr.* **21**(1), 58–69.
11. Genzel, Y., Olmer, R. M., Schafer, B., and Reichl, U. (2006) Wave microcarriers cultivation of MDCK cells for influenza virus production in serum-containing and serum-free media. *Vaccine* **24**(35–36), 6074–6087.

12. Kistner, O., Barrett, P. N., Mundt, W., Reiter, M., Schober-Bendixen, S., and Dorner, F. (1998) Development of a mammalian cell (Vero) derived candidate influenza virus vaccine. *Vaccine* **16**(9/10), 960–968.
13. Pau, M. G., Ophorst, C., Koldijk, M. H., Schouten, G., Mehtali, M., and Uytdehaag, F. (2001) The human cell line PER.C6 provides a new manufacturing system for the production of influenza vaccines. *Vaccine* **19**, 2716–2721.
14. Bahnemann, H. G. (1990) Inactivation of viral antigens for vaccine preparation with particular reference to the application of binary ethylenimine. *Vaccine* **8**(4), 299–303.
15. Budowsky, E. I. and Zalesskaya, M. A. (1991) Principles of selective inactivation of viral genome. V. Rational selection of conditions for inactivation of the viral suspension infectivity to a given extent by the action of beta-propiolactone. *Vaccine* **9**(5), 319–325.
16. Budowsky, E. I., Friedman, E. A., Zheleznova, N. V., and Noskov, F. S. (1991) Principles of selective inactivation of viral genome. VI. Inactivation of the infectivity of the influenza virus by the action of beta-propiolactone. *Vaccine* **9**(6), 398–402.
17. Mahy, B. W. J. and Kangro, H. O. (eds.) (1996) *Virology Methods Manual*. Academic Press, London, pp. 41–43.
18. "The Orange Guide": Rules and Guidance for Pharmaceutical Manufactures and Distributors (2002), Medicines Control Agency, 6th ed. London, United Kingdom, ISBN 0113225598.
19. European Pharmacopoeia (EP), Maisonneuve S. A. France, continuously updated.
20. United States Pharmacopoeia (USP), US Pharmacopeial Convention, Rockville, MD, continuously updated.
21. Points to consider in the characterization of cell lines used to produce biologicals. (1993) Department of Health and Human Services, Food and Drug Administration, Docket No. 84N-0154.
22. Guidance on viral safety evaluation of biotechnology products derived from cell lines of human and animal origin. (1998) Department of Health and Human Services, Food and Drug Administration (FDA), Docket No. 96D-0058.
23. Gregersen, J.-P. (1994) *Research and Development of Vaccines and Pharmaceuticals from Biotechnology*. VCH, Weinheim, Germany.
24. Möhler, L., Flockerzi, D., Sann, H., and Reichl, U. (2005) Mathematical model of influenza A virus production in large-scale microcarrier culture. *Biotechnol. Bioeng.* **90**(1), 46–58.

22

Retrovirus Production and Characterization

Pedro E. Cruz, Marlene Carmo, Teresa Rodrigues, and Paula Alves

Summary

Retroviral vectors are involved in more than 25% of the gene therapy trials, being the pioneering vector in this field. The production of retroviral vectors still poses some challenges, mainly because of the relatively low cell specific productivity and the low vector stability. Having clinical applications in mind, it is clear that robust production and purification processes are necessary, along with a proper vector characterization. The determination of vector titer in terms of infectious particles and transduction efficiency, combined with product quality analysis in terms of total particles and proteins present in the final preparation, are essential for a complete assessment and definition of the final product specifications.

Key Words: Retrovirus; gene therapy; production; purification; quality analysis; quantification; characterization.

1. Introduction

Gene therapy is a promising technology that may correct inherited diseases as well as treat acquired illness, such as cancer *(1–3)*. Retroviral vectors based on Moloney murine leukemia virus (MoMLV) are currently used in a wide range of clinical trials with the main advantage of having the ability to integrate into the genome, leading to long-term expression and relatively low immunogenic toxicity *(2,4)*. However, the clinical application of retroviral vectors for gene therapy is limited by the relatively low productivities of the retroviral packaging cell lines, partially because of the low stability of the vector in culture *(5,6)*. Because gene therapy applications require gene transfer to a large number of cells, high quantities of high-titer retroviral preparations are necessary *(1,6)*. To effectively produce these vectors it is therefore necessary to integrate vector production and downstream processing. Because product efficacy is essential, established protocols to assess vector titer and transduction

From: *Methods in Biotechnology, Vol. 24: Animal Cell Biotechnology: Methods and Protocols, 2nd Ed.* Edited by: R. Pörtner © Humana Press Inc., Totowa, NJ

efficiency are of crucial importance. In this sense, this chapter provides protocols for the production, purification, quantification, and characterization of retroviral vectors.

2. Materials

2.1. Cell Culture, Viral Vector Production, and Purification

2.1.1. Production of Retroviral Vectors in Cell Factory

1. Cell Factory (Nunc, Roskilde, Denmark) with 10 trays and 6320 cm^2 of total culture area, connectors, and tubing.
2. Four 2-L flasks (Schott, Mainz, Germany).
3. 1 FG50 filter 0.2 μm (Millipore, Billerica, MA), sterilized by autoclave (20 min).
4. Culture medium (*see* **Note 1**): Dulbecco's modified Eagle's medium (DMEM, Gibco-BRL, Paisley, UK) supplemented with 5% (v/v) fetal bovine serum (FBS, Gibco-BRL), 4.5 g/L glucose (Merck, Darmstadt, Germany), and 6 m*M* glutamine (Gibco-BRL), all sterile.
5. Phosphate-buffered saline (PBS), sterile: 154 m*M* NaCl (Merck), 1.34 m*M* Na_2HPO_4 (Merck), 1.54 m*M* KH_2PO_4 (Merck) in water (*see* **Note 2**), adjust to pH 7.2.
6. 0.25% (w/v) Trypsin solution containing ethylenediamine tetraacetic acid (EDTA) 1 m*M* (Gibco-BRL), sterile.

2.1.2. Purification of Retroviral Vectors

1. Buffer 1: 20 m*M* phosphate buffer, pH 7.5, 150 m*M* NaCl, sterile.
2. Buffer 2: 20 m*M* phosphate buffer, pH 7.5, 1500 m*M* NaCl, sterile.
3. Buffer 3: 20 m*M* phosphate buffer, pH 7.5, 150 m*M* NaCl, 0.5 *M* sucrose, sterile.
4. Column 1: anion exchange column, HiTrap DEAE FF (Amersham Biosciences, Uppsala, Sweden), with 1-mL bed volume.
5. Column 2: desalting column, HiPrep 26/10 Desalting (Amersham Biosciences).
6. 0.5 *N* NaOH solution.
7. 20% (v/v) Absolute ethanol (Panreac, Barcelona, Spain) in water.

2.2. Retroviral Vector Quantification

2.2.1. Infectious Particles (Using LacZ as Reporter Gene)

1. 96-Well flat-bottomed sterile plates (Starstedt, Newton, NC).
2. Culture medium DMEM supplemented with 5% (v/v) FBS, 4.5 g/L glucose, and 6 m*M* glutamine.
3. Polybrene solution 1 mg/mL (Sigma, Steinheim, Germany) in PBS, sterile.
4. 37% (v/v) Formaldehyde solution (Merck).
5. 25% (v/v) Glutaraldehyde solution (Sigma, St. Louis, MO).
6. X-gal solution, sterile: 20 mg/mL 5-bromo-4-chloro-3-indolyl-β-D-galactopyranoside (X-gal, Stratagene, La Jolla, CA) in dimethyl formamide (DMF, Riedel-de Haen, Seelze, Germany).

7. 0.5 *M* $K_3Fe(CN)_6$ solution (Merck), sterile.
8. 0.5 *M* $K_4Fe(CN)_6$ solution (Merck), sterile.
9. 0.1 *M* $MgCl_2$ solution (Merck), sterile.

2.2.2. Infectious Particles (Using GFP as Reporter Gene)

1. 24-Well flat-bottomed sterile plates (Nunc).
2. 5 mL Polystyrene round-bottomed tubes (Falcon®, Becton Dickinson, Franklin Lakes, NJ).
3. Culture medium DMEM supplemented with 5% (v/v) FBS, 4.5 g/L glucose, and 6 m*M* glutamine.
4. Polybrene solution 1 mg/mL (Sigma, Steinheim, Germany) in PBS, sterile.
5. PBS, sterile.
6. 2% (v/v) FBS solution in PBS, sterile.
7. 0.25% (w/v) Trypsin solution containing EDTA 1 m*M* (Gibco-BRL), sterile.
8. 2% (v/v) FBS solution in PBS containing 2 μg/mL propidium iodide (Sigma), sterile.

2.2.3. Total Particles (7)

1. 1.5-mL Sterile tubes.
2. LightCycler capillaries (20 μL, Roche, Mannheim, Germany).
3. DNAse I (Sigma, Steinheim, Germany).
4. First strand cDNA synthesis kit (Roche, Mannheim, Germany).
5. pSIR plasmid (Clontech, Palo Alto, CA).
6. Nuclease-free water (Promega, Madison, WI).
7. LightCycler-DNA Master SYBR Green I (Roche).
8. Forward primer (ATT GAC TGA GTC GCC CGG, *Tm* = 52.4°C, 20 μM) (TIB MOLBIOL, Berlin, Germany).
9. Reverse primer (AGC GAG ACC ACA AGT CGG AT, *Tm* = 53.6°C, 20 μ*M*) (TIB MOLBIOL).

2.3. Retroviral Vector Characterization

2.3.1. Western Blotting for MLV env and gag Proteins

1. NuPAGE 4–12% Bis-Tris gel, 1.0 × 10 well (Invitrogen, Carslbad, CA).
2. Transfer buffer: 48 m*M* Tris-base, 39 m*M* glycine, and 0.037% (v/v) sodium dodecyl sulfate (SDS) in water.
3. PVDF membrane Hybond™-P (Amersham Biosciences, Buckinghamshire, UK) and 3MM chromatography filter paper (Whatman, Maidstone, UK).
4. Dilution buffer: 0.1 *M* maleic acid and 0.15 *M* NaCl in water, adjust pH to 7.5.
5. Saturation buffer 5×: 10% (w/v) skim milk powder (Merck) in dilution buffer.
6. Wash buffer: 0.1% (v/v) Tween 20, 20 m*M* Tris-HCl, pH 8.1 and 0.5 *M* NaCl in water.
7. Antibody dilution buffer: 20% saturation buffer and 80% wash buffer 1×.

8. Primary antibody anti *gag* protein: goat polyclonal Rauscher Leukemia Virus 30 antiserum derived from purified polypeptide (Viromed, Minnetonka, MN).
9. Primary antibody anti *env* protein: rat monoclonal p69/70 antibody (83A25) kindly provided by Dr. Leonard Evans from NIAID-NIH.
10. Secondary antibody (*gag* protein): immunopure rabbit anti-goat IgG, Fc fragment peroxidase conjugated (Sigma, St. Louis, MO).
11. Secondary antibody (*env* protein): goat anti-rat IgG, horseradish peroxidase conjugated (Pierce, Rockford, IL).
12. Enhanced chemiluminescent reagents: ECL Plus Western blotting detection system (Amersham Biosciences, Buckinghamshire, UK) and Hyperfilm ECL (Amersham Biosciences).

2.3.2. Transduction Efficiency *(8)*

1. 96-well Flat-bottomed sterile plates (Nunc).
2. Culture medium DMEM supplemented with 5% (v/v) FBS, 4.5 g/L glucose, and 6 m*M* glutamine.
3. Polybrene solution 1 mg/mL (Sigma) in PBS, sterile.
4. Wash solution: 1 m*M* $MgCl_2$ in PBS.
5. Lysis buffer: 1 m*M* $MgCl_2$ and 0.5% Nonidet P-40 (Roche Diagnostics) in PBS.
6. Staining solution: 6 m*M* *o*-nitrophenyl-β-D-galactopyranoside (ONPG; Sigma, Steinheim, Germany) in lysis buffer.
7. Stop solution: 1 *M* Na_2CO_3 in water.

3. Methods

3.1. Cell Culture and Viral Vector Production

3.1.1. Production of Retroviral Vectors in Cell Factory

1. The production of retroviral vectors can be performed using several packaging cell lines. In this method a human packaging cell line, TE Fly A7, is used. This cell line derives from the TE671 cell line (ECACC no. 89071904) transformed with plasmid pMFGSnlsLacZ *(9)* and produces MLV-based vectors.
2. Prepare the 2-L flasks for inoculation and medium exchange of cell factory. Mount the 2-L flasks with the connector, tubing, and clamp. Sterilize bottles by autoclave (20 min).
3. Unpack the cell factory and place it in a laminar flow cabinet.
4. Remove the seal from one of the adaptor caps and replace it with the sterilized air filter.
5. Prepare a cell suspension with 8.5×10^4 cells/mL (2×10^4 cells/cm^2) in 1.5 L of culture medium in one of the 2-L flasks previously autoclaved (*see* **Note 3**).
6. Remove the second cap from the cell factory and replace it with the tube connector of the 2-L flask containing the cell suspension.
7. Turn the cell factory on its side, raise the flask containing the cell suspension above the cell factory level, loosen the clamp, and let the cell suspension flow into cell factory.

8. Turn the cell factory 90° so that the filling inlet faces up, and then place the cell factory in horizontal position. Remove the tube connecting to the 2-L flask and replace it with the adaptor cap, but leave the filter on.
9. Incubate cells in a humidified incubator at 37°C with 5% CO_2 for 3 d (*see* **Note 4**).
10. Connect a second 2-L flask to the cell factory using the adaptor cap port, turn the cell factory on its side, raise it above the flask level and remove cells supernatant to the flask.
11. Add 900 mL of fresh medium to the cell factory using a third flask by connecting it to the adaptor cap port and proceeding as described in **step 7**.
12. Incubate cells in a humidified incubator at 37°C with 5% CO_2 for 24 h.
13. Remove cell supernatant as described in **step 10** using a fourth 2-L flask and proceed to purification, quantification, and characterization of viral vectors.

3.1.2. Purification of Retroviral Vectors

3.1.2.1. Column Preparation and Sanitization

1. Connect columns 1 and 2 to an ÄKTA™ or FPLC system with a conductivity meter, a UV absorbance detector, pH meter, and a fraction collector. Before connecting the columns to the chromatography system, ensure there is no air in the tubing and valves. Make sure that the column inlets are filled with liquid and connect them drop-to-drop to the system.
2. Equilibrate column 1 with 20-mL of buffer 1 at a flow rate of 2 mL/min. Always ensure that the backpressure of the column does not exceed 0.3 MPa.
3. Equilibrate column 2 with 100 mL of buffer 1 at a flow rate of 5 mL/min. Always ensure that the backpressure of the column does not exceed 0.15 MPa.
4. Sanitize the whole chromatographic system with 0.5 *N* NaOH by flushing at a high flow rate (10 mL/min) all the inlet buffer and sample tubings, pumps, and outlet tubing connected to the fraction collector.
5. Sanitize column 1 with 20 mL of 0.5 *N* NaOH at a flow rate of 1 mL/min.
6. Sanitize column 2 with 60 mL of 0.5 *N* NaOH at a flow rate of 5 mL/min.
7. Equilibrate the chromatography system with buffer 1 as performed in **step 4** for sanitization.
8. Equilibrate column 1 with buffer 1 at a flow rate of 2 mL/min until the conductivity and pH of the outflow reach the values for buffer.
9. Equilibrate column 2 with buffer 3 at a flow rate of 5mL/min untill the conductivity of the outflow reaches the pH and conductivity of the buffer.

3.1.2.2. Anion Exchange Chromatography

1. To maintain viral stability and achieve higher infectious particle yields, it is best to perform the purification protocol at low temperature (4–6°C).
2. Dilute two volumes of viral supernatant, previously filtered through a 0.45-µm filter, with one volume of buffer 1 (*see* **Note 5**).
3. Equilibrate column 2 with 5 mL of buffer 1 at 2 mL/min.
4. Reset the UV detector.

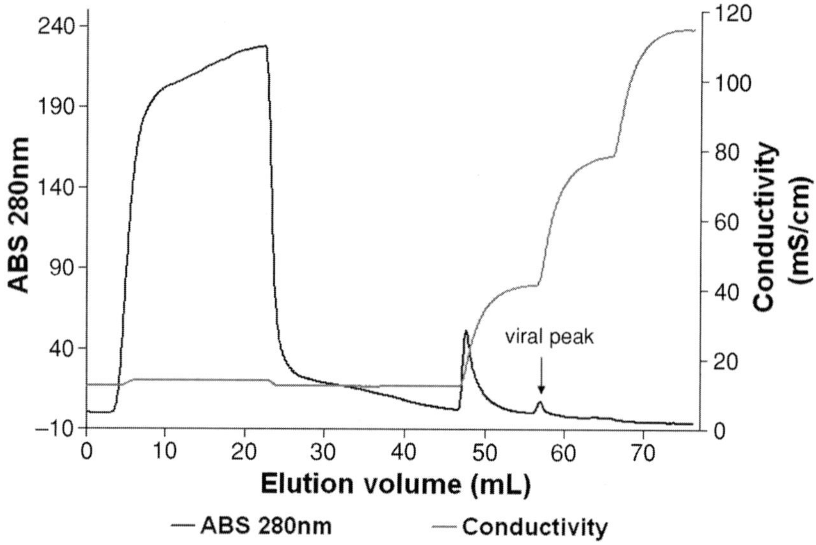

Fig. 1. Elution profile for the purification of retroviral vectors using the HiTrap DEAE FF column.

5. Load the diluted supernatant into column 1 at a flow rate of 1 mL/min (maintain this flow rate until the end of the process).
6. Wash the column with 20 mL of buffer 1.
7. Start elution of contaminant proteins with 10 mL of a mixture with 25% (v/v) of buffer 2 and 75% (v/v) of buffer 1 (*see* **Note 6**).
8. Start elution of the viral vectors with 10 mL of a mixture of 60% (v/v) of buffer 2 and 30% (v/v) of buffer 1. Collect the viral peak using the fraction collector (the elution volume will be approx 1.5–2 mL) (*see* **Note 7**). An example of an elution profile is shown in **Fig. 1**.
9. Store the recovered viral fraction at 4°C if the purification is not being performed at 4°C.
10. Regenerate the column with 10 mL of buffer 2 and 20 mL of buffer 1 afterwards (*see* **Note 8**).

3.1.2.3. Desalting of Purified Vectors

1. Connect a 2-mL sample loop to the chromatographic system (*see* **Note 9**).
2. Wash the sample loop with 10 mL of buffer 3.
3. Equilibrate column 2 with 10 mL of buffer 3 at 5 mL/min.
4. Monitor the UV absorbance at 280 nm and reset the UV detector using buffer 3.
5. Fill the sample loop with the viral peak fraction, obtained after anion exchange chromatography, using a syringe. Ensure that there are no air bubbles inside the sample loop.
6. Inject the sample loop content into column 2 at a flow rate of 2mL/min.

7. Perform an isocratic elution using buffer 3 maintaining the flow rate.
8. Collect 5-mL fractions from the injection start point using the fraction collector.
9. The viral peak will elute 10 mL after sample injection, and the peak volume will be around 4 mL (1.5- to 2-fold dilution of the initial sample volume).
10. Stop elution after appearance of a conductivity peak in the outflow, indicating elution of salts and small molecular weight species (*see* **Note 8**).
11. Filter the solution with 0.22-μm filter if sterile conditions are necessary, and store at −85°C.

3.2. Retroviral Vector Quantification

3.2.1. Infectious Particles (LacZ)

1. Prepare a target cell (HCT116) suspension with 1.65×10^5 cells/mL (5×10^4 cells/cm^2) in culture medium (*see* **Note 10**).
2. Inoculate the cell suspension in a 96-well plate, 100 μL per well. Incubate the cells for 24 h in a humidified incubator at 37°C and 5% CO_2.
3. Perform serial dilutions (depending on sample 10–10^{-7}) on the viral samples in a separate 96-well plate in culture medium containing 8 μg/mL of Polybrene.
4. Remove the medium from the seeded cell plates prepared the day before and add 20 μL of viral dilution per well in triplicates. Incubate in a humidified incubator at 37°C and 5% CO_2.
5. After 3 h of infection add 180 μL of culture medium and incubate for 48 h in a humidified incubator at 37°C and 5% CO_2.
6. Prepare the fixing solution (for 1 plate): 675 μL of glutaraldehyde 25%, 100 μL of formaldehyde 37% in 12.5 mL of PBS.
7. Prepare the staining solution (for 1 plate): 125 μL of X-Gal 20 mg/mL, 125 μL of 0.5 *M* $K_3Fe(CN)_6$, 125 μL of 0.5 *M* $K_4Fe(CN)_6$, 125 μL of 0.1 *M* $MgCl_2$ in 12 mL of PBS.
8. Remove the medium from the plates and wash the plate with PBS, 100 μL per well.
9. Add the fixing solution 100 μL per well, wait 3 min at room temperature, and remove the fixing solution.
10. Wash the plate with PBS, 100 μL per well.
11. Add the staining solution, 100 μL per well.
12. Incubate at least 24 h at 37°C (*see* **Note 11**).
13. Observe each viral dilution and count blue cells (infected cells expressing β-galactosidase) in the wells that have 10–90 blue cells per well.
14. The titer will be:

$$[\text{Titer}]\text{i.p./mL} = \frac{\text{no. of blue cells}}{0.02\,\text{mL}} \times \text{viral dilution}$$

3.2.2. Infectious Particles (GFP)

1. Prepare a target cell (NIH 3T3) suspension with 1×10^5 cells/mL in culture medium.
2. Inoculate the cell suspension in a 24-well plate, 500 μL per well. Incubate the cells for 24 h in a humidified incubator at 37°C and 5% CO_2.

3. Perform serial dilutions (depending on sample $10–10^{-7}$) to the viral samples in culture medium containing 8 μg/mL of Polybrene.
4. Remove the medium from the seeded cell plates prepared the day before and add 150 μL of viral dilution per well in triplicates. Incubate in a humidified incubator at 37°C and 5% CO_2 for 48 h.
5. At the time of infection trypsinize three wells and determine cell concentration by hemacytometer cell counting (Brand, Wertheim/Main, Germany).
6. Remove the medium from infected cells and wash each well with 500 μL of PBS, add 150 μL of trypsin to each well, and after cells detachment add 450 μL of PBS with 2% FBS and ressuspend the cells.
7. Transfer the cell suspension from each well to a centrifuge tube and centrifuge at 200*g* for 5 min in an Allegra 64R centrifuge (Beckman, Palo Alto, CA).
8. Remove the supernatant and ressuspend each pellet with 500 μL of PBS with 2% FBS and 2 μg/mL PI and transfer cell suspensions to polystyrene tubes.
9. Analyze samples for GFP fluorescence using FACScalibur flow cytometer (Becton Dickinson) (*see* **Note 12**).
10. Determine the percentage of GFP-positive viable cells using the Cellquest software (Becton Dickinson).
11. The titer will be:

$$[\text{Titer}]\ \text{i.p./mL} = \frac{\%\text{GFP}^{+}\text{viable cells}}{0.150\ \text{mL}} \times \text{viral dilution} \times \text{cell concentration}$$

3.2.3. *Total Particles* (7)

1. Incubate 50 μL of viral samples (previously filtered through a 0.45-μm filter) at 75°C for 10 min in a Thermomixer compact (Eppendorf, Hamburg, Germany).
2. Cool down the samples to room temperature and spin down the tubes in a miniSpin centrifuge (Eppendorf) in order to deposit water vapor drops on the bottom of the sample tubes.
3. Add 1 μL DNase I (1 U/μL) and 1.8 μL 25 m*M* $MgCl_2$ to the treated samples and vortex.
4. Incubate the mixture for 30 min at 25°C.
5. Incubate the mixture at 75°C for 10 min to degrade the DNase I.
6. Prepare 11.5 μL of cDNA synthesis mix for each sample to the indicated end concentrations: 2 μL of reaction buffer (1X), 4 μL $MgCl_2$ (5 m*M*), 2 μL dNTP (1 m*M*), 1.7 μL reverse primer (1 μ*M*), 1 μL RNase inhibitor (50 U), and 0.8 μL AMV reverse transcriptase (20 U) (*see* **Note 13**).
7. Add 8.5 μL of each sample to each cDNA synthesis mix, briefly vortex, and centrifuge the mixture to collect the sample at the bottom of the tube.
8. Incubate the reaction at 25°C for 10 min and then at 42°C for 60 min.
9. Incubate at 99°C for 5 min and cool at 4°C for 5 min.
10. Store the tube at 4°C for 1–2 h or –20°C for more than 2 h .
11. Mix the reagents from Sybr green PCR kit in a sterile 1.5-mL tube (master mix) to the indicated end concentrations: 2 μL LightCycler master (Fast start DNA master

SYBR Green I), 3.2 μL $MgCl_2$ (4 m*M*), 0.5 μL forward primer (0.5 μ*M*), 0.5 μL reverse primer (0.5 μ*M*), and 3.8 μL PCR grade water.

12. Vortex the master mix and add 10 μL to the necessary capillaries.
13. Add to the control capillary 10 μL of PCR-grade water and close it.
14. Make the sample dilutions in water and add 10 μL to the respective capillaries.
15. Dilute (in water) the standard pSIR plasmid (1×10^8 to 1×10^2 copies/μL) and add 10 μL to the respective capillaries.
16. Centrifuge the capillaries and start the PCR using the following LightCycler run protocol: denaturation program (95°C for 10 min); amplification and quantitation program repeated 45 times (60°C for 10 min; 72°C for 10 min with a single fluorescent measurement); melting curve program (65°–95°C for 10 min with continuous fluorescent measurement); and finally a cooling step to 40°C.
17. Analyze the data using the second derivative maximum (*see* **Note 14**).
18. The sample initial concentration is calculated according to the following formula:

$$[\text{Total particles}](\text{tp/mL}) = [\text{DNA}_{\text{PCR}}](\text{copies}/\mu\text{L}) \times \text{Dilution factor} \times 1000 \times \text{RTefficiency} \times \tfrac{1}{2}$$

19. The RT efficiency is considered to be 30% and it is also considered that each viral vector has two copies of RNA.

3.3. Retroviral Vector Characterization

3.3.1. Western Blotting for MLV env and gag Proteins (see **Note 15**)

1. Run viral samples by SDS-PAGE using NuPAGE 4–12% Bis-Tris gels.
2. The samples that have been separated by SDS-PAGE are transferred to PVDF membrane electrophoretically. These directions assume the use of a semi-dry transfer cell system (Hoefer TE70, Amersham Biosciences).
3. Incubate two PVDF membranes for 30 s in methanol.
4. Immerse the membranes and 12 filter papers in transfer buffer.
5. Place 3 filter papers on the transfer unit and above it the membrane (for each gel).
6. Disconnect the gel unit from the power supply and disassemble the gel container.
7. Remove the stacking gels and put lower gels above the PVDF membranes.
8. Above each gel place 3 filter papers and remove the air bubbles from the filter-gel-membrane sandwich.
9. Start the transfer with 0.8 × (gel area) mA and 35 V for 90 min.
10. Immerse the membranes on saturation buffer during 1 h at room temperature on a rocking platform.
11. Wash the membranes three times for 5 min with wash buffer (20 mL).
12. For *gag* protein: dilute primary antibody 1/4000 in antibody dilution buffer.
13. For *env* protein: dilute primary antibody 1/250 in antibody dilution buffer.
14. Incubate each membrane with the specific primary antibody dilution for 1 h at room temperature with agitation.
15. Wash the membranes three times for 5 min with wash buffer (20 mL).
16. For *gag* protein: dilute secondary antibody 1/5000 in antibody dilution buffer.

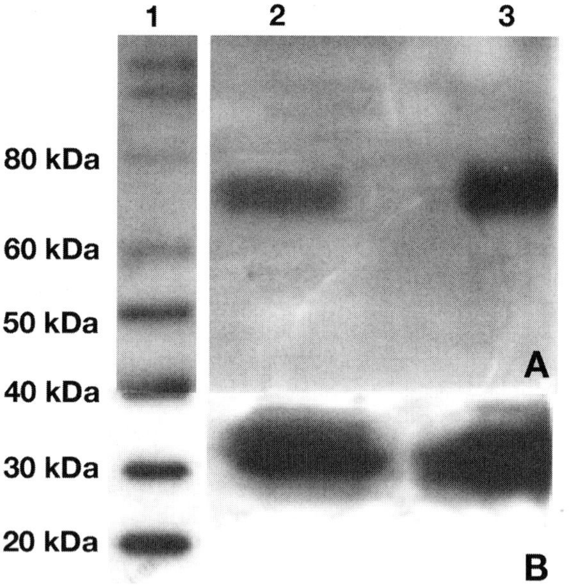

Fig. 2. Western blot for detection of *env* **(A)** and *gag* **(B)** proteins from purified vector preparations (lanes 2 and 3). Lane 1 corresponds to the molecular weight markers.

17. For *env* protein: dilute secondary antibody 1/5000 in antibody dilution buffer.
18. Incubate each membrane with the specific secondary antibody dilution for 1 h at room temperature with agitation.
19. Wash the membranes four times for 5 min with wash buffer (20 mL). The remaining steps are done in a dark room under safe light conditions.
20. Once the final wash is removed, mix the ECL plus reagents (1 mL reagent A and 25 μL reagent B) and immediately add to the blots.
21. Rotate by hand for 5 min to ensure even coverage.
22. Remove the blot from the ECL reagents, take the liquid excess out and place the membranes between the leaves of a plastic sheet that has been cut to the size of the Hypercassette™ (Amersham Biosciences).
23. Place the plastic sheet containing the membranes in the Hypercassette™ with film for a suitable exposure time (*see* **Note 16**).
24. In the Western blot to detect *env* protein, a band around 70 kDa is expected, and in the Western blot for detection of *gag* proteins, a band around 30 kDa is expected. An example of the results obtained is shown in **Fig. 2**.

3.3.2. Transduction Efficiency ***(8)***

1. Prepare a target cell (HCT116) suspension at 1.65×10^5 cells/mL (5×10^4 cells/cm^2) in culture medium.
2. Inoculate the cell suspension in a 96-well plate, 100 μL per well. Incubate cells 24 h in a humidified incubator at 37°C and 5% CO_2.

3. Perform serial dilutions (depending on sample, 1/5 to 1/250) to the viral samples in a separate 96-well plate in culture medium containing 8 μg/mL of Polybrene.
4. After 3 h of infection add 180 μL of culture medium and incubate 48 h in a humidified incubator at 37°C and 5% CO_2.
5. Remove the medium from the plates and wash the plate with wash solution, 100 μL per well.
6. Remove the wash solution, add 50 μL of lysis buffer to each well, and incubate the plate at 37°C during 30 min.
7. Add 50 μL of staining solution warmed to 37°C to each well and incubate at 37°C for 1 h (*see* **Note 17**).
8. Stop the reaction, adding 20 μL of stop solution to each well.
9. Measure the optical density at 420 nm using an absorbance plate reader spectra MAX 340 (Molecular Devices, Sunnyvale, CA); subtract nonspecific background at 650 nm.
10. Values for each point are the averages of at least triplicate wells.

4. Notes

1. Unless stated otherwise, media, additives and solutions in contact with cells are all sterilized by filtration (0.22 μm). FBS is purchased sterile and virus screened.
2. Unless stated otherwise, all solutions should be prepared in water that has a resistivity of 18.2 MΩ-cm.
3. The concentration of cells necessary to inoculate the cell factory depends on the cell type, each cell line having a specific inoculum. For this cell line (TE Fly A7), in order to prepare 1.5 L of cell suspension with 8.5×10^5 cells/mL it is necessary to tripsinize seven 175-cm^2 T-flasks.
4. Simultaneously with the inoculation of the cell factory, one 175-cm^2 T-flask should be seeded with the same cell density, 2×10^4 cells/cm^2, and used as a control to monitor cell growth. If cells do not grow as expected, the medium exchange step can be delayed or anticipated.
5. The volume of supernatant that can be used with the HiTrap column can go up to 30 mL without significant decrease in final yield (typically 40–60%). For processing larger supernatant volumes, several HiTrap columns can be connected in series or a larger column has to be packed with DEAE FF medium (Amersham Biosciences). Calculate the necessary bed volume to pack based on the criteria 10–30 mL supernatant per mL of packed medium. Use a linear flow rate that does not exceed 150 cm/h; the residence time should not be lower than 1min. The equilibration and elution volumes should be increased in proportion to the ratio of the bed volume of the two columns.
6. If the chromatographic system has no buffer mixer, prepare the buffer solutions in advance.
7. Depending on the sensitivity of the UV detector, it might be difficult to distinguish a peak from the baseline. The viral peak should elute 1.5 mL after starting to pump the elution buffer through the column.

8. At the end of the process wash the columns with 0.5 *N* NaOH and equilibrate again with buffer 1 (as for the sanitization and equilibration process). For long-term storage (>5 d) equilibrate the columns with 20% ethanol solution and store at 4°C.
9. The sample loop volume will depend on the volume of purified retroviral vectors to be desalted. With HiPrep 26/10 it is possible to desalt samples up to 15 mL. If larger volumes need to be desalted, connect several columns in series.
10. To observe β-galactosidase activity, several cell lines can be used, such as murine NIH 3T3, human HT1080, human HCT116, and human TE671. The choice of the target cell line is based on several parameters, namely the envelope of the viral vectors. In this procedure, HCT 116 cells were selected because several tests showed that these cells are reliable to quantify infectious particles of MLV vectors with amphotropic envelope with no background.
11. Plates can be stored at 4°C for at least 1 wk after staining. To store, wrap plates with parafilm in order to prevent evaporation.
12. Vortex each sample immediately before analysis in the FACS device.
13. Thaw and maintain RNase inhibitor and AMV reverse transcriptase at 4°C.
14. Examine the melting curves to check the presence of specific amplification (amplicon peak in the corresponding melting temperature, 84°C) and the absence of dimmer primers (absence of peaks at temperatures lower than that of the amplicon). The calibration curve should have a slope between –3.3 and –3.9 and an error lower than 0.1.
15. The vectors used in this procedure are MLV vectors with amphotropic envelope; for retroviral vectors with different envelopes, a primary antibody specific for the different surface protein should be used. The two Western blot methods described here can be used to detect *env* and *gag* proteins not only in purified samples but also in viral supernatants.
16. The exposure time depends on the concentration of the viral proteins as on the secondary antibody used—in this particular case 5–10 s for the detection of *gag* proteins and 1–3 min for the detection of *env* proteins.
17. The incubation time depends on the concentration of the viral samples. Incubate the plates for 15 min and observe the plate; if no yellow product has appeared, incubate again and check every 15 min.

References

1. Andreadis, S. T., Roth, C. M., Le Doux, J. M., Morgan, J. R., and Yarmush, M. L. (1999) Large-scale processing of recombinant retroviruses for gene therapy. *Biotechnol. Prog.* **15**, 1–11.
2. Mountain, A. (2000) Gene therapy: the first decade. *Trends Biotechnol.* **18**, 119–128.
3. Thomas, C. E., Ehrhardt, A., and Kay, M. A. (2003) Progress and problems with the use of viral vectors for gene therapy. *Nat. Rev. Genet.* **4**, 346–358.
4. Relph, K., Harrington, K., and Pandha, H. (2004) Recent developments and current status of gene therapy using viral vectors in the United Kingdom. *BMJ* **329**, 839–842.

5. Chuck, A. S. and Palsson, B. O. (1996) Consistent and high rates of gene transfer can be obtained using flow-through transduction over a wide range of retroviral titers. *Hum. Gene Ther.* **7**, 743–750.
6. McTaggart, S. and Al-Rubeai, M. (2000) Effects of culture parameters on the production of retroviral vectors by a human packaging cell line. *Biotechnol. Prog.* **16**, 859–865.
7. Carmo, M., Peixoto, C., Coroadinha, A. S., Alves, P. M., Cruz, P. E., and Carrondo, M. J. (2004) Quantitation of MLV-based retroviral vectors using real-time RT-PCR. *J. Virol. Methods* **119**, 115–119.
8. Le Doux, J. M., Davis, H. E., Morgan, J. R., and Yarmush, M. L. (1999) Kinetics of retrovirus production and decay. *Biotechnol. Bioeng.* **63**, 654–662.
9. Pizzato, M., Merten, O. W., Blair, E. D., and Takeuchi, Y. (2001) Development of a suspension packaging cell line for production of high titre, serum-resistant murine leukemia virus vectors. *Gene Ther.* **8**, 737–745.

23

Insect Cell Cultivation and Generation of Recombinant Baculovirus Particles for Recombinant Protein Production

Sabine Geisse

Summary

Viral transduction of eukaryotic cell lines is one possibility to efficiently generate recombinant proteins, and among all viral-based expression systems the baculovirus/insect cell expression system (BEVS) is certainly the most well known and applied. This chapter delineates the individual process steps of the system: maintenance of insect cell cultures, transfection of recombinant DNA into different insect cell lines, amplification of virus stocks, virus titer determination, and accessory experiments required to give rise to optimal production conditions.

Key Words: Insect cell lines; baculovirus; transfection; infection; recombinant protein production.

1. Introduction

The baculovirus expression vector system (BEVS) is one of the most efficient systems for the production of recombinant proteins, and consequently its application is widespread in industry as well as in academia. Since the early 1970s, when the first stable insect cell lines were established *(1)* and the infectivity of baculovirus in an in vitro culture system was demonstrated *(2,3)*, virtually thousands of reports have been published on the successful expression of proteins using this system as well as on method improvements. Thus, individual modifications described in the literature are manifold. This chapter presents in detail a standard procedure to generate recombinant proteins using the BEVS in order to introduce the system to the newcomer in the field. Additionally, alternatives and recent developments in methodology are covered to update the experienced user.

The *Autographa californica* nuclear polyhedrosis virus (AcNPV) is the prototype used for expression studies in insect cells. Its genome was completely sequenced in 1984, but its biology is as yet not fully elucidated and understood *(4)*.

From: *Methods in Biotechnology, Vol. 24: Animal Cell Biotechnology: Methods and Protocols, 2nd Ed.* Edited by: R. Pörtner © Humana Press Inc., Totowa, NJ

Alternatively, but to a much lesser extent, expression studies have also used the *Bombyx mori* (silkworm) nuclear polyhedrosis virus (BmNPV) as a carrier of transgenes *(5)*.

The set of experiments comprising the BEVS leading from the cloned gene of interest to the purified recombinant protein can be subdivided into two, albeit not independent entities: the generation of recombinant virus and the expression of recombinant protein. Both rely on state-of-the-art cultivation of insect cell lines used for transfection or infection. Without insect cells of good quality, neither sufficient viral particles can be generated nor can the protein be successfully expressed to high levels.

2. Materials

2.1. Cell Culture

1. Standard tissue culture plasticware: 6-well plates, 35-mm Petri dishes (e.g., Greiner, Becton Dickinson/Falcon), tissue culture flasks (e.g., Nunc, Becton Dickinson/ Falcon, Greiner), roller bottles (e.g., Corning, Becton Dickinson/Falcon), and (disposable) shake flasks (e.g., Corning).
2. 24-deep-well blocks (round bottom, Qiagen cat. no.19583) and AirPore™ cover sheets (Qiagen, cat. no.19571).
3. Cryovials, sterile centrifuge tubes (made of polystyrene or polypropylene).
4. Cell density determination: manually using a hemocytometer Neubauer Improved counting chamber, or automatically using, e.g., the Cedex cell counting system (Innovatis, Bielefeld/FRG) or Vi-Cell instrument (Beckman Coulter, Fullerton, CA). Staining solution: Trypan blue (Sigma, cat. no. T6146), 0.4 g dissolved in 100 mL of phosphate-buffered saline and sterile-filtered through a 0.22-μm filter to remove debris; alternatively: 0.4% ready-to-use solution (Sigma, cat. no. T8154 or Bioconcept, cat. no. 5-72F00-H).
5. Cell culture incubator, shaker incubator (e.g., from Kühner AG, Birsfelden/CH), roller bottle device (e.g., from Heraeus, HERAcell 240 or Cellroll, Integra Biosciences), cell culture microscope.
6. Special reagents: neutral red staining solution: neutral red (Sigma, cat. no. N4638) is dissolved in A.bidest at a concentration of 0.5% w/v, sterile-filtered, and stored in the dark. This stock solution is diluted 1:20 with phosphate-buffered saline prior to use. Low-melting agarose: SeaPlaque agarose (BioWhittaker cat. no. 50100), 2% w/v in A.bidest, solubilized and stored in aliquots of approx 100 mL in glass bottles with Schott caps.
7. Dimethyl sulfoxide (DMSO; Fluka, cat. no. 41640), fetal calf serum (FCS), glutamine solution (200 m*M*; Invitrogen, cat. no. 25030-024).
8. Transfection reagent: e.g., CELLFectin™ (Invitrogen, cat. no. 10362-010).

2.2. Insect Cell Lines and Cultivation Media

The establishment of numerous lepidopteran cell lines, including comparative expression experiments, has been described in the literature ***(6–8)***, but a limited

Table 1
Commercial Sources for Insect Cell Lines and Cultivation Media

Cell line	Commercial Source	Cultivation Medium
Sf21 (IPLB-Sf21-AE)	Invitrogen Orbigen BD/Pharmingen Sigma-Aldrich/ECACC	SF900 II (Invitrogen) Excell 420 (JRH Biosciences)
Sf9	Invitrogen Orbigen ATCC (CRL-1711) BD/Pharmingen Sigma-Aldrich/ECACC	SF900 II (Invitrogen), Excell 420 (JRH Biosciences)
High Five (BTI-TN-5B1-4)	Invitrogen Orbigen	SF900 II, Express Five (Invitrogen) Excell 405 (JRH Biosciences)

number of these cell lines are commonly used in conjunction with the BEVS. It is fair to say that 90% of recombinant proteins are produced by using the insect cell lines IPLB-Sf21-AE (commonly called Sf21 cells) and Sf9 cells (a clonal isolate of Sf21); both cell lines are derived from the ovarian tissue of the fall armyworm, *Spodoptera frugiperda*. The BTI-TN-5B1-4 cell line, sold under the trade name High Five™, was developed from the cabbage looper, *Trichoplusia ni* ***(9)***. These cell lines can be cultivated either adherently in stationary culture with addition of FCS or agitated under serum-free suspension culture conditions.

The development of suitable insect cell culture media has undergone significant changes in recent years. While the first cultivations were done in Grace's insect cell culture medium (TMN-FH or ILP-41) ***(6,7)***, which required supplementation by yeastolate and lipid concentrates for full functionality. The trend has now gone progressively to ready-to-use, serum-free media, which are for some applications enriched by FCS and glutamine addition.

A summary of insect cell lines and the corresponding cultivation media described in this chapter is given in **Table 1**. A more comprehensive overview of culture media and growth behavior of insect cells can be found in **ref. *10***. For special and/or less frequently employed cell lines, *see* **Note 1**.

2.3. Generation of Recombinant Baculovirus DNA for Transfection

Recombinant AcNPV virus featuring the gene of interest can be generated in several ways, because the recombination event between viral and plasmid DNA can occur either in bacteria or inside the transfected insect cells. An overview of the different approaches and their characteristics is given in **Table 2**.

Table 2
Overview on Generation of Recombinant Virus for Expression

Baculovirus expression kits and vendors	Compatible transfer vectors	Methodology for cloning foreign gene into transfer vector	Transfer of foreign gene into Baculovirus genome	Selection/ Recombination efficiency
BacPAK™ (BD Biosciences/ Clontech	Based on homologous recombination at polyhedrin locus	Ligase dependent	Homologous recombination in insect cells	≥90%
Bac-to-Bac™ (Invitrogen)	Based on site-specific transposition	Ligase-dependent, Gateway™ adapted	Site-specific transposition in bacterial cells	Selection of recombinants by blue–white selection on agar plates
BaculoDirect™ (Invitrogen)	Based on site-specific recombination	Gateway™ adapted	Site-specific recombination in Eppendorf tube	Antibiotic selection of transfectants in insect cells
flashBac™ (OET/ NextGen Sciences)	Based on homologous recombination at polyhedrin locus	Ligase dependent	Homologous recombination in insect cells	100%
BacVector™ 1000, 2000, 3000 (EMD/ Novagen)	Based on homologous recombination at polyhedrin locus	Ligase dependent	Homologous recombination in insect cells	≥95%
BaculoGold™ (BD/ Pharmingen)	Based on homologous recombination at polyhedrin locus	Ligase dependent	Homologous recombination in insect cells	≥95%
DiamondBac™ (Sigma-Aldrich)	Based on homologous recombination at polyhedrin locus	Ligase dependent	Homologous recombination in insect cells	≥95%

The protocol described in detail in **Subheading 3.** is highlighted in bold (Bac-to-Bac system/Invitrogen). For the cloning and generation of recombinant bacmid DNA, a detailed description is provided by the vendor in the manuals available on-line (Invitrogen, Carlsbad, CA: baculovirus Expression System with Gateway® Technology, Version E 11/22/2004).

3. Methods

3.1. Cultivation of Insect Cells: General Remarks and Recommendations

1. Insect cell lines are routinely cultivated at 28°C without CO_2 in tissue culture flasks, shake flasks, or roller bottles. Detailed recommendations for cultivation conditions and applications for the individual cell lines are summarized in **Table 3**.
2. Cell densities are routinely monitored manually or instrumentally. The cultures should be maintained at a continuous viability of >90% and exhibit a population doubling time of 18–24 h. Note that Sf21 and Sf9 cells require a higher seeding cell density than High Five™ cells to avoid lag phases in growth.
3. The cell cultures should be regularly checked for absence of mycoplasma. This implies, however, that the cell lines are cultivated in the absence of added antibacterial or antifungal antibiotics.
4. Insect cells should not be enzymatically detached (e.g., via trypsinization). To transfer cells, detach the cells mechanically by vigorous tapping of the flask or rinsing, because scraping of cells leads to massive cell death and low viability. Determine cell count and viability afterwards.
5. Insect cells are sensitive to shear forces and centrifugation steps; thus, the latter should be avoided whenever possible. However, for some manipulations concentrating the cells by centrifugation to obtain the required cell densities is unavoidable.
6. Maintain sufficiently large stocks of back-up cultures, e.g., uninfected cells cultivated adherently and in suspension using the conditions described in **Table 3**. These serve as starting material for all process steps described in the following.
7. Avoid handling infected and uninfected cultures simultaneously under the same laminar flow hood—this poses the risk of accidentally infecting the back-up cultures. Ideally, the cultivation and incubation of back-up cells should be done in separate, virus-free facilities.

3.2. Freezing and Thawing of Cells

It is advisable to prepare a seed stock of 20–30 cryovials from the original vial received from the vendor. The cells in culture should be regularly replaced by freshly thawed cells approximately every 2–3 mo to ensure efficient transfection and virus amplification. A cytogenetic/chromosomal instability of Sf9 cells has been demonstrated over prolonged cultivation time, possibly affecting gene expression *(13)*.

Table 3
Cultivation Conditions and Applications

Cell line/ medium	Intended purpose	Cultivation vessel	Seeding cell densities	Cultivation conditions
Sf21 in ExCell 420 + 10 % FCS	Transfection, plaque assay	6-well-plate or 35 mm petri dishes	$0.9 - 1.0 \times 10^6$ cells/2 mL medium	Standard cell culture incubator
Sf21 in ExCell 420 + 1 % FCS + 1 % glutamine	Virus amplification, generation of master and working virus stocks	Roller bottle or shake flask	3×10^5 cells/mL; for virus stocks use $7 - 10 \times 10^5$ cells/mL	Shaker incubator 94 rpm Roller bottles 10.5 rpm
Sf9 in Sf900 II or ExCell 420 media	Working virus stock generation, production	T-flask, roller bottle or shake flask, bioreactor	5×10^5 cells/mL; for production $> 1 \times 10^6$ cells/mL	Shaker incubator 94 rpm Roller bottles 10.5 rpm
High Five in SF900 II, ExCell 405 or Express Five media	Production	T-flask, shake flask or bioreactor	$2 - 3 \times 10^5$ cells/mL; for production $> 1 \times 10^6$ cells/mL	Shaker incubator 125 rpm

3.2.1. Freezing of Insect Cells

1. Label cryovials to be prepared: name of cell line, culture medium, date of freezing, amount of cells, and potentially passage number.
2. Transfer 5×10^6 to 1×10^7 cells per cryovial in the corresponding culture volume into a sterile centrifuge tube and spin down at 120*g* for 5 min.
3. Prepare freezing medium: depending on the cell line the appropriate cold insect cell culture medium is supplemented with 20% FCS and 10% DMSO (final concentrations).
4. Dissolve cell pellet gently in freezing medium; calculate 1 mL medium per cryovial.
5. Transfer mixture carefully into cryovial (1 mL per vial), close caps tightly.
6. Store vials overnight at -80°C in a styrofoam box or special freezing box (e.g., from Stratagene StrataCooler® or Nalgene Cryo 1°C Freezing Container, "Mr. Frosty") prior to transfer into the liquid nitrogen storage tank.

3.2.2. Thawing of Insect Cells

1. Remove vial from liquid nitrogen tank and transfer to the cell culture lab quickly.
2. Thaw contents of vial rapidly in a preheated water bath (37°C).

3. Transfer the cells to a sterile 15-mL centrifuge tube and dropwise add 9 mL of culture medium to gently dilute out the DMSO.
4. Centrifuge contents of tube for 5 min at 120*g*.
5. Remove supernatant and dissolve cell pellet gently in the desired volume of cell culture medium; seed cells into tissue culture flask or shake flask.
6. Monitor the culture 24 h postthawing for cell growth and viability.
7. If the cell line is to be cultivated in adherent mode in culture medium containing FCS, the diluted mixture can also be seeded directly (without centrifugation) into a tissue culture flask, followed by an exchange of medium 24 h postthawing by aspiration.

3.3. Generation of Recombinant Virus

This first step in the baculovirus expression system has undergone many refinements over time, from the early days, where the recombination event was detected by visual discrimination of nonoccluded vs occluded virus (a highly inefficient process) to methodologies that give rise to a frequency of approx 100% recombination events. Two recent reviews summarize excellently state-of-the-art methodologies currently in use ***(14,15)***.

As indicated in **Table 2**, our method of choice for the generation of recombinant viral particles is the Bac-to-Bac transfection method ***(16)***. This method offers the advantage of preselecting recombinant viral DNA in bacteria; thus, isolation of homogeneous recombinant viral populations following transfection by, e.g., plaque assay and subsequent plaque isolation is unnecessary (*see* **Note 2**).

1. Per transfection seed 7.5×10^5 Sf21 cells in 2 mL ExCell 420 medium supplemented with 10% FCS into one well of a 6-well plate. Prepare the cell suspension in the final volume prior to seeding and ensure careful mixing to achieve an even distribution of cells within the well. Allow sedimentation and attachment of the cells to the cell culture surface by incubation of the plate for at least 30 min at 28°C.
2. Prepare the DNA transfection mixture in two separate sterile polysterene tubes: in tube A resuspend 3 μg (minimal requirement) of recombinant bacmid DNA in 100 μL of ExCell 420 medium **without FCS**; in tube B add 5–10 μL of CELLFectin™ transfection reagent to 100 μL of unsupplemented ExCell 420 medium. Note: If more than 3 μg of bacmid DNA is used, take 10 μL of CELLFectin™ reagent.
3. Slowly add solution B to solution A; mix by tapping the tube (do not vortex!), and incubate the mixture for 15 min at room temperature for the lipoplexes to form. Then add 0.8 mL of unsupplemented Excell 420 medium (the total volume is now 1.0 mL).
4. Carefully remove the cell culture supernatant from the cell monolayer and wash the cells three times with 2 mL ExCell 420 medium **without FCS** per well. Immediately before adding the transfection mix, remove the entire liquid supernatant covering the cell monolayer.
5. Add the transfection mix to the cells and incubate the 6-well plate at 28°C for 5 h without agitation.

6. Remove the culture supernatant once again and replace it by addition of 2 mL ExCell 420 medium fortified with 10% FCS. Incubate the plate for 4 d at 28°C. Inspect the plate daily for signs of virus infection or bacterial contamination.
7. After 3–5 d of incubation time, clear signs of virus infection, characterized by detachment of cells and cell death as well as appearance of bloated and sausage-shaped cells, should be visible. Harvest the virus-containing supernatant in a sterile 15-mL centrifuge tube and refeed the remaining cells with 2 mL of ExCell 420 + 10% FCS. If a master virus stock is to be generated immediately, no centrifugation step is required at this stage. For storage, centrifuge the supernatant for 15 min at 2000*g* to remove the cellular debris and store the cell-free supernatant at 4°C.
8. Incubate the culture for another 4 d and recover again the virus-containing supernatant. This second harvest serves as a security back-up in case the transfection was rather inefficient giving rise to little virus in the first harvest. Prepare a cell-free supernatant as described above for subsequent storage at 4°C.

3.4. Characterization and Early Optimization of Expression

The transfection supernatants harvested at two independent time points after transfection frequently prove to be inadequate for detection of recombinant protein expression. Therefore, a first round of virus amplification is advisable prior to assessment of protein expression and viral titers.

3.4.1. Preparation of Master Virus Stock

1. This first virus amplification is done using Sf21 cells adapted to suspension culture in ExCell 420 + 1% FCS + 1% glutamine solution grown in roller bottles or shake flasks. Ensure that the cells are in logarithmic growth phase at the time of infection by passaging/diluting the culture 24 h prior to infection to a density of 1.5×10^6 cells/mL.
2. At the day of infection determine the cell count and viability of the cell culture and dilute the cells once more to a final cell density of 7×10^5 cells/mL in ExCell 420 + 1% FCS + 1% glutamine solution. The total volume of the culture should be 40 mL in either small roller bottles (Corning cat. no. 430195) or shake flasks.
3. Directly add the first 2 mL of viral supernatant recovered after transfection (first harvest) to the vessel and incubate the culture at 28°C under agitated conditions.
4. Take samples 48 and 72 h postinfection to determine cell density and viability. Harvest the entire culture 72 h postinfection by centrifugation for 15 min at 2000*g*. At this time point the viability of the cell culture should have dropped to 45–55%, and the total cell density should be less than 1.5×10^6 cells/mL.

3.4.2. Determination of Virus Titer by Plaque Assay

The determination of viral titer is commonly done by assessing the number of infectious particles reflected in the generation of "plaques," i.e. areas of lytic growth in a monolayer of unaffected insect cells, and is expressed as plaque-forming units per mL (pfu/mL).

A determination of virus titer can optionally be done on the master virus stock level, but should unequivocally be performed on the working virus stock (*see* **Subheading 3.5.1.**) to secure reproducibility of the process in case several production runs are needed necessitating repetitive virus generation.

1. Set a water bath to 37°C. An aliquot of pre-prepared 2% low-melting agarose in A.bidest is solubilized (in a boiling water bath or microwave oven) and cooled down to 37°C. Prewarm also a bottle of ExCell 420 + 10% FCS culture medium to 37°C.
2. Seed Sf21 cells cultivated in ExCell 420 + 10% FCS into 35-mm Petri dishes or wells of a 6-well plate. Each dish/well requires 1.0×10^6 cells in 2 mL of medium. A full series of dilutions comprises the neat (= undiluted) virus stock plus 10^{-1}–10^{-7} incremental dilutions, i.e., a total of 8 wells, ideally performed in duplicate (*see* **Note 3**). It is essential again to generate a homogeneous cell suspension and to plate the cells very evenly. Do not forget to label the dishes/wells accordingly.
3. Allow for settling and attachment of the cells by incubation for at least 30 min at 28°C.
4. Prepare serial 10-fold dilutions of the virus stock to be titrated by transferring 100–200 μL of virus into 0.9–1.8 mL of culture medium. If many dilutions are to be analyzed, keep the virus dilutions on ice while preparing the cells for infection.
5. Carefully (!) aspirate the cell culture supernatant above the cell monolayer of each well and add the virus dilution. Volumes of 100 μL (for Petri dishes) and 200 μL (for 6-well plate wells) are required per well to evenly cover the cell monolayer.
6. Store the plates on an agitated shaker platform at 60 rpm at room temperature for 30 min; then transfer the infected cultures to the 28°C incubator and continue the incubation for another 30 min without agitation.
7. Meanwhile, prepare the agarose overlay: mix equal volumes of prewarmed low-melting agarose with pre-armed culture medium in a sterile 50-mL Falcon centrifuge tube and return to the water bath at 37°C. Calculate 2 mL agarose overlay per infected well.
8. Carefully aspirate the virus supernatant from the cells; make sure not to disturb the cell monolayer, but work rapidly to avoid drying out of the cells.
9. Dispense 2 mL of the agarose overlay mixture into each well (*see* **Note 4**). Incubate the wells for 15 min at room temperature until the agarose has solidified. Then add 1–2 mL of Excell 420 medium + 10% FCS on top of the agarose overlay in each well.
10. Incubate the plates for 5 d at 28°C.
11. For assay coloration dilute the neutral red stock solution 1:20 with phosphate-buffered saline; calculate 1 mL of staining solution per well. Note: Wear gloves, as neutral red is toxic!
12. Add 1 mL to each well (without removing the liquid overlay); mix well by gently swirling the plates and subsequently incubate them for 4 h at 28°C.
13. Remove the entire liquid overlay of each well; invert the plate or dishes upside-down and allow for the plaques to "clear" overnight before counting (*see* **Note 5**).
14. Determine the virus titer: inspect the result of all virus dilutions tested and choose a dilution that gave rise to well-defined plaques in the range of 20–40 per well. Calculate the virus titer by multiplying the virus dilution factor times the number

of plaques counted times 10 (to backcalculate to 1 mL, if 100 μL of virus dilution was used per dish). For example: 30 plaques counted in the dish infected with 100 μL of a 10^{-6} diluted virus stock results in a final virus titer of 3×10^8 pfu/mL (*see* **Note 6**). A successfully amplified working virus stock, for instance, should result in a virus titer ranging from 5×10^7 to 1×10^9 pfu/mL.

Apart from the plaque assay outlined in detail above, several other methodologies to assess the virus titer have recently been developed. These predominantly aim at increasing the simplicity, speed, and throughput of the assay (*see* **Note 7**).

3.4.3. Expression Trials Using Different Insect Cell Lines

To assess early which insect cell line should be used for production of the desired recombinant protein, small-scale comparative expression trials using Sf21, Sf9, and HighFive™ cells can be performed. Not only do these experiments allow the choice of a suitable expression host; additionally, they shed light on the integrity of the recombinant protein produced if Western blots are used as read out. As a rule of thumb, secreted proteins are frequently expressed to higher levels using HighFive™ cells, while intracellularly expressed proteins give rise to higher protein titers in Sf21 and Sf9 cells (*see* **Notes 8** and **9**).

1. Set up in parallel the following suspension cultures: (1) Sf21 cells in ExCell 420 + 1% FCS + 1% glutamine solution (alternatively, Sf21 cells cultivated under serum-free conditions in SF900 II medium); (2) Sf9 cells cultivated in ExCell 420 or SF900 II medium; (3) HighFive™ cells cultivated in SF900 II or ExCell 405 medium. Seed each culture in 20 mL total volume at a cell density of 1×10^6 cells/mL using 125-mL Erlenmeyer shake flasks.
2. Infect each culture with 0.5 mL of the master virus stock.
3. Take 1-mL samples of the cultures after 48 and 72 h postinfection for intracellularly expressed proteins; if the recombinant protein is secreted into the culture medium, include a 96-h postinfection sampling time point.
4. Centrifuge the samples at 2000*g* to obtain a cell-free supernatant or cell pellet.
5. Analyze the samples by Coomassie-stained sodium dodecyl sulfide (SDS)-polyacrylamide gel electrophoresis (PAGE) analysis, Western blot, or detection by high-performance liquid chromatography (HPLC) or enzyme-linked immunosorbent assay (ELISA), if a protein detection tag was included in the vector construction, e.g., His_6-tag, Fc-tag, etc. (*see* **Notes 10** and **11**). An example of this type of experiment is shown in **Fig. 1**. According to the results presented, we chose to produce this protein by expression in Sf21 cells.

3.5. Generation of Working Virus Stock and Scale-Up to Production

3.5.1. Preparation of Working Virus Stock

The generation of sufficient high-titer virus stock is an essential prerequisite for subsequent production runs and as such requires again a starting cell culture in logarithmic growth phase.

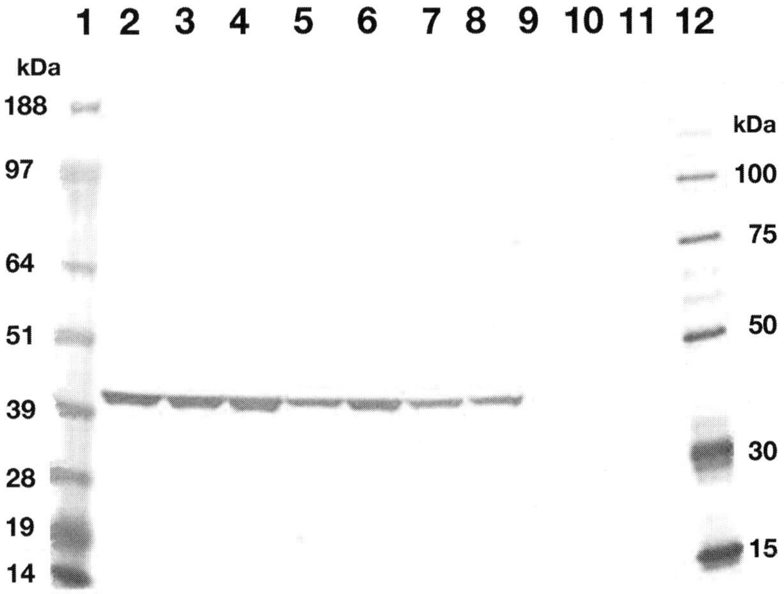

Fig. 1. The Western blot shows expression of a 41-kDa recombinant methyltransferase in different insect cell lines and culture media. The protein is tagged with a His_6-tag at the N-terminus and detected by using a monoclonal anti-his antibody (Sigma, cat. no. H1029), followed by detection with an anti-mouse IgG-HRP antibody (Sigma, cat. no. A2554). Lane 1: Protein Marker SeeBlue® Plus2 (Invitrogen); Lane 2: Sf21 cells/ExCell 401+FCS+glutamine; Lane 3: Sf21/ExCell 420; Lane 4: Sf21/SF900 II; Lane 5: Sf9/ExCell 420; Lane 6: Sf9/SF900 II; Lane 7: High Five/ExCell 405; Lane 8: High Five/SF900 II; Lane 9: Sf21 uninfected; Lane 10: Sf9 uninfected; Lane 11: High Five uninfected; Lane 12: 6xHis Protein ladder (Qiagen, cat. no. 34705).

1. Sf21 cells grown in ExCell 420 + 1% FCS + 1% glutamine solution in shake flasks are diluted 1 d prior to infection to a density of 1.5×10^6 cells/mL. Prepare a culture of at least 400 mL; 24 h later the cell density should be $2.5–3.0 \times 10^6$ cells/mL.
2. Recover 1.0×10^9 cells either by centrifugation (120*g*, 10 min at 4°C) or by directly transferring the corresponding culture volume.
3. Resuspend the cell pellet gently in fresh ExCell 420 medium plus supplements; seed two 400-mL cell suspensions at a density of 1×10^6 cells/mL into roller bottles with vented caps. Alternatively, use 3.0 l Fernbach flasks (Corning, cat. no. 431252) and prepare a culture of 800 mL at 1×10^6 cells/mL.
4. Infect the cultures by addition of 15 mL of Master virus stock per roller bottle or 15–30 mL per Fernbach shake flask (*see* **Subheading 3.4.1.**).
5. When using roller bottles, the oxygen supply may become rate limiting; thus, the culture is flushed with oxygen 48 h postinfection (0.8 bar, 4 L/min for 10 s).
6. Cell density and viability are determined at 24-h intervals, starting at 72 h postinfection.

7. When the cell viability has dropped to 55–75%—usually 3–4 days after infection—the working virus stock solution of 2 × 400 mL is harvested by centrifugation (3000*g*, 30 min, 4°C). At harvest the total cell density should be less than 3.5 × 10^6 cells/mL.
8. The supernatant is now ready for virus titer determination and storage until use (*see* **Note 12**). At this stage an accurate assessment of virus titer is mandatory or at least highly recommended to validate the quality of the virus stock generated (*see* **Subheading 3.4.2.** and **Notes 3–7**).

3.5.2. Defining Production Conditions by Kinetic Experiments

This final preproduction experiment aims at determination of the cell density of the culture at infection (commonly termed "time point of infection," or TOI), the amount of virus required to infect the culture, expressed as virus particles per cell ("multiplicity of infection," or MOI), and the precise time point of harvest of the culture postinfection. All three parameters impact on optimal yields and the highest degree of solubility and integrity of the recombinant protein produced.

Considering the multitude of proteins previously expressed using the BEVS, it is certainly advisable to perform a literature search for the protein of interest, which may result in useful clues and advice. However, in the case of novel genes or as-yet-unpublished candidates, this experiment should be performed. Recently protocols have been developed to minimize the efforts and expenditure of virus by using deep-well blocks as cultivation vessels for this assay ***(14,42–44)***. If a cell line evaluation has been done already at an earlier stage of the process (*see* **Subheading 3.4.3.**), the experiment can be limited to addressing the three parameters outlined above.

1. Seed Sf21, Sf9, or High Five™ insect cells into a 24-deep-well block in 6 mL of total culture volume at cell densities of 1.2 × 10^6 cells/mL and 1.8 × 10^6 cells/mL (two different TOIs). Prepare a total of 4–8 deep wells per TOI.
2. Infect the cells in half of the wells with recombinant virus at a MOI of 0.1, the other half at a MOI of 1 (two different MOIs).
3. Seal the deep-well block with an AirPore™ cover sheet to minimize losses during incubation because of evaporation.
4. Incubate the block at 28°C on a shaker platform, shaking at 250 rpm. For High Five™ cells, use a shaking speed of 350 rpm.
5. Take samples at 48 and 72 h postinfection, in case of secreted proteins also at 96 h postinfection.
6. Analyze for optimal expression conditions by protein analyses (SDS-PAGE, Western blot, HPLC, ELISA).
7. Additional variation in cell densities and MOIs for infection may require analysis if the protein proves to be not well expressed using the above-listed assay conditions **(45)**.

3.5.3. Large-Scale Recombinant Protein Production Using the BEVS

Once all the above-outlined process steps have been successfully performed, the scale-up to the desired production volume is fairly straightforward. In brief, parameters evaluated prior to production which are now applied comprise:

- Assessment of the appropriate production cell line.
- Optimal density of cells at infection (TOI).
- Optimal ratio of virus particles per cell (MOI).
- Optimal time point of harvest of the culture.

The scale at which the production run is then performed is highly dependent on the available equipment of the laboratory and ranges from milliliter to liter scale (using cells cultivated in T-flasks, roller bottles, or shake flasks) to large-scale bioreactors, such as conventional stirred tank reactors or more recently the Wave™ bioreactor ***(46,47)***. Detailed descriptions on how to perform these large-scale runs would exceed the scope of this chapter and can be found elsewhere in the literature (*see*, e.g., **ref. *10*** and references therein).

3.5.4. Summary and Outlook

The wealth of existing literature on the baculovirus expression system proves that this system has become the most popular and successful eukaryotic expression system for the generation of recombinant (tool) proteins over the past three decades. Method improvements to the system have accompanied this development, on the individual laboratory scale as well as on the commercial side. One can assume that this process has not reached its final stage yet: most recent trends focus on automation of individual process steps to further increase the throughput, and first instruments—mainly liquid handling stations—such as the 'BaculoWorkstation™' (NextGen Sciences, Huntington, UK) and 'Piccolo'/'Sonata' (The Automation Partnership, Royston, UK) are available on the market.

4. Notes

1. Another cell line, **expresSF+**, which represents a selected subclone of Sf9 cells, is marketed by Protein Sciences Corp. (http://www.proteinsciences.com). The company claims improved growth properties and yields vs standard Sf9 cells. A license agreement/fee is requested by Protein Sciences for its use.
   Mimic cells, also a derivative of Sf9 cells, which were genetically modified to express several mammalian glycosyltransferases, are claimed to give rise to more complex *N*-glycosylation patterns than normally produced by insect cells ***(11,12)***. These are sold by Invitrogen (cat. no. 12552-014).
2. In our hands comparative analyses of bacmid-derived clonal virus populations have revealed no difference in expression rates; thus, purifying the virus by plaque isolation/purification appears not to be necessary. As a caveat, however, it should

be mentioned that bacmid-derived virus appears to be less stable over time; spontaneous deletion of the transposon has been described over prolonged passaging (i.e., reamplification) of the virus stock *(17)*.

3. The minimal requirement with respect to assessment of an accurate virus titer is to analyze the 10^{-3}–10^{-7} dilutions of a given virus stock, plus one well for undiluted virus as control. If Petri dishes are used, it is recommended to place them on moistened tissues/paper towels onto solid support, e.g., a 500-cm^2 tissue culture plate, to minimize evaporation during incubation.
4. This is the most critical step of the plaque assay, which requires some practice and skill. If pipetting of the agarose mixture proceeds too slowly, the agarose may start to solidify in the pipet; if done too quickly, the result is an uneven overlay and ill-shaped plaque formation. Make sure that the agarose solution has the right temperature; too high a temperature causes cell damage, too low premature settling of the agarose during pipetting.
5. Upon removal of the liquid overlay plaques will become visible almost instantly; however, a more accurate result is obtained after the so-called "clearing" of plaques overnight. If the coloration step results in a grainy appearance of the overlay, the neutral red solution has deteriorated—make sure to always store this solution at 4°C in the dark. If no plaques are visible at all, inspect the cell monolayer for viable cells. If all cells appear dead, either the handling time for addition of virus or agarose was too long (cells died by drying out) or the agarose overlay did not have the desired temperature, killing the cells by heat shock.
6. If virus dilutions greater than 200 μL in volume are added to the wells (e.g., 1 mL to facilitate the pipetting of dilutions), we realized that the observed virus titer is rather low, presumably because not all virus particles present in 1-mL volume per dilution are able to sediment to the cell monolayer within 1-h incubation time. If one wants to work with 1-mL volumes for infection, a correction factor of 5 (observed titer vs real titer) should be introduced in correctly calculating the virus titer.
7. A summary outlining alternative methods for virus titer determination is presented below (for details see literature references):
   - End-point dilution assay ***(18,19)***.
   - Co-expression of quantifiable reporter proteins such as β-galactosidase or green fluorescent protein ***(20–23)***.
   - Antibody-based approaches directed against baculoviral proteins ***(24)***, also available as commercial immunostaining kits (FastPlax Titer Kit, Novagen, cat. no. 70814-3 and BD BacPAK™ baculovirus Rapid Titer Kit, Clontech cat. no. K1599-1).
   - Quantitation by flow cytometry ***(25)***.
   - Quantitation by real-time polymerase chain reaction ***(26)***.
   - Assays assessing the attenuation of cell growth upon virus infection, e.g., proliferation assays using MTT ***(27)*** or alamarBlue ***(28)*** as reagents, or measuring the increase in cell diameter induced by virus infection ***(29)***.
8. One of the major secreted proteins produced by baculovirus-infected insect cells is the enyzme chitinase (*chiA*), responsible for liquefaction of infected larvae in

conjunction with v-cathepsin ***(30)***. The huge overproduction of this nonessential, virus-encoded protein, along with the shut-down of host protein synthesis upon virus infection, may lead to compromised secretion of recombinant proteins. Thus, AcNPV virus DNA featuring a deleted *chiA* gene has been developed recently to overcome this problem (*see* **Note 9** for details).

9. Another problem that may become obvious during the small-scale expression experiment (or during the assessment of production conditions; *see* **Subheading 3.5.2.**) is that the desired protein is prone to proteolytic degradation. This observation, encountered not infrequently, finds its cause in the lytic nature of the viral expression system accompanied by release of proteases from the insect cell or transcribed from the viral DNA itself. To date, several proteases have been identified and (partially) characterized: a cathepsin-like cysteine protease (*v-cath*) is encoded on the viral genome and accounts for 65% of the proteolysis of protein observed. Apart from *v-cath*, insect-cell-derived proteases of the cysteine and aspartic acid protease families were found (for review, *see* **refs. *10,31***).

   To circumvent proteolytic degradation of the target protein during the production phase, addition of protease inhibitors to the culture can be useful. E-64 (*N*-[*N*-(L-3-*trans*-carboxyoxirane-2-carbonyl)-L-leucyl]-agmatin, Roche Molecular Diagnostics, cat. no. 10874523001), leupeptin, and pepstatin A were described as being the most effective with respect to protease inhibition while exhibiting simultaneously low cell cytotoxicity ***(32,33)***.

   Alternatively, viral DNA constructs featuring deletions of *chiA* and *v-cath* genes have been described and are now being marketed (e.g., AcΔCC constructed at the University of Wageningen, NL ***[34]***, flashBac system [NextGen Sciences], BacVector3000 sold by EMD/Novagen, Multi-Bac vector ***[35]***).

10. If the solubility of an intracellularly expressed recombinant protein is to be addressed, the cell pellet obtained after the first centrifugation step is resuspended in lysis buffer (50 m*M* Tris-HCl, pH 8.0, 100 μg/mL polymethylsulfonyl fluoride, 10 m*M* β-mercaptoethanol, dissolved in 0.9% NaCl solution; this corresponds to a twofold concentrated buffer). Choose a volume that corresponds to 10–20% of the culture volume, e.g., if 1 mL of culture was centrifuged, add 100–200 μL of lysis buffer to the pellet. Carefully resuspend the cell pellet by vortexing, followed by 15-min incubation on ice and subsequent centrifugation at 10.000*g* for 2 min. The supernatant now contains the soluble fraction of the protein expressed.

    The remaining pellet harbors the insoluble fraction of the protein. Upon removal of the supernatant, this pellet is dissolved again in 100–200 μL of the following buffer: 20 m*M* hydroxyethyl piperazine ethane sulfonate, pH 7.4, 100 m*M* NaCl, 1 m*M* β-mercaptoethanol, 0.1% SDS, and Complete™ protease inhibitor cocktail (Roche, w/o EDTA cat. no. 1873580, one tablet per 40 mL of buffer). The mixture is sonicated twice for 15 s (amplitude 41% = 20 μ) prior to further analysis by SDS-PAGE and Western blot.

11. In some instances a given protein proves to be well expressed but entirely insoluble within the insect cell. Remedies for this aggregation phenomenon may include (1)

changes in choice and positioning of protein tag/fusion partner in the vector construct, (2) co-expression of an interacting protein partner or chaperone ***(36,37)***, (3) very careful monitoring of the expression conditions (time point of harvest! *see* **Subheading 3.5.2.**), or (4) employing a weaker promoter active earlier in the infectious cell cycle, e.g., the AcNPV basic protein promoter ***(38,39)***. The latter solution may, however, affect protein yields negatively.

12. Early culture supernatants as well as master and working virus stocks should be stored at 4°C until further use; it is estimated that the virus titer remains stable for approx 3–6 mo. For long-term storage virus stocks can also be frozen at –80°C; storage at –20°C leads to rapid loss of infectivity. However, if storage of viral supernatants derived from serum-free cultivations is envisaged, 0.1–1% of FCS or bovine serum albumin should be added for stabilization ***(40)***. Alternatively, DMSO (2.5–10%), glycerol (2.5–10%), or sucrose (0.25–1 *M*) can also be added to improve the stability of the virus stocks ***(41)***.

We routinely keep two volumes (1 mL) of the master virus stock frozen in cryovials at –80°C. Alternatively, the cell pellets of infected cells of master and/or working virus stock can also be stored at –80°C; for revival of the virus present intracellularly, the pellet is thawed and fresh medium is added (see http://www.baculovirus.com for a protocol). Significant loss in virus titer appears to occur through exposure to light ***(40)***; thus, it is recommended to protect the recombinant baculovirus stocks from light as much as possible.

Acknowledgments

I would like to thank my colleagues Mirjam Buchs and Yann Pouliquen for excellent technical assistance in developing these protocols as well as for their support during the preparation of this manuscript.

References

1. Vaughn, J. L., Goodwin, R. H., Tompkins, G. J., and McCawley, P. (1977) The establishment of two cell lines from the insect *Spodoptera frugiperda* (Lepidoptera: Noctuidae). *In Vitro Cell. Dev. Biol.* **13**, 213–217.
2. Smith, G. E., Summers, M. D., and Fraser, M. J. (1983) Production of human interferon in insect cells infected with a baculovirus expression vector. *Mol. Cell. Biol.* **3**, 2156–2165.
3. Pennock, G. D., Shoemaker, C., and Miller, L. K. (1984) Strong and regulated expression of *Escherichia coli* β-galactosidase in insect cells with a baculovirus vector. *Mol. Cell. Biol.* **4**, 399–406.
4. Ayres, M. D., Howard, S. C., Kuzio, J., Lopez-Ferber, M., and Possee, R. D. (1984) The complete DNA sequence of *Autographa californica* nuclear polyhedrosis virus. *Virology* **202**, 586–605.
5. Maeda, S., Kawan, T., Obinata, M., et al. (1988) Production of human interferon in silkworm using a baculovirus vector. *Nature* **315**, 592–594.

6. Hink, W. F., Thomsen, D. R., Davidson, D. J., Meyer, A. L., and Castellino, F. J. (1991) Expression of three recombinant proteins using baculovirus vectors in 23 insect cell lines. *Biotechnol. Prog.* **7**, 9–14.
7. Davis, T. R., Wickham, T. J., McKenna, K. A., Granados, R. R., Shuler, M. L., and Wood, H. A. (1993) Comparative recombinant protein production of eight insect cell lines. *In Vitro Cell. Dev. Biol.* **29A**, 388–390.
8. Chen, Y. P., Gundersen-Rindal, D. E., and Lynn, D. E. (2005) baculovirus-based expression of an insect viral protein in 12 different insect cell lines. *In Vitro Cell. Dev. Biol. Anim.* **41**(1), 43–49.
9. Wickham, T. J. and Nemerov, G. R. (1993) Optimization of growth methods and recombinant protein production in BTI-Tn5B1-4 insect cell using the baculovirus expression vector. *Biotechnol. Prog.* **9**, 25–30.
10. Ikonomou, L., Schneider, Y.-J., and Agathos, S. N. (2003) Insect cell culture for industrial production of recombinant proteins. *Appl. Microbiol. Biotechnol.* **62**, 1–20.
11. Hollister, J., Grabenhorst, E., Nimtz, M., Conradt, H., and Jarvis, J. L. (2002) Engineering the protein glycosylation pathway in insect cells for production of biantennary, complex N-glycans. *Biochemistry* **41**, 15093–15104.
12. Jarvis, D. L. (2003) Developing baculovirus-insect cell expression systems for humanized recombinant glycoprotein production. *Virology* **310**, 1–7.
13. Jarman-Smith, R. F., Armstrong, S. J., Mannix, C. J., and Al-Rubeai, M. (2002) Chromosome instability in *Spodoptera frugiperda* Sf-9 cell line. *Biotechnol. Progr.* **18**, 623–628.
14. Hunt, I. (2005) From gene to protein: a review of new and enabling technologies for multi-parallel protein expression. *Protein Expr. Purif.* **40**, 1–22.
15. Kost, T. A., Condreay, J. P., and Jarvis, D. L. (2005) baculovirus as versatile vectors for protein expression in insect and mammalian cells. *Nat. Biotechnol.* **23**, 567–575.
16. Luckow, V. A., Lee, S. C., Barry, G. F., and Olins, P. O. (1993) Efficient generation of infectious recombinant baculoviruses by site-specific transposon-mediated insertion of foreign genes into a baculovirus genome propagated in *Escherichia coli*. *J. Virol.* **67**, 4566–4579.
17. Piljman, G. P., van Schijndel, J. E., and Vlak, J. M. (2003) Spontaneous excision of BAC vector sequences from bacmid-derived baculovirus expression vectors upon passage in insect cells. *J. Gen. Virol.* **84**, 2669–2678.
18. O'Reilly, D. R., Miller, L. K., and Luckow, V. A. (eds.) (1992) *Baculovirus Expression Vectors; A Laboratory Manual*. W.H. Freeman and Company, New York.
19. Lynn, D. E. (1992) Improved efficiency in determining the titer of the *Autographa californica* baculovirus nonoccluded virus. *BioTechniques* **13**(2), 282–285.
20. Yahata, T., Andriole, S., Isselbacher, K. J., and Shioda, T. (2000) Estimation of baculovirus titer by β-galactosidase activity assay of virus preparations. *BioTechniques* **29**, 214–215.
21. Cha, H. J., Gotoh, T., and Bentley, W. E. (1997) Simplification of titer determination for recombinant baculovirus by green fluorescent protein marker. *BioTechniques* **23**, 782–786.

22. Malde, V. and Hunt, I. (2004) Calculation of baculovirus titer using a microfluidic-based bioanalyzer. *BioTechniques* **36**, 942–946.
23. Philipps, B., Forstner, M., and Mayr, L. M. (2004) baculovirus expression system for magnetic sorting of infected cells and enhanced titer determination. *BioTechniques* **36**, 80–83.
24. Kwon, M. S., Dojima, T., Toriyama, M., and Park, E. Y. (2002) Development of an antibody-based assay for determination of baculovirus titers in 10 h. *Biotechnol.Prog.* **18**(3), 647–651.
25. Shen, C. F., Meghrous, J., and Kamen, A. (2002) Quantitation of baculovirus particles by flow cytometry. *J.Virol.Methods* **105**, 321–330.
26. Lo, H. R. and Chao, Y. C. (2004) Rapid titer determination of baculovirus by quantitative real-time polymerase chain reaction. *Biotechnol. Prog.* **20**(1), 354–360.
27. Mena, J.A., Ramirez, O.T., and Palomares, L.A. (2003) Titration of non-occluded baculovirus using a cell viability assay. *BioTechniques* **34**, 260–264.
28. Pouliquen, Y., Kolbinger, F., Geisse, S., and Mahnke, M. (2005) Automated baculovirus titration assay based on viable cell growth monitoring using AlamarBlue™. BioTechniques, in press.
29. Janakiraman, V., Forrest, W. F., Chow, B., and Seshagiri, S. (2005) A rapid method for estimation of baculovirus titer based on viable cell size. *J. Virol. Methods*, in press, available on-line.
30. Hawtin, R. E., Zarkowska, T., Arnold, K., et al. (1997) Liquefaction of *Autographa californica* nucleopolyhedrosis-infected insects is dependent on the integrity of virus-encoded chitinase and cathepsin genes. *Virology* **238**, 243–253.
31. Gotoh, T., Miyazaki, Y., Kikuchi, K. I., and Bentley, W. E. (2001) Investigation of sequential behaviour of carboxyl protease and cysteine protease activities in virus-infected Sf9 insect cell culture by inhibition assay. *Appl. Microbiol. Biotechnol.* **56**, 742–749.
32. Grosch, H.-W. and Hasilik, A. (1998) Protection of proteolysis-prone recombinant proteins in baculovirus expression systems. *BioTechniques* **24**, 930–934.
33. Martensen, P. M. and Justesen, J. (2001) Specific inhibitors prevent proteolytic degradation of recombinant proteins expressed in HighFive™ cells. *BioTechniques* **30**, 782–792.
34. Kaba, S. A., Salcedo, A. M., Wafula, P. O., Vlak, J. M., and van Oers, M. M. (2004) Development of a chitinase and v-cathepsin negative bacmid for improved integrity of secreted recombinant proteins. *J. Virol. Methods* **122**, 113–118.
35. Berger, I., Fitzgerald, D. J., and Richmond, T. J. (2004) baculovirus expression system for heterologous multiprotein complexes. *Nat. Biotechnol.* **22**, 1583–1587.
36. Hsu, T., Watson, S., Eiden, J. J., and Betenbaugh, M. J. (1996) Rescue of immunoglobulins from insolubility is facilitated by PDI in the baculovirus expression system. *Protein Expr. Purif.* **7**, 281–288.
37. Martinez-Torrecuadrada, J. L., Romero, S., Nunez, A., Alfonso, P., Sanchez-Cespedes, M., and Casal, J. I. (2005) An efficient expression system for the production of functionally active human LKB1. *J. Biotechnol.* **115**, 23–34.
38. Bonning, B. C., Roelvink, P. W., Vlak, J. M., Possee, R. D., and Hammock, B. D. (1994) Superior expression of juvenile hormone esterase and β-galactosidase from

the basic protein promoter of *Autographa californica* nuclear polyhedrosis virus compared to the p10 protein and polyhedrin promoters. *J. Gen. Virol.* **75**, 1551–1556.
39. Lawrie, A. M., King, L. A., and Ogden, J. E. (1995) High level synthesis and secretion of human urokinase using a late gene promoter of the *Autographa californica* nuclear polyhedrosis virus. *J. Biotechnol.* **39**, 1–8.
40. Jarvis, D. L. and Garcia, A., Jr. (1994) Long-term stability of baculoviruses stored under various conditions. *BioTechniques* **16**, 508–513.
41. Jorio, H., Tran, R., and Kamen A. (2005) Stability of serum-free and purified baculovirus stocks under various storage conditions. *Biotechnol. Progr.* DOI:10.1021/bp050218v
42. Chambers, S. P., Austen, D. A., Fulghum, J. R., and Kim, W. M. (2004) High-throughput screening for soluble recombinant expressed kinases in *Escherichia coli* and insect cells. *Protein Expr. Purif.* **36**, 40–47.
43. McCall, E. J., Danielsson, A., Hardern, I. M., et al. (2005) Improvements to the throughput of recombinant protein expression in the baculovirus/insect cell system. *Protein Expr.Purif.* **42**, 29–36.
44. Bahia, D., Cheung, R., Buchs, M., Geisse, S., and Hunt, I. (2005) Optimisation of insect cell growth in deep-well blocks: development of a high-throughput insect cell expression system. *Protein Expr. Purif.* **39**, 61–70.
45. Zhang, Y. H., Enden, G., and Merchuk, J. C. (2005) Insect cells-baculovirus system: factors affecting growth and low MOI infection. *Biochem. Eng. J.* **27**(1), 8–16.
46. Singh, V. (1999) Disposable bioreactor for cell culture using wave-induced agitation. *Cytotechnology* **30**, 149–158.
47. Weber, W., Weber, E., Geisse, S. and Memmert, K. (2002) Optimisation of protein expression and establishment of the Wave bioreactor for baculovirus/insect cell culture. *Cytotechnology* **38**, 77–85.

Index